ELECTRONIC COMMERCE

Tenth Edition

ELECTRONIC COMMERCE

Tenth Edition

Gary P. Schneider, Ph.D., CPA
Quinnipiac University

COURSE TECHNOLOGY
CENGAGE Learning·

Australia • Brazil • Japan • Korea • Mexico • Singapore • Spain • United Kingdom • United States

COURSE TECHNOLOGY
CENGAGE Learning

Electronic Commerce, Tenth Edition
Gary P. Schneider, Ph.D., CPA

Editor-in-Chief: Joe Sabatino

Senior Acquisitions Editor:
Charles McCormick, Jr.

Senior Product Manager: Kate Mason

Development Editor: Amanda Brodkin

Marketing Director: Keri Witman

Marketing Manager: Adam Marsh

Senior Marketing Communications Manager:
Libby Shipp

Marketing Coordinator: Eileen Corcoran

Content Project Management:
PreMediaGlobal

Media Editor: Chris Valentine

Art Direction and Cover Design:
PreMediaGlobal

Cover Credit: ©Baris Simsek, iStock;
©VectorForever, Shutterstock

Manufacturing Planner: Julio Esperas

Compositor: PreMediaGlobal

For product information and technology assistance, contact us at
Cengage Learning Customer & Sales Support, 1-800-354-9706
For permission to use material from this text or product,
submit all requests online at **www.cengage.com/permissions**.
Further permissions questions can be e-mailed to
permissionrequest@cengage.com.

Library of Congress Control Number: 2012934926

Student Edition
ISBN-13: 978-1-133-52682-7
ISBN-10: 1-133-52682-9

Instructor Edition
ISBN-13: 978-1-133-59615-8
ISBN-10: 1-133-59615-0

Course Technology
20 Channel Center Street
Boston, MA 02210
USA

Printed in the United States of America
1 2 3 4 5 6 7 16 15 14 13 12

BRIEF CONTENTS

Preface xvii

Part 1: Introduction

Chapter 1
Introduction to Electronic Commerce 3

Chapter 2
Technology Infrastructure: The Internet and the World Wide Web 53

Part 2: Business Strategies for Electronic Commerce

Chapter 3
Selling on the Web 105

Chapter 4
Marketing on the Web 153

Chapter 5
Business-to-Business Activities: Improving Efficiency and Reducing Costs 205

Chapter 6
Social Networking, Mobile Commerce, and Online Auctions 245

Chapter 7
The Environment of Electronic Commerce: Legal, Ethical, and Tax Issues 281

Part 3: Technologies for Electronic Commerce

Chapter 8
Web Server Hardware and Software   333

Chapter 9
Electronic Commerce Software   371

Chapter 10
Electronic Commerce Security   407

Chapter 11
Payment Systems for Electronic Commerce   461

Part 4: Integration

Chapter 12
Planning for Electronic Commerce   497

Glossary   529

Index   561

TABLE OF CONTENTS

Preface   xvii

Part 1: Introduction

Chapter 1 *Introduction to Electronic Commerce*   3
  Electronic Commerce: Into the Third Wave   5
    Electronic Commerce and Electronic Business   5
    Categories of Electronic Commerce   6
    The Development and Growth of Electronic Commerce   8
    The Dot-Com Boom, Bust, and Rebirth   10
    The Second Wave of Electronic Commerce   11
    The Third Wave Begins   14
  Business Models, Revenue Models, and Business Processes   15
    Focus on Specific Business Processes   16
    Role of Merchandising   16
    Product/Process Suitability to Electronic Commerce   17
  Electronic Commerce: Opportunities, Cautions, and Concerns   18
    Opportunities for Electronic Commerce   18
    Electronic Commerce: Cautions and Concerns   20
  Economic Forces and Electronic Commerce   22
    Transaction Costs   23
    Markets and Hierarchies   24
    Using Electronic Commerce to Reduce Transaction Costs   26
    Network Economic Structures   27
    Network Effects   28
  Identifying Electronic Commerce Opportunities   29
    Strategic Business Unit Value Chains   29
    Industry Value Chains   31
    SWOT Analysis: Evaluating Business Unit Opportunities   33
  International Nature of Electronic Commerce   34
    Trust Issues on the Web   35
    Language Issues   36
    Cultural Issues   37
    Culture and Government   38
    Infrastructure Issues   40
  Summary   43
  Key Terms   43
  Review Questions   44
  Exercises   45
  Cases   45
  For Further Study and Research   49

Chapter 2   *Technology Infrastructure: The Internet and
the World Wide Web*   53

The Internet and the World Wide Web   55
  Origins of the Internet   55
  New Uses for the Internet   56
  Commercial Use of the Internet   57
  Growth of the Internet   57
Packet-Switched Networks   58
  Routing Packets   59
  Public and Private Networks   60
  Virtual Private Networks (VPNs)   61
  Intranets and Extranets   61
Internet Protocols   62
  TCP/IP   62
  IP Addressing   63
  Electronic Mail Protocols   64
  Web Page Request and Delivery Protocols   65
Emergence of the World Wide Web   66
  The Development of Hypertext   66
  Graphical Interfaces for Hypertext   67
  The World Wide Web   67
  The Deep Web   68
  Domain Names   69
Markup Languages and the Web   70
  Markup Languages   72
  Hypertext Markup Language   72
  Extensible Markup Language (XML)   78
  HTML and XML Editors   83
Internet Connection Options   84
  Connectivity Overview   84
  Voice-Grade Telephone Connections   85
  Broadband Connections   85
  Leased-Line Connections   87
  Wireless Connections   87
Internet2 and the Semantic Web   91
Summary   93
Key Terms   93
Review Questions   95
Exercises   96
Cases   98
For Further Study and Research   99

Part 2: Business Strategies for Electronic Commerce

Chapter 3   *Selling on the Web*   105
Revenue Models for Online Business   106
  Web Catalog Revenue Models   107
  Fee-for-Content Revenue Models   112
  Advertising as a Revenue Model Element   114

Fee-for-Transaction Revenue Models	119
Fee-for-Service Revenue Models	125
Free for Many, Fee for a Few	126
Changing Strategies: Revenue Models in Transition	127
Subscription to Advertising-Supported Model	127
Advertising-Supported to Advertising-Subscription Mixed Model	127
Advertising-Supported to Fee-for-Services Model	128
Advertising-Supported to Subscription Model	128
Multiple Changes to Revenue Models	128
Revenue Strategy Issues for Online Businesses	130
Channel Conflict and Cannibalization	130
Strategic Alliances	131
Luxury Goods Strategies	132
Overstock Sales Strategies	132
Creating an Effective Business Presence Online	132
Identifying Web Presence Goals	133
Web Site Usability	136
How the Web Is Different	136
Meeting the Needs of Web Site Visitors	136
Trust and Loyalty	139
Usability Testing	140
Customer-Centric Web Site Design	140
Using the Web to Connect with Customers	142
The Nature of Communication on the Web	142
Summary	145
Key Terms	145
Review Questions	146
Exercises	146
Cases	147
For Further Study and Research	150
Chapter 4 *Marketing on the Web*	153
Web Marketing Strategies	155
The Four Ps of Marketing	155
Product-Based Marketing Strategies	156
Customer-Based Marketing Strategies	157
Communicating with Different Market Segments	158
Trust, Complexity, and Media Choice	158
Market Segmentation	160
Market Segmentation on the Web	162
Offering Customers a Choice on the Web	162
Beyond Market Segmentation: Customer Behavior and Relationship Intensity	163
Segmentation Using Customer Behavior	163
Customer Relationship Intensity and Life-Cycle Segmentation	165
Acquisition, Conversion, and Retention of Customers	167
Customer Acquisition, Conversion, and Retention: The Funnel Model	169
Advertising on the Web	170
Banner Ads	171
Text Ads	173
Other Web Ad Formats	174
Mobile Device Advertising	175

Site Sponsorships 175
Online Advertising Cost and Effectiveness 176
Effectiveness of Online Advertising 177
E-Mail Marketing 178
Permission Marketing 179
Combining Content and Advertising 180
Outsourcing E-Mail Processing 180
Technology-Enabled Customer Relationship Management 180
CRM as a Source of Value in the Marketspace 181
Creating and Maintaining Brands on the Web 183
Elements of Branding 183
Emotional Branding vs. Rational Branding 184
Affiliate Marketing Strategies 185
Viral Marketing Strategies and Social Media 186
Search Engine Positioning and Domain Names 188
Search Engines and Web Directories 188
Paid Search Engine Inclusion and Placement 189
Web Site Naming Issues 191
Summary 194
Key Terms 194
Review Questions 196
Exercises 196
Cases 197
For Further Study and Research 201

Chapter 5   *Business-to-Business Activities: Improving Efficiency and Reducing Costs* 205
Purchasing, Logistics, and Business Support Processes 208
Purchasing Activities 208
Direct vs. Indirect Materials Purchasing 211
Logistics Activities 212
Business Process Support Activities 213
E-Government 215
Network Model of Economic Organization in Purchasing: Supply Webs 217
Electronic Data Interchange 217
Early Business Information Interchange Efforts 218
Emergence of Broader Standards: The Birth of EDI 219
How EDI Works 220
Value-Added Networks 223
EDI Payments 225
Supply Chain Management Using Internet Technologies 226
Value Creation in the Supply Chain 226
Increasing Supply Chain Efficiencies 228
Materials-Tracking Technologies 229
Creating an Ultimate Consumer Orientation in the Supply Chain 231
Building and Maintaining Trust in the Supply Chain 232
Electronic Marketplaces and Portals 232
Independent Industry Marketplaces 232
Private Stores and Customer Portals 234
Private Company Marketplaces 234
Industry Consortia-Sponsored Marketplaces 235

Summary   237
Key Terms   237
Review Questions   238
Exercises   238
Cases   239
For Further Study and Research   242

Chapter 6   *Social Networking, Mobile Commerce, and Online Auctions*   245
From Virtual Communities to Social Networks   246
    Virtual Communities   247
    Early Web Communities   247
    Social Networking Emerges   248
    Revenue Models for Social Networking Sites   251
Mobile Commerce   255
    Mobile Operating Systems   255
    Mobile Apps   257
    Tablet Devices   258
    Mobile Payment Apps   258
Online Auctions   259
    Auction Basics   259
    Online Auctions and Related Businesses   263
    Auction-Related Services   270
Summary   272
Key Terms   272
Review Questions   273
Exercises   273
Cases   274
For Further Study and Research   277

Chapter 7   *The Environment of Electronic Commerce:*
Legal, Ethical, and Tax Issues   281
The Legal Environment of Electronic Commerce   283
    Borders and Jurisdiction   283
    Jurisdiction on the Internet   286
    Conflict of Laws   290
    Contracting and Contract Enforcement in Electronic Commerce   291
Use and Protection of Intellectual Property in Online Business   296
    Copyright Issues   297
    Patent Issues   299
    Trademark Issues   300
    Domain Names and Intellectual Property Issues 300
    Protecting Intellectual Property Online   302
    Defamation   303
    Deceptive Trade Practices   304
    Advertising Regulation   304
Online Crime, Terrorism, and Warfare   306
    Online Crime: Jurisdiction Issues   306
    New Types of Crime Online   308
    Online Warfare and Terrorism   309
Ethical Issues   310
    Ethics and Online Business Practices   310

Privacy Rights and Obligations 311
Communications with Children 315
Taxation and Electronic Commerce 317
Nexus 317
U.S. Income Taxes 318
U.S. State Sales Taxes 319
Import Tariffs 320
European Union Value Added Taxes 320
Summary 321
Key Terms 322
Review Questions 323
Exercises 323
Cases 324
For Further Study and Research 327

Part 3: Technologies for Electronic Commerce

Chapter 8 *Web Server Hardware and Software* 333
Web Server Basics 335
Dynamic Content Generation 336
Multiple Meanings of "Server" 337
Web Client/Server Architectures 338
Software for Web Servers 340
Operating Systems for Web Servers 340
Web Server Software 341
Finding Web Server Software Information 342
Electronic Mail (E-Mail) 342
E-Mail Benefits 343
E-Mail Drawbacks 343
Spam 343
Solutions to the Spam Problem 344
Web Site Utility Programs 352
Finger and Ping Utilities 352
Tracert and Other Route-Tracing Programs 352
Telnet and FTP Utilities 353
Indexing and Searching Utility Programs 354
Data Analysis Software 354
Link-Checking Utilities 354
Remote Server Administration 355
Web Server Hardware 355
Server Computers 355
Web Servers and Green Computing 356
Web Server Performance Evaluation 357
Web Server Hardware Architectures 358
Summary 362
Key Terms 362
Review Questions 363
Exercises 364
Cases 364
For Further Study and Research 368

Chapter 9 *Electronic Commerce Software* 371
  Web Hosting Alternatives 373
  Basic Functions of Electronic Commerce Software 374
    Catalog Display Software 375
    Shopping Cart Software 377
    Transaction Processing 380
  How Electronic Commerce Software Works with Other Software 381
    Databases 382
    Middleware 382
    Enterprise Application Integration 383
    Integration with ERP Systems 384
    Web Services 384
  Electronic Commerce Software for Small and Midsize Companies 387
    Basic Commerce Service Providers 387
    Mall-Style Commerce Service Providers 388
    Estimating Operating Expenses for a Small Web Business 389
  Electronic Commerce Software for Midsize to Large Businesses 390
    Web Site Development Tools 391
  Electronic Commerce Software for Large Businesses 392
    Enterprise-Class Electronic Commerce Software 393
    Content Management Software 395
    Knowledge Management Software 395
    Supply Chain Management Software 396
    Customer Relationship Management Software 396
    Cloud Computing 398
  Summary 399
  Key Terms 399
  Review Questions 400
  Exercises 400
  Cases 402
  For Further Study and Research 404

Chapter 10 *Electronic Commerce Security* 407
  Online Security Issues Overview 408
    Origins of Security on Interconnected Computer Systems 409
    Computer Security and Risk Management 410
    Elements of Computer Security 411
    Establishing a Security Policy 411
  Security for Client Computers 413
    Cookies and Web Bugs 413
    Active Content 415
    Java Applets 416
    JavaScript 417
    ActiveX Controls 417
    Graphics and Plug-Ins 418
    Viruses, Worms, and Antivirus Software 419
    Digital Certificates 424
    Steganography 427
    Physical Security for Clients 428
    Client Security for Mobile Devices 428
  Communication Channel Security 429
    Secrecy Threats 429

Integrity Threats	431
Necessity Threats	431
Threats to the Physical Security of Internet Communications Channels	432
Threats to Wireless Networks	432
Encryption Solutions	433
Using a Hash Function to Create a Message Digest	440
Converting a Message Digest into a Digital Signature	440
Security for Server Computers	441
Web Server Threats	441
Database Threats	443
Other Programming Threats	443
Threats to the Physical Security of Web Servers	444
Access Control and Authentication	445
Firewalls	447
Organizations that Promote Computer Security	449
CERT	449
Other Organizations	449
Computer Forensics and Ethical Hacking	450
Summary	451
Key Terms	451
Review Questions	453
Exercises	453
Cases	454
For Further Study and Research	456

Chapter 11 *Payment Systems for Electronic Commerce*	461
Online Payment Basics	463
Micropayments and Small Payments	463
Online Payment Methods	464
Payment Cards	465
Advantages and Disadvantages of Payment Cards	466
Payment Acceptance and Processing	466
Electronic Cash	471
Privacy and Security of Electronic Cash	471
Holding Electronic Cash: Online and Offline Cash	472
Advantages and Disadvantages of Electronic Cash	472
Digital Wallets	474
Software-Only Digital Wallets	475
Hardware-Based Digital Wallets	476
Stored-Value Cards	476
Magnetic Strip Cards	476
Smart Cards	477
Internet Technologies and the Banking Industry	478
Check Processing	478
Mobile Banking	480
Criminal Activity and Payment Systems: Phishing and Identity Theft	480
Phishing Attacks	480
Using Phishing Attacks for Identity Theft	483
Phishing Attack Countermeasures	485
Summary	486
Key Terms	486
Review Questions	487

Exercises   488
Cases   489
For Further Study and Research   491

Part 4: Integration

Chapter 12   *Planning for Electronic Commerce*   497
   Identifying Benefits and Estimating Costs of Electronic Commerce Initiatives   498
      Identifying Objectives   499
      Linking Objectives to Business Strategies   499
      Identifying and Measuring Benefits   500
      Identifying and Estimating Costs   502
      Funding Online Business Startups   504
      Comparing Benefits to Costs   505
      Return on Investment (ROI)   506
   Strategies for Developing Electronic Commerce Web Sites   507
      Internal Development vs. Outsourcing   508
      New Methods for Implementing Partial Outsourcing   511
   Managing Electronic Commerce Implementations   513
      Project Management   513
      Project Portfolio Management   514
      Staffing for Electronic Commerce   514
      Postimplementation Audits   517
      Change Management   518
   Summary   519
   Key Terms   519
   Review Questions   520
   Exercises   521
   Cases   522
   For Further Study and Research   524

Glossary   529
Index   561

Electronic Commerce, Tenth Edition provides complete coverage of the key business and technology elements of electronic commerce. The book does not assume that readers have any previous electronic commerce knowledge or experience.

In 1998, having spent several years doing electronic commerce research, consulting, and corporate training, I began developing undergraduate and graduate business school courses in electronic commerce. Although I had used a variety of books and other materials in my corporate training work, I was concerned that those materials would not work well in university courses because they were written at widely varying levels and did not have the organization and pedagogic features, such as review questions, that are so important to students.

After searching for a textbook that offered balanced coverage of both the business and technology elements of electronic commerce, I concluded that no such book existed. The first edition of *Electronic Commerce* was written to fill that void. Since that first edition, I have worked to improve the book and keep it current with the rapid changes in this dynamic field.

New to this Edition

This edition includes the usual updates to keep the content current with the rapidly occurring changes in electronic commerce. The tenth edition also includes new material on the following topics:

- Introduces the emergence of a third wave of electronic commerce (Chapter 1)
- Censorship issues in China (Chapter 1)
- Use of social media during the Arab Spring (Chapter 1)
- Mobile commerce using smartphones and tablet devices (Chapters 1, 3, 6)
- Rapid growth in online business in Asia driven by smartphone usage (Chapter 2)
- New online film and television programming distribution channels (Chapter 3)
- New strategies for selling luxury goods online (Chapter 3)
- Using social media to create viral marketing strategies (Chapter 4)
- Impact sourcing as an offshoring strategy (Chapter 5)
- Social commerce (Chapter 6)
- Cloud computing (Chapter 8)
- Content management software and social media (Chapter 9)
- Security for mobile devices (Chapter 10)
- New major viruses and security threats (Chapter 10)
- Mobile payment-processing technologies (Chapter 11)

ORGANIZATION AND COVERAGE

Electronic Commerce: Tenth edition introduces readers to both the theory and practice of conducting business over the Internet and World Wide Web. The book is organized into four sections: an introduction, business strategies, technologies, and integration.

Introduction

The book's first section includes two chapters. Chapter 1, "Introduction to Electronic Commerce," defines electronic commerce and describes how companies use it to create new products and services, reduce the cost of existing business processes, and improve the efficiency and effectiveness of their operations. The concept of the second wave of electronic commerce is presented and developed in this chapter. Chapter 1 also describes the history of the Internet and the Web, explains the international environment in which electronic commerce exists, provides an overview of the economic structures in which businesses operate, and describes how electronic commerce fits into those structures. Two themes are introduced in this chapter and recur throughout later chapters: examining a firm's value chain can suggest opportunities for electronic commerce initiatives, and reductions in transaction costs are important elements of many electronic commerce initiatives.

Chapter 2, "Technology Infrastructure: The Internet and the World Wide Web," introduces the technologies used to conduct business online, including topics such as Internet infrastructure, protocols, and packet-switched networks. Chapter 2 also describes the markup languages used on the Web (HTML and XML) and discusses Internet connection options and tradeoffs, including wireless technologies.

Business Strategies for Electronic Commerce

The second section of the book includes five chapters that describe the business strategies that companies and other organizations are using to do business online. Chapter 3, "Selling on the Web," describes revenue models that companies are using on the Web and explains how some companies have changed their revenue models as the Web has matured. The chapter explains important concepts related to revenue models, such as cannibalization and coordinating multiple marketing channels. The chapter also describes how firms that understand the nature of communication on the Web can identify and reach the largest possible number of qualified customers.

Chapter 4, "Marketing on the Web," provides an introduction to Internet marketing and online advertising. It includes coverage of market segmentation, technology-enabled customer relationship management, rational branding, contextual advertising, localized advertising, viral marketing, and permission marketing. The chapter also explains how online businesses can share and transfer brand benefits through affiliate marketing and cooperative efforts among brand owners.

Chapter 5, "Business-to-Business Activities: Improving Efficiency and Reducing Costs" explores the variety of methods that companies are using to improve their purchasing and logistics primary activities with Internet and Web technologies. Chapter 5 also provides an overview of EDI and describes how companies are outsourcing some of their business processes to less-developed countries. Chapter 5 describes how businesses are using technologies such as e-procurement, radio-frequency identification, and reverse auctions in the practice of supply chain management online.

Chapter 6, "Social Networking, Mobile Commerce, and Online Auctions," explains how companies now use the Web to do things that they have never done before, such as creating social networks, engaging in mobile commerce, and operating auction sites.

The chapter describes how businesses are developing social networks and using existing social networking Web sites to increase sales and do market research. The emergence of mobile commerce in meaningful volumes after many years of anticipation is outlined. The chapter also explains how companies are using Web auction sites to sell goods to their customers and generate advertising revenue.

Chapter 7, "The Environment of Electronic Commerce: Legal, Ethical, and Tax Issues," discusses the legal and ethical aspects of intellectual property usage and the privacy rights of customers. Online crime, terrorism, and warfare are covered as well. The chapter also explains that the large number of government units that have jurisdiction and power to tax makes it essential that companies doing business on the Web understand the potential liabilities of doing business with customers in those jurisdictions.

Technologies for Electronic Commerce

The third section of the book includes four chapters that describe the technologies of electronic commerce and explains how they work. Chapter 8, "Web Server Hardware and Software," describes the computers, operating systems, e-mail systems, utility programs, and Web server software that organizations use in the operation of their electronic commerce Web sites, including cloud computing technologies. The chapter describes the problem of unsolicited commercial e-mail (UCE, or spam) and outlines both technical and legal solutions to the problem.

Chapter 9, "Electronic Commerce Software," describes the basic functions that all electronic commerce Web sites must accomplish and explains the various software options used to perform those functions by companies of various sizes. This chapter includes an overview of Web services, database management, shopping cart, cloud computing, and other types of software used in electronic commerce. The chapter also includes a discussion of Web hosting options for online businesses of various sizes.

Chapter 10, "Electronic Commerce Security," discusses security threats and counter-measures that organizations can use to ensure the security of client computers (and smartphones and tablet devices), communications channels, and Web servers. The chapter emphasizes the importance of a written security policy and explains how encryption and digital certificates work. The chapter also includes an update on the most recent computer viruses, worms, and other threats.

Chapter 11, "Payment Systems for Electronic Commerce," presents a discussion of electronic payment systems, including mobile banking, electronic cash, electronic wallets, and the technologies used to make stored-value cards, credit cards, debit cards, and charge cards work. The chapter describes how payment systems operate, including approval of transactions and disbursements to merchants, and describes how banks are using Internet technologies to improve check clearing and payment-processing operations. The use of mobile technologies for making payments and doing online banking is outlined. The chapter also includes a discussion of the threats that phishing attacks and identity theft crimes pose for individuals and online businesses.

Integration

The fourth and final section of the book includes one chapter that integrates the business and technology strategies used in electronic commerce. Chapter 12, "Planning for Electronic Commerce," presents an overview of key elements that are typically included in business plans for electronic commerce implementations, such as the setting of objectives and estimating project costs and benefits. The chapter describes outsourcing strategies used in electronic commerce and covers the use of project management and project port-folio management as formal ways to plan and control tasks and resources used in electronic commerce implementations. This chapter includes a discussion of change management and outlines specific jobs available in organizations that conduct electronic commerce.

FEATURES

The tenth edition of *Electronic Commerce* includes a number of features and offers additional resources designed to help readers understand electronic commerce. These features and resources include:

- **Business Case Approach** The introduction to each chapter includes a real business case that provides a unifying theme for the chapter. The case provides a backdrop for the material described in the chapter. Each case illustrates an important topic from the chapter and demonstrates its relevance to the current practice of electronic commerce.

- **Learning From Failures** Not all electronic commerce initiatives have been successful. Each chapter in the book includes a short summary of an electronic commerce failure related to the content of that chapter. We all learn from our mistakes—this feature is designed to help readers understand the missteps of electronic commerce pioneers who learned their lessons the hard way.

- **Summaries** Each chapter concludes with a Summary that concisely recaps the most important concepts in the chapter.

- **Web Links** The Web Links are a set of Web pages maintained by the publisher for readers of this book. The Web Links complement the book and link to Web sites referred to in the book and to other online resources that further illustrate the concepts presented. The Web is constantly changing, and the Web Links are continually monitored and updated for those changes so that its links continue to lead to useful Web resources for each chapter. You can find the Web Links for this book by visiting Course Technology's Web site at *www.cengage.com/mis* and searching for Electronic Commerce.

- **Web Links References in Text** Throughout each chapter, there are Web Links references that indicate the name of a link included in the Web Links. Text set in bold, green, sans-serif letters (Metabot Pro) indicates a like-named link in the Web Links. The links are organized under chapter and subchapter headings that correspond to those in the book. The Web Links also contains many supplemental links to help students explore beyond the book's content.

- **Review Questions and Exercises** Each chapter concludes with meaningful review materials including both conceptual discussion questions and hands-on exercises. The review questions are ideal for use as the basis for class discussions or as written homework assignments. The exercises give students hands-on experiences that yield computer output or a written report.

- **Cases** Each chapter concludes with two comprehensive cases. One case uses a fictitious setting to illustrate key learning objectives from that chapter. The other case gives students an opportunity to apply what they have learned from the chapter to an actual situation that a real company or organization has faced. The cases offer students a rich environment in which they can apply what they have learned and provide motivation for doing further research on the topics.

- **For Further Study and Research** Each chapter concludes with a comprehensive list of the resources that were consulted during the writing of the chapter. These references to publications in academic journals, books, and the IT industry and business press provide a sound starting point for readers who want to learn more about the topics contained in the chapter.

- **Key Terms and Glossary** Terms within each chapter that may be new to the student or have specific subject-related meaning are highlighted by boldface type. The end of each chapter includes a list of the chapter's key terms. All of the book's key terms are compiled, along with definitions, in a Glossary at the end of the book.

TEACHING TOOLS

When this book is used in an academic setting, instructors may obtain the following teaching tools from Course Technology:

- **Instructor's Manual** The Instructor's Manual has been carefully prepared and tested to ensure its accuracy and dependability. The Instructor's Manual is available through the Course Technology Instructor Downloads. (Call your customer service representative to obtain your username and password.)
- **ExamView©** This textbook is accompanied by ExamView, a powerful testing software package that allows instructors to create and administer printed, computer (LAN-based), and Internet exams. ExamView includes hundreds of questions that correspond to the topics covered in this text, enabling students to generate detailed study guides that include page references for further review. The computer-based and Internet testing components allow students to take exams at their computers and also save the instructor time by grading each exam automatically.
- **PowerPoint Presentations** Microsoft PowerPoint slides are included for each chapter as a teaching aid for classroom presentations, to make available to students on a network for chapter review, or to be printed for classroom distribution. Instructors can add their own slides for additional topics they introduce to the class. The presentations are included on the Instructor's CD.
- **WebTutor** Whether you want to Web-enable your class or teach entirely online, WebTutor provides customizable, rich, text-specific content that can be used with both WebCT and Blackboard. WebTutor allows instructors to easily blend, add, edit, reorganize, or delete content. Each WebTutor product provides media assets, quizzing, Web Links, discussion topics, and more.

ACKNOWLEDGMENTS

I owe a great debt of gratitude to my good friends at Cengage who made this book possible. Cengage remains the best publisher with which I have ever worked. Everyone at Cengage put forth tremendous effort to publish this edition on a very tight schedule. My heartfelt thanks go to Charles McCormick, Jr., Senior Acquisitions Editor; Kate Mason and Aimee Poirier, who shared the job of Product Manager; and Divya Divakaran, Production Project Manager, for their tireless work and dedication. I am deeply indebted to Amanda Brodkin, Development Editor extraordinaire, for her outstanding contributions to all 10 editions of this book. Amanda performed the magic of turning my manuscript drafts into a high-quality textbook and was always ready with encouragement and fresh ideas when I was running low on them. Many of the best elements of this book resulted from Amanda's ideas and inspirations. In particular, I want to thank Amanda for contributing the Dutch auction example in Chapter 6 and the ideas for the cases in Chapters 7 and 8.

I want to thank the following reviewers for their insightful comments and suggestions on previous editions:

Paul Ambrose	University of Wisconsin, Milwaukee
Kirk Arnett	Mississippi State University
Tina Ashford	Macon State College
Rafael Azuaje	Sul Ross State University
Robert Chi	California State University-Long Beach
Chet Cunningham	Madisonville Community College
Roland Eichelberger	Baylor University
Mary Garrett	Michigan Virtual High School
Barbara Grabowski	Benedictine University
Milena Head	McMaster University
Perry M. Hidalgo	Gwinnett Technical Institute
Brent Hussin	University of Wisconsin, Green Bay
Cheri L. Kase	Legg Mason Corporate Technology
Joanne Kuzma	St. Petersburg College
Rick Lindgren	Graceland University
Victor Lipe	Trident Technical College
William Lisenby	Alamo Community College
Diane Lockwood	Albers School of Business and Economics, Seattle University
Jane Mackay	Texas Christian University
Michael P. Martel	Culverhouse School of Accountancy, University of Alabama
William E. McTammany	Florida State College at Jacksonville
Leslie Moore	Jackson State Community College
Martha Myers	Kennesaw State University
Pete Partin	Forethought Financial Services
Andy Pickering	University of Maryland University College
David Reavis	Texas A&M University
George Reynolds	Strayer University
Barbara Warner	University of South Florida
Gene Yelle	Megacom Services

Special thanks go to reviewer A. Lee Gilbert of Nanyang Technological University in Singapore, who provided extremely detailed comments and many useful suggestions for improving Chapter 12. My thanks also go to the many professors who have used the previous editions in their classes and who have sent me suggestions for improving the text. In particular, I want to acknowledge the detailed recommendations made by David Bell of Pacific Union College regarding the coverage of IP addresses in Chapter 2.

The University of San Diego provided research funding that allowed me to work on the first edition of this book and gave me fellow faculty members who were always happy to discuss and critically evaluate ideas for the book. Of these faculty members, my thanks go first to Jim Perry for his contributions as co-author on the first two editions of this book. Tom Buckles, now a professor of marketing at Biola University, provided many useful suggestions, pointed out a number of valuable research resources, and was willing to sit and discuss ideas for this book long after everyone else had left the building. Rahul Singh, now teaching at the University of North Carolina-Greensboro, provided suggestions regarding the book's coverage of electronic commerce infrastructure. Carl Rebman made recommendations on a number of networking, telecommunications, and security topics. The University of San Diego School of Business Administration also provided the research assistance of many graduate students who helped me with work on the first seven editions of this book. Among those research assistants were Sebastian Ailioaie, a Fulbright Fellow who did substantial work on the Web Links, and Anthony Coury, who applied his considerable legal knowledge to reviewing Chapter 7 and suggesting many improvements.

Many of my graduate students provided helpful suggestions and ideas. My special thanks go to two of those students, Dima Ghawi and Dan Gordon. Dima shared her significant background research on reverse auctions and helped me develop many of the ideas presented in Chapters 5 and 6. Dan gave me the benefit of his experiences as manager of global EDI operations for a major international firm and provided an in-depth review of Chapter 5. I am also grateful to Robin Lloyd for her help with the Lonely Planet case (in Chapter 3) and to Zu-yo Wang for his help with the Alibaba.com case (Chapter 6). Other students who provided valuable suggestions include Maximiliano Altieri, Adrian Boyce, Karl Flaig, Kathy Glaser, Emilie Johnson Hersh, Chad McManamy, Dan Mulligan, Firat Ozkan, Suzanne Phillips, Susan Soelaiman, Carolyn Sturz, and Leila Worthy.

Finally, I want to express my deep appreciation for the support and encouragement of my wife, Cathy Cosby. Without her support and patience, writing this book would not have been possible.

DEDICATION

To the memory of my father, Anthony J. Schneider.

ABOUT THE AUTHOR

Gary Schneider is the William S. Perlroth Professor of Accounting at Quinnipiac University. His prior teaching appointments include the University of San Diego, the University of Tennessee, and Xavier University. He has won a number of teaching and research awards. He served as academic director of the University of San Diego's graduate programs in electronic commerce and information systems. Gary has published more than 50 books and 100 research papers on a variety of accounting, information systems, and management topics. His books have been translated into Chinese, French, Italian, Korean, and Spanish. Gary's research has been funded by the Irvine Foundation and the U.S. Office of Naval Research. His work has appeared in the *Journal of Information Systems, Interfaces, Issues in Accounting Education*, and the *Information Systems Audit & Control Journal*. He has served as editor of the *Business Studies Journal* and the *Accounting Systems and Technology Reporter,* as accounting discipline editor of *Advances in Accounting, Finance and Economics*, as associate editor of the *Journal of Global Information Management*, and on the editorial boards of the *Journal of Information Systems,* the *Journal of Electronic Commerce in Organizations*, the *Journal of Database Management*, and the *Information Systems Audit & Control Journal*. Gary has lectured on electronic commerce topics at universities and businesses in the United States, Europe, South America, and Asia. He has provided consulting and training services to a number of major clients, including Gartner, Gateway, Honeywell, the National Science Foundation, Qualcomm, and the U.S. Department of Commerce. In 1999, he was named a Fellow of the Gartner Institute. In 2003, he was awarded the Clarence L. Steber professorship by the University of San Diego. Gary is a licensed CPA in Ohio, where he practiced public accounting for 14 years. He holds a Ph.D. in accounting information systems from the University of Tennessee, an M.B.A. in accounting from Xavier University, and a B.A. in economics from the University of Cincinnati.

PART 1

INTRODUCTION

CHAPTER 1
Introduction to Electronic Commerce, 3

CHAPTER 2
Technology Infrastructure: The Internet and the World Wide Web, 53

INTRODUCTION TO ELECTRONIC COMMERCE

LEARNING OBJECTIVES

In this chapter, you will learn about:

- What electronic commerce is and how it has evolved into a second wave of growth
- Why companies concentrate on revenue models and the analysis of business processes instead of business models when they undertake electronic commerce initiatives
- How economic forces have created a business environment that is fostering the continued growth of electronic commerce
- How businesses use value chains and SWOT analysis to identify electronic commerce opportunities
- The international nature of electronic commerce and the challenges that arise in engaging in electronic commerce on a global scale

INTRODUCTION

In the late 1990s, electronic commerce was still emerging as a new way to do business; at that time, most companies were doing very little buying or selling online. They still were selling products in physical stores or taking orders over the telephone and by mail. However, a few companies had established solid footholds online. Amazon.com was a rapidly growing bookseller and eBay had taken the lead as a profitable auction site. The business of providing search tools for finding information online was dominated by a few well-established sites, including AltaVista, HotBot, Lycos, and

Yahoo!. Most industry observers at that time believed that any new search engine Web site would find it very difficult to compete against these established operations.

Search engines of the late 1990s provided results based on the number of times a search term appeared on Web pages. Pages that included the greatest number of occurrences of a user's search term would be more highly ranked and would thus appear near the top of the search results list. By 1998, two Stanford University students, Lawrence Page and Sergey Brin, had been working on a search engine research project for two years. Page and Brin believed that a search ranking based on the relationships between Web sites would give users better and more useful results. They developed search algorithms based on the number of links a particular Web page had to and from other highly relevant pages. In 1998, they started Google (*Note*: This typeface indicates a corresponding link to a related Web page in the book's Web Links Google's URL is http://www.google.com) in a friend's garage with about $1.1 million of seed money invested by a group of Stanford graduates and local businesspersons.

Most industry observers agree that Google's page ranking system, which has been continually improved since its introduction, consistently provides users with more relevant results than other search engines. Internet users flocked to Google, which became one of the most popular sites on the Internet. The site's popularity allowed Google to charge increasingly higher rates for advertising space on its Web pages. Marketing staff at Google noticed that another search engine, Goto.com (now owned by Yahoo! and operated as Yahoo! Search Marketing), was selling ad space on Web sites by allowing advertisers to bid on the price of keywords and then charging based on the number of users who clicked the ads. For example, a car dealer could bid on the price of the keyword "car." If the car dealer were the high bidder at 12 cents, then the car dealer would pay for the ad at a rate of 12 cents times the number of site visitors who clicked the ad. Google adopted this keyword bidding model in 2000 and has used it since then to sell small text ads that appear on search results pages.

This approach to selling advertising was extremely successful. Combined with the highly relevant search results provided by the page ranking system, it led to Google's continued growth. When the company went public in 2004 (raising $1.67 billion), its market valuation was nearly $23 billion. Today, Google is one of the most successful online companies in the world. The Web provides a quick path to potential customers for any businessperson with a unique product or service. Google's improved page ranking system was available to anyone in the world the day it was introduced online.

ELECTRONIC COMMERCE: INTO THE THIRD WAVE

The business phenomenon that we now call electronic commerce has had an interesting history. From humble beginnings in the mid-1990s, electronic commerce grew rapidly until 2000, when a major downturn occurred. The popular media published endless news stories describing how the "dot-com boom" had turned into the "dot-com bust." Between 2000 and 2003, many industry observers were writing obituaries for electronic commerce. Just as the unreasonable expectations for immediate success had fueled unwarranted high expectations during the boom years, overly gloomy news reports colored perceptions during this time.

Beginning in 2003, electronic commerce began to show signs of new life. Companies that had survived the downturn were not only seeing growth in sales again, but many of them were showing profits. As the economy grew, electronic commerce grew also, but at a more rapid pace. Thus, electronic commerce gradually became a larger part of the total economy. In the general economic recession that started in 2008, electronic commerce was hurt less than most of the economy. Even in the face of recession, the second wave of electronic commerce continued forward. The technologies that underlie the future expansion of electronic commerce continue to be developed. Today's handheld devices, including mobile telephones and tablet computers, offer the potential for a third wave in the evolution of online business. This section defines electronic commerce and describes its evolution from first wave into the second wave and outlines prospects for movement into a third wave of development.

Electronic Commerce and Electronic Business

To many people, the term "electronic commerce" means shopping on the part of the Internet called the World Wide Web (the Web). However, **electronic commerce (or e-commerce)** also includes many other activities, such as businesses trading with other businesses and internal processes that companies use to support their buying, selling, hiring, planning, and other activities. Some people use the term **electronic business (or e-business)** when they are talking about electronic commerce in this broader sense. For example, IBM defines electronic business as "the transformation of key business processes through the use of Internet technologies." Most people use the terms "electronic commerce" and "electronic business" interchangeably. In this book, the term electronic commerce (or e-commerce) is used in its broadest sense and includes all business activities that use Internet technologies. Internet technologies include the Internet, the World Wide Web, and other technologies such as wireless transmissions on mobile telephone networks. Companies that operate only online are often called **dot-com** or **pure dot-com** businesses to distinguish them from companies that operate in physical locations (solely or together with online operations).

Categories of Electronic Commerce

Categorizing electronic commerce by the types of entities participating in the transactions or business processes is a useful and commonly accepted way to define online business. The five general electronic commerce categories are business-to-consumer, business-to-business, transactions and business processes, consumer-to-consumer, and business-to-government. The three categories that are most commonly used are:

- Consumer shopping on the Web, often called **business-to-consumer** (or **B2C**)
- Transactions conducted between businesses on the Web, often called **business-to-business** (or **B2B**)
- Transactions and business processes in which companies, governments, and other organizations use Internet technologies to support selling and purchasing activities

A single company might participate in activities that fall under multiple e-commerce categories. Consider a company that manufactures stereo speakers. The company might sell its finished product to consumers on the Web, which would be B2C electronic commerce. It might also purchase the materials it uses to make the speakers from other companies on the Web, which would be B2B electronic commerce. Businesses often have entire departments devoted to negotiating purchase transactions with their suppliers. These departments are usually named **supply management** or **procurement**. Thus, B2B electronic commerce is sometimes called **e-procurement**.

In addition to buying materials and selling speakers, the company must also undertake many other activities to convert the purchased materials into speakers. These activities might include hiring and managing the people who make the speakers, renting or buying the facilities in which the speakers are made and stored, shipping the speakers, maintaining accounting records, obtaining customer feedback, purchasing insurance, developing advertising campaigns, and designing new versions of the speakers. An increasing number of these transactions and business processes can be done on the Web. Manufacturing processes (such as the fabrication of the speakers) can be controlled using Internet technologies within the business. All of these communication, control, and transaction-related activities have become an important part of electronic commerce. Some people include these activities in the B2B category; others refer to them as underlying or supporting business processes.

For more than 80 years, business researchers have been studying the ways people behave in businesses. This research has helped managers better understand how workers do their jobs and what motivates them to work more effectively. The research results have helped managers, and more recently, the workers themselves, improve job performance. By changing the nature of jobs, managers and workers can, as the saying goes, "work smarter, not harder." An important part of doing these job studies is to learn what activities each worker performs. In this setting, an **activity** is a task performed by a worker in the course of doing his or her job.

For a much longer time—centuries, in fact—business owners have kept records of how well their businesses are performing. The formal practice of accounting, or recording transactions, dates back to the Middle Ages. A **transaction** is an exchange of value, such as a purchase, a sale, or the conversion of raw materials into a finished product. By recording transactions, accountants help business owners keep score and measure how well they are doing. All transactions involve at least one activity, and some transactions involve many activities. Not all activities result in measurable (and therefore recordable) transactions. Thus, a transaction always has one or more activities associated with it, but an activity might not be related to a transaction.

The group of logical, related, and sequential activities and transactions in which businesses engage are often collectively referred to as **business processes**. Transferring funds, placing orders, sending invoices, and shipping goods to customers are all types of activities or transactions. For example, the business process of shipping goods to customers might include a number of activities (or tasks, or transactions), such as inspecting the goods, packing the goods, negotiating with a freight company to deliver the goods, creating and printing the shipping documents, loading the goods onto the truck, and sending payment to the freight company. One important way that the Web is helping people work more effectively is by enabling employees of many different kinds of companies to work at home or from other locations (such as while traveling). In this arrangement, called **telecommuting** or **telework**, the employee logs in to the company network through the Internet instead of traveling to an office.

Figure 1-1 shows the three main categories of electronic commerce. The figure presents a rough approximation of the relative sizes of these elements. In terms of dollar volume and number of transactions, B2B electronic commerce is much greater than B2C electronic commerce. However, the number of supporting business processes is greater than the number of all B2C and B2B transactions combined.

The large oval in Figure 1-1 that represents the business processes that support selling and purchasing activities is the largest element of electronic commerce.

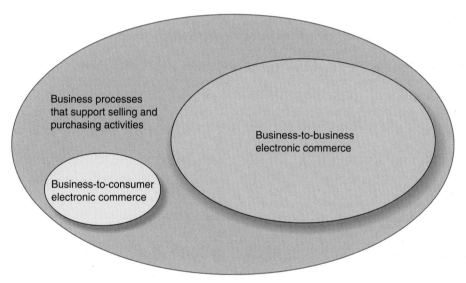

© Cengage Learning

FIGURE 1-1 Elements of electronic commerce

Some researchers define a fourth category of electronic commerce, called **consumer-to-consumer** (or **C2C**), which includes individuals who buy and sell items among themselves. For example, C2C electronic commerce occurs when a person sells an item through a Web auction site to another person. In this book, C2C sales are included in the B2C category because the person selling the item acts much as a business would for purposes of the transaction.

Finally, some researchers also define a category of electronic commerce called **business-to-government** (or **B2G**); this category includes business transactions with government agencies, such as paying taxes and filing required reports. An increasing number of states have Web sites that help companies do business with state government agencies. In this book, B2G transactions are included in the discussions of B2B electronic commerce. Figure 1-2 summarizes these five categories of electronic commerce.

Category	Description	Example
Business-to-consumer (B2C)	Businesses sell products or services to individual consumers.	Walmart.com sells merchandise to consumers through its Web site.
Business-to-business (B2B)	Businesses sell products or services to other businesses.	Grainger.com sells industrial supplies to large and small businesses through its Web site.
Business processes that support buying and selling activities	Businesses and other organizations maintain and use information to identify and evaluate customers, suppliers, and employees. Increasingly, businesses share this information in carefully managed ways with their customers, suppliers, employees, and business partners.	Dell Computer uses secure Internet connections to share current sales and sales forecast information with suppliers. The suppliers can use this information to plan their own production and deliver component parts to Dell in the right quantities at the right time.
Consumer-to-consumer (C2C)	Participants in an online marketplace can buy and sell goods to each other. Because one party is selling, and thus acting as a business, this book treats C2C transactions as part of B2C electronic commerce.	Consumers and businesses trade with each other in the eBay.com online marketplace.
Business-to-government (B2G)	Businesses sell goods or services to governments and government agencies. This book treats B2G transactions as part of B2C electronic commerce.	CA.gov procurement site allows businesses to sell online to the state of California.

FIGURE 1-2 Electronic commerce categories

The Development and Growth of Electronic Commerce

Over the thousands of years that people have engaged in commerce with one another, they have adopted the tools and technologies that became available. For example, the advent of sailing ships in ancient times opened new avenues of trade to buyers and sellers. Later innovations, such as the printing press, steam engine, and telephone, have changed the way people conduct commerce activities. The Internet has changed the way people buy, sell, hire, and organize business activities in more ways and more rapidly than any other technology in the history of business.

Electronic Funds Transfers (EFTs)

Although the Web has made online shopping possible for many businesses and individuals, in a broader sense, electronic commerce has existed for many years. For more than 40 years, banks have been using **electronic funds transfers** (**EFTs**, also called **wire transfers**), which are electronic transmissions of account exchange information over private communications' networks.

Electronic Data Interchange (EDI)

Businesses have also been engaging in a type of electronic commerce, known as electronic data interchange, for many years. **Electronic data interchange (EDI)** occurs when one business transmits computer-readable data in a standard format to another business. In the 1960s, businesses realized that many of the documents they exchanged were related to the shipping of goods; for example, invoices, purchase orders, and bills of lading. These documents included the same set of information for almost every transaction. Businesses also realized that they were spending a good deal of time and money entering this data into their computers, printing paper forms, and then reentering the data on the other side of the transaction. Although the purchase order, invoice, and bill of lading for each transaction contained much of the same information—such as item numbers, descriptions, prices, and quantities—each paper form usually had its own unique format for presenting the information. By creating a set of standard formats for transmitting the information electronically, businesses were able to reduce errors, avoid printing and mailing costs, and eliminate the need to reenter the data.

Businesses that engage in EDI with each other are called **trading partners**. The standard formats used in EDI contain the same information that businesses have always included in their standard paper invoices, purchase orders, and shipping documents. Firms such as General Electric, Sears, and Wal-Mart have been pioneers in using EDI to improve their purchasing processes and their relationships with suppliers. The U.S. government, which is one of the largest EDI trading partners in the world, was also instrumental in bringing businesses into EDI.

One problem that EDI pioneers faced was the high cost of implementation. Until the late 1990s, doing EDI meant buying expensive computer hardware and software and then either establishing direct network connections (using leased telephone lines) to all trading partners or subscribing to a value-added network. A **value-added network (VAN)** is an independent firm that offers connection and transaction-forwarding services to buyers and sellers engaged in EDI. Before the Internet came into existence as we know it today, VANs provided the connections between most trading partners and were responsible for ensuring the security of the data transmitted. VANs usually charged a fixed monthly fee plus a per-transaction charge, adding to the already significant expense of implementing EDI. Many smaller firms could not afford to participate in EDI and lost business to their larger competitors who could afford EDI.

In the late 1990s, many industry observers believed that the Internet would provide smaller companies with an alternative to EDI. Many articles in the trade press announced that the death of EDI was imminent. However, EDI continued to thrive because it was well entrenched in large companies. They had invested large amounts of money in their EDI systems and had built many of their sales, purchasing, and accounting systems around EDI. And the Internet, as an inexpensive communications medium, gave smaller companies a way to participate in EDI. The companies that operated VANs gradually moved EDI traffic to the Internet, and new companies developed other ways to help smaller businesses conduct EDI transactions on the Internet. These movements of EDI traffic to the Internet have dramatically reduced the cost of participating in EDI and have made it possible for even the smallest suppliers to do business with large customers who require its use. As a result, EDI continues to be a large portion of B2B electronic commerce and is growing steadily every year in number of transactions and dollar volume. You will learn more about EDI, VANs, and new B2B transaction technologies in Chapter 5.

The Dot-Com Boom, Bust, and Rebirth

Between 1997 and 2000, more than 12,000 Internet-related businesses were started with more than $100 billion of investors' money. In an extended burst of optimism, and what many later described as irrational exuberance, investors feared that they might miss the money-making opportunity of a lifetime. As more investors competed for a fixed number of good ideas, the price of those ideas increased. Many good ideas suffered from poor implementation. Worse, a number of bad ideas were proposed and funded. More than 5,000 of these Internet start-up firms went out of business or were acquired in the downturn that began in 2000. The media coverage of the "dot-com bust" was extensive. However, between 2000 and 2003, more than $200 billion was invested in purchasing electronic commerce businesses that were in trouble and starting new online ventures, according to the industry research firm WebMergers. This second wave of financial investment was not reported extensively in either the general or business media, but these investments quietly fueled a rebirth of growth in online business activity. This second wave provided another chance at success for many online business ideas that were poorly implemented in the early days of the Internet.

Despite the many news stories that appeared between 2000 and 2002 proclaiming the death of electronic commerce, the growth in online B2C sales actually had continued through that period, although at a slower pace than during the boom years of the late 1990s. Thus, the "bust" that was so widely reported in the media was really more of a slowdown than a collapse. After four years of doubling or tripling every year, growth in online sales slowed to an annual rate of 20 to 30 percent starting in 2001. This growth rate continued through the recession of 2008–2009.

The 2008–2009 global recession devastated many traditional retailers, particularly in the United States and Europe. Large Asian economies, such as those in China and India, were affected less and continued to expand. Around the globe, online sales overall continued to grow during that period, although at a lower rate than the 20 to 30 percent annual rates achieved earlier in the decade. As many traditional businesses remain mired in the aftereffects of that recession, online business activity has picked up and appears to be at the leading edge of economic growth. Online business growth in Asia continued at relatively high rates throughout the recessionary period, which boosted global online sales totals. In fact, online retail sales in China exceeded those in the United States for the first time in 2010. Most experts expect to see global online business grow at a sustained rate of 15 to 25 percent through 2015.

One force driving the growth in global online sales to consumers is the ever-increasing number of people who have access to the Internet. Today, billions of people around the world still do not have computers and, therefore, do not have computer access to the Internet. The predictions for continued global online business are based in part on the growing numbers of people using inexpensive devices such as mobile phones and tablet computers to access the Internet.

In addition to the growth in the B2C sector, B2B sales online have been increasing steadily for almost two decades. The dollar total of B2B online sales has been greater than B2C sales because B2B incorporates EDI, a technology that accounted for more than $400 billion per year in transactions in 1995, when Internet-based electronic commerce was just beginning. This book defines B2B sales as including companies' transactions with other businesses, with their employees, and with governmental agencies (for example, when they pay their taxes) because these business processes are all candidates for the application of Internet technologies.

The dollar amount of these B2B transactions is substantial. Intel is one example of a company that sells its products to other businesses rather than to consumers. Intel accepts more than 98 percent of its orders (more than $38 billion per year) through the Internet. Intel also purchases billions of dollars' worth of supplies and raw materials on the Web each year. The total volume of all worldwide business activities on the Web is expected to exceed $11.9 trillion by 2013. Figure 1-3 summarizes the growth of actual and estimated global online sales for the B2C and B2B categories.

Year	B2C Sales: Actual and Estimated $ Billions	B2B Sales (including EDI): Actual and Estimated $ Billions
2013	963	11,900
2012	821	10,600
2011	681	9,500
2010	573	8,600
2009	487	7,500
2008	453	6,500
2007	426	5,600
2006	361	4,800
2005	255	4,100
2004	179	2,800
2003	103	1,600
2002	91	900
2001	73	730
2000	52	600
1999	26	550
1998	11	520
1997	5	490
1996	Less than 1	460

Adapted from reports by ClickZ Network (http://www.clickz.com/stats/stats_toolbox/); eMarketer (http://www.emarketer.com/); Forrester Research (http://www.forrester.com); Internet Retailer (http://www.internetretailer.com), the Statistical Abstract of the United States, 2008, Washington: U.S. Census Bureau, and the Statistical Abstract of the United States, 2011, Washington: U.S. Census Bureau.

FIGURE 1-3   Actual and estimated online sales in B2C and B2B categories

The Second Wave of Electronic Commerce

Many researchers have noted that electronic commerce is a major change in the way business is conducted and compare it to other historic changes in economic organization such as the Industrial Revolution. However, the Industrial Revolution was not a single event, but a series of developments that took place over a 50- to 100-year period. Economists Chris Freeman and Francisco Louçã describe four distinct waves (or phases) that occurred in the Industrial Revolution in their book *As Time Goes By* (see the For Further Study and Research section at the end of this chapter). In each wave, different business strategies were successful. Electronic commerce and the information revolution brought about by the Internet will likely go through a series of waves, too. Researchers agree that the second wave of electronic commerce is well under way. This section outlines the defining characteristics of the first wave of electronic commerce and describes how the second wave is different. Later, you will learn about the third wave that is taking shape.

The first wave of electronic commerce was predominantly a U.S. phenomenon. Web pages were primarily in English, particularly on commerce sites. The second wave is characterized by its international scope, with sellers doing business in many countries and in many languages. Language translation and currency conversion have been two impediments to the efficient conduct of global business in the second wave. You will learn more about the issues that arise in global electronic commerce later in this chapter, in Chapter 7, which concerns legal issues, and in Chapter 11, which concerns online payment systems.

In the first wave, easy access to start-up capital led to an overemphasis on creating new large enterprises to exploit electronic commerce opportunities. Investors were excited about electronic commerce and wanted to participate, no matter how much it cost or how weak the underlying ideas were. In the second wave, established companies are using their own internal funds to finance gradual expansion of electronic commerce opportunities. These measured and carefully considered investments are helping electronic commerce grow more steadily, though more slowly.

The Internet technologies used in the first wave, especially in B2C commerce, were slow and inexpensive. Most consumers connected to the Internet using dial-up modems. The increase in broadband connections in homes is a key element in the B2C component of the second wave. In 2004, the number of U.S. homes with broadband connections began to increase rapidly. Most industry estimates showed that about 12 percent of U.S. homes had broadband connections in early 2004. By late 2011, those estimates were ranging between 80 and 85 percent. Other countries, such as South Korea, subsidize their citizens' Internet access and have an even higher rate of broadband usage. The increased use of home Internet connections to transfer large audio and video files is generally seen as the reason large numbers of people spent the extra money required to obtain a broadband connection. The increased speed of broadband not only makes Internet use more efficient, but it also can alter the way people use the Web. For example, a broadband connection allows a user to watch movies and television programs online—something that is impossible to do with a dial-up connection. This opens up more opportunities for businesses to make online sales. It also changes the way that online retailers can present their products to Web site visitors. Although business customers, unlike retail customers, have had fast connections to the Internet for many years, the increasing availability of wireless Internet connections has increased the volume and nature of B2B electronic commerce. Salespeople using laptop computers can stay in touch with customers, prepare quotes, and check on orders being fulfilled from virtually anywhere they happen to be. You will learn more about different types of connections in Chapter 2 and how connection speed can affect consumers' online shopping experiences in Chapters 3 and 4. You will learn how the pervasiveness of computers (laptops and tablets) and mobile phones that can access the Internet is changing B2B electronic commerce in Chapter 5.

Electronic mail (or e-mail) was used in the first wave as a tool for relatively unstructured communication. In the second wave, both B2C and B2B sellers began using e-mail as an integral part of their marketing and customer contact strategies. You will learn about e-mail technologies in Chapter 2 and e-mail marketing in Chapter 4.

Online advertising was the main intended revenue source of many failed dot-com businesses in the first wave. After a two-year dip in online advertising activity and revenues, companies began the second wave with a renewed interest in making the Internet work as an effective advertising medium. Some categories of online advertising, such as employment services (job wanted ads) are growing rapidly and are replacing traditional advertising outlets. Companies such as Google have devised ways of delivering specific ads to Internet users who are most likely to be interested in the products or services offered by those ads. You will learn about second-wave advertising strategies in Chapter 4.

The sale of digital products was fraught with difficulties during the first wave of electronic commerce. The music recording industry was unable (or, some would say, unwilling) to devise a way to distribute digital music on the Web. This created an environment in which digital piracy—the theft of musical artists' intellectual property—became rampant. The promise of electronic books was also unfulfilled. The second wave is fulfilling the promise of available technology by supporting the legal distribution of music, video, and other digital products on the Web. Apple Computer's **iTunes** Web site is an example of a second-wave digital product distribution business that is meeting the needs of consumers and its industry. You will learn more about digital product distribution strategies in Chapter 3 and about the related legal issues in Chapter 7.

Another group of technologies have emerged that have combined to make new businesses possible on the Web. The general term for these technologies is **Web 2.0**, and they include software that allows users of Web sites to participate in the creation, editing, and distribution of content on a Web site owned and operated by a third party. Sites such as Wikipedia, YouTube, and Facebook use Web 2.0 technologies. Customer relationships management software that runs from the Web, such as Salesforce.com, also uses Web 2.0 technologies. You will learn about Web 2.0 business opportunities throughout this book and you will learn about the technologies used to implement them in Chapter 9.

In the first wave of electronic commerce, many companies and investors believed that being the first Web site to offer a particular type of product or service would give them an opportunity to be successful. This strategy is called the **first-mover advantage**. As business researchers studied companies who had tried to gain a first-mover advantage, they learned that being first did not always lead to success (see the Suarez and Lanzolla article reference in the For Further Study and Research section at the end of this chapter). First movers must invest large amounts of money in new technologies and make guesses about what customers will want when those technologies are functioning. The combination of high uncertainty and the need for large investments makes being a first mover very risky. As many business strategists have noted, "It is the second mouse that gets the cheese."

First movers that were successful tended to be large companies that had an established reputation (or brand) and that also had marketing, distribution, and production expertise. First movers that were smaller or that lacked the expertise in these areas tended to be unsuccessful. Also, first movers that entered highly volatile markets or in those industries with high rates of technological change often did not do well. In the second wave, fewer businesses rely on a first-mover advantage when they take their businesses online. A good example of a company that was successful in the second wave by not being a first mover is illustrated in the opening case for this chapter about Google.

Figure 1-4 shows a summary of some key characteristics of the first and second wave of electronic commerce. This list can never be complete because every day brings new technologies and combinations of existing technologies that make additional second-wave opportunities possible.

Electronic Commerce Characteristic	First Wave	Second Wave
International character of electronic commerce	Dominated by U.S. companies	Global enterprises in many countries participating in electronic commerce
Languages	Most electronic commerce Web sites in English	Many electronic commerce Web sites available in multiple languages
Funding	Many new companies started with outside investor money	Established companies funding electronic commerce initiatives with their own capital
Connection technologies	Many electronic commerce participants used slow Internet connections	Rapidly increasing use of broadband technologies for Internet connections
E-mail contact with customers	Unstructured e-mail communication with customers	Customized e-mail strategies now integral to customer contact
Advertising and electronic commerce integration	Reliance on simple forms of online advertising as main revenue source	Use of multiple sophisticated advertising approaches and better integration of electronic commerce with existing business processes and strategies
Distribution of digital products	Widespread piracy due to ineffective distribution of digital products	New approaches to the sale and distribution of digital products
First-mover advantage	Rely on first-mover advantage to ensure success in all types of markets and industries	Realize that first-mover advantage leads to success only for some companies in certain specific markets and industries

FIGURE 1-4   Key characteristics of the first two waves of electronic commerce

The Third Wave Begins

Since about 2001, industry analysts have been predicting the emergence of mobile telephone-based commerce (often called **mobile commerce** or **m-commerce**) every year. And year after year, they were surprised that the expected development of mobile commerce did not occur. The limited capabilities of mobile telephones were a major impediment until very recently.

Mobile commerce is finally taking off with the increasingly widespread use of mobile phones that allow Internet access and smart phones. **Smart phones** are mobile phones that include a Web browser, a full keyboard, and an identifiable operating system that allows users to run various software packages. These phones are available with usage plans that include very high or even unlimited data transfers at a fixed monthly rate.

Another technological development was the introduction of tablet computers. These handheld devices are larger than a smart phone but smaller than a laptop computer. Most tablet computers (and smart phones) can connect to the Internet through a wireless phone service carrier or a local wireless network. This flexibility is important, especially if the wireless data plan restricts the amount of data that can be downloaded. The availability of these devices and the low cost of Internet connectivity have made mobile commerce possible on a large scale for the first time. Leading online business research firms, including Forrester, Coda, and ABI Research, estimate mobile commerce to be about $1 billion today but expect rapid growth to levels between $10 billion and $30 billion by 2015.

One of the most important changes brought about by fully operational handheld devices is that the Internet becomes truly available everywhere. This constant availability can change buyer behavior in many ways (discussed in Chapters 3 and 4) and it can provide new opportunities for online businesses that could not exist without such broad-based connectivity. You will learn about these opportunities for mobile commerce in Chapter 6.

In the first two waves, Internet technologies were integrated into B2B transactions and internal business processes by using bar codes and scanners to track parts, assemblies, inventories, and production status. These tracking technologies were not well integrated. Also, companies sent transaction information to each other using a patchwork of communication methods, including fax, e-mail, and EDI. In the third wave, Radio Frequency Identification (RFID) devices and smart cards are being combined with biometric technologies, such as fingerprint readers and retina scanners, to control more items and people in a wider variety of situations. These technologies are increasingly integrated with each other and with communication systems that allow companies to communicate with each other and share transaction, inventory level, and customer demand information effectively. You will learn more about how these technologies are integrated with B2B electronic commerce in Chapter 5.

The Web 2.0 technologies that enabled part of the growth in electronic commerce that occurred in the second wave will play a major role in the third wave. For example, Web sites such as Facebook and technologies such as Twitter can be used to engage in social commerce. **Social commerce** is the use of interpersonal connections online to promote or sell goods and services. Because a handheld device connected to the Internet can put a user online virtually all the time, social interactions can be used to advertise, promote, or suggest specific products or services. Internet Retailer notes that current social commerce sales are under $1 billion but expects volume to increase to $14 billion by 2015. You will learn more about social commerce in Chapter 6.

Large businesses—both existing businesses and new businesses that had obtained large amounts of capital early on—dominated the first wave. The second wave saw a major increase in the participation of small businesses (those with fewer than 200 employees) in the online economy. Still, more than 30 percent of small businesses in the United States do not have Web sites. In other parts of the world, this percentage is much higher. The third wave of electronic commerce will include the participation of a significantly larger proportion of these smaller businesses. Providing services that help smaller companies use electronic commerce will also be a substantial area of growth.

Not all of the future of electronic commerce is based on second and third wave developments. Some of the most successful first-wave companies, such as Amazon.com, eBay, and Yahoo!, continue to grow by offering increasingly innovative products and services. The third wave of electronic commerce will provide new opportunities for these businesses, too.

BUSINESS MODELS, REVENUE MODELS, AND BUSINESS PROCESSES

A **business model** is a set of processes that combine to achieve a company's primary goal, which is typically to yield a profit. In the first wave of electronic commerce, many investors tried to find start-up companies that had new, Internet-driven business models. These investors expected that the right business model would lead to rapid sales growth and market dominance. If a company was successful using a new "dot-com" business model, investors would clamor to copy that model or find a start-up company that planned to use a similar business model. This strategy led the way to many business failures, some of them quite dramatic.

In the wake of the dot-com debacle that ended the first wave of electronic commerce, many business researchers analyzed the efficacy of this "copy a successful business model" approach and began to question the advisability of focusing great attention on a company's business model. One of the main critics, Harvard Business School professor Michael Porter, argued that business models not only did not matter, they probably did not exist. (You can read more about Porter's criticisms of the business model approach in the articles cited in the For Further Study and Research section at the end of this chapter.)

Today, most companies realize that copying or adapting someone else's business model is neither an easy nor wise road map to success. Instead, companies should examine the elements of their business; that is, they should identify business processes that they can streamline, enhance, or replace with processes driven by Internet technologies.

Companies and investors do use the idea of a **revenue model**, which is a specific collection of business processes used to identify customers, market to those customers, and generate sales to those customers. The revenue model idea is helpful for classifying revenue-generating activities for communication and analysis purposes. The details of revenue models that are used on the Web are presented in Chapter 3.

Focus on Specific Business Processes

In addition to the revenue model grouping of business processes, companies think of the rest of their operations as specific business processes. Those processes include purchasing raw materials or goods for resale, converting materials and labor into finished goods, managing transportation and logistics, hiring and training employees, managing the finances of the business, and many other activities.

An important function of this book is to help you learn how to identify those business processes that firms can accomplish more effectively by using electronic commerce technologies. In some cases, business processes use traditional commerce activities very effectively, and technology cannot improve them. Products that buyers prefer to touch, smell, or examine closely can be difficult to sell using electronic commerce. For example, customers might be reluctant to buy items that have an important element of tactile feel or condition such as high-fashion clothing (you cannot touch it online and subtle color variations that are hard to distinguish on a computer monitor can make a large difference) or antique jewelry (for which elements of condition that require close inspection can be critical to value) if they cannot closely examine the products before agreeing to purchase them.

This book will help you learn how to use Internet technologies to improve existing business processes and identify new business opportunities. An important aspect of electronic commerce is that firms can use it to help them adapt to change. The business world is changing more rapidly than ever before. Although much of this book is devoted to explaining technologies, the book's focus is on the business of electronic commerce; the technologies only enable the business processes.

Role of Merchandising

Retail merchants have years of traditional commerce experience in creating store environments that help convince customers to buy. This combination of store design, layout, and product display knowledge is called **merchandising**. In addition, many salespeople have developed skills that allow them to identify customer needs and find products or services that meet those needs.

The skills of merchandising and personal selling can be difficult to practice remotely. However, companies must be able to transfer their merchandising skills to the Web for their Web sites to be successful. Some products are easier to sell on the Internet than others because the merchandising skills related to those products are easier to transfer to the Web. You will learn more about how merchandising can be accomplished online in Chapters 3 and 4.

Product/Process Suitability to Electronic Commerce

Some products, such as books or CDs, are good candidates for electronic commerce because customers do not need to experience the physical characteristics of the particular item before they buy it. Because one copy of a new book is identical to other copies, and because the customer is not concerned about fit, freshness, or other such qualities, customers are usually willing to order a title without examining the specific copy they will receive. The advantages of electronic commerce, including the ability of one site to offer a wider selection of titles than even the largest physical bookstore, can outweigh the advantages of a traditional bookstore—for example, the customer's ability to browse the pages of the books. In later chapters, you will learn how to evaluate the advantages and disadvantages of using electronic commerce for specific business processes. Figure 1-5 lists examples of business processes categorized by suitability for electronic commerce and traditional commerce. As technologies develop, many processes that were strictly handled through traditional commerce have become more suitable for electronic commerce. This trend will likely continue. You will learn more about transitions of this type in Chapter 3.

Well Suited to Electronic Commerce	Suited to a Combination of Electronic and Traditional Commerce Strategies	Well Suited to Traditional Commerce
Sale/purchase of books and CDs	Sale/purchase of automobiles	Sale/purchase of impulse items for immediate use
Sale/purchase of goods that have strong brand reputations	Banking and financial services	Sale/purchase of used, unbranded goods
Online delivery of software and digital content, such as music and movies	Roommate-matching services	
Sale/purchase of travel services	Sale/purchase of residential real estate	
Online shipment tracking	Sale/purchase of high-value jewelry and antiques	
Sale/purchase of investment and insurance products		

FIGURE 1-5   Business process suitability to type of commerce

One business process that is especially well-suited to electronic commerce is the selling of commodity items. A **commodity item** is a product or service that is hard to distinguish from the same products or services provided by other sellers; its features have become standardized and well known. The only difference a buyer perceives when shopping for a commodity item is its price. Gasoline, office supplies, soap, computers, and airline transportation are all examples of commodity products or services, as are the books and CDs sold by Amazon.com.

Not all commodity items are good candidates for electronic commerce. They must have an attractive shipping profile to be sold online. A product's **shipping profile** is the collection of attributes that affect how easily that product can be packaged and delivered. A high value-to-weight ratio can help by making the overall shipping cost a small fraction of the selling price. A DVD is an excellent example of an item that has a high value-to-weight ratio. Products that are consistent in size, shape, and weight can make warehousing and shipping much simpler and less costly. Commodity items that have an attractive shipping profile include books, clothing, shoes, kitchen accessories, and many other small household items.

A product that has a strong brand reputation—such as a Kodak camera—is easier to sell on the Web than an unbranded item, because the brand's reputation reduces the buyer's concerns about quality when buying that item sight unseen. Expensive jewelry has a high value-to-weight ratio, but many people are reluctant to buy it without examining it in person unless the jewelry is sold under a well-known brand name and with a generous return policy.

Other items that are well-suited to electronic commerce are those that appeal to small, but geographically dispersed, groups of customers. Collectible comic books are an example of this kind of product.

Traditional commerce, rather than electronic commerce, can be a better way to sell items that rely on personal selling skills. For example, sales of commercial real estate involve large amounts of money and a high degree of interpersonal trust. Even if commercial real estate is listed online, it will usually require personal contact to negotiate the deal. Many businesses are using a combination of personal contact enhanced by an online presence to sell items such as high-fashion clothing, antiques, or specialized food items.

A combination of electronic and traditional commerce strategies works best when the business process includes both commodity and personal inspection elements. For example, most people find information on the Web about new and used automobiles and do considerable research on specific makes and models before they visit a dealership to buy. In the case of used cars, electronic commerce provides a good way for buyers to obtain information about available models, features, reliability, prices, and dealerships, and also helps buyers find specific vehicles that meet their exact requirements. The range of conditions of used cars makes the traditional commerce component of personal inspection a key part of the transaction negotiation.

ELECTRONIC COMMERCE: OPPORTUNITIES, CAUTIONS, AND CONCERNS

Electronic commerce has changed the way business is conducted in many industries. However, not every business process is suitable for electronic commerce. As technologies advance, more and more types of business processes become candidates for electronic commerce. This section outlines some opportunities and points out some cautions that businesses should consider in evaluating opportunities to engage in online business activities.

Opportunities for Electronic Commerce

Electronic commerce is attractive to businesses because, quite simply, it can help increase profits. It can do this because electronic commerce can increase sales and decrease business costs. Advertising done well on the Web can get even a small firm's promotional message out to potential customers in every country in the world. A firm can use electronic commerce to reach small groups of customers that are geographically scattered. The Web is particularly useful in creating virtual communities that become ideal target markets for specific types of products or services. A **virtual community** is a gathering of people who share a common interest, but instead of this gathering occurring in the physical world, it takes place on the Internet. In recent years, virtual communities have taken advantage of Web 2.0 technologies to make their activities more accessible and interesting to community members. Thomas Petzinger has written extensively in his *Wall Street Journal* newspaper columns and his book, *The New Pioneers*, about new patterns of work and commerce that have evolved from these virtual communities. You will learn about Web sites (called **social networking sites**) that individuals and businesses use to conduct social interactions online and the business opportunities they present in Chapter 6.

Just as electronic commerce increases sales opportunities for the seller, it also increases purchasing opportunities for the buyer. Businesses can use electronic commerce to identify new suppliers and business partners. Negotiating price and delivery terms is easier in electronic commerce because the Internet can help companies efficiently obtain competitive bid information. Electronic commerce increases the speed and accuracy with which businesses can exchange information, which reduces costs on both sides of transactions. Many companies are reducing their costs of handling sales inquiries, providing price quotes, and determining product availability by using electronic commerce in their sales support and order-taking processes.

Cisco Systems, a leading manufacturer of computer networking equipment, currently sells almost all its products online. Because no customer service representatives are involved in making these sales, Cisco operates very efficiently. In 1998, the first year in which its online sales initiative was operational, Cisco made 72 percent of its sales on the Web. Cisco avoided handling 500,000 calls per month and saved $500 million in that first year. Today, Cisco conducts more than 99 percent of its purchase and sales transactions online.

Electronic commerce provides buyers with a wider range of choices than traditional commerce because buyers can consider many different products and services from a wider variety of sellers. This wide variety is available for consumers to evaluate 24 hours a day, every day. Some buyers prefer a great deal of information in deciding on a purchase; others prefer less. Electronic commerce provides buyers with an easy way to customize the level of detail in the information they obtain about a prospective purchase. Instead of waiting days for the mail to bring a catalog or product specification sheet, or even minutes for a fax transmission, buyers can have instant access to detailed information on the Web. Allowing customers to create their own ideal information environment saves money and provides an opportunity for increased sales.

Most digital products, such as software, music, video, or images, can be delivered through the Internet to reduce the time buyers must wait to begin using their purchases. The ability to deliver digital products online is not just a cost-reduction strategy; it can provide an opportunity for increased sales. Intuit sells its TurboTax income tax preparation software online and lets customers download the software immediately if they wish. Intuit sells a considerable amount of TurboTax software late in the evening on April 14 each year. (April 15 is the deadline for filing personal income tax returns in the United States.)

The benefits of electronic commerce extend to the general welfare of society. Electronic payments of tax refunds, public retirement, and welfare support cost less to issue and arrive securely and quickly when transmitted over the Internet. Furthermore, electronic payments can be easier to audit and monitor than payments made by check, providing protection against fraud and theft losses. To the extent that electronic commerce enables people to telecommute, everyone benefits from the reduction in commuter-caused traffic and pollution. Electronic commerce can also make products and services available in remote areas. For example, distance learning makes it possible for people to learn skills and earn degrees no matter where they live or which hours they have available for study.

Electronic Commerce: Cautions and Concerns

Some business processes might never lend themselves to electronic commerce. For example, perishable foods and high-cost, unique items such as custom-designed jewelry can be very difficult to inspect adequately from a remote location, regardless of any technologies that might be devised in the future. Most of the cautions and concerns regarding electronic commerce today, however, stem from the rapidly developing pace of the underlying technologies and the reluctance of people to change the way they do things. These barriers have disappeared for many types of online business and will continue to disappear as electronic commerce matures and becomes more generally accepted.

The Need for a Critical Mass

Some products and services require that a critical mass of potential buyers be equipped and willing to buy through the Internet. For example, online grocers such as **Peapod** initially offered their delivery services only in a few cities. As more of Peapod's potential customers became connected to the Internet and felt comfortable with purchasing online, the company was able to expand slowly and carefully into more geographic areas. After more than 10 years of operation, Peapod operates in fewer than 20 U.S. metropolitan areas. Most online grocers focus their sales efforts on packaged goods and branded items. Perishable grocery products, such as fruit and vegetables, are much harder to sell online because customers want to examine and select specific items for freshness and quality. Peapod is a good example of how challenging it can be to build a business in an industry that requires this kind of critical mass. Although it was one of the first online grocery stores, Peapod has had a difficult time staying in business, and was even offline for a short time in 2000. Peapod was subsequently acquired by Royal Ahold, a European firm that was willing to invest additional cash to keep it in operation. Two of Peapod's major competitors, WebVan and HomeGrocer, were unable to stay in business long enough to attract a sufficient customer base.

Established traditional grocery chains in the United States such as **Safeway** also offer online ordering and delivery services in a second wave of using Internet technologies in the grocery business. By using their existing infrastructure (including warehouses, purchasing systems, and physical stores in multiple locations), they are able to avoid having to make the large capital investment in facilities that led to the demise of first-wave dot-com grocers such as WebVan and HomeGrocer.

One online grocer that has successfully implemented an updated version of the WebVan and HomeGrocer operational approach is **FreshDirect**. By limiting its service area to the densely populated region in and around New York City, FreshDirect has found the right combination of operating scale and market. The company started in 2002 and achieved profitability in 2004 with sales of $90 million. This is a much smaller sales volume than either WebVan or HomeGrocer would have needed to be profitable.

Outside the United States, online grocers have done quite well. Three of the most successful online grocery efforts in the world are **Grocery Gateway** in Toronto, **Disco Virtual** in Buenos Aires, and **Tesco** in the United Kingdom. Grocery Gateway and Disco Virtual operate in densely populated urban environments that offer sufficiently large numbers of customers within relatively small geographic areas, which make their delivery routes profitable. Tesco started its operations in London, which offers a similar densely populated urban area. However, Tesco has also expanded its operations to selected rural areas that are near a Tesco supermarket.

Predictability of Costs and Revenues

Businesses often calculate return-on-investment numbers before committing to any new technology. This has been difficult to do for investments in electronic commerce because the costs and benefits are often hard to quantify or predict with any degree of accuracy. Costs that are a function of technology can change dramatically even during a short-lived online business implementation project because the underlying technologies are changing so rapidly.

Many firms have had trouble recruiting and retaining employees with the technological, design, or business process skills needed to take their business online. Larger firms often try to use existing personnel who are steeped in traditional ways of doing business. These employees often have difficulty adapting what they have learned about the business to an online environment in which the risks and benefits are often very different. You will learn more about return-on-investment calculations and employee recruitment and retention issues in Chapter 12.

Technology Integration Issues

Another problem facing firms that want to do business on the Internet is the difficulty of integrating existing databases and transaction-processing software designed for traditional commerce into the software that enables electronic commerce. Although a number of companies offer software design and consulting services that promise to tie existing systems into new online business systems, these services can be expensive. The outcome of any systems integration effort can be highly uncertain as well. You will learn more about how companies deal with these software issues in Chapter 9.

Cultural and Legal Concerns

In addition to technology and software issues, many businesses face cultural and legal obstacles to conducting all types of electronic commerce. B2C electronic commerce must deal with the fact that many consumers are still fearful of sending their credit card numbers over the Internet and having online merchants—merchants they have never met—know so much about them. Other consumers are simply resistant to change and are uncomfortable viewing merchandise on a computer screen rather than in person.

B2B electronic commerce is also affected by cultural and legal considerations. The details of business transactions are often not specified; businesses frequently rely on a long history of doing business a particular way. These established business practices can vary greatly from country to country and making assumptions when engaging in international commerce can be disastrous. You will learn more about electronic commerce security, privacy issues, and payment systems later in this book.

The legal environment in which electronic commerce is conducted is full of unclear and conflicting laws. In many cases, government regulators have not kept up with technologies. As you will learn in Chapter 7, laws that govern commerce were written when signed documents were a reasonable expectation in any business transaction. However, as more businesses and individuals find the benefits of electronic commerce to be compelling, many of these technology- and culture-related disadvantages will be resolved or seem less problematic.

Pets.com

In February 1999, Pets.com launched its Web site with the hopes of making substantial sales to the 60 percent of U.S. households that own pets and spend more than $20 billion each year feeding, entertaining, and caring for them. More than 10,000 stores sold pet supplies. These stores included small retail outlets, grocery stores, discount retailers (such as Wal-Mart and Costco), and a new generation of pet superstores. Pets.com had acquired an excellent domain name and intended to exploit the opportunities presented by high levels of investor interest in funding electronic commerce companies. The plan for Pets.com was to spend heavily to develop a brand and a Web presence that would rapidly make the company the premier online source for pet-related products.

After launching the site, Pets.com raised $110 million from private investors in 1999, and another $80 million in a public sale of stock in early 2000. Pets.com spent more than $100 million of the money on advertising during its short life. It also spent significant sums to create a Web store that offered more than 12,000 different products. In November 2000—less than two years after launching its Web site—Pets.com went out of business.

Pets.com had created an electronic commerce initiative in an industry in which online business offered few advantages over traditional commerce. The products had a very low value-to-weight ratio. The shipping costs for pet food, one of the company's best-selling product categories, caused it to lose money on every sale. Pet products come in all shapes, sizes, and weights, and are, therefore, difficult to pack and ship efficiently. Pets.com was also spending money rapidly at a time when investors were beginning to question the long-run viability of all electronic commerce businesses. The lesson here is that Pets.com could not develop any sustainable advantage over traditional pet stores. Without such an advantage, the business was doomed.

In the years following the Pets.com failure, a number of companies such as PETCO and PetFoodDirect.com began selling pet food and related items online. These companies were more careful than Pets.com was about what they offered for sale. By selling only items that had an appropriate shipping profile, many of these companies have now become successful. For example, veterinarians who formulate foods that meet the needs of specific pet diets are finding they can charge enough for those products to make online sales profitable.

ECONOMIC FORCES AND ELECTRONIC COMMERCE

Economics is the study of how people allocate scarce resources. One important way that people allocate resources is through commerce (the other major way is through government actions, such as taxes or subsidies). Many economists are interested in how people organize their commerce activities. One way people do this is to participate in markets. Economists use a formal definition of **market** that includes two conditions: first, that the potential sellers of a good come into contact with potential buyers, and second, that a medium of exchange is available. This medium of exchange can be currency or barter. Most economists agree that markets are strong and effective mechanisms for allocating scarce resources. Thus, one would expect most business transactions to occur within markets. However, much business activity today occurs within large **hierarchical business organizations**, which economists generally refer to as **firms**, or **companies**.

Most hierarchical organizations are headed by a top-level president or chief operating officer. Reporting to the president are a number of executives who, in turn, have a larger number of middle managers who report to them, and so on. An organization can have a relatively flat hierarchy, in which there are only a few levels of management, or it can have many reporting levels. In either case, the bottom level includes the largest number of employees and is usually made up of production workers or service providers. Thus, the hierarchical organization always has a pyramid-shaped structure.

These large firms often conduct many different business activities entirely within the organizational structure of the firm and participate in markets only for purchasing raw materials and selling finished products. If markets are indeed highly effective mechanisms for allocating scarce resources, these large corporations should participate in markets at every stage of their production and value-generation processes. Nobel laureate Ronald Coase wrote an essay in 1937 in which he questioned why individuals who engaged in commerce often created firms to organize their activities. He was particularly interested in the hierarchical structure of these business organizations. Coase concluded that transaction costs were the main motivation for moving economic activity from markets to hierarchically structured firms.

Transaction Costs

Transaction costs are the total of all costs that a buyer and seller incur as they gather information and negotiate a purchase-and-sale transaction. Although brokerage fees and sales commissions can be a part of transaction costs, the cost of information search and acquisition is often far larger. Another significant component of transaction costs can be the investment a seller makes in equipment or in the hiring of skilled employees to supply the product or service to the buyer.

To understand better how transaction costs occur in markets, consider the following example: A sweater dealer could obtain sweaters by engaging in market transactions with a number of independent sweater knitters. Each knitter could sell sweaters to one or several dealers. Transaction costs incurred by the dealer would include the costs of identifying the independent knitters, visiting them to negotiate the purchase price, arranging for delivery of the sweaters, and inspecting the sweaters on arrival. The knitters would also incur costs, such as the purchase of knitting supplies. Because individual knitters could not know whether any sweater dealer would ever buy sweaters from them, the investments they make to enter the sweater-knitting business have an uncertain yield. This risk is a significant transaction cost for the knitters.

After purchasing the sweaters, sweater dealers take them to a different market in which sweater dealers meet and do business with the retail shops that sell sweaters to the consumer. The dealers can learn which colors, patterns, and styles are in demand from price and quantity negotiations with the retail shops in this market. The sweater dealers can then use that information to negotiate price and other terms in the knitters' market. A diagram of this set of markets appears in Figure 1-6.

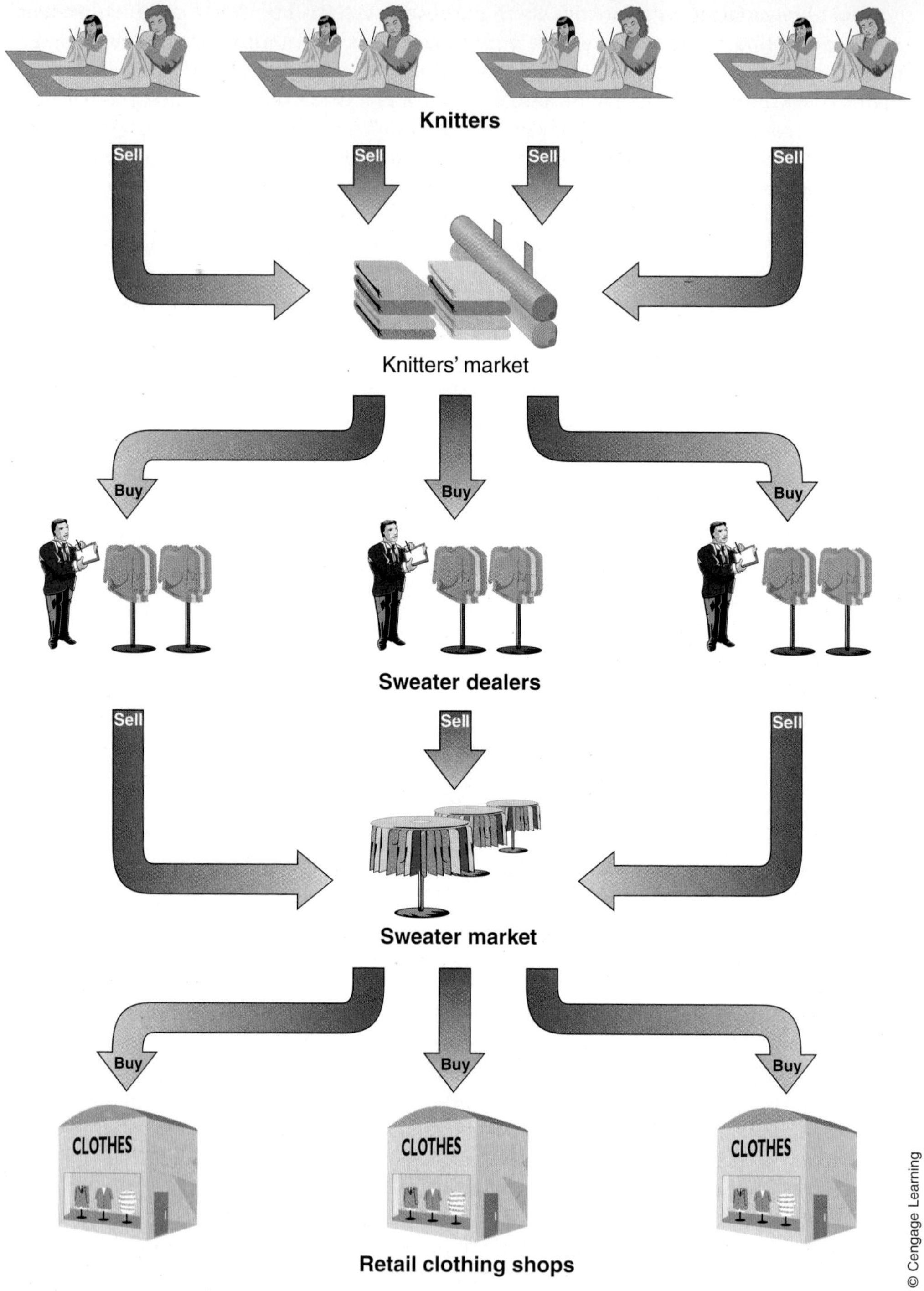

FIGURE 1-6   Market form of economic organization

© Cengage Learning

Markets and Hierarchies

Coase reasoned that when transaction costs were high, businesspeople would form organizations to replace market-negotiated transactions. These organizations would be hierarchical and would include strong supervision and worker-monitoring elements.

Instead of negotiating with individuals to purchase sweaters they had knit, a hierarchical organization would hire knitters, and then supervise and monitor their work activities. This supervision and monitoring system would include flows of monitoring information from the lower levels to the higher levels of the organization. It would also have control of information flowing from the upper levels of the organization to the lower levels. Although the costs of creating and maintaining a supervision and monitoring system are high, they can be lower than transaction costs in many instances.

In the sweater example, the sweater dealer would hire knitters, supply them with yarn and knitting tools, and supervise their knitting activities. This supervision could be done mainly by first-line supervisors, who might be drawn from the ranks of the more skilled knitters. The practice of an existing firm replacing one or more of its supplier markets with its own hierarchical structure for creating the supplied product is called **vertical integration**. Figure 1-7 shows how the sweater example would look after the knitters and the individual sweater dealers were vertically integrated into the hierarchical structure of a single sweater dealer.

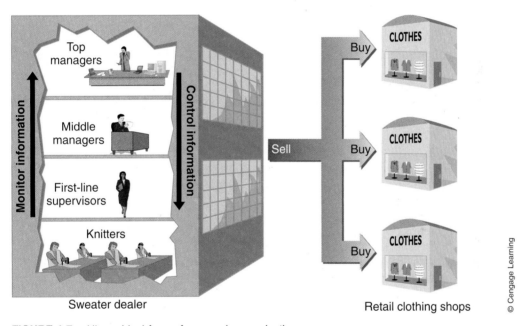

FIGURE 1-7   Hierarchical form of economic organization

Oliver Williamson, an economist who extended Coase's analysis, noted that firms in industries with complex manufacturing and assembly operations tended to be hierarchically organized and vertically integrated. Many of the manufacturing and administrative innovations that occurred in businesses during the twentieth century increased the efficiency and effectiveness of hierarchical monitoring activities. Assembly lines and other mass production technologies allowed work to be broken down into small, easily supervised procedures. The advent of computers brought tremendous increases in the ability of upper-level managers to monitor and control the detailed activities of their subordinates. Some of these direct measurement techniques are even more effective than the first-line supervisors on the shop floor.

During the years from the Industrial Revolution through the present, improvements in monitoring became commonplace and the size and level of vertical integration of firms have increased. In some very large organizations, however, monitoring systems have not kept pace with the organization's increase in size. This has created problems because the economic viability of a firm depends on its ability to track operational activities effectively at the lowest levels of the firm. These firms have instituted decentralization programs that allow business units to function as separate organizations, negotiating transactions with other business units as if they were operating in a market rather than as part of the same firm. Economists argue that large companies decentralize because they have grown too large to be managed effectively as hierarchical structures, so their managers need the information provided by market mechanisms.

To expose their decentralized operations to market mechanisms, these companies allow their divisions to operate as independent business units. A **strategic business unit (SBU)**, or simply **business unit**, is an autonomous part of a company that is large enough to manage itself but small enough to respond quickly to changes in its business environment. SBUs have their own mission and objectives; therefore, they have their own strategies for marketing, product development, purchasing, and long-term growth. General Electric, one of the largest companies in the world, has used SBUs to handle its diverse business operations since the 1960s. For example, General Electric makes both jet engines and light bulbs. These two businesses have different products, distribution channels, and customer types; therefore, they require different objectives, product development strategies, marketing plans, and manufacturing operations. General Electric's Jet Engine Division and Light Bulb Division operate as separate SBUs. Although an SBU operates as a participant in a market (rather than as part of the hierarchical structure of the owning company), the SBU itself is organized internally as a hierarchy.

Exceptions to the general trend toward hierarchies do exist. Many commodities, such as wheat, sugar, and crude oil, are still traded in markets. The commodity nature of the products traded in these markets significantly reduces transaction costs. There are a large number of potential buyers for an agricultural commodity such as wheat, and farmers do not make any special investment in customizing or modifying the product for particular customers. Thus, neither buyers nor sellers in commodity markets experience significant transaction costs.

Using Electronic Commerce to Reduce Transaction Costs

Businesses and individuals can use electronic commerce to reduce transaction costs by improving the flow of information and increasing the coordination of actions. By reducing the cost of searching for potential buyers and sellers and increasing the number of potential market participants, electronic commerce can change the attractiveness of vertical integration for many firms.

To see how electronic commerce can change the level and nature of transaction costs, consider an employment transaction. The agreement to employ a person has high transaction costs for the seller—the employee who sells his or her services. These transaction costs include a commitment to forego other employment and career development opportunities. Individuals make a high investment in learning and adapting to the culture of their employers. If accepting the job involves a move, the employee can incur very high costs, including actual costs of the move and related costs, such as the loss of a spouse's job. Much of the employee's investment is specific to a particular job and location; the employee cannot transfer the investment to a new job.

If a sufficient number of employees throughout the world can telecommute, then many of these transaction costs could be reduced or eliminated. Instead of uprooting a spouse and family to move, a worker could accept a new job by simply logging on to a different company server!

Network Economic Structures

Some researchers argue that many companies and strategic business units operate today in an economic structure that is neither a market nor a hierarchy. In this **network economic structure**, companies coordinate their strategies, resources, and skill sets by forming long-term, stable relationships with other companies and individuals based on shared purposes. These relationships are often called **strategic alliances** or **strategic partnerships**, and when they occur between or among companies operating on the Internet, these relationships are also called **virtual companies**.

In some cases, these entities, called **strategic partners**, come together as a team for a specific project or activity. The team dissolves when the project is complete; however, the partners maintain contact with each other through the ensuing period of inactivity. When the need for a similar project or activity arises, the same organizations and individuals build teams from their combined resources. In other cases, the strategic partners form many intercompany teams to undertake a variety of ongoing activities. Later in this book, you will see many examples of strategic partners creating alliances of this sort on the Web. In a hierarchically structured business environment, these types of strategic alliances would not last very long because the larger strategic partners would buy out the smaller partners and form a larger single company.

Network organizations are particularly well suited to technology industries that are information intensive. In the sweater example, the knitters might organize into networks of smaller organizations that specialize in certain styles or designs. Some of the particularly skilled knitters might leave the sweater dealer to form their own company to produce custom-knit sweaters. Some of the sweater dealer's marketing employees might form an independent firm that conducts market research on what the retail shops plan to buy in the upcoming months. This firm could sell its research reports to both the sweater dealer and the custom-knitting firm. As market conditions change, these smaller and more nimble organizations could continually reinvent themselves and take advantage of new opportunities that arise in the sweater markets. An illustration of such a network organization appears in Figure 1-8.

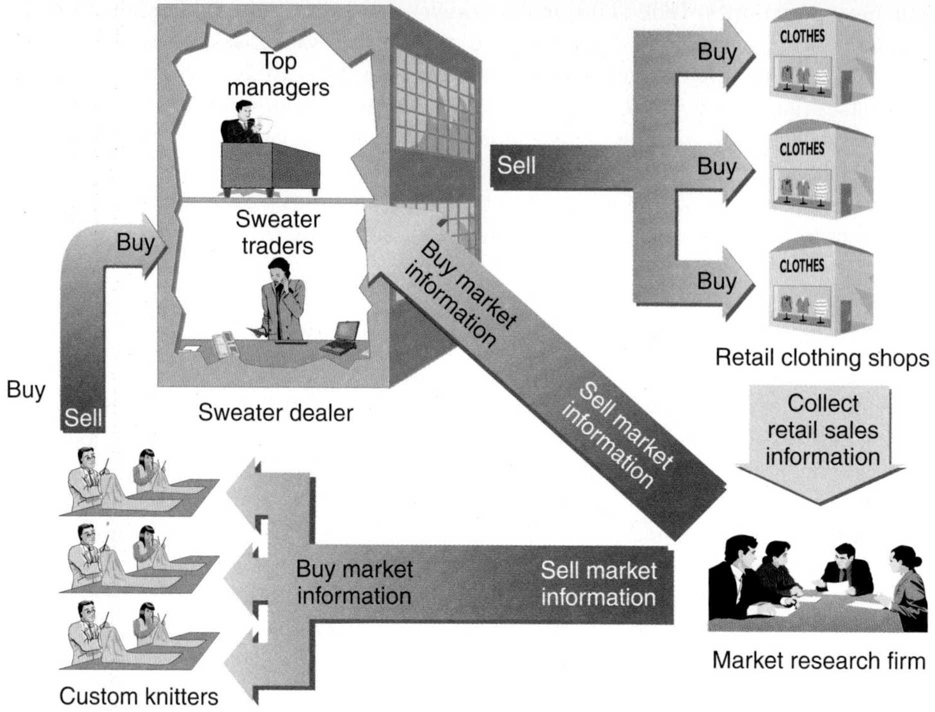

FIGURE 1-8 Network form of economic organization

Electronic commerce can make such networks, which rely extensively on information sharing, much easier to construct and maintain. Some researchers believe that these network forms of organizing commerce will become predominant in the near future. One of these researchers, Manuel Castells, even predicts that economic networks will become the organizing structure for all social interactions among people.

Network Effects

Economists have found that most activities yield less value as the amount of consumption increases. For example, a person who consumes one hamburger obtains a certain amount of value from that consumption. As the person consumes more hamburgers, the value provided by each hamburger decreases. Few people find the fifth hamburger as enjoyable as the first. This characteristic of economic activity is called the **law of diminishing returns**. In networks, an interesting exception to the law of diminishing returns occurs. As more people or organizations participate in a network, the value of the network to each participant increases. This increase in value is called a **network effect**.

To understand how network effects work, consider an early user of the telephone in the 1800s. When telephones were first introduced, few people had them. The value of each telephone increased as more people had them installed. As the network of telephones grew, the capability of each individual telephone increased because it could be used to communicate with more people. This increase in the value of each telephone as more and more telephones are able to connect to each other is the result of a network effect. Imagine how much less useful (and therefore, less valuable) your mobile phone today would be if you could only use it to talk with other people who had the same mobile phone carrier.

Your e-mail account, which gives you access to a network of other people with e-mail accounts, is another example of a network effect. If your e-mail account were part of a small network, it would be less valuable than it is. Most people today have e-mail accounts that are part of the Internet (a global network of computers, about which you will learn more in Chapter 2). In the early days of e-mail, most e-mail accounts only connected people in the same company or organization. Internet e-mail accounts are far more valuable than single-organization e-mail accounts because of the network effect.

Regardless of how businesses in a particular industry organize themselves—as markets, hierarchies, or networks—you need a way to identify business processes and evaluate whether electronic commerce is suitable for each process. The next section presents one useful structure for examining business processes.

IDENTIFYING ELECTRONIC COMMERCE OPPORTUNITIES

Internet technologies can be used to improve so many business processes that it can be difficult for managers to decide where and how to use them. One way to focus on specific business processes as candidates for electronic commerce is to break the business down into a series of value-adding activities that combine to generate profits and meet other goals of the firm. In this section, you will learn one popular way to analyze business activities as a sequence of activities that create value for the firm.

Commerce is conducted by firms of all sizes. Smaller firms might focus on one product, distribution channel, or type of customer. Larger firms often sell many different products and services through a variety of distribution channels to several types of customers. In these larger firms, managers organize their work around the activities of strategic business units. Multiple business units owned by a common set of shareholders make up a firm, or company, and multiple firms that sell similar products to similar customers make up an **industry**.

Strategic Business Unit Value Chains

In his 1985 book, *Competitive Advantage*, Michael Porter introduced the idea of value chains. A **value chain** is a way of organizing the activities that each strategic business unit undertakes to design, produce, promote, market, deliver, and support the products or services it sells. In addition to these **primary activities**, Porter also includes **supporting activities**, such as human resource management and purchasing, in the value chain model. Figure 1-9 shows a value chain for a strategic business unit, including both primary and supporting activities. These value chain activities will occur in some form in any strategic business unit.

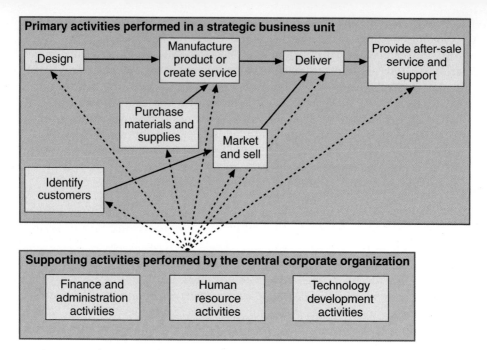

FIGURE 1-9 Value chain for a strategic business unit

The left-to-right flow in Figure 1-9 does not imply a strict time sequence for these processes. For example, a business unit might engage in marketing activities before purchasing materials and supplies. Each strategic business unit conducts the following primary activities:

- *Design*: activities that take a product from concept to manufacturing, including concept research, engineering, and test marketing
- *Identify customers*: activities that help the firm find new customers and new ways to serve existing customers, including market research and customer satisfaction surveys
- *Purchase materials and supplies*: procurement activities, including vendor selection, vendor qualification, negotiating long-term supply contracts, and monitoring quality and timeliness of delivery
- *Manufacture product or create service*: activities that transform materials and labor into finished products, including fabricating, assembling, finishing, testing, and packaging
- *Market and sell*: activities that give buyers a way to purchase and that provide inducements for them to do so, including advertising, promoting, managing salespeople, pricing, and identifying and monitoring sales and distribution channels

- *Deliver*: activities that store, distribute, and ship the final product or provide the service, including warehousing, handling materials, consolidating freight, selecting shippers, and monitoring timeliness of delivery
- *Provide after-sale service and support*: activities that promote a continuing relationship with the customer, including installing, testing, maintaining, repairing, fulfilling warranties, and replacing parts

The importance of each primary activity depends on the product or service the business unit provides and to which customers it sells. Each business unit must also have support activities that provide the infrastructure for the unit's primary activities. The central corporate organization typically provides the support activities that appear in Figure 1-9. These activities include the following:

- *Finance and administration activities*: providing the firm's basic infrastructure, including accounting, paying bills, borrowing funds, reporting to government regulators, and ensuring compliance with relevant laws
- *Human resource activities*: coordinating the management of employees, including recruiting, hiring, training, compensation, and managing benefits
- *Technology development activities*: improving the product or service that the firm is selling and that helps improve the business processes in every primary activity, including basic research, applied research and development, process improvement studies, and field tests of maintenance procedures

Industry Value Chains

Porter's book also identifies the importance of examining where the strategic business unit fits within its industry. Porter uses the term **value system** to describe the larger stream of activities into which a particular business unit's value chain is embedded. However, many subsequent researchers and business consultants have used the term **industry value chain** when referring to value systems. When a business unit delivers a product to its customer, that customer might use the product as purchased materials in its value chain. By becoming aware of how other business units in the industry value chain conduct their activities, managers can identify new opportunities for cost reduction, product improvement, or channel reconfiguration.

Every product or service has an industry value chain that can be identified and analyzed for these opportunities. To create an industry value chain, start with the inputs to your SBU and work backward to identify your suppliers' suppliers, then the suppliers of those suppliers, and so on. Then start with your customers and work forward to identify your customers' customers, then the customers of those customers, and so on.

An example of an industry value chain appears in Figure 1-10. This value chain is for a wooden chair and traces the life of the product from its inception as trees in a forest to its grave in a landfill or at a sawdust recycler.

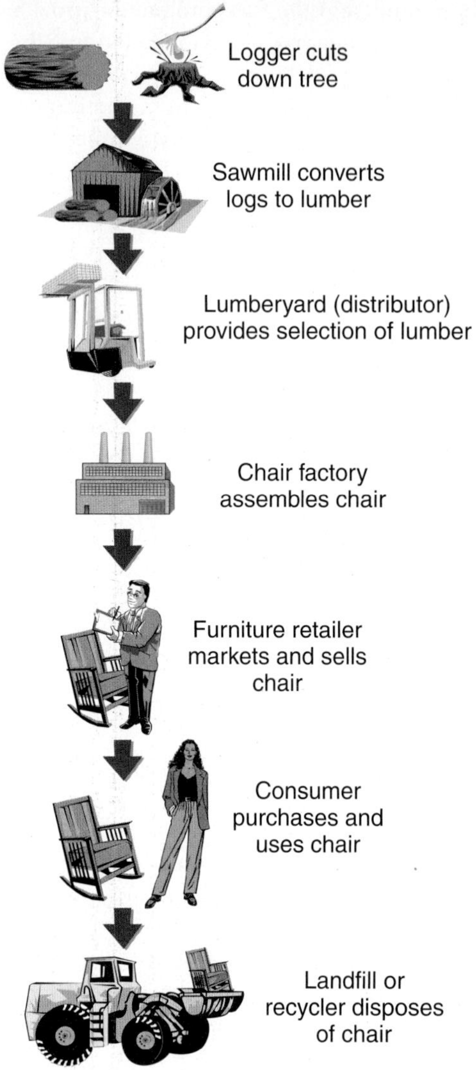

FIGURE 1-10   Industry value chain for a wooden chair

Each business unit (logger, sawmill, lumberyard, chair factory, retailer, consumer, and recycler) shown in Figure 1-10 has its own value chain. For example, the sawmill purchases logs from the tree harvester and combines them in its manufacturing process with inputs, such as labor and saw blades, from other sources. Among the sawmill customers are the chair factory, shown in Figure 1-10, and other users of cut lumber. Examining this industry value chain could be useful for the sawmill that is considering entering the tree-harvesting business or the furniture retailer who is thinking about partnering with a trucking line. The industry value chain identifies opportunities up and down the product's life cycle for increasing the efficiency or quality of the product.

As they examine their industry value chains, many managers are finding that they can use electronic commerce and Internet technologies to reduce costs, improve product quality, reach new customers or suppliers, and create new ways of selling existing products. For example, a software developer who releases annual updates to programs might consider removing the software retailer from the distribution channel for software updates by offering to send the updates through the Internet directly to the consumer.

This change would modify the software developer's industry value chain and would provide an opportunity for increasing sales revenue (the software developer could retain the margin that a retailer would have added to the price of the update), but it would not appear as part of the software developer business unit value chain. By examining elements of the value chain outside the individual business unit, managers can identify many business opportunities, including those that can be exploited using electronic commerce.

The value chain concept is a useful way to think about business strategy in general. When firms are considering electronic commerce, the value chain can be an excellent way to organize the examination of business processes within their business units and in other parts of the product's life cycle. Using the value chain reinforces the idea that electronic commerce should be a business solution, not a technology implemented for its own sake.

SWOT Analysis: Evaluating Business Unit Opportunities

Now that you have learned about industry value chains and SBUs, you can learn one popular technique for analyzing and evaluating business opportunities. Most electronic commerce initiatives add value by either reducing transaction costs, creating some type of network effect, or a combination of both. In **SWOT analysis** (the acronym is short for strengths, weaknesses, opportunities, and threats), the analyst first looks into the business unit to identify its strengths and weaknesses. The analyst then reviews the environment in which the business unit operates and identifies opportunities presented by that environment and the threats posed by that environment. Figure 1-11 shows questions that an analyst would ask in conducting a SWOT analysis for any company or SBU.

Strengths	Weaknesses
• What does the company do well?	• What does the company do poorly?
• Is the company strong in its market?	• What problems could be avoided?
• Does the company have a strong sense of purpose and the culture to support that purpose?	• Does the company have serious financial liabilities?

Opportunities	Threats
• Are industry trends moving upward?	• What are competitors doing well?
• Do new markets exist for the company's products/services?	• What obstacles does the company face?
• Are there new technologies that the company can exploit?	• Are there troubling changes in the company's business environment (technologies, laws, and regulations)?

© Cengage Learning

FIGURE 1-11　SWOT analysis questions

By considering all of the issues that it faces in a systematic way, a business unit can formulate strategies to take advantage of its opportunities by building on its strengths, avoiding any threats, and compensating for its weaknesses.

In the mid-1990s, **Dell Computer** used a SWOT analysis to create a business strategy that helped it become a strong competitor in its industry value chain. Dell identified its strengths in selling directly to customers and in designing its computers and other products to reduce manufacturing costs. It acknowledged the weakness of having no relationships with local computer dealers. Dell faced threats from competitors such as Compaq (now a part of Hewlett-Packard) and IBM, both of which had much stronger brand names and reputations for quality at that time. Dell identified an opportunity by noting that its customers were becoming more knowledgeable about computers and could specify exactly what they wanted without having Dell salespeople answer questions or develop configurations for them. It also saw the Internet as a potential marketing tool. Dell carefully considered and answered the SWOT analysis questions shown in Figure 1-11. The results of Dell's SWOT analysis appear in Figure 1-12.

Strengths	Weaknesses
• Sell directly to consumers • Keep costs below competitors' costs	• No strong relationships with computer retailers

Opportunities	Threats
• Consumer desire for one-stop shopping • Consumers know what they want to buy • Internet could be a powerful marketing tool	• Competitors have stronger brand names • Competitors have strong relationships with computer retailers

© Cengage Learning

FIGURE 1-12   Results of Dell's SWOT analysis

The strategy that Dell followed after doing the analysis took all four of the SWOT elements into consideration. Dell decided to offer customized computers built to order and sold over the phone, and eventually, over the Internet. Dell's strategy capitalized on its strengths and avoided relying on a dealer network. The brand and quality threats posed by Compaq and IBM were lessened by Dell's ability to deliver higher perceived quality because each computer was custom made for each buyer. Ten years later, Dell observed that the environment of personal computer sales had changed and did start selling computers through dealers.

INTERNATIONAL NATURE OF ELECTRONIC COMMERCE

Because the Internet connects computers all over the world, any business that engages in electronic commerce instantly becomes an international business, with exposure to potential customers in other countries and cultures. When companies use the Web to improve a business process, they are automatically operating in a global environment. The first wave of electronic commerce was dominated by U.S. businesses. In the second wave, European and Asian businesses expanded online. Today, a rapidly increasing proportion of online business activity is based outside the United States. Figure 1-13 shows the proportions of online B2C sales that arise in the main geographic regions of the world.

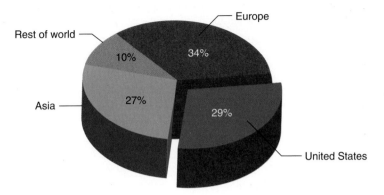

Source: Internet Retailer report of Goldman Sachs estimates
http://www.internetretailer.com/trends/sales/

FIGURE 1-13   Proportion of online B2C sales by geographic region, 2010

Asian online markets are growing at the most rapid pace, with sales expected to double by 2014. Although much of the online sales activity in each of the world regions depicted in the figure is intraregion, an increasing proportion of online business involves companies making sales across multiple international boundaries. The key issues that a company faces when it conducts international commerce include trust, culture, language, government, and infrastructure. These topics are covered in the following sections. The related issues of international law and currency conversion are covered in Chapter 7.

Trust Issues on the Web

It is important for all businesses to establish trusting relationships with their customers. Companies with established reputations in the physical world often create trust by ensuring that customers know who they are. These businesses can rely on their established brand names to create trust on the Web. New companies that want to establish online businesses face a more difficult challenge because a kind of anonymity exists for companies trying to establish a Web presence.

For example, a U.S. bank can establish a Web site that offers services throughout the world. No potential customer visiting the site can determine just how large or well established the bank is simply by browsing through the site's pages. Because Web site visitors will not become customers unless they trust the company behind the site, a plan for establishing credibility is essential. Sellers on the Web cannot assume that visitors will know that the site is operated by a trustworthy business.

Customers' inherent lack of trust in "strangers" on the Web is logical and to be expected; after all, people have been doing business with their neighbors—not strangers—for thousands of years. When a company grows to become a large corporation with multinational operations, its reputation grows commensurately. Before a company can do business in dozens of countries, it must prove its trustworthiness by satisfying customers for many years as it grows. Businesses on the Web must find ways to overcome this well-founded tradition of distrusting strangers, because today a company can incorporate one day and, through the Web, be doing business the next day with people all over the world. For businesses to succeed on the Web, they must find ways to quickly generate the trust that traditional businesses take years to develop.

Language Issues

Most companies realize that the only way to do business effectively in other cultures is to adapt to those cultures. The phrase "think globally, act locally" is often used to describe this approach. The first step that a Web business usually takes to reach potential customers in other countries, and thus in other cultures, is to provide local language versions of its Web site. This may mean translating the Web site into another language or regional dialect. Researchers have found that customers are far more likely to buy products and services from Web sites in their own language, even if they can read English well. Only about 400 million of the world's 7 billion people learned English as their native language.

Researchers estimate that about 50 percent of the content available on the Internet today is in English, but more than half of current Internet users do not read English. Industry analysts estimate that by 2015, more than 90 percent of Internet users will be outside the United States, and 70 percent of electronic commerce transactions will involve at least one party located outside the United States.

Some languages require multiple translations for separate dialects. For example, the Spanish spoken in Spain is different from that spoken in Mexico, which is different from that spoken elsewhere in Latin America. People in parts of Argentina and Uruguay use yet a fourth dialect of Spanish. Many of these dialect differences are spoken inflections, which are not important for Web site designers (unless, of course, their sites include audio or video elements); however, a significant number of differences occur in word meanings and spellings. You might be familiar with these types of differences, because they occur in the U.S. and British dialects of English. The U.S. spelling of *gray* becomes *grey* in Great Britain, and the meaning of *bonnet* changes from a type of hat in the United States to an automobile hood in Great Britain. Chinese has two main systems of writing: simplified Chinese, which is used in mainland China, and traditional Chinese, which is used in Hong Kong and Taiwan.

Most companies that translate their Web sites choose to translate all of their pages. However, as Web sites grow larger, companies are becoming more selective in their translation efforts. Some sites have thousands of pages with much targeted content; the businesses operating those sites can find the cost of translating all pages to be prohibitive.

The decision whether to translate a particular page should be made by the corporate department responsible for each page's content. The home page should have versions in all supported languages, as should all first-level links to the home page. Beyond that, pages that are devoted to marketing, product information, and establishing brand should be given a high translation priority. Some pages, especially those devoted to local interests, might be maintained only in the relevant language. For example, a weekly update on local news and employment opportunities at a company's plant in Frankfurt probably needs to be maintained only in German.

Links to the Web sites of firms that provide Web page translation services and translation software for companies are included in the Additional Resources section of the Web Links under the heading Language Translation Services. These firms translate Web pages and maintain them for a fee that is usually between 25 and 90 cents per word for translations done by skilled human translators. Languages that are complex or that are spoken by relatively few people are generally more expensive to translate than other languages.

Different approaches can be appropriate for translating the different types of text that appear on an electronic commerce site. For key marketing messages, the touch of a human translator can be essential to capture subtle meanings. For more routine transaction-processing functions, automated software translation may be an acceptable alternative. Software translation, also called **machine translation**, can reach speeds of 400,000 words per hour, so even if the translation is not perfect, businesses might find it preferable to a human who can translate only about 500 words per hour. Many of the companies in this field are working to develop software and databases of previously translated material that can help human translators work more efficiently and accurately.

The translation services and software manufacturers that work with electronic commerce sites do not generally use the term "translation" to describe what they do. They prefer the term **localization**, which means a translation that considers multiple elements of the local environment, such as business and cultural practices, in addition to local dialect variations in the language. The cultural element is very important because it can affect—and sometimes completely change—the user's interpretation of text.

Cultural Issues

An important element of business trust is anticipating how the other party to a transaction will act in specific circumstances. A company's brand conveys expectations about how the company will behave; therefore, companies with established brands can build online businesses more quickly and easily than a new company without a reputation. For example, a potential buyer might like to know how the seller would react to a claim by the buyer that the seller misrepresented the quality of the goods sold. Part of this knowledge derives from the buyer and seller sharing a common language and common customs. Buyers are more comfortable doing business with sellers they know are trustworthy.

The combination of language and customs is often called **culture**. Most researchers agree that culture varies across national boundaries and, in many cases, varies across regions within nations. For example, the concept of private property is an important cultural value and underlies laws in many European and North American countries. Asian cultures do not value private property in the same way, so laws and business practices in those countries can be quite different. All companies must be aware of the differences in language and customs that make up the culture of any region in which they intend to do business. The Additional Resources section of the Web Links includes links to Web sites that provide detailed information on cultural issues for specific countries under the heading Global Trust and Culture.

Managers at Virtual Vineyards (now a part of Wine.com), a company that sells wine and specialty food items on the Web, were perplexed. The company was getting an unusually high number of complaints from customers in Japan about short shipments. Virtual Vineyards sold most of its wine in case (12 bottles) or half-case quantities. Thus, to save on operating costs, it stocked shipping materials only in case, half-case, and two-bottle sizes. After an investigation, the company determined that many of its Japanese customers ordered only one bottle of wine, which was shipped in a two-bottle container. To these Japanese customers, who consider packaging to be an important element of a high-quality product such as wine, it was inconceivable that anyone would ship one bottle of wine in a two-bottle container. They were e-mailing to ask where the other bottle was, notwithstanding the fact that they had ordered only one bottle.

Some errors stemming from subtle language and cultural standards have become classic examples that are regularly cited in international business courses and training sessions. For example, General Motors' choice of name for its Chevrolet Nova automobile amused people in Latin America—*no va* means "it will not go" in Spanish. Pepsi's "Come Alive" advertising campaign fizzled in China because its message came across as "Pepsi brings your ancestors back from their graves."

Another story that is widely used in international business training sessions is about a company that sold baby food in jars adorned with the picture of a very cute baby. The jars sold well everywhere they had been introduced except in parts of Africa. The mystery was solved when the manufacturer learned that food containers in those parts of Africa always carry a picture of their contents. This story is particularly interesting because it never happened. However, it illustrates a potential cultural issue so dramatically that it continues to appear in marketing textbooks and international business training materials.

Designers of Web sites for international commerce must be very careful when they choose icons to represent common actions. For example, in the United States, a shopping cart is a good symbol to use when building an electronic commerce site. However, many Europeans use shopping *baskets* when they go to a store and may never have seen a shopping *cart*. In Australia, people would recognize a shopping cart image but would be confused by the text "shopping cart" if it were used with the image. Australians call them shopping *trolleys*. In the United States, people often form a hand signal (the index finger touching the thumb to create a circle) that indicates "OK" or "everything is just fine." A Web designer might be tempted to use this hand signal as an icon to indicate that the transaction is completed or the credit card is approved, unaware that in some countries, including Brazil, this hand signal is an obscene gesture.

The cultural overtones of simple design decisions can be dramatic. In India, for example, it is inappropriate to use the image of a cow in a cartoon or other comical setting. Potential customers in Muslim countries can be offended by an image that shows human arms or legs uncovered. Even colors or Web page design elements can be troublesome. For example, white, which denotes purity in Europe and the Americas, is associated with death and mourning in China and many other Asian countries. A Web page that is divided into four segments can be offensive to a Japanese visitor because the number four is a symbol of death in that culture.

Japanese shoppers have resisted the U.S. version of electronic commerce because they generally prefer to pay in cash or by cash transfer instead of by credit card, and they have a high level of apprehension about doing business online. Softbank, a major Japanese firm that invests in Internet companies, devised a way to introduce electronic commerce to a reluctant Japanese population. Softbank created a joint venture with 7-Eleven, Yahoo! Japan, and Tohan (a major Japanese book distributor) to sell books and CDs on the Web. This venture, called eS-Books, allows customers to order items on the Internet, and then pick them up and pay for them in cash at the local 7-Eleven convenience store. By adding an intermediary that satisfies the needs of the Japanese customer, Softbank has been highly successful in bringing business-to-consumer electronic commerce to Japan.

Culture and Government

Some parts of the world have cultural environments that are extremely inhospitable to the type of online discussion that occurs on the Internet. These cultural conditions, in some cases, lead to government controls that can limit electronic commerce development. The Internet is a very open form of communication. This type of unfettered communication is not desired or even considered acceptable in some cultures. For example, a **Human Rights Watch** report stated that many countries in the Middle East and North Africa do not allow their citizens unrestricted access to the Internet. The report notes that many governments in this part of the world regularly prevent free expression by their citizens and have taken specific steps to prevent the exchange of information outside of state controls. For instance, Saudi Arabia, Yemen, and the United Arab Emirates all filter the Web content that is available in their countries. An organization devoted to the international promotion of democracy and civil liberties, **Freedom House**, offers a number of downloadable publications on its site, including in-depth reports on Internet censorship activities of governments throughout the world.

In many North African and Middle Eastern countries, officials have publicly denounced the Internet as a medium that helps distribute materials that are sexually explicit, anti-Islam, or that cast doubts on the traditional role of women in their societies. In many of these countries, uncontrolled use of Internet technologies is so at odds with existing traditions, cultures, and laws that electronic commerce is unlikely to exist locally at any significant level in the near future. In contrast, other Islamic jurisdictions in that part of the world, including Algeria, Morocco, and the Palestinian Authority, do not limit online access or content.

A number of restrictive governments in the world control Internet access as a way to prevent the formation and growth of internal independent political activist organizations. By limiting access or monitoring all Internet traffic, the planners of rebellions against the government can be thwarted. During the Arab Spring of 2011, young people in Egypt and Tunisia used social media to share information and coordinate protest locations and activities. The Egyptian authorities were so concerned that they made several (unsuccessful) attempts to steal every Facebook password in the country. One of the first acts of the Libyan rebels after they overthrew Muammar Qaddafi was to restore the country's Internet connection, which had been cut at the start of the rebellion. They also sent a text message to millions of Libyan mobile phone users saying, "Long live free Libya," and added $40 worth of calling credit to each individual phone account.

The censorship of Internet content and communications restricts electronic commerce because it prevents certain types of products and services from being sold or advertised. Further, it reduces the interest level of many potential participants in online activities. If large numbers of people in a country are not interested in being online, businesses that use the Internet as an information and product delivery channel will not develop in those countries.

Other countries, such as the People's Republic of China and Singapore, are wrestling with the issues presented by the growth of the Internet as a vehicle for doing business. These countries have a tradition of controlling their citizens' access to information from outside the country, but they want their economies to reap the benefits of electronic commerce. China created a complex set of registration requirements and regulations that govern any business that engages in electronic commerce. These regulations are enforced by the Public Security Bureau, which is a branch of the state police, not an independent administrative agency. For example, companies in China that sell Internet services must register all of their customers with the Public Security Bureau and must retain copies of all e-mail messages and chat room conversations for 60 days. Chinese citizens entering a chat room at **Sohu.com**, one of China's leading portal sites ("sohu" means "search fox" in Chinese), are greeted with a Web page containing the following text (translated here from the original Chinese):

> Warning! Please take note that the following issues are prohibited according to Chinese law: 1) Criticism of the People's Republic of China Constitution. 2) Revealing State secrets, and discussion about overthrowing the Communist government.
> 3) Topics which damage the reputation of the State.

The Chinese government regularly conducts reviews of ISPs and their records. Every year, the Chinese Public Security Bureau shuts down thousands of Internet cafes for failing to keep adequate records and requires many others to suspend operations while they implement required electronic record-keeping procedures. Operators of Web sites in China are required to monitor all content that appears on their sites. Blogbus, a Chinese site that allowed visitors to post essays, was shut down in 2004 because one posting (out of 15,000) contained an essay that included what the government deemed to be "forbidden content." Hundreds of people have been jailed in China for posting "subversive" content on Web pages.

More recently, China has required the installation of censoring software on all computers used in schools and Internet cafes. This software, called the Green Dam Youth Escort, blocks any Web sites on a government banned list and tracks details of the use of the computer on which it is installed. A requirement that all computers sold in China have this software installed was withdrawn in 2009; however, other government efforts to limit access to the Internet are in place. For example, China's Golden Shield Project is an $800 million effort to limit its citizens' access to information on the Internet that it deems to be forbidden. The Chinese government actively monitors developments in the world to determine what it will censor. For example, Chinese human rights activist Liu Xiaobo became a forbidden topic when he won the 2010 Nobel Peace Prize.

North Korea, Singapore, and a number of Middle Eastern countries have also adopted rules and policies that restrict their citizens' use of the Internet. These countries will continue to face difficult policy choices as they maintain their attempts to control individuals' use of the Internet while at the same time trying to encourage growth in online business transactions.

Some countries, although they do not ban electronic commerce entirely, have strong cultural requirements that have found their way into the legal codes that govern business conduct. In France, an advertisement for a product or service must be in French. Thus, a business in the United States that advertises its products on the Web and is willing to ship goods to France must provide a French version of its pages if it intends to comply with French law. Many U.S. electronic commerce sites include in their Web pages a list of the countries from which they will accept orders through their Web sites.

Infrastructure Issues

Businesses that successfully meet the challenges posed by trust, language, and culture issues still face the challenges posed by variations and inadequacies in the infrastructure that supports the Internet throughout the world. Internet infrastructure includes the computers and software connected to the Internet and the communications networks over which the message packets travel. In many countries other than the United States, the telecommunications industry is either government-owned or heavily regulated by the government. In many cases, regulations in these countries have inhibited the development of the telecommunications infrastructure or limited the expansion of that infrastructure to a size that cannot reliably support Internet traffic.

Local connection costs through the existing telephone networks in many developing countries are very high compared to U.S. costs for similar access. This can have a profound effect on the behavior of electronic commerce participants. For example, in countries where Internet connection costs are high, few businesspeople would spend time surfing the Web to shop for a product. They would use a Web browser only to navigate to a specific site that they know offers the product they want to buy. Thus, to be successful in selling to businesses in such countries, a company would need to advertise its Web presence in traditional media instead of relying on Web search engines to deliver customers to their Web sites.

More than half of all businesses on the Web turn away international orders because they do not have the processes in place to handle such orders. Some of these companies are losing millions of dollars' worth of international business each year. This problem is global; not only are U.S. businesses having difficulty reaching their international markets, but businesses in other countries are having similar difficulties reaching the U.S. market.

The paperwork and often-convoluted processes that accompany international transactions are targets for technological solutions. Most firms that conduct business internationally rely on a complex array of freight-forwarding companies, customs brokers, international freight carriers, bonded warehouses, and importers to navigate the maze of paperwork that must be completed at every step of the transaction to satisfy government and insurance requirements. A **freight forwarder** is a company that arranges shipping and insurance for international transactions. A **customs broker** is a company that arranges the payment of tariffs and compliance with customs laws for international shipments. A number of companies combine these two functions and offer a full range of export management services. A **bonded warehouse** is a secure location where incoming international shipments can be held until customs requirements are satisfied or until payment arrangements are completed. The multiple flows of information and transfers of physical objects that occur in a typical international trade transaction are illustrated in Figure 1-14.

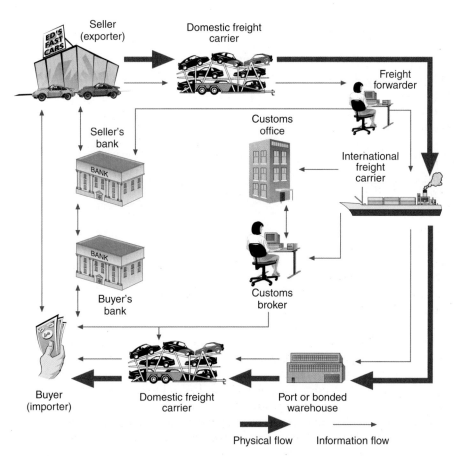

FIGURE 1-14 Parties involved in a typical international trade transaction

As you can see in Figure 1-14, the information flows can be complex. Domestic transactions usually include only the seller, the buyer, their respective banks, and one freight carrier. International transactions almost always require physical handling of goods by several freight carriers, storage in a freight forwarder's facility before international shipment, and storage in a port or bonded warehouse facility in the destination country. This handling and storage require monitoring by government customs offices in addition to the monitoring by seller and buyer that occurs in domestic transactions. International transactions usually require the coordinated efforts of customs brokers and freight forwarding agencies because the regulations and procedures governing international transactions are so complex. You will learn more about how businesses transfer money in international transactions in Chapter 11.

Industry experts estimate that the annual cost of handling paperwork for international transactions is $700 billion. Companies sell software that can automate some of the paperwork; however, many countries have their own paper-based forms and procedures with which international shippers must comply. To further complicate matters, some countries that have automated some procedures use computer systems that are incompatible with those of other countries.

Some governments provide assistance to companies that want to do international business on the Web. The Argentine government operates the **Fundación Invertir** Web site to provide information to companies that want to do business in Argentina. The **U.S. Commercial Service** (an agency of the U.S. Department of Commerce) operates the **BuyUSA** site, a portal for U.S. companies that want to sell abroad and non-U.S. companies that want to buy from U.S. companies.

Infrastructure issues will continue to prevent international business from reaching its full potential until technology is adapted to overcome barriers instead of being a part of those barriers.

Summary

In this chapter, you learned that electronic commerce is the application of new technologies, particularly Internet and Web technologies, to help individuals, businesses, and other organizations conduct business more effectively. Electronic commerce is being adopted in waves of change. The first wave of electronic commerce ended in 2000. The second wave, with new approaches to integrating Internet technologies into business processes, is under way. In the second wave, businesses are focusing less on overall business models and more on improving specific business processes. A third wave of electronic commerce is just now beginning that will capitalize on the availability of mobile devices such as smart phones and tablet computers. These devices, along with increasing use of social media Web sites, will extend the reach of the Internet to new customers and locations, opening new avenues of electronic commerce for companies around the world.

Using electronic commerce, some businesses have been able to create new products and services, and others have improved the promotion, marketing, and delivery of existing offerings. Firms have also found many ways to use electronic commerce to improve purchasing and supply activities; identify new customers; and operate their finance, administration, and human resource management activities more efficiently. You learned that electronic commerce can help businesses reduce transaction costs or create network economic effects that can lead to greater revenue opportunities.

You examined an overview of markets, hierarchies, and networks—the economic structures in which businesses operate—and learned how electronic commerce fits into those structures. Porter's ideas about value chains at the business unit and industry levels were presented, and you learned how to use value chains and SWOT analysis as ways to understand business processes and analyze their suitability for electronic commerce implementation.

The inherently global nature of electronic commerce leads to many opportunities and a few challenges. You learned that companies engaged in international electronic commerce must understand the trust, cultural, language, and legal issues that arise when doing business across national borders.

Key Terms

Activity	Electronic business (e-business)
Bonded warehouse	Electronic commerce (e-commerce)
Business model	Electronic data interchange (EDI)
Business processes	Electronic funds transfer (EFT)
Business unit	Firm
Business-to-business (B2B)	First-mover advantage
Business-to-consumer (B2C)	Freight forwarder
Business-to-government (B2G)	Hierarchical business organization
Commodity item	Industry
Company	Industry value chain
Consumer-to-consumer (C2C)	Law of diminishing returns
Culture	Localization
Customs broker	Machine translation
Dot-com	Market
E-procurement	Merchandising

Mobile commerce (m-commerce)

Network economic structure

Network effect

Primary activities

Procurement

Pure dot-com

Revenue model

Shipping profile

Smart phone

Social commerce

Social networking site

Strategic alliance

Strategic business unit (SBU)

Strategic partners

Strategic partnership

Supply management

Supporting activities

SWOT analysis

Telecommuting

Telework

Trading partners

Transaction

Transaction costs

Value-added network (VAN)

Value chain

Value system

Vertical integration

Virtual community

Virtual company

Web 2.0

Wire transfer

Review Questions

1. Briefly describe the technologies that are leading businesses into the third wave of electronic commerce.

2. Figure 1-5 lists roommate-matching services as a type of business that is well suited to a combination of electronic and traditional commerce. In one paragraph, describe the elements of this service that would be best handled using traditional commerce, and explain why.

3. Briefly describe the specific activities that a motorcycle manufacturer might include in B2B electronic commerce for its supply management or procurement operations.

4. What are the main functions of a value-added network?

5. Many business analysts have discussed the concept of the first-mover advantage. What are some of the disadvantages of being a first mover?

6. What is a shipping profile, and why is it an important consideration for firms making online sales?

7. What are transaction costs, and why are they important?

8. Provide one example of how electronic commerce could help change an industry's economic structure from a hierarchy to a network.

9. Why would a strategic business unit have its own mission and objectives?

10. How might a university use SWOT analysis to identify new degree programs that it could offer online?

11. Briefly describe the function a customs broker might play in the delivery of online that were purchased online.

1. Companies that sell luxury goods, such as Chanel, Lilly Pulitzer, and Vera Wang, were reluctant to offer their products for sale on their Web sites for many years. These businesses preferred to use their Web sites to display information about their products only and to sell their products through exclusive retail stores. Summarize the reasons these luxury goods producers might have been hesitant to sell online and speculate why they might have changed their thinking.

2. You have decided to buy a new color printer for your home office. You have not decided whether an ink-jet or laser printer would be best for you. List specific activities that you must undertake as you gather information about printer capabilities and features. Use the HPshopping.com, Office Depot, Best Buy, OfficeMax, and Staples Web sites to gather information. Write a short summary of the process you undertook to serve as a model for others who plan to undertake a similar task.

3. Choose one of the Web sites listed in the previous question and identify three ways the company has reduced its transaction costs by using a Web site to provide information about printers. List these three transaction cost-reduction elements and write a paragraph in which you discuss one transaction cost-reduction opportunity that you believe the company missed.

4. Create a diagram (similar to the diagram in Figure 1-10) that describes the industry value chain for a stainless steel water bottle. Identify stages of the chain in which a company might use electronic commerce and explain how the company might use it in those stages.

5. Read the following business messages and come up with a list of words or phrases in each message that you believe might be troublesome for automated translation software. Then use either the Yahoo! Babel Fish or the FreeTranslation Web site to translate the messages from English to one of the foreign languages available on that site. Translate each message back into English. Write a short memo that compares the problems you anticipated with those that occurred in the automated translation. The business messages are the following:

   a. The flight has been delayed for several hours and your shipment of components will not arrive as scheduled.

   b. We would be happy to bid on your proposal; however, we will need the drawings of subassembly #24 and the supervising mechanical engineer's quality control report by next Thursday.

   c. Our company offers the latest and greatest hot deals on wheels. We would love to send you a brochure that explains why our brakes, wheels, and suspension components will do the job for you effectively and economically.

Cases

C1. Amazon.com

In 1994, a 29-year-old financial analyst and fund manager named Jeff Bezos became intrigued by the rapid growth of the Internet. Looking for a way to capitalize on this hot new marketing tool, he made a list of 20 products that might sell well on the Internet. After some intense analysis, he determined that books were at the top of that list. Although Bezos liked the name Abracadabra, he decided to call his online bookshop Amazon.com. Today, Amazon.com has more than 100 million customers and sells billions of dollars' worth of all types of merchandise.

When he started, Bezos had no experience in the bookselling business, but he realized that books had an ideal shipping profile for online sales. He believed that many customers would be willing to buy books without inspecting them in person and that books could be impulse purchase items if properly promoted on a Web site. By accepting orders on its Web site, Bezos believed that Amazon.com could reduce transaction costs in the sale to the customer.

Several million book titles are in print at any one time throughout the world, and more than a million of those are in English. However, the largest physical bookstore cannot stock more than 200,000 books and carries even fewer titles because bookstores stock more than one copy of each title. Having a wide selection was important because Bezos believed it would help create a network economic effect. People would visit Amazon.com whenever they wanted to buy a book because it would be the most likely store (physical or online) to have a particular title. After becoming satisfied customers, people would return to Amazon.com to buy more books and would eventually stop looking elsewhere.

The structure of the supply side of the book business was equally important to Amazon.com's success. Music CDs, which were second on Bezos' list, were produced by a few major recording companies who could easily control Amazon.com's supply. In contrast, there were a large number of book publishers, none of which held a dominant position in the book-selling marketplace. Thus, it was unlikely that a single supplier could restrict Bezos' supply of books or enter his market as a competitor. He decided to locate his firm in Seattle, close to a large pool of programming talent and near one of the largest book distribution warehouses in the world. These supply factors were important because Bezos wanted to develop efficiencies that would allow Amazon.com to reduce transaction costs for its purchases as well as its sales transactions.

Bezos encouraged early customers to submit reviews and ratings of books, which he posted with the publisher's information about the book and with reviews written by Amazon.com employees. This customer participation served as a substitute for the corner bookshop staff's friendly advice and recommendations. Bezos saw the power of the Internet in reaching small, highly focused market segments, but he realized that his comprehensive bookstore could not be all things to all people. Therefore, he created a sales associate program in which Web sites devoted to a particular topic, such as model railroading, could provide links to Amazon.com books that related to that topic. In return, Amazon.com remits a percentage of the referred sales to the owner of the referring site.

Although Bezos' original vision was to create an online bookstore with the world's best selection, Amazon has moved into other product lines where opportunities for network economic effects and transaction cost reductions looked promising. In 1998, Amazon.com began selling music CDs and videos, first on VHS tape, and then later on DVD. More recently, Amazon added MP3 music downloads. Today, Amazon offers thousands of products in dozens of categories.

By paying attention to every process involved in buying, promoting, selling, and shipping consumer goods, and by working to improve each process continually, Bezos and Amazon.com became one of the first highly visible success stories in electronic commerce. In fact, Amazon.com now generates significant revenue by supplying other sellers of consumer goods with the technology to sell those goods online. One of its first partnerships was with Toys"R"Us, a company that had experienced difficulties in selling online and making deliveries on time in the 1999 holiday shopping season. Toys"R"Us signed an agreement with Amazon.com in 2000 that placed Toys"R"Us products on the Amazon.com Web site. Amazon.com would accept the orders on its Web site and would ship products to customers for Toys"R"Us in exchange for a percentage of each sale. Amazon.com also agreed not to sell toys itself or on behalf of other partners for whom it might provide online sales services in the future. For example, when

CHAPTER 2

TECHNOLOGY INFRASTRUCTURE: THE INTERNET AND THE WORLD WIDE WEB

LEARNING OBJECTIVES

In this chapter, you will learn:

- About the origin, growth, and current structure of the Internet
- How packet-switched networks are combined to form the Internet
- How Internet, e-mail, and Web protocols work
- About Internet addressing and how Web domain names are constructed
- About the history and use of markup languages on the Web
- How HTML tags and links work
- About technologies people and businesses use to connect to the Internet
- About Internet2 and the Semantic Web

INTRODUCTION

Most people who use the Internet today do so using a computer. However, a growing number of

Internet users, especially in developing countries, use an Internet-capable mobile phone as their

primary means of accessing the Internet. Although the first Internet-capable mobile phones were

developed in the late 1990s, a number of technological issues prevented them from being very useful

as a way to browse the Internet. Their screens were small and lacked color, they did not have alphanumeric keyboards, their ability to store information was limited, and the networks through which they connected to the Internet were slow and unreliable.

In 2001, Handspring introduced its Treo phones, and Research in Motion (RIM) introduced its BlackBerry phones. These mobile phones included small alphanumeric keyboards, significantly larger memory capacities than other phones of the time, and were designed for quick access to e-mail. Nokia was quick to follow with smart phones that had similar features. By 2009, every major phone manufacturer offered a range of smart phones and Internet-capable mobile phones. Most of these phones were too expensive for markets in developing countries; however, by 2011, a variety of Internet-capable mobile phones were being sold in these countries. Nokia has been a leader in developing lower-cost phones for these markets.

Although many companies have created Web pages for their mobile users that are designed to be used without a mouse and that are readable on the relatively small screens of phones, most have not. Without mobile-ready interfaces, it can be a challenge for users to fully implement their smart phones as tools of electronic commerce. As more online businesses realize that mobile phone users are potential customers, more Web sites will be redesigned to give mobile users a better experience.

In the developed industrial countries, Internet-capable phones and tablet devices are tools of convenience; they provide continual access to e-mail and the Web for busy people who work from multiple locations. In the rest of the world, they are often the only affordable way to access the Internet. For example, about 85 percent of the U.S. population (about 340 million people) had regular access to the Internet in 2011. Although many of these people could access the Internet through Internet-capable phones, very few of them relied on their phones as their only Internet access. In 2011 in China, only 32 percent of the population had Internet access and more than half of that access was through Internet-capable phones. In India, a mere 8 percent of the population (about 96 million people) had Internet access; slightly less than half of it through Internet-capable phones.

As you learned in Chapter 1, rapid growth in the use of Internet-capable phones is expected to continue in developing countries. As their Internet access increases and their economies develop, many observers expect vast increases in online business activity to follow.

THE INTERNET AND THE WORLD WIDE WEB

A **computer network** is any technology that allows people to connect computers to each other. An **internet** (small "i") is a group of computer networks that have been interconnected. In fact, "internet" is short for "interconnected network." One particular internet, which uses a specific set of rules and connects networks all over the world to each other, is called the **Internet** (capital "i"). Networks of computers and the Internet that connects them to each other form the basic technological structure that underlies virtually all electronic commerce.

This chapter introduces you to many of the hardware and software technologies that make electronic commerce possible. First, you will learn how the Internet and the World Wide Web work. Then, you will learn about other technologies that support the Internet, the Web, and electronic commerce. In this chapter, you will be introduced to several complex networking technologies. If you are interested in learning more about how computer networks operate, you can consult one of the computer networking books cited in the For Further Study and Research section at the end of this chapter, or you can take courses in data communications and networking.

The part of the Internet known as the **World Wide Web**, or, more simply, the **Web**, is a subset of the computers on the Internet that are connected to one another in a specific way that makes them and their contents easily accessible to each other. The most important thing about the Web is that it includes an easy-to-use standard interface. This interface makes it possible for people who are not computer experts to use the Web to access a variety of Internet resources.

Origins of the Internet

In the early 1960s, the U.S. Department of Defense became concerned about the possible effects of nuclear attack on its computing facilities. The Defense Department realized that the weapons of the future would require powerful computers for coordination and control. The powerful computers of that time were all large mainframe computers.

The Defense Department began examining ways to connect these computers to each other and also to connect them to weapons installations distributed all over the world. Employing many of the best communications technology researchers, the Defense Department funded research at leading universities and institutes. The goal of this research was to design a worldwide network that could remain operational, even if parts of the network were destroyed by enemy military action or sabotage. These researchers determined that the best path to accomplishing their goals was to create networks that did not require a central computer to control network operations.

The computer networks that existed at that time used leased telephone company lines for their connections. These telephone company systems established a single connection between sender and receiver for each telephone call, and then that connection carried all data along a single path. When a company wanted to connect computers it owned at two different locations, the company placed a telephone call to establish the connection, and then connected one computer to each end of that single connection.

The Defense Department was concerned about the inherent risk of this single-channel method for connecting computers, and its researchers developed a different method of sending information through multiple channels. In this method, files and messages are broken into packets that are labeled electronically with codes for their origins, sequences, and destinations. You will learn more about how packet networks operate later in this chapter.

In 1969, Defense Department researchers in the Advanced Research Projects Agency (ARPA) used this direct connection network model to connect four computers—one each at the University of California at Los Angeles, SRI International, the University of California at Santa Barbara, and the University of Utah—into a network called the ARPANET. The ARPANET was the earliest of the networks that eventually combined to become what we now call the Internet. Throughout the 1970s and 1980s, many researchers in the academic community connected to the ARPANET and contributed to the technological developments that increased its speed and efficiency. At the same time, researchers at other universities were creating their own networks using similar technologies.

New Uses for the Internet

Although the goals of the Defense Department network were to control weapons systems and transfer research files, other uses for this vast network began to appear in the early 1970s. E-mail was born in 1972 when Ray Tomlinson, a researcher who used the network, wrote a program that could send and receive messages over the network. This new method of communicating became widely used very quickly. The number of network users in the military and education research communities continued to grow. Many of these new participants used the networking technology to transfer files and access computers remotely.

The first e-mail mailing lists also appeared on these military and education research networks. A **mailing list** is an e-mail address that forwards any message it receives to any user who has subscribed to the list. In 1979, a group of students and programmers at Duke University and the University of North Carolina started **Usenet**, an abbreviation for **User's News Network**. Usenet allows anyone who connects to the network to read and post articles on a variety of subjects. Usenet survives on the Internet today, with more than 1000 different topic areas that are called **newsgroups**.

Although the people using these networks were developing many creative applications, use of the networks was limited to those members of the research and academic communities who could access them. Between 1979 and 1989, these network applications were improved and tested by an increasing number of users. The Defense Department's networking software became more widely used in academic and research institutions as these organizations recognized the benefits of having a common communications network. As the number of people in different organizations using these networks increased, security concerns arose; these concerns continue to be problematic. You will learn more about these network security issues in Chapter 10. The explosion of personal computer use during the 1980s also helped more people become comfortable with computers. During the 1980s, other independent networks (such as Bitnet) were developed by academics worldwide and researchers in specific countries other than the United States (such as the United Kingdom's academic research network, Janet). In the late 1980s, these independent academic and research networks from all over the world merged into what we now call the Internet.

Commercial Use of the Internet

As personal computers became more powerful, affordable, and available during the 1980s, companies increasingly used them to construct their own internal networks. Although these networks included e-mail software that employees could use to send messages to each other, businesses wanted their employees to be able to communicate with people outside their corporate networks. The Defense Department network and most of the academic networks that had teamed up with it were receiving funding from the National Science Foundation (NSF). The NSF prohibited commercial network traffic on its networks, so businesses turned to commercial e-mail service providers to handle their e-mail needs. Larger firms built their own networks that used leased telephone lines to connect field offices to corporate headquarters.

In 1989, the NSF permitted two commercial e-mail services, MCI Mail and CompuServe, to establish limited connections to the Internet for the sole purpose of exchanging e-mail transmissions with users of the Internet. These connections allowed commercial enterprises to send e-mail directly to Internet addresses, and allowed members of the research and education communities on the Internet to send e-mail directly to MCI Mail and CompuServe addresses. The NSF justified this limited commercial use of the Internet as a service that would primarily benefit the Internet's noncommercial users. As the 1990s began, people from all walks of life—not just scientists or academic researchers—started thinking of these networks as the global resource that we now know as the Internet. Although this network of networks had grown from four Defense Department computers in 1969 to more than 300,000 computers on many interconnected networks by 1990, the greatest growth of the Internet was yet to come.

Growth of the Internet

In 1991, the NSF further eased its restrictions on commercial Internet activity and began implementing plans to privatize the Internet. The privatization of the Internet was substantially completed in 1995, when the NSF turned over the operation of the main Internet connections to a group of privately owned companies. The new structure of the Internet was based on four **network access points (NAPs)** located in San Francisco, New York, Chicago, and Washington, D.C., each operated by a separate telecommunications company. As the Internet grew, more companies opened more NAPs in more locations. These companies, known as **network access providers**, sell Internet access rights directly to larger customers and indirectly to smaller firms and individuals through other companies, called **Internet service providers (ISPs)**.

The Internet was a phenomenon that had truly sneaked up on an unsuspecting world. The researchers who had been so involved in the creation and growth of the Internet just accepted it as part of their working environment. However, people outside the research community were largely unaware of the potential offered by a large interconnected set of computer networks. Figure 2-1 shows the consistent and dramatic growth in the number of **Internet hosts**, which are computers directly connected to the Internet.

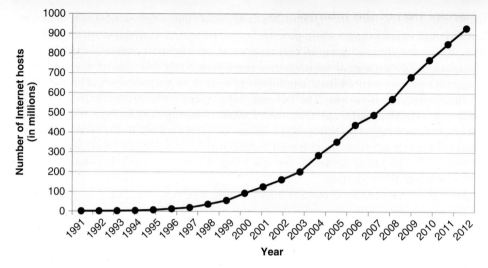

Source: Internet Software Consortium (http://www.isc.org/) and author's estimates

FIGURE 2-1 Growth of the Internet

In 40 years, the Internet has grown to become one of the most significant technological and social accomplishments of the last millennium. Millions of people, from elementary school students to research scientists, now use this complex, interconnected network of computers. These computers run thousands of different software packages. The computers are located in almost every country of the world. Every year, billions of dollars change hands over the Internet in exchange for all kinds of products and services. All of this activity occurs with no central coordination point or control, which is especially ironic given that the Internet began as a way for the military to maintain control of weapons systems while under attack.

The Internet is a set of interconnected networks. Thus, to understand the technologies used to build the Internet, you must first learn about the structure of its component networks.

PACKET-SWITCHED NETWORKS

A network of computers that are located close together—for example, in the same building—is called a **local area network (LAN)**. Networks of computers that are connected over greater distances are called **wide area networks (WANs)**.

The early models (dating back to the 1950s) for WANs were the circuits of the local and long-distance telephone companies of the time, because the first early WANs used leased telephone company lines for their connections. A telephone call establishes a single connection path between the caller and receiver. Once that connection is established, data travels along that single path. Telephone company equipment (originally mechanical, now electronic) selects specific telephone lines to connect to one another by closing switches. These switches work like the switches you use to turn lights on and off in your home, except that they open and close much faster, and are controlled by mechanical or electronic devices instead of human hands.

The combination of telephone lines and the closed switches that connect them to each other is called a **circuit**. This circuit forms a single electrical path between caller and receiver. This single path of connected circuits switched into each other is maintained for the entire length of the call. This type of centrally controlled, single-connection model is known as **circuit switching**.

Although circuit switching works well for telephone calls, it does not work as well for sending data across a large WAN or an interconnected network like the Internet. The Internet was designed to be resistant to failure. In a circuit-switched network, a failure in any one of the connected circuits causes the connection to be interrupted and data to be lost. Instead, the Internet uses packet switching to move data between two points. In a **packet-switched** network, files and e-mail messages are broken down into small pieces, called **packets**, that are labeled electronically with their origins, sequences, and destination addresses. Packets travel from computer to computer along the interconnected networks until they reach their destinations. Each packet can take a different path through the interconnected networks, and the packets may arrive out of order. The destination computer collects the packets and reassembles the original file or e-mail message from the pieces in each packet.

Routing Packets

As an individual packet travels from one network to another, the computers through which the packet travels determine the most efficient route for getting the packet to its destination. The most efficient route changes from second to second, depending on how much traffic each computer on the Internet is handling at each moment. The computers that decide how best to forward each packet are called **routing computers**, **router computers**, **routers**, **gateway computers** (because they act as the gateway from a LAN or WAN to the Internet), or **border routers** (because they are located at the border between the organization and the Internet). The programs on router computers that determine the best path on which to send each packet contain rules called **routing algorithms**. The programs apply their routing algorithms to information they have stored in **routing tables** or **configuration tables**. This information includes lists of connections that lead to particular groups of other routers, rules that specify which connections to use first, and rules for handling instances of heavy packet traffic and network congestion.

Individual LANs and WANs can use a variety of different rules and standards for creating packets within their networks. The network devices that move packets from one part of a network to another are called hubs, switches, and bridges. Routers are used to connect networks to other networks. You can take a data communications and networking class to learn more about these network devices and how they work.

When packets leave a network to travel on the Internet, they must be translated into a standard format. Routers usually perform this translation function. As you can see, routers are an important part of the infrastructure of the Internet. When a company or organization becomes part of the Internet, it must connect at least one router to the other routers (owned by other companies or organizations) that make up the Internet. Figure 2-2 is a diagram of a small portion of the Internet that shows its router-based architecture. The figure shows only the routers that connect each organization's WANs and LANs to the Internet, not the other routers that are inside the WANs and LANs or that connect them to each other within the organization.

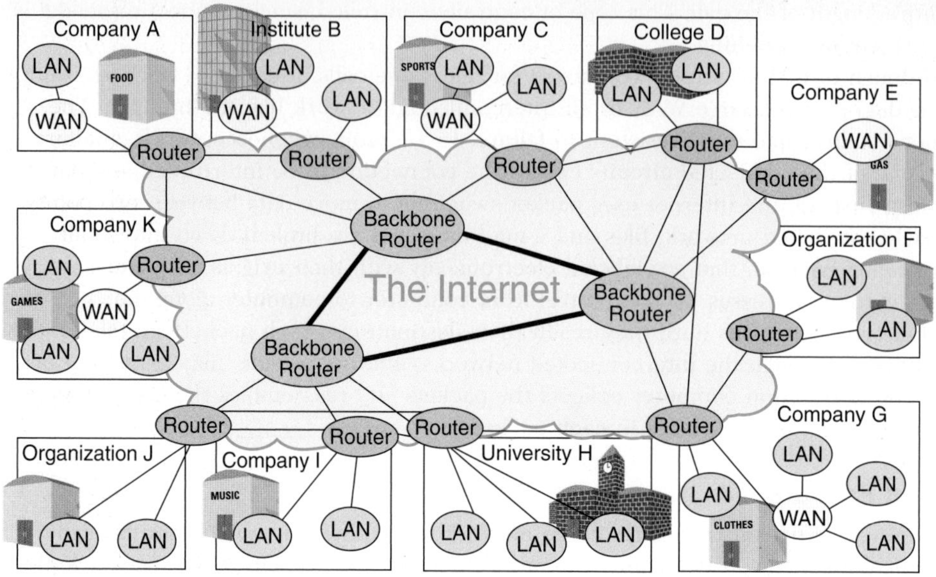

FIGURE 2-2   Router-based architecture of the Internet

The Internet also has routers that handle packet traffic along the Internet's main connecting points. These routers and the telecommunications lines connecting them are collectively referred to as the **Internet backbone**. These routers, sometimes called **backbone routers**, are very large computers that can each handle more than 3 billion packets per second. You can see in the figure that a router connected to the Internet always has more than one path to which it can direct a packet. By building in multiple packet paths, the designers of the Internet created a degree of redundancy in the system that allows it to keep moving packets, even if one or more of the routers or connecting lines fails.

Public and Private Networks

A **public network** is any computer network or telecommunications network that is available to the public. The Internet is one example of a public network. Public networks such as the Internet, as you will learn in later chapters, do not provide much security as part of their basic structures.

A **private network** is a leased-line connection between two companies that physically connects their computers and/or networks to one another. A **leased line** is a permanent telephone connection between two points. Unlike the telephone circuit connection you create when you dial a telephone number, a leased line is always active. The advantage of a leased line is security. Only the two parties that lease the line to create the private network have access to the connection.

The largest drawback to a private network is the cost of the leased lines, which can be quite expensive. Every pair of companies wanting a private network between them requires a separate line connecting them. For instance, if a company wants to set up private network connections with seven other companies, the company must pay the cost of seven leased lines, one for each company. Although the cost of leasing these lines has dropped significantly over the past two decades, it can still be substantial, especially for organizations that need to connect many offices or other locations to each other.

Virtual Private Networks (VPNs)

A **virtual private network (VPN)** is a connection that uses public networks and their protocols to send data in a way that protects the data as well as a private network would, but at a lower cost. VPN software must be installed on the computers at both ends of the transmission. The technology that most VPN software uses is called IP tunneling or encapsulation.

IP tunneling creates a private passageway through the public Internet that provides secure transmission from one computer to another. The passageway is created by VPN software that encrypts the packet content and then places the encrypted packets inside another packet in a process called **encapsulation**. The outer packet is called an **IP wrapper**. The Web server sends the encapsulated packets to their destinations over the Internet, which is a public network. The computer that receives the packet unwraps it and decrypts the message using VPN software that is the same as, or is compatible with, the VPN software used to encrypt and encapsulate the packet at the sending end.

The word "virtual" is used as part of VPN because, although the connection appears to be a permanent connection, it is actually temporary. The VPN is created, carries out its work over the Internet, and is then terminated.

The VPN is like a separate, covered commuter lane on a highway (the Internet) in which the passengers cannot be seen by vehicles traveling in the other lanes. Company employees in remote locations can send sensitive information to company computers using the VPN private tunnels established on the Internet. You will learn more about VPNs, firewalls, and other network security devices in Chapter 10.

Intranets and Extranets

In the early days of the Internet, the distinction between private and public networks was clear. Organizations could have one or more private networks that they operated internally. They could also participate in public networks with other organizations. The Internet was one such public network. However, as networking (and inter-networking) technologies became less expensive and easier to deploy, organizations began building more and more internets (small "i"), or interconnected networks. Some of these internets did not extend beyond the boundaries of the building organization.

The term **intranet** describes an internet that does not extend beyond the organization that created it. In the past, most intranets were constructed by interconnecting a number of private networks; however, organizations today can create secure intranets using VPN technologies. If security is not an issue, they can even build intranets using public networks. Similarly, an **extranet** was originally defined as an intranet that had been extended to include specific entities outside the boundaries of the organization, such as business partners, customers, or suppliers. Extranets were used to save money and increase efficiency by replacing traditional communication tools such as fax, telephone, and overnight express document carriers. To maintain security within extranets, almost all organizations that created them did so by interconnecting private networks.

As the Web became more widely used, many organizations began using the Internet, the public network on which the Web operates, as part of their extranets (and, in some cases, intranets). The addition of VPN technologies allowed organizations to use the Internet (a public network), yet have the same level of security over their data that had been provided by their use of private networks in the past.

This evolution of technologies over time has led to some confusion today when people use the terms public network, private network, VPN, intranet, and extranet. Remember that "intranet" is used when the internet does not extend beyond the boundaries of a particular organization; "extranet" is used when the internet extends beyond the boundaries of an organization and includes networks of other organizations. The technologies used (public networks, private networks, or VPNs) are independent of organizational boundaries. For example, an intranet could use private networks, VPNs, or even public networks (if security is not an issue).

INTERNET PROTOCOLS

A **protocol** is a collection of rules for formatting, ordering, and error checking data sent across a network. For example, protocols determine how the sending device indicates that it has finished sending a message and how the receiving device indicates that it has received (or not received) the message. A protocol also includes rules about what is allowed in a transmission and how it is formatted. Computers that communicate with each other must use the same protocol for data transmission. As you learned earlier in this chapter, the first packet-switched network, the ARPANET, connected only a few universities and research centers. Following its inception in 1969, this experimental network grew during the next few years and began using the **Network Control Protocol (NCP)**. In the early days of computing, each computer manufacturer created its own protocol, so computers made by different manufacturers could not be connected to each other. This practice was called **proprietary architecture** or **closed architecture**. NCP was designed so it could be used by any computer manufacturer and was made available to any company that wanted it. This **open architecture** philosophy that was developed for the evolving ARPANET, which later became the core of the Internet, included the use of a common protocol for all computers connected to the Internet and four key rules for message handling:

- Independent networks should not require any internal changes to be connected to the network.
- Packets that do not arrive at their destinations must be retransmitted from their source network.
- Router computers act as receive-and-forward devices; they do not retain information about the packets that they handle.
- No global control exists over the network.

The open architecture approach has contributed to the success of the Internet because computers manufactured by different companies (Apple, Dell, Hewlett-Packard, Sun, etc.) can be interconnected. The ARPANET and its successor, the Internet, use routers to isolate each LAN or WAN from the other networks to which they are connected. Each LAN or WAN can use its own set of protocols for packet traffic within the LAN or WAN, but must use a router (or similar device) to move packets onto the Internet in its standard format (or protocol). Following these simple rules makes the connections between the interconnected networks operate effectively.

TCP/IP

The Internet uses two main protocols: the **Transmission Control Protocol (TCP)** and the **Internet Protocol (IP)**. Developed by Internet pioneers Vinton Cerf and Robert Kahn, these protocols are the rules that govern how data moves through the Internet and how network connections are established and terminated. The acronym **TCP/IP** is commonly used to refer to the two protocols.

The TCP controls the disassembly of a message or a file into packets before it is transmitted over the Internet, and it controls the reassembly of those packets into their original formats when they reach their destinations. The IP specifies the addressing details for each packet, labeling each with the packet's origination and destination addresses. Soon after the new TCP/IP protocol set was developed, it replaced the NCP that ARPANET originally used.

In addition to its Internet function, TCP/IP is used today in many LANs. The TCP/IP protocol is provided in most personal computer operating systems commonly used today, including Linux, Macintosh, Microsoft Windows, and UNIX.

IP Addressing

The version of IP that has been in use since 1981 on the Internet is **Internet Protocol version 4 (IPv4)**. It uses a 32-bit number to identify computers connected to the Internet. This address is called an **IP address**. Computers do all of their internal calculations using a **base 2** (or **binary**) number system in which each digit is either a 0 or a 1, corresponding to a condition of either off or on. IPv4 uses a 32-bit binary number that allows for more than 4 billion different addresses (2^{32} = 4,294,967,296).

When a router breaks a message into packets before sending it onto the Internet, the router marks each packet with both the source IP address and the destination IP address of the message. To make them easier to read, IP numbers (addresses) appear as four numbers separated by periods. This notation system is called **dotted decimal** notation. An IPv4 address is a 32-bit number, so each of the four numbers is an 8-bit number ($4 \times 8 = 32$). In most computer applications, an 8-bit number is called a **byte**; however, in networking applications, an 8-bit number is often called an **octet**. In binary, an octet can have values from 00000000 to 11111111; the decimal equivalents of these binary numbers are 0 and 255, respectively.

Because each of the four parts of a dotted decimal number can range from 0 to 255, IP addresses range from 0.0.0.0 (written in binary as 32 zeros) to 255.255.255.255 (written in binary as 32 ones). Although some people find dotted decimal notation to be confusing at first, most do agree that writing, reading, and remembering a computer's address as 216.115.108.245 is easier than 11011000011100110110110011110101, or its full decimal equivalent, which is 3,631,433,189.

Today, IP addresses are assigned by three not-for-profit organizations: the **American Registry for Internet Numbers (ARIN)**, the **Reséaux IP Européens (RIPE)**, and the **Asia-Pacific Network Information Center (APNIC)**. These registries assign and manage IP addresses for various parts of the world: ARIN for North America, South America, the Caribbean, and sub-Saharan Africa; RIPE for Europe, the Middle East, and the rest of Africa; and APNIC for countries in the Asia-Pacific area.

You can use the **ARIN Whois** page at the ARIN Web site to search the IP addresses owned by organizations in North America. Enter an organization name into the search box on the page, then click the Search WHOIS button, and the Whois server returns a list of the IP addresses owned by that organization. For example, performing a search on the word *Carnegie* displays the IP address blocks owned by Carnegie Bank, Carnegie Mellon University, and a number of other organizations whose names begin with Carnegie. You can also enter an IP address and find out who owns that IP address. If you enter "3.0.0.0" (without the quotation marks), you will find that General Electric owns the entire block of IP addresses from 3.0.0.0 to 3.255.255.255. General Electric can use these addresses, which number approximately 16.7 million, for its own computers, or it can lease them to other companies or individuals to whom it provides Internet access services.

In the early days of the Internet, the 4 billion addresses provided by the IPv4 rules certainly seemed to be more addresses than an experimental research network would ever need. However, about 2 billion of those addresses today are either in use or unavailable for use because of the way blocks of addresses were assigned to organizations. The new kinds of devices on the Internet's many networks, such as wireless personal digital assistants and smart phones, promise to keep demand high for IP addresses.

Network engineers have devised a number of stopgap techniques to stretch the supply of IP addresses. One of the most popular techniques is **subnetting**, which is the use of reserved private IP addresses within LANs and WANs to provide additional address space. **Private IP addresses** are a series of IP numbers that are not permitted on packets that travel on the Internet. In subnetting, a computer called a **Network Address Translation (NAT) device** converts those private IP addresses into normal IP addresses when it forwards packets from those computers to the Internet.

The Internet Engineering Task Force (IETF) worked on several new protocols that could solve the limited addressing capacity of IPv4, and in 1997, approved **Internet Protocol version 6 (IPv6)** as the protocol that will replace IPv4. The new IP is being implemented gradually because the two protocols are not directly compatible. The process of switching the Internet over to IPv6 completely will take many years; however, network engineers have devised ways to run both protocols in parallel on interconnected networks. In 2011, the Internet Society conducted a 24-hour worldwide test of IPv6 that included more than 1000 Web sites; however, fewer than 10 percent of all Web hosts currently support the new protocol. The chief technology officer of the Internet Society has set a target of 20 percent deployment for the protocol by the end of 2012.

The major advantage of IPv6 is that it uses a 128-bit number for addresses instead of the 32-bit number used in IPv4. The number of available addresses in IPv6 (2^{128}) is 34 followed by 37 zeros—billions of times larger than the address space of IPv4. The new IP also changes the format of the packet itself. Improvements in networking technologies over the past 20 years have made many of the fields in the IPv4 packet unnecessary. IPv6 eliminates those fields and adds fields for security and other optional information.

IPv6 has a shorthand notation system for expressing addresses, similar to the IPv4 dotted decimal notation system. However, because the IPv6 address space is much larger, its notation system is more complex. The IPv6 notation uses eight groups of 16 bits ($8 \times 16 = 128$). Each group is expressed as four hexadecimal digits and the groups are separated by colons; thus, the notation system is called **colon hexadecimal** or **colon hex**. A **hexadecimal (base 16)** numbering system uses 16 characters (0, 1, 2, 3, 4, 5, 6, 7, 8, 9, a, b, c, d, e, and f). An example of an IPv6 address expressed in this notation is: CD18:0000:0000:AF23:0000:FF9E:61B2:884D. To save space, the zeros can be omitted, which reduces this address to: CD18:::AF23::FF9E:61B2:884D.

Electronic Mail Protocols

Electronic mail, or **e-mail**, that is sent across the Internet must also be formatted according to a common set of rules. Most organizations use a client/server structure to handle e-mail. The organization has a computer called an **e-mail server** that is devoted to handling e-mail. Software running on the e-mail server stores and forwards e-mail messages. People in the organization might use a variety of programs, called **e-mail client software**, to read and send e-mail. These programs include Microsoft Outlook, Mozilla Thunderbird, and many others. The e-mail client software communicates with the e-mail server software on the e-mail server computer to send and receive e-mail messages.

Many people also use e-mail on their computers at home. In most cases, the e-mail servers that handle their messages are operated by the companies that provide their connections to the Internet. An increasing number of people use e-mail services that are offered by Web sites such as Yahoo! Mail, Microsoft's Hotmail, or Google's Gmail. In these cases, the e-mail servers and the e-mail clients are operated by the owners of the Web sites. The individual users only see the e-mail client software (and not the e-mail server software) in their Web browsers when they log on to the Web mail service.

With so many different e-mail client and server software choices, standardization and rules are very important. If e-mail messages did not follow standard rules, an e-mail message created by a person using one e-mail client program could not be read by a person using a different e-mail client program. As you have already learned in this chapter, rules for computer data transmission are called protocols.

SMTP and POP are two common protocols used for sending and retrieving e-mail. **Simple Mail Transfer Protocol (SMTP)** specifies the format of a mail message and describes how mail is to be administered on the e-mail server and transmitted on the Internet. An e-mail client program running on a user's computer can request mail from the organization's e-mail server using the **Post Office Protocol (POP)**. A POP message can tell the e-mail server to send mail to the user's computer and delete it from the e-mail server; send mail to the user's computer and not delete it; or simply ask whether new mail has arrived. POP provides support for **Multipurpose Internet Mail Extensions (MIME)**, which is a set of rules for handling binary files, such as word-processing documents, spreadsheets, photos, or sound clips that are attached to e-mail messages.

The **Interactive Mail Access Protocol (IMAP)** is a newer e-mail protocol that performs the same basic functions as POP, but includes additional features. For example, IMAP can instruct the e-mail server to send only selected e-mail messages to the client instead of all messages. IMAP also allows the user to view only the header and the e-mail sender's name before deciding to download the entire message. POP requires users to download e-mail messages to their computers before they can search, read, forward, delete, or reply to those messages. IMAP lets users create and manipulate e-mail folders (also called mailboxes) and individual e-mail messages while the messages are still on the e-mail server; that is, the user does not need to download e-mail before working with it.

The tools that IMAP provides are important to the large number of people who access their e-mail from different computers at different times. IMAP lets users manipulate and store their e-mail on the e-mail server and access it from any computer. The main drawback to IMAP is that users' e-mail messages are stored on the e-mail server. As the number of users increases, the size of the e-mail server's disk drives must also increase. In general, server computers use faster (and thus, more expensive) disk drives than desktop computers. Therefore, it is more expensive to provide disk storage space for large quantities of e-mail on a server computer than to provide that same disk space on users' desktop computers. As the price of all disk storage continues to decrease, these cost concerns become less important. You can learn more about IMAP at the University of Washington's IMAP Connection Web site.

Web Page Request and Delivery Protocols

The Web is software that runs on computers that are connected to each other through the Internet. **Web client computers** run software called **Web client software** or **Web browser software**. Examples of popular Web browser software include Google Chrome, Microsoft Internet Explorer, and Mozilla Firefox. Web browser software sends requests for Web page files to other computers, which are called Web servers. A Web server computer runs software called **Web server software**. Web server software receives requests from many different Web clients and responds by sending files back to those Web client computers. Each Web client computer's Web client software renders those files into a Web page. Thus, the purpose of a Web server is to respond to requests for Web pages from Web clients. This combination of client computers running Web client software and server computers running Web server software is an example of a **client/server architecture**.

The set of rules for delivering Web page files over the Internet is in a protocol called the **Hypertext Transfer Protocol (HTTP)**, which was developed by Tim Berners-Lee in 1991. When a user types a domain name (for example, www.yahoo.com) into a Web browser's address bar, the browser sends an HTTP-formatted message to a Web server computer at Yahoo! that stores Web page files. The Web server computer at Yahoo! then responds by sending a set of files (one for the Web page and one for each graphic object, sound, or video clip included on the page) back to the client computer. These files are sent within a message that is HTTP formatted.

To initiate a Web page request using a Web browser, the user types the name of the protocol, followed by the characters "//:" before the domain name. Thus, a user would type http://www.yahoo.com to go to the Yahoo! Web site. Most Web browsers today automatically insert the http:// if the user does not include it. The combination of the protocol name and the domain name is called a **Uniform Resource Locator (URL)** because it lets the user locate a resource (the Web page) on another computer (the Web server).

EMERGENCE OF THE WORLD WIDE WEB

At a technological level, the Web is nothing more than software that runs on computers that are connected to the Internet. The network traffic generated by Web software is the largest single category of traffic on the Internet today, outpacing e-mail, file transfers, and other data-transmission traffic. But the ideas behind the Web developed from innovative ways of thinking about and organizing information storage and retrieval. These ideas go back many years. Two important ideas that became key technological elements of the Web are hypertext and graphical user interfaces.

The Development of Hypertext

In 1945, Vannevar Bush, who was director of the U.S. Office of Scientific Research and Development, wrote an article in *The Atlantic Monthly* about ways that scientists could apply the skills they learned during World War II to peacetime activities. The article included a number of visionary ideas about future uses of technology to organize and facilitate efficient access to information. Bush speculated that engineers would eventually build a machine that he called the Memex, a memory extension device that would store all of a person's books, records, letters, and research results on microfilm. Bush's Memex would include mechanical aids, such as microfilm readers and indexes, that would help users quickly and flexibly consult their collected knowledge.

In the 1960s, Ted Nelson described a similar system in which text on one page links to text on other pages. Nelson called his page-linking system **hypertext**. Douglas Engelbart, who also invented the computer mouse, created the first experimental hypertext system on one of the large computers of the 1960s. In 1987, Nelson published *Literary Machines*, a book in which he outlined project Xanadu, a global system for online hypertext publishing and commerce. Nelson used the term *hypertext* to describe a page-linking system that would interconnect related pages of information, regardless of where in the world they were stored.

In 1989, Tim Berners-Lee was trying to improve the laboratory research document-handling procedures for his employer, CERN: European Laboratory for Particle Physics. CERN had been connected to the Internet for two years, but its scientists wanted to find better ways to circulate their scientific papers and data among the high-energy physics research community throughout the world. Berners-Lee proposed a hypertext development project intended to provide this data-sharing functionality.

Over the next two years, Berners-Lee developed the code for a hypertext server program and made it available on the Internet. A **hypertext server** is a computer that stores files written in **Hypertext Markup Language (HTML)**, the language used for the creation of Web pages. The hypertext server is connected through the Internet to other computers that can connect to the hypertext server and read those HTML files. Hypertext servers used on the Web today are usually called **Web servers**. HTML, which Berners-Lee developed from his original hypertext server program, is a language that includes a set of codes (or tags) attached to text. These codes describe the relationships among text elements. For example, HTML includes tags that indicate which text is part of a header element, which text is part of a paragraph element, and which text is part of a numbered list element. One important type of tag is the hypertext link tag. A **hypertext link**, or **hyperlink**, points to another location in the same or another HTML document. The details of HTML and other markup languages are covered later in this chapter.

Graphical Interfaces for Hypertext

Several different types of software are available to read HTML documents, but most people use a Web browser such as Mozilla Firefox or Microsoft Internet Explorer. A **Web browser** is a software interface that lets users read (or browse) HTML documents and move from one HTML document to another through text formatted with hypertext link tags in each file. If the HTML documents are on computers connected to the Internet, you can use a Web browser to move from an HTML document on one computer to an HTML document on any other computer on the Internet.

An HTML document differs from a word-processing document in that it does not specify how a particular text element will appear. For example, you might use word-processing software to create a document heading by setting the heading text font to Arial, its font size to 14 points, and its position to centered. The document displays and prints these exact settings whenever you open the document in that word processor. In contrast, an HTML document simply includes a heading tag with the heading text. Many different browser programs can read an HTML document. Each program recognizes the heading tag and displays the text in whatever manner each program normally displays headings. Different Web browser programs might each display the text differently, but all of them display the text with the characteristics of a heading.

A Web browser presents an HTML document in an easy-to-read format in the browser's graphical user interface. A **graphical user interface (GUI)** is a way of presenting program control functions and program output to users and accepting their input. It uses pictures, icons, and other graphical elements instead of displaying just text. Almost all personal computers today use a GUI such as Microsoft Windows or the Macintosh user interface.

The World Wide Web

Berners-Lee called his system of hyperlinked HTML documents the World Wide Web. The Web caught on quickly in the scientific research community, but few people outside that community had software that could read the HTML documents. In 1993, a group of students led by Marc Andreessen at the University of Illinois wrote Mosaic, the first GUI program that could read HTML and use HTML hyperlinks to navigate from page to page on computers anywhere on the Internet. Mosaic was the first Web browser that became widely available for personal computers, and some Web surfers still use it today.

Programmers quickly realized that a system of pages connected by hypertext links would provide many new Internet users with an easy way to access information on the Internet. Businesses recognized the profit-making potential offered by a worldwide network of easy-to-use computers. In 1994, Andreessen and other members of the University of Illinois Mosaic team joined with James Clark of Silicon Graphics to found Netscape Communications (which is now owned by Time Warner). Its first product, the Netscape Navigator Web browser program based on Mosaic, was an instant success. Netscape became one of the fastest-growing software companies ever. Microsoft created its Internet Explorer Web browser and entered the market soon after Netscape's success became apparent. Today, Internet Explorer is the most widely used Web browser in the world. Its main competitor, Mozilla Firefox, is a descendant of Netscape Navigator.

The number of Web sites has grown even more rapidly than the Internet itself. The number of Web sites is currently estimated at more than 250 million, and individual Web pages number more than 50 billion because each Web site might include hundreds or even thousands of individual Web pages. Therefore, nobody really knows how many Web pages exist. Figure 2-3 shows the overall rapid growth rate of the Web. Other than a brief consolidation period during the 2001–2002 economic downturn, the Web has grown at a consistently rapid rate.

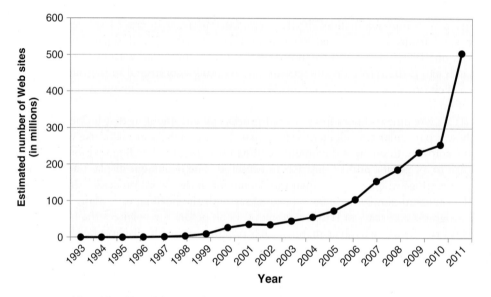

Adapted from Netcraft Computer Surveys (http://www.netcraft.com) and author's estimates

FIGURE 2-3   Growth of the World Wide Web

Noteworthy is the increase from 2010 to 2011, a year in which the number of Web sites doubled. This exceptional growth was driven in part by the large number of new Web sites opening in developing countries, primarily in Asia and Eastern Europe.

The Deep Web

In addition to Web pages that are specifically programmed to exist in a permanent form, the Web provides access to customized Web pages that are created in response to a particular user's query. Such Web pages pull their content from databases. For example, if you visit Amazon.com and search for a book about "online business," computers at Amazon.com query their databases of information about books and create a Web page that is a customized response to your search. The Web page that lists your search results never existed before your visit. This store of information that is available though the Web is called the **deep Web**. Researchers estimate the number of possible pages in the deep Web to be in the trillions.

Domain Names

The founders of the Internet were concerned that users might find the dotted decimal notation difficult to remember. To make the numbering system easier to use, they created an alternative addressing method that uses words. In this system, an address such as www.cengage.com is called a domain name. **Domain names** are sets of words that are assigned to specific IP addresses. Domain names can contain two or more word groups separated by periods. The rightmost part of a domain name is the most general. Each part of the domain name becomes more specific as you move to the left.

For example, the domain name www.sandiego.edu contains three parts separated by periods. Beginning at the right, the name "edu" indicates that the computer belongs to an educational institution. The institution, University of San Diego, is identified by the name "sandiego." The "www" indicates that the computer is running software that makes it a part of the World Wide Web. Most, but not all, Web addresses follow this "www" naming convention. For example, the group of computers that operate the Yahoo! Games service is named games.yahoo.com.

The rightmost part of a domain name is called a **top-level domain (TLD)**. For many years, these domains have included a group of generic domains—such as .edu, .com, and .org—and a set of country domains. Since 1998, the Internet Corporation for Assigned Names and Numbers (ICANN) has had the responsibility of managing domain names and coordinating them with the IP address registrars. ICANN is also responsible for setting standards for the router computers that make up the Internet. Since taking over these responsibilities, ICANN has added a number of new TLDs. Some of these TLDs are **generic top-level domains (gTLDs)**, which are available to specified categories of users. ICANN is itself responsible for the maintenance of gTLDs. Other new domains are **sponsored top-level domains (sTLDs)**, which are TLDs for which an organization other than ICANN is responsible. The sponsor of a specific sTLD must be a recognized institution that has expertise regarding and is familiar with the community that uses the sTLD. For example, the .aero sTLD is sponsored by SITA, an air transport industry association that has expertise in and is familiar with airlines, airports, and the aerospace industry. Individual countries are permitted to maintain their own TLDs, which their residents can use alone or in combination with other TLDs. For example, the URL of the University of Queensland in Brisbane, Australia is www.uq.edu.au, which combines .edu with .au to indicate that it is an educational institution in Australia. Figure 2-4 presents a list of some commonly used TLDs, including gTLDs and some of the more frequently used country TLDs.

TLD	Use
.com	U.S. commercial
.edu	Four-year educational institution
.gov	U.S. federal government
.mil	U.S. military
.net	U.S. general use
.org	U.S. not-for-profit organization
.us	U.S. general use
.asia	Companies, individuals, and organizations based in Asian–Pacific regions
.biz	Businesses
.info	General use
.name	Individual persons
.pro	Licensed professionals (such as accountants, lawyers, physicians)
.au	Australia
.ca	Canada
.de	Germany
.fi	Finland
.fr	France
.jp	Japan
.se	Sweden
.uk	United Kingdom

Source: Internet Assigned Numbers Authority Root Zone Database, http://www.iana.org/domains/root/db/

FIGURE 2-4   Commonly used domain names

Although ICANN has always chosen new gTLDs after much deliberation and careful consideration, many people have been highly critical of the selections (see, for example, the **ICANNWatch** Web site). In 2011, ICANN decided to stop managing the addition of new gTLDs so tightly. Starting in 2012, individuals and businesses can petition for just about any TLD they would like to have. This has generated some controversy; you can learn more about the related issues on the Web sites of the **Internet Governance Project** and the **Convergence Center**, both at Syracuse University. Increases in the number of TLDs can make it more difficult for companies to protect their corporate and product brand names, as you will learn in Chapter 7.

MARKUP LANGUAGES AND THE WEB

Web pages can include many elements, such as graphics, photographs, sound clips, and even small programs that run in the Web browser. Each of these elements is stored on the Web server as a separate file. The most important parts of a Web page, however, are the structure of the page and the text that makes up the main part of the page. The page structure and text are stored in a text file that is formatted, or marked up, using a text markup language. A **text markup language** specifies a set of tags that are inserted into the text. These **markup tags**, or **tags**, provide formatting instructions that Web client software can understand. The Web client software uses those instructions as it renders the text and page elements contained in the other files into the Web page that appears on the screen of the client computer.

The markup language most commonly used on the Web is HTML, which is a subset of a much older and far more complex text markup language called **Standard Generalized Markup Language (SGML)**. Figure 2-5 shows how HTML, XML, and XHTML have descended from the original SGML specification. SGML was used for many years by the publishing industry to create documents that needed to be printed in various formats and that were revised frequently. In addition to its role as a markup language, SGML is a **metalanguage**, which is a language that can be used to define other languages. Another markup language that was derived from SGML for use on the Web is **Extensible Markup Language (XML)**, which is increasingly used to mark up information that companies share with each other over the Internet. The X in XML comes from the word extensible; you might see the word extensible shown as eXtensible. XML is also a meta language because users can create their own markup elements that extend the usefulness of XML (which is why it is called an "extensible" language).

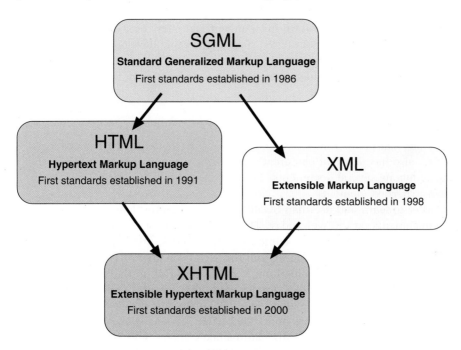

© Cengage Learning

FIGURE 2-5   Development of markup languages

The **World Wide Web Consortium (W3C)**, a not-for-profit group that maintains standards for the Web, presented its first draft form of XML in 1996; the W3C issued its first formal version recommendation in 1998. Thus, it is a much newer markup language than HTML. In 2000, the W3C released the first version of a recommendation for a new markup language called **Extensible Hypertext Markup Language (XHTML)**, which is a reformulation of HTML version 4.0 as an XML application. The Web Links include a link to the W3C XHTML Version 1.0 Specification.

Markup Languages

Since the 1960s, publishers have used markup languages to create documents that can be formatted once, stored electronically, and then printed many times in various layouts that each interpret the formatting differently. U.S. Department of Defense contractors also used early markup languages to create manuals and parts lists for weapons systems. These documents contained many information elements that were often reprinted in different versions and formats. Using electronic document storage and programs that could interpret the formats to produce different layouts saved a tremendous amount of retyping time and cost.

A **Generalized Markup Language (GML)** emerged from these early efforts to create standard formatting styles for electronic documents. In 1986, after many elements of the standard had been in use for years, the International Organization for Standardization (ISO) adopted a version of GML called SGML. SGML offers a system of marking up documents that is independent of any software application. Many organizations, such as the Association of American Publishers, Hewlett-Packard, and Kodak, use SGML because they have complex document-management requirements.

SGML is nonproprietary and platform independent and offers user-defined tags. However, it is not well suited to certain tasks, such as the rapid development of Web pages. SGML is costly to set up and maintain, requires the use of expensive software tools, and is hard to learn. Creating document-type definitions in SGML can be expensive and time consuming.

Hypertext Markup Language

HTML includes tags that define the format and style of text elements in an electronic document. HTML also has tags that can create relationships among text elements within one document or among several documents. The text elements that are related to each other are called **hypertext elements**.

HTML is easier to learn and use than SGML. HTML is the prevalent markup language used to create documents on the Web today. The early versions of HTML let Web page designers create text-based electronic documents with headings, title bar titles, bullets, lines, and ordered lists. As the use of HTML and the Web itself grew, HTML creator Berners-Lee turned over the job of maintaining versions of HTML to the W3C. Later versions of HTML included tags for tables, frames, and other features that helped Web designers create more complex page layouts. The W3C maintains detailed information about HTML versions and related topics on its W3C HTML Working Group page.

The process for approval of new HTML features takes a long time, so Web browser software developers created some features, called **HTML extensions**, that would only work in their browsers. At various times during the history of HTML, both Microsoft and Netscape enabled their Web browsers to use these HTML extension tags before those tags were approved by the W3C. In some cases, these tags were enabled in one browser and not the other. In other cases, the tags used were never approved by the W3C or were approved in a different form than the one implemented in the Web browser software. Web page designers who wanted to use the latest available tags were often frustrated by this inconsistency. Many of these Web designers had to create separate sets of Web pages for the different types of browsers, which was inefficient and expensive. Most of these tag difference issues were resolved when the W3C issued the specification for HTML version 4.0 in 1997, although enough of them remained to cause regular problems for Web designers.

After HTML 4.0 was finalized in 1999, development on new versions of HTML slowed. Browser developers worked on adding new features to their software and the W3C directed its efforts to other matters. In 2007, three browser developers (Apple, Opera, and the Mozilla Foundation) began working on an updated version of HTML that would include features such as audio and video within the markup language itself. Audio and video elements in Web pages have always required the use of add-on software. The current working draft of HTML version 5.0 is authorized to be active until 2014, but it could be finalized before then. You can learn more about this latest HTML version by visiting the W3C HTML 5 page.

HTML Tags

An HTML document contains document text and elements. The tags in an HTML document are interpreted by the Web browser and used by it to format the display of the text enclosed by the tags. In HTML, the tags are enclosed in angle brackets (<>). Most HTML tags have an **opening tag** and a **closing tag** that format the text between them. The closing tag is preceded by a slash within the angle brackets (</>). The general form of an HTML element is:

```
<tagname properties>Displayed information affected by tag</tagname>
```

Two good examples of HTML tag pairs are the strong character-formatting tags and the emphasis character-formatting tags. For example, a Web browser reading the following line of text:

```
<strong>A Review of the Book <em>HTML Is Fun!</em></strong>
```

would recognize the <strong> and </strong> tags as instructions to display the entire line of text in bold and the <em> and </em> tags as instructions to display the text enclosed by those tags in italics. The Web browser would display the text as:

A Review of the Book *HTML Is Fun!*

Some Web browsers allow the user to customize the interpretations of the tags, so that different Web browsers might display the tagged text differently. For example, one Web browser might display text enclosed by strong tags in a blue color instead of displaying the text as bold. Tags are generally written in lowercase letters; however, older versions of HTML allowed the use of either case and you might still see Web pages that include uppercase (or mixed case) HTML tags. Although most tags are two-sided (they use both an opening and a closing tag), some are not. Tags that only require opening tags are known as one-sided tags. The tag that creates a line break (</br>) is a common one-sided tag. Some tags, such as the paragraph tag (<p>...</p>), are two-sided tags for which the closing tag is optional. Designers sometimes omit the optional closing tags, but this practice is poor markup style.

In a two-sided tag set, the closing tag position is very important. For example, if you were to omit the closing bold tag in the preceding example, any text that followed the line would be bolded. Sometimes an opening tag contains one or more property modifiers that further refine how the tag operates. A tag's property might modify a text display, or it might designate where to find a graphic element. Figure 2-6 (on the next page) shows some sample text marked up with HTML tags and Figure 2-7 (on page 75) shows this text as it appears in a Web browser. The tags in these two figures are among the most common HTML tags in use today on the Web.

```
<html>

    <head>

        <title>HTML Tag Examples</title>

    </head>

    <body>

    <h1>This text is set in Heading one tags</h1>
    <h2>This text is set in Heading two tags</h2>
    <h3>This text is set in Heading three tags</h3>

    <p>
    This text is set within Paragraph tags. It will appear as one paragraph: the
    text will wrap at the end of each line that is rendered in the Web browser no
    matter where the typed text ends. The text inside Paragraph tags is rendered
    without regard to extra spaces typed in the text, such as these:
    Character formatting can also be applied within Paragraph tags. For
    example, <strong>the Strong tags will cause this text to appear bolded in
    most Web browsers</strong> and <em>the emphasis tags will cause this to
    appear italicized in most Web browsers</em>.
    </p>

    <pre>
    HTML includes tags that instruct the Web browser to render the text
    Exactly      the     way      it     is       typed,
    as in this example.
    </pre>

    <p>
    HTML includes tags that instruct the Web browser to place text in bulleted or
    numbered lists:
    </p>

    <ul>
        <li>Bulleted list item one</li>
        <li>Bulleted list item two</li>
        <li>Bulleted list item three</li>
    </ul>

    <ol>
        <li>Numbered list item one</li>
        <li>Numbered list item two</li>
        <li>Numbered list item three</li>
    </ol>

    <p>
    The most important tag in HTML is the Anchor Hypertext Reference tag,
    which is the tag that provides a link to another Web page (or another location
    in the same Web page). For example, the underlined text
    <a href="http://www.w3c.org/">World Wide Web Consortium</a>
    is a link to the not-for-profit organization that develops Web technologies.
    </p>

    </body>

</html>
```

FIGURE 2-6　Text marked up with HTML tags

This text is set in Heading one tags

This text is set in Heading two tags

This text is set in Heading three tags

This text is set within Paragraph tags. It will appear as one paragraph: the text will wrap at the end of each line that is rendered in the Web browser no matter where the typed text ends. The text inside Paragraph tags is rendered without regard to extra spaces typed in the text, such as these: Character formatting can also be applied within Paragraph tags. For example, **the Strong tags will cause this text to appear bolded in most Web browsers** and *the emphasis tags will cause this to appear italicized in most Web browsers.*

```
HTML includes tags that instruct the Web browser to render the text
Exactly     the      way      it      is       typed,
as in this example.
```

HTML includes tags that instruct the Web browser to place text in bulleted or numbered lists:

- Bulleted list item one
- Bulleted list item two
- Bulleted list item three

1. Numbered list item one
2. Numbered list item two
3. Numbered list item three

The most important tag in HTML is the Anchor Hypertext Reference tag, which is the tag that provides a link to another Web page (or another location in the same Web page). For example, the underlined text World Wide Web Consortium is a link to the not-for-profit organization that develops Web technologies.

FIGURE 2-7 Text marked up with HTML tags as it appears in a Web browser

Other frequently used HTML tags (not shown in the figures) let Web designers include graphics on Web pages and format text in the form of tables. The text and HTML tags that form a Web page can be viewed when the page is open in a Web browser by clicking the Page button and selecting View source in Internet Explorer or by selecting View, Page Source from the context menu in Firefox. A number of online sources (such as the W3C Getting Started with HTML page) and textbooks are available that describe HTML tags and their uses, and you can consult them for an in-depth look at HTML.

HTML Links

The Web organizes interlinked pages of information residing on sites around the world. Hyperlinks on Web pages form a "web" of those pages. A user can traverse the interwoven pages by clicking hyperlinked text on one page to move to another page in the web of pages. Users can read Web pages in serial order or in whatever order they prefer by following hyperlinks. Figure 2-8 illustrates the differences between reading a paper catalog in a linear way and reading a hypertext catalog in a nonlinear way.

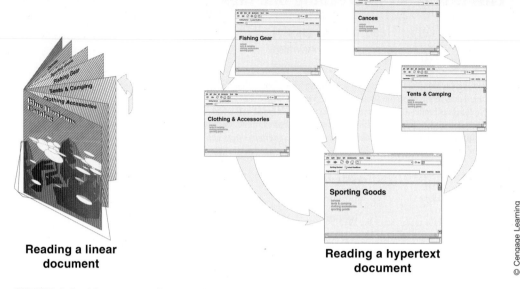

Reading a linear document

Reading a hypertext document

© Cengage Learning

FIGURE 2-8 Linear vs. nonlinear paths through documents

Web sites can use links to direct customers to pages on the company's Web server. The way links lead customers through pages can affect the usefulness of the site and can play a major role in shaping customers' impressions of the company. Two commonly used link structures are linear and hierarchical. A **linear hyperlink structure** resembles conventional paper documents in that the reader begins on the first page and clicks the Next button to move to the next page in a serial fashion. This structure works well when customers fill out forms prior to a purchase or other agreement. In this case, the customer reads and responds to page one, and then moves on to the next page. This process continues until the entire form is completed. The only Web page navigation choices the user typically has are Back and Next.

Another link arrangement is called a hierarchical structure. In a **hierarchical hyperlink structure**, the Web user opens an introductory page called a **home page** or **start page**. This page contains one or more links to other pages, and those pages, in turn, link to other pages. This hierarchical arrangement resembles an inverted tree in which the root is at the top and the branches are below it. Hierarchical structures are good for leading customers from general topics or products to specific product models and quantities. A company's home page might contain links to help, company history, company officers, order processing, frequently asked questions, and product catalogs.

Many sites that use a hierarchical structure include a page on the Web site that contains a map or outline listing of the Web pages in their hierarchical order. This page is called a **site map**. Of course, hybrid designs that combine linear and hierarchical structures are also possible. Figure 2-9 illustrates these three common Web page organization structures.

Linear structure

Hierarchical structure

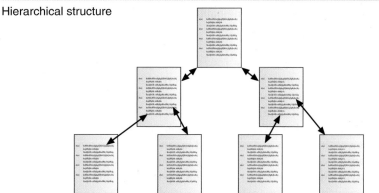

Hybrid structure

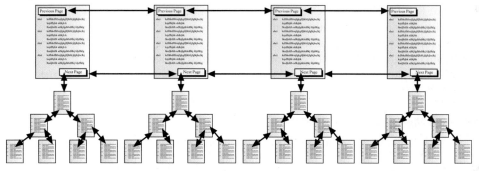

© Cengage Learning

FIGURE 2-9 Three common Web page organization structures

In HTML, hyperlinks are created using the HTML **anchor tag**. Whether you are linking to text within the same document or to a document on a distant computer, the anchor tag has the same basic form:

```
<a href="address">Visible link text</a>
```

Anchor tags have opening and closing tags. The opening tag has a hypertext reference (HREF) property, which specifies the remote or local document's address. Clicking the text following the opening link transfers control to the HREF address, wherever that happens to be. A person creating an electronic résumé on the Web might want to make a university's name and address under the Education heading a hyperlink instead of plain text. Anyone viewing the résumé can click the link, which leads the reader to the university's home page. The following example shows the HTML code to create a hyperlink to another Web server:

```
<a href="http://www.gsu.edu">Georgia State University </a>
```

Similarly, the résumé could include a local link to another part of the same document with the following marked-up text:

```
<a href="#references">References are found here</a>
```

In both of these examples, the text between the anchors appears on the Web page as a hyperlink. Most browsers display the link in blue and underline it. In most browser software, the action of moving the mouse pointer over a hyperlink causes the mouse pointer to change from an arrow to a pointing hand.

Scripting Languages and Style Sheets

Versions of HTML released by the W3C after 1997 include an HTML tag called the object tag and include support for **Cascading Style Sheets (CSS)**. Web designers can use the object tag to embed scripting language code on HTML pages. You will learn more about Web page scripting techniques in Chapter 8.

CSS are sets of instructions that give Web developers more control over the format of displayed pages. Similar to document styles in word-processing programs, CSS lets designers define formatting styles that can be applied to multiple Web pages. The set of instructions, called a **style sheet**, is usually stored in a separate file and is referenced using the HTML style tag; however, it can be included as part of a Web page's HTML file. The term *cascading* is used because designers can apply many style sheets to the same Web page, one on top of the other, and the styles from each style sheet flow (or cascade) into the next. For example, a three-stage cascade might include one style sheet with formatting instructions for text within heading 1 tags, a second style sheet with formatting instructions for text within heading 2 tags, and a third style sheet with formatting instructions for text within paragraph tags. A designer who later decides to change the formatting of heading 2 text can just replace the second style sheet with a different one. Those changes would cascade into the third style sheet.

Extensible Markup Language (XML)

As the Web grew, HTML continued to provide a useful tool for Web designers who wanted to create attractive layouts of text and graphics on their pages. However, as companies began to conduct electronic commerce on the Web, the need to present large amounts of data on Web pages also became important. Companies created Web sites that contained lists of inventory items, sales invoices, purchase orders, and other business data. The need to keep these lists updated was also important and posed a new challenge for many Web designers. The tool that had helped these Web designers create useful Web pages, HTML, was not such a good tool for presenting or maintaining information lists.

In the late 1990s, companies began turning to XML to help them maintain Web pages that contained large amounts of data. XML uses paired start and stop tags in much the same way as database software defines a record structure. For example, a company that sells products on the Web might have Web pages that contain descriptions and photos of the products it sells. The Web pages are marked up with HTML tags, but the product information elements themselves, such as prices, identification numbers, and quantities on hand, are marked up with XML tags. The XML document is embedded within the HTML document.

XML includes data-management capabilities that HTML cannot provide. To better understand the strengths of XML and weaknesses of HTML in data-management tasks, consider the simple example of a Web page that includes a list of countries and some basic facts about each country. A Web designer might decide to use HTML tags to show each fact the same way for each country. Each fact would use a different tag. Assume that the Web designer in this case decided to use the HTML heading tags to present the data. Figure 2-10 shows the data and the HTML heading tags for four countries (this is only an example; the actual list would include more than 150 countries). The first item in the list provides the definitions for each tag. Figure 2-11 (on the next page) shows this HTML document as it appears in a Web browser.

```
<html>

  <head>

    <title>Countries</title>

  </head>

  <body>

    <h1>Countries</h1>

    <h2>CountryName</h2>
    <h3>CapitalCity</h3>
    <h4>AreaInSquareKilometers</h4>
    <h5>OfficialLanguage</h5>
    <h6>VotingAge</h6>

    <h2>Argentina</h2>
    <h3>Buenos Aires</h3>
    <h4>2,766,890</h4>
    <h5>Spanish</h5>
    <h6>18</h6>

    <h2>Austria</h2>
    <h3>Vienna</h3>
    <h4>83,858</h4>
    <h5>German</h5>
    <h6>19</h6>

    <h2>Barbados</h2>
    <h3>Bridgetown</h3>
    <h4>430</h4>
    <h5>English</h5>
    <h6>18</h6>

    <h2>Belarus</h2>
    <h3>Minsk</h3>
    <h4>207,600</h4>
    <h5>Byelorussian</h5>
    <h6>18</h6>

  </body>

</html>
```

FIGURE 2-10 Country list data marked up with HTML tags

Countries

CountryName

CapitalCity

AreaInSquareKilometers

OfficialLanguage

Voting Age

Argentina

Buenos Aires

2,766,890

Spanish

18

Austria

Vienna

83,858

German

19

Barbados

Bridgetown

430

English

18

Belarus

Minsk

207,600

Byelorussian

18

FIGURE 2-11   Country list data as it appears in a Web browser

These figures reveal some of the shortcomings of using HTML to present a list of items when the meaning of each item in the list is important. The Web designer in this case used HTML heading tags. HTML has only six levels of heading tags; thus, if the individual items had additional information elements than shown in this example (such as population and continent), this approach would not work at all. The Web designer could use various combinations of text attributes such as size, font, color, bold, or italics to distinguish among items, but none of these tags would convey the meaning of the individual data elements. The only information about the meaning of each country's listing appears in the first list item, which includes the definitions for each element. In the late 1990s, Web professionals began to consider XML as a list-formatting alternative to HTML that would more effectively communicate the meaning of data.

XML differs from HTML in two important respects. First, XML is not a markup language with defined tags. It is a framework within which individuals, companies, and other organizations can create their own sets of tags. Second, XML tags do not specify how text appears on a Web page; the tags convey the meaning (the semantics) of the information included within them. To understand this distinction between appearance and semantics, consider the list of countries example again. In XML, tags can be created for each fact that define the meaning of the fact. Figure 2-12 shows the countries data marked up with XML tags. Some browsers, such as Internet Explorer, can render XML files directly without additional instructions. Figure 2-13 (on the next page) shows the country list XML file as it would appear in an Internet Explorer browser window.

```
declaration —<?xml version="1.0"?>

root
element —<CountriesList>
       <Country Name = "Argentina">
         <CapitalCity>Buenos Aires</CapitalCity>
         <AreaInSquareKilometers>2,766,890</AreaInSquareKilometers>
         <OfficialLanguage>Spanish</OfficialLanguage>
         <VotingAge>18</VotingAge>
       </Country>

       <Country Name = "Austria">
         <CapitalCity>Vienna</CapitalCity>
         <AreaInSquareKilometers>83,858</AreaInSquareKilometers>
         <OfficialLanguage>German</OfficialLanguage>
         <VotingAge>19</VotingAge>
       </Country>

       <Country Name = "Barbados">
         <CapitalCity>Bridgetown</CapitalCity>
         <AreaInSquareKilometers>430</AreaInSquareKilometers>
         <OfficialLanguage>English</OfficialLanguage>
         <VotingAge>18</VotingAge>
       </Country>

       <Country Name = "Belarus">
         <CapitalCity>Minsk</CapitalCity>
         <AreaInSquareKilometers>207,600</AreaInSquareKilometers>
         <OfficialLanguage>Byelorussian</OfficialLanguage>
         <VotingAge>18</VotingAge>
       </Country>

     </CountriesList>
```

© Cengage Learning

FIGURE 2-12 Country list data marked up with XML tags

```
<?xml version="1.0" ?>
<CountriesList>
- <Country Name="Argentina">
    <CapitalCity>Buenos Aires</CapitalCity>
    <AreaInSquareKilometers>2,766,890</AreaInSquareKilometers>
    <OfficialLanguage>Spanish</OfficialLanguage>
    <VotingAge>18</VotingAge>
  </Country>
- <Country Name="Austria">
    <CapitalCity>Vienna</CapitalCity>
    <AreaInSquareKilometers>83,858</AreaInSquareKilometers>
    <OfficialLanguage>German</OfficialLanguage>
    <VotingAge>19</VotingAge>
  </Country>
- <Country Name="Barbados">
    <CapitalCity>Bridgetown</CapitalCity>
    <AreaInSquareKilometers>430</AreaInSquareKilometers>
    <OfficialLanguage>English</OfficialLanguage>
    <VotingAge>18</VotingAge>
  </Country>
- <Country Name="Belarus">
    <CapitalCity>Minsk</CapitalCity>
    <AreaInSquareKilometers>207,600</AreaInSquareKilometers>
    <OfficialLanguage>Byelorussian</OfficialLanguage>
    <VotingAge>18</VotingAge>
  </Country>
</CountriesList>
```

FIGURE 2-13   Country list data marked up with XML tags as it would appear in Internet Explorer

The first line in the XML file shown in Figures 2-12 and 2-13 is the declaration, which indicates that the file uses version 1.0 of XML. XML markup tags are similar in appearance to SGML markup tags, thus the declaration can help avoid confusion in organizations that use both. The second line and the last line are the root element tags. The root element of an XML file contains all of the other elements in that file and is usually assigned a name that describes the purpose or meaning of the file.

The other elements are called child elements; for example, Country is a child element of CountriesList. Each of the other attributes is, in turn, a child element of the Country element. Unlike an HTML file, when an XML file is displayed in a browser, the tags are visible. The names of these child elements were created specifically for use in this file. If programmers in another organization were to create a file with country information, they might use different names for these elements (for example, "Capital" instead of "CapitalCity"), which would make it difficult for the two organizations to share information. Thus, the greatest strength of XML, that it allows users to define their own tags, is also its greatest weakness.

To overcome that weakness, many companies have agreed to follow common standards for XML tags. These standards, in the form of data-type definitions (DTDs) or XML schemas, are available for a number of industries, including **Extensible Business Reporting Language (XBRL)** for accounting and financial information standards, **LegalXML** for information in the legal profession, and **MathML** for mathematical and scientific information.

A number of industry groups have formed to create standard XML tag definitions that can be used by all companies in that industry. **RosettaNet** is an example of such an industry group. In 2001, the W3C released a set of rules for XML document interoperability that many researchers believe will help resolve incompatibilities between different sets of XML tag definitions. A set of XML tag definitions is sometimes called an **XML vocabulary**. Hundreds of publicly defined XML vocabularies have been developed or are currently circulating. You can find links to many of them on the Oasis **Cover Pages: XML Applications and Initiatives** Web page. You can learn more about XML by reading the **W3C XML Pages**.

Although it is possible to display XML files in some Web browsers, XML files are not intended to be displayed in a Web browser. They are designed to be translated using another file that contains formatting instructions or to be read by a program. Formatting instructions are often written in the **Extensible Stylesheet Language (XSL)**, and the programs that read or transform XML files are usually written in the Java programming language. These programs, sometimes called **XML parsers**, can format an XML file so it can appear on the screen of a computer, a smart phone, an Internet-capable mobile phone, or some other device. A diagram showing one way that a Web server might process HTTP requests for Web pages generated from an XML database in different formats for different Web browsing devices appears in Figure 2-14.

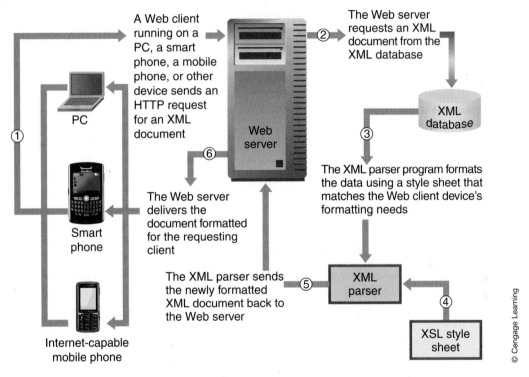

FIGURE 2-14 Processing requests for Web pages from an XML database

HTML and XML Editors

Web designers can create HTML documents in any general-purpose text editor or word processor. However, a special-purpose HTML editor can help Web designers create Web pages much more easily. HTML editors are also included as part of more sophisticated programs that are sometimes called Web site design tools. With these programs, Web designers can create and manage complete Web sites, including features for database access, graphics, and fill-in forms. These programs display the Web page as it will appear in a Web browser in one window and display the HTML-tagged text in another window. The designer can edit in either window and changes are reflected in the other window.

For example, the designer can drag and drop objects such as graphics onto the Web browser view page and the program automatically generates the HTML tags to position the graphics.

Web site design programs also include features that allow the designer to create a Web site on a PC and then upload the entire site (HTML documents, graphics files, and so on) to a Web server computer. When the site needs to be edited later, the designer can edit the copy of the site on the PC, and then instruct the program to synchronize those changes on the copy of the site that resides on the Web server. The most widely used Web site design tool is **Adobe Dreamweaver**.

XML files, like HTML files, can be created in any text editor. However, programs designed to make the task of designing and managing XML files easier are also available. These programs provide tag validation and XML creation capabilities in addition to making the job of marking up text with XML tags more efficient. An example of a leading XML editing program is **XML Spy**.

INTERNET CONNECTION OPTIONS

The Internet is a set of interconnected networks. Most organizations have their computers connected to each other using a network. Many families have their home computers connected to each other in a network. Mobile phones are connected to the wireless phone service provider's network. These networks can be connected to the Internet in a number of ways, as described in this section. Companies that provide Internet access to individuals, businesses, and other organizations, called **Internet access providers (IAPs)** or ISPs, usually offer several connection options. This section briefly describes current connection choices and presents their advantages and disadvantages.

Connectivity Overview

ISPs offer several ways to connect to the Internet. The most common connection options are voice-grade telephone lines, various types of broadband connections, leased lines, and wireless. One of the major distinguishing factors between various ISPs and their connection options is the bandwidth they offer. **Bandwidth** is the amount of data that can travel through a communication medium per unit of time. The higher the bandwidth, the faster data files travel and the faster Web pages appear on your screen. Each connection option offers different bandwidths, and each ISP offers varying bandwidths for each connection option. Traffic on the Internet and at your local service provider greatly affects **net bandwidth**, which is the actual speed that information travels. When few people are competing for service from an ISP, net bandwidth approaches the carrier's upper limit. On the other hand, users experience slowdowns during high-traffic periods.

Bandwidth can differ for data traveling to or from the ISP depending on the user's connection type. Connection types include:

- **Symmetric connections** that provide the same bandwidth in both directions.
- **Asymmetric connections** that provide different bandwidths for each direction.

Bandwidth refers to the amount of data that travels and the rate at which it travels. The two bandwidth types in an asymmetric connection are as follows:

- **Upstream bandwidth**, also called **upload bandwidth**, is a measure of the amount of information that can travel from the user to the Internet in a given amount of time.

- **Downstream bandwidth**, also called **download** or **downlink bandwidth**, is a measure of the amount of information that can travel from the Internet to a user in a given amount of time (for example, when a user receives a Web page from a Web server).

Voice-Grade Telephone Connections

In the early days of the Web, most individuals connected to their ISPs through a modem connected to their local telephone service providers. **Plain old telephone service (POTS)** uses existing telephone lines and an analog modem to provide a bandwidth of between 28 and 56 Kbps. Today, most people use other connection methods, including a higher grade of telephone service called **Digital Subscriber Line (DSL)** protocol. DSL connection methods do not use a modem. They use a piece of equipment that is a form of network switch, but most people call this piece of equipment (incorrectly) a "DSL modem." **Integrated Services Digital Network (ISDN)** was the first technology developed to use the DSL protocol suite and has been available in parts of the United States since 1984. ISDN is more expensive than regular telephone service and offers bandwidths of between 128 and 256 Kbps.

Broadband Connections

Connections that operate at speeds of greater than about 200 Kbps are called **broadband** services. One of the newest technologies that uses the DSL protocol to provide service in the broadband range is **asymmetric digital subscriber line (ADSL**, usually abbreviated **DSL**). It provides transmission bandwidths from 100 to 640 Kbps upstream and from 1.5 to 9 Mbps (million bits per second) downstream. For businesses, a **high-speed DSL (HDSL)** connection service can provide more than 768 Kbps of symmetric bandwidth.

Cable modems—connected to the same broadband coaxial cable that serves a television—typically provide transmission speeds between 300 Kbps and 1 Mbps from the client to the server. The downstream transmission rate can be as high as 10 Mbps. In the United States alone, more than 160 million homes have broadband cable service available, and more than 110 million homes subscribe to cable television. The latest estimates indicate that there are more than 11 million cable modem subscribers in the United States. In recent years, DSL monthly fees have been slightly lower than those of cable companies in markets where they compete. Today, about 13 million households have broadband DSL connections. Virtually all companies and organizations of any size have some type of broadband Internet connection.

DSL is a private line with no competing traffic. Unlike DSL, cable modem connection bandwidths vary with the number of other subscribers competing for the shared resource. Transmission speeds can decrease dramatically in heavily subscribed neighborhoods at prime times—in neighborhoods where many people are using cable modems simultaneously.

Connection options based on cable or telephone line connections are wonderful for urban and suburban Web users, but those living in rural areas often have limited telephone service and no cable access at all. The telephone lines used to cover the vast distances between rural customers are usually **voice-grade lines**, which cost less than telephone lines designed to carry data, are made of lower-grade copper, and were never intended to carry data. These lines can carry only limited bandwidth—usually less than 14 Kbps. Telephone companies have wired most urban and suburban areas with **data-grade lines** (made more carefully and of higher-grade copper than voice-grade lines) because the short length of the lines in these areas makes it less expensive to install than in rural areas where connection distances are much longer.

LEARNING FROM FAILURES

NorthPoint Communications

In 1997, Michael Malaga was a successful telecommunications executive with an idea. He wanted to sell broadband Internet access to small businesses in urban areas. DSL technology was just gaining acceptance, and leased telephone lines were available from telephone companies. He wanted to avoid residential customers because they would soon have inexpensive cable modem access to meet their broadband needs. He also wanted to avoid suburban and rural businesses to keep the telephone line leasing costs low (lease charges are higher for longer distances). He and five friends started NorthPoint Communications with $500,000 of their combined savings and raised another $11 million within a few months. After six months, the company had raised more money from investors and had acquired 1500 customers, but it was posting a net loss of $30 million. On the strength of its number of customers, the company began the task of raising the $100 million that Malaga estimated it would need to create the network infrastructure.

Independent DSL providers such as NorthPoint were pressed by customers to install service rapidly, but had to rely on local telephone companies to ensure that their lines would support DSL. In many cases, the telephone companies had to install switches and other equipment to make DSL work on a particular line. The telephone companies often were in no rush to do this because they also sold DSL service, and speedy service would be helping a competitor. The delays led to unpredictable installation holdups and many unhappy NorthPoint customers. Customers with problems after the service was installed were often bounced from the telephone company to NorthPoint, without obtaining satisfactory or timely resolutions of their problems.

Although NorthPoint was unable to make its relationship with each customer profitable, Malaga and his team were rapidly raising money in the hot capital markets of the time. The company raised $162 million before its first stock offering in 1999, which brought in an additional $387 million. At that time, the company had 13,000 customers, which means that NorthPoint had raised more than $42,000 from outside investors for each customer. Considering that each customer would generate revenue of about $1,000 per year, the economics of the business did not look good. By the end of 1999, NorthPoint had spent $300 million of the cash it had raised to build its network infrastructure and reported an operating loss of $184 million. At this point, NorthPoint was operating in 28 cities.

During the next year, the company continued to raise additional funds, gain more customers, and lose money on each customer. In August 2000, the telephone company Verizon agreed to purchase 55 percent of the company for $800 million paid in installments. The total funding that NorthPoint had obtained by the end of 2000, including the partial payments received from Verizon, added up to $1.2 billion. By the end of the year, NorthPoint was in 109 cities and needed to spend $66 million in cash per month just to stay in business. Verizon withdrew from the purchase agreement, the stock plunged, and the layoffs began.

NorthPoint filed for bankruptcy in January 2001 and sold its networking hardware to AT&T in March for $135 million. AT&T was not interested in continuing the DSL business (it just wanted the hardware), so NorthPoint's 87,000 small business customers lost their Internet service overnight. In many of the cities that NorthPoint had served, there were no competitors to pick up the service. Because the capital markets of the late 1990s were so eager to invest in anything that appeared to be connected with the Internet, NorthPoint was able to raise incredible amounts of money. However, NorthPoint sold Internet access to customers for less than it cost to provide the service. No amount of investor money could overcome that basic business mistake.

Leased-Line Connections

Large firms with large amounts of Internet traffic can connect to an ISP using higher bandwidth connections that they can lease from telecommunications carriers. These connections use a variety of technologies and are usually classified by the equivalent number of telephone lines they include. (The connection technologies they use were originally developed to carry large numbers of telephone calls.)

A telephone line designed to carry one digital signal is called DS0 (digital signal zero, the name of the signaling format used on those lines) and has a bandwidth of 56 Kbps. A **T1** line (also called a DS1) carries 24 DS0 lines and operates at 1.544 Mbps. **T3** service (also called DS3) offers 44.736 Mbps (the equivalent of 30 T1 lines or 760 DS0 lines). All of these leased telephone line connections are much more expensive than POTS, ISDN, or DSL connections.

Large organizations that need to connect hundreds or thousands of individual users to the Internet require very high bandwidth. NAPs use T1 and T3 lines. NAPs and the computers that perform routing functions on the Internet backbone also use technologies such as **frame relay** and **asynchronous transfer mode (ATM)** connections and **optical fiber** (instead of copper wire) connections with bandwidths determined by the class of fiber-optic cable used. An OC3 (optical carrier 3) connection provides 156 Mbps, an OC12 provides 622 Mbps, an OC48 provides 2.5 Gbps (gigabits, or 1 billion bits per second), and an OC192 provides 10 Gbps.

Wireless Connections

For many people in rural areas, satellite microwave transmissions have made connections to the Internet possible for the first time. In the first satellite technologies, the customer placed a receiving dish antenna on the roof or in the yard and pointed it at the satellite. The satellite sent microwave transmissions to handle Internet downloads at speeds of around 500 Kbps. Uploads were handled by a POTS modem connection. For Web browsing, this was not too bad, since most of the uploaded messages were small text messages (e-mails and Web page requests). People who wanted to send large e-mail attachments or transfer files over the Internet found the slow upload speeds unsatisfactory.

Today, companies offer satellite Internet connections that do not require a POTS modem connection for uploads. These connections use a microwave transmitter for Internet uploads. This transmitter provides upload speeds as high as 150 Kbps. Initially, the installation charges were much higher than for other residential Internet connection services because a professional installer was needed to carefully aim the transmitter's dish antenna at the satellite. Recently, the accuracy of the antennas improved, and some of these companies now offer a self-installation option that drastically reduces the initial cost. For installations in North America, the antennas must have a clear line of sight into the southwestern sky. This requirement can make these services unusable for many people living in large cities or on the wrong side of an apartment building. In the United States, about 2 million homes are connected to the Internet through a satellite broadband service.

Although satellite connections were the only wireless Internet access media for many years, many types of wireless networks are available now. People today use Internet-capable mobile phones, smart phones, game consoles, and notebook computers equipped with wireless network cards to connect to a variety of wireless networks that, in turn, are connected to the Internet. More than half of U.S. Internet users have used a wireless device to access the Internet.

Bluetooth and Ultra Wideband (UWB)

One of the first wireless protocols, designed for personal use over short distances, is called **Bluetooth**. (The protocol was developed in Norway and is named for Harald Bluetooth, a 10th century Scandinavian king.) Bluetooth operates reliably over distances of up to 35 feet and can be a part of up to 10 networks of eight devices each. It is a low-bandwidth technology, with speeds of up to 722 Kbps. Bluetooth is useful for tasks such as wireless synchronization of laptop computers with desktop computers and wireless printing from laptops or mobile phones. These small Bluetooth networks are called **personal area networks (PANs)** or **piconets**.

One major advantage of Bluetooth technology is that it consumes very little power, which is an important consideration for mobile devices. Another advantage is that Bluetooth devices can discover one another and exchange information automatically. For example, a person using a laptop computer in a temporary office can print to a local Bluetooth-enabled printer without logging in to the network or installing software on either device. The printer and the laptop computer electronically recognize each other as Bluetooth devices and can immediately begin exchanging information.

Another wireless communication technology, **Ultra Wideband (UWB)**, provides wide bandwidth (up to about 480 Mbps in current versions) connections over short distances (30 to 100 feet). UWB was developed for short-range secure communications in military applications during the 1960s. Many observers believe that UWB technologies will be used in future personal area networking applications such as home media centers (for example, a PC could beam stored video files to a nearby television) and in linking mobile phones to the Internet. UWB is faster and more reliable than the wireless Ethernet technologies now used for these purposes.

Wireless Ethernet (Wi-Fi)

The most common wireless connection technology for use on LANs is called **Wi-Fi**, **wireless Ethernet**, or **802.11b** (802.11 is the number of the technology's **network specification**, which is the set of rules that equipment connected to the network must follow). Wireless networking specifications are created by the **IEEE** (originally an acronym for an organization named the Institute of Electrical and Electronic Engineers, the letters are now used as the title of the organization and are pronounced eye-triple-E). A computer equipped with a Wi-Fi network card can communicate through a wireless access point connected to a LAN to become a part of that LAN. A **wireless access point (WAP)** is a device that transmits network packets between Wi-Fi-equipped computers and other devices that are within its range. The user must have authorization to connect to the LAN and might be required to perform a login procedure before the laptop can access the LAN through the WAP.

Wi-Fi that uses the 802.11b specification has a potential bandwidth of 11 Mbps and a range of about 300 feet. In actual installations, the achieved bandwidth and range can be dramatically affected by the construction material of the objects (such as walls, floors, doors, and windows) through which the signals must pass. For example, reinforced concrete walls and certain types of tinted glass windows greatly reduce the effective range of Wi-Fi. Despite these limitations, organizations can make Wi-Fi a key element of their LAN structures by installing a number of WAPs throughout their premises.

In 2002, an improved version of Wi-Fi, called **802.11a** (the 802.11b protocol was easier to implement, thus it was introduced first) was introduced. The 802.11a protocol is capable of transmitting data at speeds up to 54 Mbps, but it is not compatible with 802.11b devices. Later in 2002, the **802.11g** protocol, which has the 54-Mbps speed of 802.11a and is compatible with 802.11b devices, was introduced. Because of its compatibility with the many 802.11b devices that were in use, 802.11g was an immediate success. In 2003, work began on the **802.11n** standard, which was completed in 2009. The 802.11n wireless networking products provide significantly higher actual bandwidths (300–450 Mbps) than any earlier Wi-Fi standard products.

Wi-Fi devices are capable of **roaming**, that is, shifting from one WAP to another, without requiring intervention by the user. Some organizations, including airports, convention centers, and hotels, operate WAPs that are open to the public. These access points are called **hot spots**. Some organizations allow free access to their hot spots; others charge an access fee. A number of restaurants and fast food retailers, such as McDonald's, Panera, and Starbucks, offer hot spots. Hotels and office buildings have found that installing a WAP can be cheaper and easier than running network cable, especially in older buildings. Some hotels offer wireless access free; others charge a small fee. Users of fee-based networks authorize a connection charge when they log in. There are Web sites that offer hot spot directories that show hot spots by location, but these sites tend to open and close frequently, so these directories become out of date rather quickly. The best way to find hot spots (or a hot spot directory) is to use your favorite search engine.

Some communities have installed wireless networks that can be accessed from anywhere in the area. For example, the city of Grand Haven, Michigan, installed a metropolitan area Wi-Fi network. Grand Haven is a growing town on the shores of Lake Michigan. The company that built the network, Ottawa Wireless, sells network access to residents and businesses throughout the area. The company offers access not only on land, but on boats up to 15 miles out on Lake Michigan. Several small company owners use this network to conduct their online business while sailing!

Fixed-Point Wireless

In a growing number of rural areas that do not have cable TV service or telephone lines with the high-grade wires necessary to provide Internet bandwidths, some small companies have begun to offer fixed-point wireless service as an inexpensive alternative to satellite service. One version of **fixed-point wireless** uses a system of repeaters to forward a radio signal from the ISP to customers. The **repeaters** are transmitter–receiver devices (also called **transceivers**) that receive the signal and then retransmit it toward users' roof-mounted antennas and to the next repeater, which receives the signal and passes it on to the next repeater, which can be up to 20 miles away. The users' antennas are connected to a device that converts the radio signals into Wi-Fi packets that are sent to the users' computers or wireless LANs. Another version of fixed-point wireless directly transmits Wi-Fi packets through hundreds, or even thousands, of short-range transceivers that are located close to each other. This approach is called **mesh routing**. As Wi-Fi technologies improve, the number and variety of options for wireless connections to the Internet should continue to increase.

Mobile Telephone Networks

By the end of 2012, industry experts estimate that about 7 billion mobile phones and other devices that use mobile telephone networks will be in operation around the world, which is roughly one for every person on earth (although many people in the world do not have a mobile phone, many more own multiple mobile phones and devices). These phones are sometimes called cellular (or cell) phones because they broadcast signals to (and receive signals from) antennas that are placed about 3 miles apart in a grid, and the hexagonal area that each antenna covers within this grid is called a cell.

Many mobile phones have a small screen and can be used to send and receive short text messages using a protocol called **short message service (SMS)**. As you learned at the beginning of this chapter, Internet-enabled mobile phones and smart phones are very popular in highly developed countries as convenient ways to stay connected while on the go. But more important, mobile phones are giving large numbers of people in developing countries their first access to the online world.

In addition to mobile phones, a variety of other devices now use mobile telephone networks. These devices include small computers called **netbooks** and tablet devices. **Tablet devices** are larger than a mobile phone but smaller than most computers (including most netbooks). Most netbooks, tablet devices, and many mobile phones have the ability to use either a mobile telephone network or a locally available wireless network. These devices almost all have the ability to switch automatically to a wireless network when one is available. Using a local wireless network can be less expensive than using a mobile telephone network.

Although mobile phones were originally designed to handle voice communications, they have always been able to transmit data. However, their data transmission speeds were very low, ranging from 10 to 384 Kbps. Most mobile telephone networks today use one of a series of technologies called **third-generation (3G) wireless technology** that offer download speeds up to 2 Mbps and upload speeds up to 800 Kbps. However, the major U.S. wireless carriers are rapidly introducing newer technologies, including **Long Term Evolution (LTE)** and **Worldwide Interoperability for Microwave Access (WiMAX)**, that are generally referred to as **fourth-generation (4G) wireless technology**. These 4G technologies offer download speeds up to 12 Mbps and upload speeds up to 5 Mbps.

The newer mobile phone technologies are not yet widely available around the world. They are available in the United States, Japan, South Korea, Taiwan, and much of Europe. In other parts of the world, the cost of Internet-capable phones can be prohibitive. In China, for example, about 60 percent of all people who have Internet access (that number is about 450 million) have it through Internet-capable mobile phones. China has 900 million mobile phones in the country, and about 400 million of those are mobile phones without Internet access, which suggests that considerable future potential exists in China for increased Internet access through mobile phones.

In India, about 860 million people have mobile phones, but only about 40 million of them have reliable Internet access through their phones. Only about 96 million Indian citizens have any form of regular Internet access. In recent years, India's telecom companies have been building infrastructure that will allow them to offer better Internet access to their phone customers. Industry analysts expect that Internet access in India through mobile phones will increase rapidly in the 2012–2015 time frame, providing the country with increased online business opportunities through mobile commerce.

As you learned in Chapter 1, companies have seen great potential for these wireless networks and the devices connected to them in the development of mobile commerce. You will learn more about revenue models that use wireless technologies in Chapter 3 and cost-reduction strategies that use wireless technologies in Chapter 5. Chapter 6 includes an overview of the development of mobile commerce to date and an outline of future directions. In Chapter 11, you will learn how some companies are using these technologies to process online payments for goods and services. Figure 2-15 summarizes speed and cost information for the most commonly available wired and wireless options for connecting a home or business to the Internet.

Service	Upstream Speed (Kbps)	Downstream Speed (Kbps)	Capacity (Number of Simultaneous Users)	One-time Startup Costs	Continuing Monthly Costs
Residential-Small Business Services					
POTS	28–56	28–56	1	$0–$20	$9–$20
Wireless 3G network	10–800	10–2000	1	$0–$120	$30–150
ISDN	128–256	128–256	1–3	$60–$300	$50–$90
ADSL	100–640	500–9000	4–20	$50–$100	$200–$500
Cable	300–1500	500–10,000	4–10	$0–$100	$40–$300
Satellite	125–150	400–500	1–3	$0–$800	$40–$100
Fixed-point wireless	250–1500	500–3000	1–4	$0–$350	$50–$150
Wireless 4G network	500–5000	1000–12,000	1	$0–$200	$80–$200
Business Services					
Leased digital line (DS0)	64	64	1–10	$50–$200	$40–$150
Fixed-point wireless	500–10,000	500–10,000	5–1000	$0–$500	$300–$5000
T1 leased line	1544	1544	100–200	$100–$2000	$300–$1600
T3 leased line	44,700	44,700	1000–10,000	$1000–$9000	$3000–$12,000
Large Organization					
OC3 leased line	156,000	156,000	1000–50,000	$3000–$12,000	$9000–$22,000
OC12 leased line	622,000	622,000	Backbone	Negotiated	$25,000–$100,000
OC48 leased line	2,500,000	2,500,000	Backbone	Negotiated	Negotiated
OC192 leased line	10,000,000	10,000,000	Backbone	Negotiated	Negotiated

FIGURE 2-15   Internet connection options

INTERNET2 AND THE SEMANTIC WEB

At the high end of the bandwidth spectrum, a group of network research scientists from nearly 200 universities and a number of major corporations joined together in 1996 to recapture the original enthusiasm of the ARPANET with an advanced research network called Internet2. When the National Science Foundation turned over the Internet backbone to commercial interests in 1995, many scientists felt that they had lost a large, living laboratory. **Internet2** is the replacement for that laboratory. An experimental test bed for new networking technologies that is separate from the original Internet, Internet2 has achieved bandwidths of 10 Gbps and more on parts of its network.

Internet2 is also used by universities to conduct large collaborative research projects that require several supercomputers connected at very fast speeds, or that use multiple video feeds—things that would be impossible on the Internet given its lower bandwidth limits. For example, doctors at medical schools that are members of Internet2 regularly use its technology to do live videoconference consultations during complex surgeries. Internet2 serves as a proving ground for new technologies and applications of those technologies that will eventually find their way to the Internet. In 2008, CERN (the birthplace of the original Web in Switzerland) began using Internet2 to share data generated by its new particle accelerator with a research network of 70 U.S. universities. Every few weeks, each university downloads about two terabytes (a terabyte is one thousand gigabytes) of data within a four-hour time period.

The Internet2 project is focused mainly on technology development. In contrast, Tim Berners-Lee began a project in 2001 that has a goal of blending technologies and information into a next-generation Web. This **Semantic Web** project envisions words on Web pages being tagged (using XML) with their meanings. The Web would become a huge machine-readable database. People could use intelligent programs called **software agents** to read the XML tags to determine the meaning of the words in their contexts. For example, a software agent given the instruction to find an airline ticket with certain terms (date, cities, cost limit) would launch a search on the Web and return with an electronic ticket that meets the criteria. Instead of a user having to visit several Web sites to gather information, compare prices and itineraries, and make a decision, the software agent would automatically do the searching, comparing, and purchasing.

For software agents to perform these functions, Web standards must include XML, a resource description framework, and an ontology. You have already seen how XML tags can describe the semantics of data elements. A **resource description framework (RDF)** is a set of standards for XML syntax. It would function as a dictionary for all XML tags used on the Web. An **ontology** is a set of standards that defines, in detail, the relationships among RDF standards and specific XML tags within a particular knowledge domain. For example, the ontology for cooking would include concepts such as ingredients, utensils, and ovens; however, it would also include rules and behavioral expectations, such as that ingredients can be mixed using utensils, that the resulting product can be eaten by people, and that ovens generate heat within a confined area. Ontologies and the RDF would provide the intelligence about the knowledge domain so that software agents could make decisions as humans would.

The development of the Semantic Web is expected to take many years. The first step in this project is to develop ontologies for specific subjects. Thus far, several areas of scientific inquiry have begun developing ontologies that will become the building blocks of the Semantic Web in their areas. Biology, genomics, and medicine have all made progress toward specific ontologies. These fields can benefit greatly from a tool like the Semantic Web, which can increase the speed with which research results, experimental data, and new procedures can be made available to all researchers in the field. Thus, these fields have a high incentive to collaborate on the hard work involved in creating ontologies.

Other sciences, such as climatology, hydrology, and oceanography have similar incentives (as many researchers around the world work on common problems such as global warming) and scientists are developing ontologies for their disciplines. The government of the United Kingdom is also developing an ontology for data it collects with the hope that it will be useful to a wide range of researchers.

Although many researchers involved in the Semantic Web project have expressed frustration at its slow progress, a number of important users of the Semantic Web have developed important ontologies that will allow the project to continue moving forward. You can learn more about the current status of this project by following the Web Links to the **W3C Semantic Web** pages.

Summary

In this chapter, you learned about the history of the Internet and the Web, including how these technologies emerged from research projects and grew to be the supporting infrastructure for electronic commerce today. You learned about intranets and extranets and that they can be implemented using public network, private network, or virtual private network technologies.

You also learned about the protocols, programs, languages, and architectures that support the Internet and the World Wide Web. TCP/IP is the protocol suite used to create and transport information packets across the Internet. IP addresses identify computers on the Internet. Domain names such as www.amazon.com also identify computers on the Internet, but those names are translated into IP addresses by the routing computers on the Internet. HTTP is the set of rules for transferring Web pages and requests for those Web pages on the Internet. POP, SMTP, and IMAP are protocols that help manage e-mail.

Hypertext Markup Language (HTML) was derived from the more generic meta language SGML. HTML defines the structure and content of Web pages using markup symbols called tags. Over time, HTML has evolved to include a large number of tags that accommodate graphics, Cascading Style Sheets, and other Web page elements. Hyperlinks are HTML tags that contain a URL. The URL can be a local or remote computer. HTML editors facilitate Web page construction with helpful tools and drag-and-drop capabilities. Extensible Markup Language (XML) is also derived from SGML. However, unlike HTML, XML uses markup tags to describe the meaning, or semantics, of the text, rather than its display characteristics. XML offers businesses hope for a common language that they will be able to use to describe products, services, and even business processes to each other in common, shared data-bases. XML could help companies dramatically reduce the costs of handling intercompany information flows.

Internet service providers offer many different types of connections to the Internet. Basic telephone connections are the most economical and easiest to install, but they are the slowest. Broadband cable, satellite microwave transmission, and DSL services provide Internet access at relatively high speeds. Other, more expensive options provide the bandwidth that larger businesses need. A variety of wireless connection options are becoming available, including fixed-point wireless. The wireless connection options available through mobile phones show promise in creating new opportunities for revenue generation, cost reduction, and payment-processing applications.

Internet2 is an experimental network built by a consortium of research universities and businesses that provides a test bed for creating and perfecting the high-speed networking technologies of tomorrow. The Semantic Web project is moving slowly toward its goal of making research data widely available and enabling many user interactions with the Web to be handled by intelligent software agents.

Key Terms

802.11a	Asynchronous transfer mode (ATM)
802.11b	Backbone router
802.11g	Bandwidth
802.11n	Base 2
Anchor tag	Binary
Asymmetric connection	Bluetooth
Asymmetric digital subscriber line (ADSL, DSL)	Border router

Broadband
Byte
Cascading Style Sheets (CSS)
Circuit
Circuit switching
Client/server architecture
Closed architecture
Closing tag
Colon hex
Colon hexadecimal
Computer network
Configuration table
Data-grade lines
Deep Web
Digital Subscriber Line (DSL)
Domain name
Dotted decimal
Downlink bandwidth
Download
Downstream bandwidth
Electronic mail
E-mail
E-mail client software
E-mail server
Encapsulation
Extensible Hypertext Markup Language
(XHTML)
Extensible Markup Language (XML)
Extensible Stylesheet Language (XSL)
Extranet
Fixed-point wireless
Fourth-generation (4G) wireless technology
Frame relay
Gateway computer
Generalized Markup Language (GML)
generic top-level domain (gTLD)
Graphical user interface (GUI)
Hexadecimal (base 16)
Hierarchical hyperlink structure
High-speed DSL (HDSL)
Home page
Hot spot
HTML extensions

Hyperlink
Hypertext
Hypertext element
Hypertext link
Hypertext Markup Language (HTML)
Hypertext server
Hypertext Transfer Protocol (HTTP)
IEEE
Integrated Services Digital Network (ISDN)
Interactive Mail Access Protocol (IMAP)
Internet
internet
Internet access provider (IAP)
Internet backbone
Internet host
Internet Protocol (IP)
Internet Protocol version 4 (IPv4)
Internet Protocol version 6 (IPv6)
Internet service provider (ISP)
Internet2
Intranet
IP address
IP tunneling
IP wrapper
Leased line
Linear hyperlink structure
Local area network (LAN)
Long Term Evolution (LTE)
Mailing list
Markup tags
Mesh routing
Metalanguage
Multipurpose Internet Mail Extensions (MIME)
Net bandwidth
Netbook
Network access points (NAPs)
Network access provider
Network Address Translation (NAT) device
Network Control Protocol (NCP)
Network specification
Newsgroup
Octet
Ontology

Open architecture	T3
Opening tag	Tablet device
Optical fiber	Tag
Packet	TCP/IP
Packet-switched	Text markup language
Personal area network (PAN)	Third-generation (3G) wireless technology
Piconet	Top-level domain (TLD)
Plain old telephone service (POTS)	Transceiver
Post Office Protocol (POP)	Transmission Control Protocol (TCP)
Private IP address	Ultra Wideband (UWB)
Private network	Uniform Resource Locator (URL)
Proprietary architecture	Upload bandwidth
Protocol	Upstream bandwidth
Public network	Usenet
Repeater	User's News Network
Resource description framework (RDF)	Virtual private network (VPN)
Roaming	Voice-grade lines
Router	Web
Router computers	Web browser
Routing algorithm	Web browser software
Routing computer	Web client computer
Routing table	Web client software
Semantic Web	Web server
Short message service (SMS)	Web server software
Simple Mail Transfer Protocol (SMTP)	Wi-Fi
Site map	Wide area network (WAN)
Software agents	Wireless access point (WAP)
sponsored top-level domains (sTLDs)	Wireless Ethernet
Standard Generalized Markup Language (SGML)	World Wide Web
	World Wide Web Consortium (W3C)
Start page	Worldwide Interoperability for Microwave Access (WiMAX)
Style sheet	
Subnetting	XML parser
Symmetric connection	XML vocabulary
T1	

Review Questions

1. In one or two paragraphs, describe how the Internet changed from a government research project into a technology for business users.

2. In two paragraphs, outline how the ideas of Vannevar Bush and Ted Nelson became key elements of the World Wide Web.

3. In about 100 words, describe the function of the Internet Corporation for Assigned Names and Numbers. Include a discussion of the differences between gTLDs and sTLDs in your answer.

4. The Web uses a client/server architecture. In about 100 words, describe the client and server elements of this architecture, including specific examples of software and hardware that are used to form the Web.

5. In about 100 words, explain the difference between an extranet and an intranet. In your answer, describe when you might use a VPN in either.

6. In about 100 words, explain how markup tags work in HTML, and describe the function of the HTML anchor tag. Explain the importance of the anchor tag in the evolution of electronic commerce activity on the Web.

7. In about 200 words, define "markup languages." Include overviews of HTML and XML in your definition. As part of your answer, provide examples of at least two situations in which an organization would use XML and two situations in which an organization would use HTML.

8. In about 100 words, describe the differences between symmetric and asymmetric connections. Include a discussion of why one might be preferable to the other in a specific situation.

9. In about 100 words, describe how ontologies and resource description frameworks could help software agents provide useful services on the Semantic Web.

Exercises

1. In 2003, ICANN and the major domain name registries began offering a five-day grace period for new domain registrations. The idea was to give registrants time to correct typographical errors and misspellings in the names they registered. If a registrant found an error in that five-day period, they could cancel their registration and, presumably, re-register a corrected domain name. This policy led to a problem called "domain tasting" that required considerable effort and cooperation to resolve nearly six years after the policy was implemented. Using your library or your favorite search engine, learn more about domain tasting. Prepare a report of about 300 words that defines domain tasting, outlines its negative effects on Web users, and describes how the problem of domain tasting was resolved.

2. Bridgewater Engineering Company (BECO), a privately held machine shop, makes heavy-duty machinery for factory assembly lines. It sells its presses, grinders, and milling equipment using a few inside salespeople and telephones. It buys its raw materials and supplies from a variety of steel mills and small-parts fabricators located around the world. BECO's president, Tom Dalton, has hired you as a consultant and would like your advice regarding how best to share information with the company's suppliers. Tom would like to connect his network of computers into their ordering systems so he can order supplies quickly when he needs them. He is interested in learning more about how he can use the Internet to set up such connections. Use the Web and the book's accompanying Web Links to locate information about extranets and VPNs. Write a report that briefly describes how companies use extranets to link their systems with those of their suppliers, and then write an evaluation of at least two companies (using information you have gathered in your Web searches) that could help develop an extranet that would work for Tom. Close the report with an overview of how BECO could use VPN technologies in this type of extranet. The three parts of your report should total about 500 words.

3. Tanya Trago is the IT manager for Greenway Enterprises, a large landscaping company with hundreds of home and commercial customers. She is interested in finding ways to reduce the costs of maintaining the company's tree trimming and lawn maintenance equipment. Greenway runs its own repair and maintenance facility because it operates a large number of mowers, cranes, backhoes, and similar machinery. The facility purchases replacement parts and repair supplies for all of this equipment. Tanya is interested in creating a database to track these parts and supplies. She would also like to integrate that database with information provided by the vendors that sell those parts and supplies to Greenway. Several of these vendors use XML tags to describe their inventory, but no common standard tag system has been adopted in the industry. Use the Web Links, the Web, and your library to conduct research on the use of XML in the landscaping equipment and machinery industry, summarize your findings, and prepare a report of about 300 words in which you give Tanya advice on the advantages and disadvantages of using XML tags as descriptors in this situation.

4. As you learned in this chapter, XML allows users to define their own markup tags. You also learned that this flexibility can lead to problems when IT professionals who have developed tag sets for their own organizations share information with other organizations that are using other tag sets. One way organizations can avoid this problem is to agree to follow common standards. A common standard for financial information is XBRL. Accountants and financial analysts around the world have agreed to use XBRL to format financial statements and other reports. In about 300 words, outline the advantages that companies and financial analysts can obtain by using the XBRL standard. You can research this subject in your school library or online using your favorite search engine and the links provided for this exercise in the Web Links.

5. You are the assistant to Yin Chan, the service manager of Quick Fix Repair Systems. Quick Fix offers repair and maintenance services to homeowners throughout the tri-state area. Quick Fix service technicians can each do minor plumbing, electrical, and carpentry work. Yin wants to equip each service technician with the technology they need to report their time and materials usage on each job. Today, the service technicians carry a notebook computer for recording this information at job sites. Many of them also carry a smart phone to stay in touch with the office, order parts that they do not have with them, and keep track of their schedule. Yin would like to ensure that service technicians have access to the Quick Fix main computers while they are on the job site so they can check supplies inventory and access Quick Fix service guides that help them make repairs in the most effective ways possible. Yin asks you to investigate various options for giving her service technicians wireless remote access to the Quick Fix main computers. She wants you to consider options that use the technicians' notebook computers or that use their smart phones. Prepare a report for Yin in which you discuss the advantages and disadvantages of having the technicians use notebook computers and smart phones for access to the Quick Fix main computers. The smart phones are connected through a corporate wireless phone plan that provides unlimited data transfers each month; you are to briefly review at least three options for connecting the notebook computers, writing no more than two paragraphs for each option. Then choose the best option and write a one-page evaluation of your choice's strengths and weaknesses. Use the Web Links and your favorite Web search engines to do your research.

Cases

C1. Internet Access in Hyderabad

Hyderabad is the fourth-largest city in India, with a population of nearly 7 million in the city itself and more than 14 million in the metropolitan area. It is the capital of the State of Andhra Pradesh, which has 84 million people and a $100 billion-per-year economy. Hyderabad itself accounts for $60 billion of that annual activity, much of it in information technology (the city houses the Indian headquarters of Amazon.com, Google, and Microsoft, for example) and pharmaceuticals. With more than a dozen universities, the city is a leader in education and research.

Like the rest of the country, however, citizens in the Hyderabad metropolitan area are less connected to the Internet than the city's strong presence in the information technology industry would suggest. Overall, only about 1 million residents (about 8 percent of the population) have regular online access. About 425 million (about 3 percent of the population) of those with regular access use an Internet-capable phone as their primary access device. Approximately 72 percent of Hyderabad citizens own mobile phones, but most of them are not Internet-capable.

Internet access in Hyderabad lags far behind the United States, where 85 percent of the population has regular online access, virtually all of it using computers as the primary access device. It also lags behind online access in China, where about 32 percent of the population has regular online access (about 19 percent of the population uses Internet-capable phones as their primary access devices).

Required:

1. What are the implications of the low Internet access rates for the citizens of Hyderabad as they become active participants in the world economy over the next five to 10 years? Summarize your thoughts in about 100 words.

2. Using the Web Links for this case, your library, or your favorite search engine, identify current trends in the growth of Internet-capable phones and other online access devices in India. In a report of about 200 words, evaluate the prospects for significant changes in online access rates over the next few years.

3. Use what you learned about online access technologies in this chapter to outline several alternatives that the government of Hyderabad should consider developing (perhaps in partnership with local information technology companies) that might increase online access rates for its citizens. Prepare a report of about 200 words in which you discuss at least two of these alternatives.

Note: Your instructor might assign you to a group to complete this case, and might ask you to prepare a formal presentation of your results to your class.

C2. Portable Fun Instruments

Yash Gupta is the founder and president of Portable Fun Instruments (PFI), a company that has had great success in the handheld game market. Its first products were dedicated handheld devices that each offered a specific game, such as backgammon, checkers, or chess. As the power of microprocessors for handheld devices grew, and the size and cost of those microprocessors shrank, PFI was able to build better and more complex games into its devices.

Today, PFI offers a wide variety of dedicated handheld devices on which users can play card games, adventure games, and sports simulations, and solve various kinds of puzzles. Most of the elements in the game displays are graphics, not words. This helps PFI sell the devices in many different markets around the world without having to build separate interfaces for each language. PFI's game devices have retail prices that range between $20 and $40, but the retailers and distributors buy them from PFI for prices that range between $4 and $18.

PFI is profitable because Yash has worked hard to keep development and production costs low. Most of the programming is done in Bangalore, India, and the devices are built in production facilities located in Xixiang, China, and Penang, Malaysia. Although Yash has been successful in controlling production costs, he worries about continuing to operate the company with a long-term strategy that requires PFI to build a new physical device for each sale. The large retail chains that have become PFI's main customers are always asking for discounts and reduced prices on new orders, and production costs are creeping upward even though the facilities are located in some of the lowest-cost areas in the world.

Yash wants to explore the potential PFI has for moving its games to other platforms. PFI has translated some of its games to run on smart phones, but the results have been disappointing. Until recently, most smart phone users have been businesspeople who use their smart phones for e-mail, appointments, address books, travel expenses, and other data-management functions. These users are not avid game players, and sales of PFI's games for these platforms have not been strong.

Some of PFI's marketing team members have been telling Yash about the success of Apple's iPhone and the online store for software that runs on that phone (called Applications for iPhone). Apple shares the revenue earned from software sales on its site with the developers of that software. Other team members have mentioned Google's Android operating system for smart phones built by a variety of manufacturers. Software for those phones sells in the Android Marketplace, which operates in much the same way as Apple's software store, sharing revenue with software providers.

Yash has hired you as a consultant to investigate the Apple iPhone and Android Marketplace as options for selling versions of PFI's games that will work on smart phones.

Required:

1. Use the links in the Web Links for this case, your favorite search engine, and resources in your library to learn more about Android and Apple as program delivery systems for smart phones. Prepare a 300-word executive summary for Yash that describes each delivery system you identify and outlines the current or likely near-term availability of each system for content providers such as PFI.

2. Prepare a report for Yash and the PFI executive team in which you outline and analyze the strengths and weaknesses of each content delivery system you have identified. Your report should conclude with a specific recommendation regarding the suitability of each content delivery system for PFI's games. This report should be about 300 words in length.

Note: Your instructor might assign you to a group to complete this case, and might ask you to prepare a formal presentation of your results to your class.

For Further Study and Research

Arthur, C. 2009. "China's Internet Users Surpass U.S. Population," *The Guardian*, July 16. (http://www.guardian.co.uk/technology/2009/jul/16/china-internet-more-users-us-population)

BBC News. 2010. "Over 5 Billion Mobile Phone Connections Worldwide," *BBC News*, July 9. (http://www.bbc.co.uk/news/10569081)

Bellman, E. 2009. "Rural India Snaps Up Mobile Phones," *The Wall Street Journal*, February 9, B1, B5.

Belson, K. 2007. "Unlike U.S., Japanese Push Fiber Over Profit," *The New York Times*, October 3. (http://www.nytimes.com/2007/10/03/business/worldbusiness/03broadband.html)

Bergman, M. 2001. *The Deep Web: Surfacing Hidden Value*. Sioux Falls, SD: BrightPlanet.com. (http://brightplanet.com/technology/deepweb.asp)

Boles, C. 2007. "States Step In to Close Broadband Gap," *The Wall Street Journal*, November 1, B3.

Bonson, E., V. Cortijo, and T. Escobar. 2009. "Toward the Global Adoption of XBRL Using International Financial Reporting Standards (IFRS)," *International Journal of Accounting Information Systems*, 10(1), March, 46–60.

Bosak, J. and T. Bray. 1999. "How XML Will Fix the Web: Tags Categorizing Facts, Not Formats, Speed Up Transactions," *Scientific American*, 280(5), May, 89.

Brewin, B. 2004. "Michigan City Turns on Citywide Wi-Fi," *Computerworld*, July 30. (http://www.computerworld.com/mobiletopics/mobile/wifi/story/0,10801,94928,00.html)

Bruno, A. 2009. "Call of the iPhone," *Billboard*, April 4, 24–28.

Campbell, T. 1998. "The First E-Mail," *Pretext Magazine*, March. (http://www.pretext.com/mar98/features/story2.htm)

Chao, L., J. Ye, and Y. Kane. 2009. "Apple, Facing Competition, Readies iPhone for Launch in Giant China Market," *The Wall Street Journal*, August 27, B1–B2.

Chester, J. 2006. "The End of the Internet?" *The Nation*, February 1. (http://www.thenation.com/doc/20060213/chester)

Cramer, J. 2009. "The Biggest Thing Since E-mail: Why the Smart Phone Market Is Only Just Beginning to Take Off," *New York*, August 24, 36–38.

Dipert, B. 2009. "802.11n: Complicated and About to Become Even Messier," *EDN*, May 28. (http://www.edn.com/article/CA6659414.html)

Dominque, J., D. Fensel, and J. Hendler. 2011. *Handbook of Semantic Web Technologies*, London: Springer.

Dyck, T. 2002. "Going Native: XML Databases," *PC Magazine*, 21(12), June 30, 136–139.

The Economist, 2008. "India: 3G at Last," *The Economist Intelligence Unit Country Monitor*, 16(28), July 28, 1.

EContent. 2009. "XML for the Masses," 32(3), April, 45.

Einhorn, B. 2009. "Will China Pick Up the OPhone?" *BusinessWeek*, September 7, 20.

Ely, A. 2008. "Where in the World is IPv6?" *InformationWeek*, December 22, 43–44.

Fensel, D., J. Hendler, H. Lieberman, and W. Wahlster. 2002. *Spinning the Semantic Web: Bringing the World Wide Web to Its Full Potential*. Cambridge, MA: MIT Press.

Garbellotto, G. 2009. "XBRL Implementation Strategies: The Built-in Approach," *Strategic Finance*, 91(2), August, 56–57.

Goldfarb, C. 1981. "A Generalized Approach to Document Markup," *ACM Sigplan Notices*, (16)6, June, 68–73.

Hannon, N. and M. Willis. 2005. "Combating Everyday Data Problems with XBRL," *Strategic Finance*, 87(1), July, 57–59.

Hannon, N. and M. Willis. 2005. "Combating Everyday Data Problems with XBRL, Part 2," *Strategic Finance*, 87(2), August, 59–61.

Hawn, C. 2001. "Management By Stock Market: NorthPoint Rode the Web Wave," *Forbes*, 167(10), April 30, 52–53.

Henschen, D. 2005. "XBRL Offers a Faster Route to Intelligence," *Intelligent Enterprise*, 8(8), August, 12.

Horrigan, J. 2009. *Wireless Internet Use*. Washington, DC: Pew Internet & American Life Project. (http://pewinternet.org/Reports/2009/12-Wireless-Internet-Use.aspx)

Horrigan, J. 2009. *Home Broadband Adoption 2009*. Washington, DC: Pew Internet & American Life Project. (http://pewinternet.org/Reports/2009/10-Home-Broadband-Adoption-2009.aspx)

Horrocks, I. 2008. "Ontologies and the Semantic Web," *Communications of the ACM*, 21(12), December, 58–67.

International Telecommunications Union (ITU). 2011. *Measuring the Information Society, 2011 Edition*. Geneva: ITU.

Internet Society. 2011. "World IPv6 Day." (http://www.worldipv6day.org/)

Kim, H., W. Kim, and M. Lee. 2009. "Semantic Web Constraint Language and its Application to an Intelligent Shopping Agent," *Decision Support Systems*, 46(4), March, 882–894.

Kim, W. 2009. "Mobile WiMAX: The Leader of the Mobile Internet Era," *IEEE Communications Magazine*, 47(6), June, 10–12.

Kisiel, R. 2009. "Dealership Web sites shrink to fit on phones," *Automotive News*, March 16, 30.

Kristof, N. 2005. "When Pigs Wi-Fi," *The New York Times*, August 7, 13.

Kumaravel, K. 2011. "Comparative Study of 3G and 4G in Mobile Technology," *International Journal of Computer Science Issues*, 8(5), September, 256–263.

Lawton, C. 2009. "Making the Connection," *The Wall Street Journal*, April 20, R4.

Lawton, C. and S. Silver. 2009. "Smart Phones are Edging Out Other Gadgets," *The Wall Street Journal*, March 24, D1–D3.

Lawton, G. 2011. "4G: Engineering Versus Marketing," *IEEE Computer*, 44(3), March, 14–16.

Lee, M. 2008. "HTML 5 Comes to Fruition," *InformationWeek*, March 31, 48–49.

Liebman, L. 2001. "XML's Tower Of Babel," *InternetWeek*, April 30, 25–26.

Luk, L. and J. Scheck. 2009. "Dell Developing Phones for China," *The Wall Street Journal*, August 18, B4.

Malnig, A. 2005. "XBRL: Deep Drilling for Financials," *Seybold Report: Analyzing Publishing Technologies*, 5(4), May 18, 11–14.

Marriot, M. 2006. "Hey Neighbor, Stop Piggybacking on My Wireless," *The New York Times*, March 5. (http://www.nytimes.com/2006/03/05/technology/05wireless.html)

McCracken, H. 2009. "Smart Phone OS Smackdown," *PC World*, 27(2), February, 54–58.

Nelson, T. 1987. *Literary Machines*. Swarthmore, PA: Nelson.

Nielsen, J. 2003. "Mobile Devices: One Generation From Useful," *Alertbox*, August 18. (http://www.useit.com/alertbox/20030818.html)

Nielsen, J. 2009. "Mobile Usability," *Alertbox*, July 20. (http://www.useit.com/alertbox/mobile-usability.html)

Panigrahi, S. and S. Biswas. 2011. "Next Generation Semantic Web and Its Application," *International Journal of Computer Science Issues*, 8(2), March, 385–392.

Panko, R. and J. Panko. 2011. *Business Data Networks and Telecommunications*. Eighth Edition. Upper Saddle River, NJ: Prentice Hall.

Paulraj A. 2011. "Evolution of Indian Wireless Networks," *IETE Technical Review*, 28(5), 375–380.

Poppcuviu, C. 2009. "Implementing IPv6," *Broadcast Engineering*, 51(7), July, 38–40.

Pringle, D. 2005. "Wi-Fi Woes: Wireless Networks Are Great—If You Can Figure Out How to Set Them Up," *The Wall Street Journal*, July 18, R11.

Rodenbaugh, M. 2009. "Abusive Domain Registrations: ICANN Policy Developments (or Lack Thereof?), *Computer & Internet Lawyer*, 26(5), May, 17–22.

Saint-Andre, P. 2009. "XMPP: Lessons Learned from Ten Years of XML Messaging," *IEEE Communications Magazine*, 47(4), April, 92–96.

Shadbolt, N., T. Berners-Lee, and W. Hall. 2006. "The Semantic Web Revisited," *IEEE Intelligent Systems*, 21(3), 96–101.

Sharma, A. and D. Thoppil. 2011. "Google Sees India Web Explosion," *The Wall Street Journal*, September 16, B7.

Strategic Finance. 2009. "XBRL Reporting Is Now Mandatory," 90(7), January, 61.

Sullivan, M. 2011. "4G Wireless Speed Tests: Which Is Really the Fastest?" March 13. (http://www.pcworld.com/article/221931/4g_wireless_speed_tests_which_is_really_the_fastest.html)

Telecommunications Reports. 2011. "ICANN Approves Expansion of gTLDs," 77(13), July 1, 22.

Tie, R. 2005. "XBRL: It's Unstoppable: Interview With Charles Hoffman," *Journal of Accountancy*, August, 32–35.

Thurm, S. 2002. "Cisco Profit Exceeds Expectations," *The Wall Street Journal*, May 8, A3.

Vance, A. 2011. "The Cloud: Battle of the Tech Titans," *Bloomberg Businessweek*, March 3. (http://www.businessweek.com/magazine/content/11_11/b4219052599182.htm)

Weinberg, N. 2008. "802.11n: It's MIMO Time-O," *Network World*, January 14, 36.

Weinberger, D. 2009. "The Dream of the Semantic Web," *KM World*, 18(3), March, 1–3.

White, C. 2011. *Data Communications and Computer Networks: A Business User's Approach*. 6th Edition. Cincinnati: South-Western.

White, M. and B. Briggs. 2011. *Tech Trends 2011: The Natural Convergence of Business and IT*. New York: Deloitte Consulting LLP.

Wood, G. 2011. "IPv6: Making Room for the World on the Future Internet," *IEEE Internet Computing*, 15(4), July–August, 88–89.

Zhang, M. and R. Wolff. 2004. "Crossing the Digital Divide: Cost-Effective Broadband Wireless Access for Rural and Remote Areas," *IEEE Communications Magazine*, 42(2), February, 99–105.

Zhang, Z., G. Dong, Z. Peng, and Z. Yan. 2011. "A Framework for Incremental Deep Web Crawler Based on URL Classification," *Lecture Notes in Computer Science*, 6988, 302–310.

Zhu, H. and H. Wu. 2011. "Interoperability of XBRL Financial Statements in the U.S." *International Journal of E-Business Research*, 7(2), April–June, 19–33.

PART 2

BUSINESS STRATEGIES FOR ELECTRONIC COMMERCE

CHAPTER 3
Selling on the Web, 105

CHAPTER 4
Marketing on the Web, 153

CHAPTER 5
Business-to-Business Activities: Improving Efficiency and Reducing Costs, 205

CHAPTER 6
Social Networking, Mobile Commerce, and Online Auctions, 245

CHAPTER 7
The Environment of Electronic Commerce: Legal, Ethical, and Tax Issues, 281

SELLING ON THE WEB

INTRODUCTION

In the 1980s, **Progressive** was a relatively small auto insurance company that specialized in writing policies for people who had poor driving records and could not qualify for regular policies sold by other insurers. Progressive charged higher premiums for these policies, which the insurance industry calls substandard policies. Often, other insurers who could not write standard polices for customers would refer those customers to Progressive. The combination of high premiums and the lower cost of its smaller sales force enabled Progressive to earn good profits on the substandard business. Eventually, other insurers noticed Progressive's success and began to offer their own substandard policies.

To respond to the increased competition, Progressive improved its claim service and was one of the first insurance companies to offer 24/7 service every day of the year. During the 1990s,

Progressive developed a full line of auto insurance products for all types of drivers and worked hard to make sure that it offered the lowest prices in every market. Progressive's marketing mentions the quality of its service, but it always emphasizes its low prices.

Progressive was the first auto insurance company to launch a Web site (in 1995) and was the first to sell policies online (in 1997). Knowing that most potential insurance buyers shop multiple Web sites to find the best rate, the company began showing its competitors' rates on its Web site in 2002, allowing potential customers to compare prices without leaving Progressive's site. The site displays these rates even when Progressive's rate is higher than a competitor's rate on a particular policy.

By providing these competitive quotes, Progressive hopes to convince shoppers that their Web site is an important one to visit early in their search because it can save them time. The practice of displaying competitors' quotes also creates an impression of openness and honesty. Progressive believes that people prefer to buy insurance from honest companies who offer the best prices. Its Web site conveys its belief and provides a consistent corporate message to potential customers. In 2008, Progressive introduced a female character, "Flo," who embodies openness, honesty, and a devotion to low prices. Flo appears in the company's television and radio ads, and is featured prominently on its Web site. In fact, the character often appears in television ad vignettes that tout the price comparison feature of the Web site. The comparative quotes feature of the Web site and its use of the Flo character are examples of how companies can successfully integrate their Web presence into their overall brand positioning strategy and reinforce the message they want to deliver to customers and potential customers.

REVENUE MODELS FOR ONLINE BUSINESS

As you learned in Chapter 1, a useful way to think about electronic commerce implementations is to consider how they can generate revenue. Not all electronic commerce initiatives have the goal of providing revenue; some are undertaken to reduce costs or improve customer service. You will learn about those types of initiatives in Chapter 5. In this chapter, you will learn about various models that online businesses currently use to generate revenue, including Web catalog, digital content, advertising-supported, advertising-subscription mixed, and fee-based models. These approaches can work for both business-to-consumer (B2C) and business-to-business (B2B) electronic commerce. Many companies create one Web site to handle both B2C and B2B sales. Even when companies create separate sites (or separate pages within one site), they often use the same revenue model for both types of sales.

Web Catalog Revenue Models

Many companies sell goods and services on the Web using an adaptation of a revenue model that is more than 100 years old. In 1872, a traveling salesman named Aaron Montgomery Ward started selling dry goods to farmers through a one-page list. Richard Sears and Alvah Roebuck began mailing catalogs to farmers and small-town residents in 1895. Both Montgomery Ward and Sears, Roebuck & Company grew to become dominant retailers in the United States by the 1950s, with retail stores serving urban markets and the catalog business well established in serving rural and small-town markets. The general acceptance of the mail order catalog business built a solid base for the Web-based version that would evolve from it in the 1990s.

In the traditional catalog-based retail revenue model, the seller establishes a brand image, and then uses the strength of that image to sell through printed information mailed to prospective buyers, who place orders by mail or telephone. For more than a century, this revenue model, called the **mail-order** or **catalog model**, has been successful for a wide variety of consumer items, including apparel, computers, electronics, housewares, and gifts. Other companies that succeeded as mail-order businesses in the twentieth century include J. C. Penney, LL Bean, and Hickory Farms.

Many companies have adapted this revenue model to the online world by replacing or supplementing their print catalogs with information on their Web sites. This revenue model is called the **Web catalog revenue model**. Most customers today place orders through the Web site, but in the early years of electronic commerce, many shoppers used the Web to obtain information about products and compare prices and features, and then made their purchases by telephone. Types of retail businesses that use the Web catalog revenue model include sellers of computers, consumer electronics, books, music, videos, jewelry, clothing, flowers, and gifts. Many general merchandisers also use the Web catalog revenue model. B2B sellers have also been avid adopters of the Web catalog model. Items such as tools, electrical and plumbing parts, and every imaginable industrial supply item from sandpaper to valve gaskets are now offered for sale online.

Many of the most successful online businesses using the Web catalog revenue model are firms that were already operating in the mail-order business and simply extended their operations to the Web. Other companies that use the Web catalog revenue model adopted it after realizing that the products they sold in their physical stores could also be sold on the Web. This additional sales outlet did not require them to build additional stores, yet provided access to new customers throughout the world.

Discount Retailers: Getting a Great Deal Online

A number of discounters, such as Overstock.com, began their first retail operations online. Borrowing a concept from the physical world's Wal-Marts and discount club stores, these discounters sell merchandise at extremely low prices.

Traditional discount retailers, such as Costco, Kmart, Target, and Wal-Mart, were reluctant to implement online sales on their Web sites, which they used originally for general information distribution. They had huge investments in their physical stores, were making large amounts of sales in those stores, and did not really understand the world of online retailing. However, after some false starts and learning challenges, all of these major retailers now use the Web catalog revenue model in their online sales operations.

LEARNING FROM FAILURES

Walmart.Com

Wal-Mart is the world's largest retailer, with thousands of stores and annual sales exceeding $400 billion. Founded in 1962 by retailing legend Sam Walton, the company has won numerous awards for business innovation. However, Wal-Mart's moves into online retailing have been troubled.

Wal-Mart launched its first Web site in July 1996. Like most company sites of that time, it contained some information about the company, but did not offer any products for sale. Wal-Mart did little to develop the Web site over the next three years, but it did add a Web store—just in time to participate in the disastrous 1999 holiday shopping season.

Wal-Mart was not the only Web retailer to have trouble in 1999. Many companies found that they were ill-prepared for the large number of customers who decided to try electronic commerce in that year's holiday season. Lost orders, unfilled orders, and shipments that failed to arrive until January 2000 were common for many Web retailers that year. Wal-Mart was noted as an industry leader in shipping and logistics management; however, the announcement on its Web site that it could not promise Christmas delivery for items ordered after December 14 was particularly embarrassing.

To make matters worse, Wal-Mart was in the middle of developing a new Web site that it had hoped to launch before the holiday season. The project, which industry analysts estimate cost more than $100 million, ran months late and did not operate until January 2000.

After eight months of operating the new Web site, Wal-Mart found itself with low levels of customer traffic (well below those of its major rivals J.C. Penney, Sears, Kmart, and Target) and high levels of criticism from Web site design experts who found the site slow, difficult to use, and lacking customer service features.

In October 2000, Wal-Mart closed the site completely for four weeks. Earlier in the year, it had created Walmart.com, a joint venture with Accel Partners to develop a new Web site, but the new site was not ready to launch until November. Industry analysts widely criticized Wal-Mart's decision to completely shut down its Web operations for such a long time period at the beginning of the holiday shopping season.

The new Web site was a vast improvement over the old site; much better organized with improved browsing and search functions. The site offered about the same number of items as the previous site (about 500,000 items; several times more than what the physical stores carry); however, the newer site had more consumer electronics, toys, and sporting goods with fewer offerings of consumable products. Wal-Mart also created a separate distribution center to serve Walmart.com exclusively.

A decade later, Wal-Mart's online operations were once again in the news. In 2011, industry analysts estimated that Walmart.com was the sixth-largest online retailer in North America and noted that this was not a particularly impressive showing for the world's largest retailer. In August, the company announced that it was ending sales of music downloads after failing to compete successfully against Apple's iTunes store and that two of its top online executives were leaving the company. In the wake of these developments, Wal-Mart announced a major reorganization of its online operations in North America, the United Kingdom, and Japan that it hopes will better integrate online and physical store operations. Online business managers will now report to retail operations directors in each country rather than to a global e-commerce director. After more than 10 years, it is reversing its 2000 strategy of separating these operations.

Wal-Mart's experience is a testament to how difficult it can be to get Web retailing right. Success eluded the largest retailer in the world for years. Wal-Mart is estimated to have spent hundreds of millions of dollars on various Web implementations and product distribution strategies over the past 15 years. And now it appears to be starting over once again.

Using Multiple Marketing Channels

Having more than one way to reach customers is often a good idea for companies, as Montgomery Ward and Sears found out many years ago. They used one channel (retail stores) to reach urban customers and another channel (mail order catalog) to reach rural customers. Each different pathway to customers is called a **marketing channel**. Companies find that having several marketing channels lets them reach more customers at less cost. For example, it is expensive to stock a large number of different items in a physical store, so a company such as Best Buy will stock the most popular items in its stores but will sell a wider variety of items (including those that are not in high demand at every one of its retail locations) on its Web site. Customers who want to have physical contact with a product (putting fingers on a laptop computer's keyboard, for example) before buying can visit the retail location. A customer who wants a high-end and expensive home theater system can find it on the Web site. By having two marketing channels (retail store and Web site), Best Buy reaches more customers and offers more products than it could by using either channel alone. Like many other retailers, Home Depot encourages online sales by offering an option to have online orders shipped free to a nearby physical store location for the customer to pick up. This is an especially attractive option for large or heavy items.

Some retailers, such as Talbots, combine the benefits of these two marketing channels by offering in-store online ordering. This allows customers to examine a product in the store, and then find their exact size or the color they like by placing an order on the retailer's Web site from the store.

Similarly, a retailer that mails print catalogs might include a product's general description and photo in the catalog, but refer customers to the retailer's Web site for detailed specifications or more information about the product. Mailed catalogs (or newspaper advertising inserts) continue to be an effective marketing tool because they inform customers of products they might not otherwise know about. The catalog arrives in the mail (or the newspaper insert arrives with the newspaper) to inform them. In contrast, a Web site only delivers the marketing message if the customer visits the Web site.

Using multiple marketing channels to reach the same set of customers can be an effective strategy for retailers. Figure 3-1 shows two examples (there are many other possibilities) of how retailers might combine two marketing channels.

Retailer with physical stores and Web site

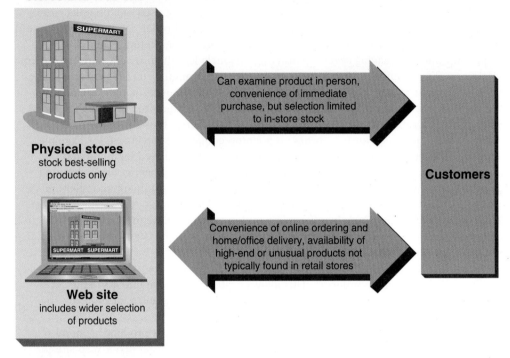

Physical stores
stock best-selling products only

Web site
includes wider selection of products

Can examine product in person, convenience of immediate purchase, but selection limited to in-store stock

Convenience of online ordering and home/office delivery, availability of high-end or unusual products not typically found in retail stores

Customers

Retailer with mailed catalogs and Web site

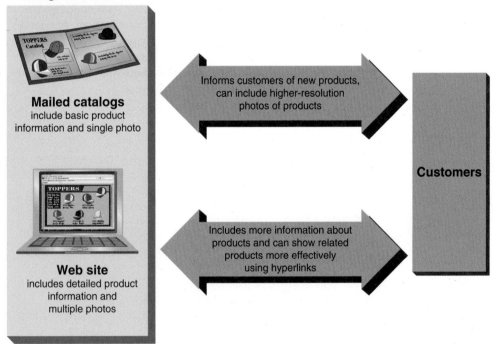

Mailed catalogs
include basic product information and single photo

Web site
includes detailed product information and multiple photos

Informs customers of new products, can include higher-resolution photos of products

Includes more information about products and can show related products more effectively using hyperlinks

Customers

FIGURE 3-1 Combining marketing channels: Two retailer examples

Adding the Personal Touch

A number of apparel sellers have adapted their catalog sales model to the Web. These Web stores display photos of casual and business clothing categorized by style and described with prices, sizes, colors, and tailoring details. Their intent is to have customers examine the clothing and place orders through the Web site. Lands' End pioneered the idea of online Web shopping assistance with its Lands' End Live feature in 1999. A Web customer with a question can initiate a text chat with a customer service representative or click a button on the Web page to have the representative call. In addition to answering questions, the representative can offer suggestions by pushing Web pages to the customer's browser.

Today, many Web sites offer a chat feature that is activated by the Web site visitor clicking a button on the Web page. Some sites activate a chat window when a visitor remains on a particular Web page longer than a certain time interval. These chat windows simulate the experience of having a helpful salesperson approach the customer in a physical retail store. Other online general apparel retailers have added online chat, personal shopper, and virtual model features to their sites. Some of these sites include a feature that lets two shoppers browse the Web site together from different computers. Only one of the shoppers can purchase items, but either shopper can select items to view. The selected items appear in both Web browsers. Web sites can buy this technology from vendors such as **DecisionStep** (its product is called ShopTogether).

Lands' End also was a pioneer in adding personal shopper and virtual model features to its site. The **personal shopper** is an intelligent agent program that learns the customer's preferences and makes suggestions. The **virtual model** is a graphic image built from customer measurements and descriptions on which customers can try clothes. Lands' End found that the dollar amount of orders placed by customers who use the virtual model is significantly higher than other orders. The Canadian company that developed this Web site feature, **My Virtual Model**, has sold the technology to a number of other clothing retailers. The My Virtual Model Web site stores an individual's virtual model details and makes that information available through any other clothing retailer site that offers the service.

One problem that the Web presents for clothing retailers of all types is that the color settings on computer monitors vary widely. It is difficult for customers to get an accurate idea of what the product's color will look like when it arrives. Most online clothing stores will send a fabric swatch on request. The swatch also gives the customer a sense of the fabric's texture—an added benefit not provided by catalogs. Most Web catalog retailers also have generous return policies that allow customers to return unused merchandise for any reason.

In addition to text chat, some online retailers use video to communicate with customers who have Webcams attached to or built into their computers. ITSRx is an online pharmacy that fills prescriptions for patients who have serious and chronic diseases such as cancer or multiple sclerosis. Many of these prescription drugs are injectables and a number of ITSRx's customers are new users of these medications. In the past, ITSRx used e-mail, telephone, or fax communications to answer customer questions about the drugs. Today, the pharmacy uses text chat and online video to instruct customers and to watch them as they administer their injections. The use of video in this case is much more efficient and safer than relying on e-mails or even text-based chat interactions.

Fee-for-Content Revenue Models

Firms that own written information (words or numbers) or rights to that information have embraced the Web as a highly efficient distribution mechanism. Many of these companies use a **digital content revenue model**; that is, they sell rights to access the information they own. Many companies sell subscriptions that give customers the right to access all or a specified part of the information; others sell the right to access individual items. A number of companies combine these two approaches and sell both subscriptions and individual access rights.

Legal, Academic, Business, and Technical Content

Many digital content providers specialize in legal, academic research, business, or technical material; however, all types of content are now available online. Whether you are an engineer who needs to find out if an idea you have has already been patented by someone else or a physician checking on a potential prescription interaction, you can find a digital content provider online who wants to fulfill your need.

LexisNexis offers a variety of information services for lawyers and law enforcement officials, court cases, public records, and resources for law libraries. In the past, law firms had to subscribe to and install expensive dedicated computer systems to obtain access to this information, but the Web has given LexisNexis customers much more flexibility in how they access their subscriptions.

Many academic and professional organizations, such as the American Psychological Association and the Association for Computing Machinery, sell subscriptions and individual access rights to their journals and other publications online. Academic publishing has always been a difficult business in which to make a profit because the base of potential subscribers is so small. Even highly regarded academic journals might have fewer than 2000 subscribers. To break even, academic journals must often charge each subscriber hundreds or even thousands of dollars per year. Electronic publishing eliminates the high costs of paper, printing, and delivery, and makes dissemination of research results less expensive and more timely.

A number of academic information aggregation services, such as ProQuest and EBSCO Information Services, purchase the rights to academic journals, newspapers, and other publications and resell those rights in subscription packages to schools, libraries, companies, and not-for-profit institutions.

Dow Jones, a business-focused publisher of newspapers such as *The Wall Street Journal* and *Barron's*, was one of the first publishers to create a Web site for selling subscriptions to digitized newspaper, magazine, and journal content. Today, Dow Jones operates an online content management and integration service called Factiva, which gives companies the ability to manage internal information and integrate it with external information to track company and industry news, perform analysis of acquisition candidates, and manage the company's risk in a dynamic business environment. Factiva also sells subscriptions to individuals who want to do research on businesses for employment searches or investment analysis.

Electronic Books

Another type of digital content sold online is the electronic book. Companies such as Audible and Books-on-Tape (now both owned by Amazon.com) sold digital audio editions of books for many years, first as cassette tapes, then as CDs, and later as various types of digital files. Today, the market leaders in electronic books are Amazon.com's Kindle products, Barnes & Noble's Nook products, and Google's eBookstore. Sales of electronic books are steadily growing; for example, Amazon.com announced that in 2011 it would sell more electronic books than paper books.

Books, magazines, and newspapers sold by these services as digital content are available for the physical readers (the Kindles and the Nooks) and for related applications that run on computers, tablet devices, and smart phones. Books are sold individually for these devices/applications; magazines and newspapers are sold on a subscription basis.

Online Music

The recording industry was slow to embrace online distribution of music because audio files are digital products that can be easily copied once purchased. Following a period of several years during which audio files were illegally shared among thousands of users, much of the recording industry finally stopped resisting digital sales of audio files. Starting around 2006, the recording companies began to identify ways they could capture some of the market for music files by selling their audio tracks online.

The largest online music stores include Amazon MP3, Apple's iTunes, eMusic, Google Music, Microsoft's MSN Music, and Rhapsody. These sites sell single songs (tracks) for about a dollar each and sell albums at various prices (most are between $5 and $12). Although some sites offer subscription plans, most of the sales revenue on these sites is generated from the sale of individual songs or digital albums.

The online music market has been complicated because no single store offers all of the music that is available in digital format and because many of the stores try to promote their own music file formats. Artists and recording companies sometimes only offer their music through one store and some refuse to offer their music online at all. By promoting their own file formats, stores are trying to encourage music consumers to use one store exclusively. Some online music sellers require buyers to download and install software, called **Digital Rights Management (DRM)** software, that limits the number of copies that can be made of each audio file. This does not prevent illegal copying, but it does make copying somewhat more difficult and the sellers hope that the extra effort required will discourage some of this copying. However, each store has different rules about how many copies are permitted and on which devices the files can be played. Consumers, especially those who buy music from more than one store, have found these varying restrictions confusing.

In 2007, the Amazon MP3 store was the first major online retailer to offer music tracks from several major recording companies in DRM-free MP3 format. Since then, other major retailers have followed Amazon's lead and most of them now offer some or all of their music in DRM-free, compatible file formats. Also, without DRM, it is now easier to convert files from one format to another. A report published by Strategy Analytics in 2011 predicts that online music sales will reach $2.8 billion in 2012, surpassing the declining total of music CD sales (estimated to be $2.7 billion) for the first time.

Online Video

Digital video can be sold or rented online as either a file download or as a streaming video. DRM software provides control over the number of copies that can be made of the downloaded video, the devices on which the video can be installed, and restrictions on how long the video remains available for watching. Videos offered for sale online include previously released movies, television shows, and programming that is developed specifically for the online market.

In the past, video sales have been limited by three main issues: the large size of video files (which can make download times long and streaming feeds uneven), concern that such sales might impair other sales of the video, and technological barriers that prevent downloaded videos from being played on a variety of devices. Online businesses have been working to overcome these issues and have had some success in addressing all three issues.

First, videos are still the largest types of files that are regularly transmitted on the Internet, but companies are continually experimenting with technologies that improve the delivery of large files and video streams. You will learn more about these content delivery technologies, pioneered by companies such as Akamai, Amazon.com, and Google, in Chapter 8.

Second, the companies that produce media are learning more about how online distribution fits into their overall revenue strategy. Movies traditionally have been released by the major Hollywood studios (20th Century Fox, Paramount, Sony, Walt Disney, Warner Brothers, and Universal) into different markets in a well-defined serial pattern. Movies were first distributed to theaters, which paid a high price for the right to show the movie first. After its initial theater run, the movie might then have been sold to airlines for in-flight showings and to premium cable channels such as HBO or Starz. Next, the movie was released on DVD and became available for purchase or rental through retail video stores. Eventually, the movie was sold to broadcast television stations and basic cable channels. This serial release pattern was designed to provide the movie's creators with the highest revenue obtainable at each point in the life of the product. Media producers released movies in this pattern for years, out of fear that any online distribution might steal sales away from one of their traditional outlets. These media producers now are experimenting with alternative distribution strategies. Some are now releasing movies online and on DVD simultaneously. As the number of online content distributors that charge either a subscription or a per-view fee for movies increases, media producers will be more amenable to releasing their product online because they can get paid for it.

Finally, video delivery technologies are becoming more transparent. For example, HTML 5 allows the delivery of movies through a standard Web browser without requiring plug-ins or external software. The availability of Web browsers on devices other than computers (for example, smart phones and tablet devices) has reduced concerns about technology barriers to video delivery on multiple devices.

Amazon.com sells the right to view movies and television shows on its Web site. The struggling video rental chain **Blockbuster** sells and rents access to video downloads, as does **Netflix**, which includes online access to movies on its Web site as part of its DVD rental subscription plans. Apple's **iTunes** service includes video offerings for rent or purchase in addition to its many free video downloads.

Television programs are also available online. Three of the major U.S. broadcast networks (ABC, Fox, and NBC) formed a joint venture to operate **Hulu**, which offers video clips of popular television programs and movies. Hulu offers much of its content free (using an advertising-supported revenue model) but also offers a monthly subscription, which makes premium content available. The other major U.S. broadcast network, CBS, operates **TV.com**, which offers free selected CBS-owned content, using an advertising-supported revenue model. Premium cable channel providers such as HBO and Showtime offer online access to their content for customers who have subscriptions to their services through their local cable company.

Advertising as a Revenue Model Element

Instead of charging a fee or subscription for content, many online businesses display advertising on their Web sites. The fees they charge advertisers are used to support the operation of the Web site and pay for the development or purchase of its content. Some sites rely entirely on advertising for their revenue, others use it only to provide part of their revenue. In this section, you will learn how advertising revenue is incorporated into the revenue models of various content-providing online businesses.

Advertising-Supported Revenue Models

The **advertising-supported revenue model** is the one used by broadcast network television in the United States. Broadcasters provide free programming to an audience along with advertising messages. The advertising revenue is sufficient to support the operations of the network and the creation or purchase of the programs. With the exception of the overall Web growth slowdown during 2000–2002, which you learned about in Chapter 1, Web advertising has increased steadily since the mid-1990s. As you will learn in Chapter 4, online advertising is now well established as an important component of the advertising mix used by businesses of all types. As online advertising grows, more and more Web sites can use it as a revenue source, either alone or in combination with other revenue sources.

The use of online advertising as the sole revenue source for a Web site has faced two major challenges. First, there has been little consensus on how to measure and charge for site visitor views, even after almost 20 years of experience with the medium. Because Web sites can take multiple measurements, such as number of visitors, number of unique visitors, number of click-throughs, and can measure other attributes of visitor behavior, Web advertisers have struggled to develop standards for advertising charges. In addition to the number of visitors or page views, stickiness is a critical element in creating a presence that attracts advertisers. The **stickiness** of a Web site is its ability to keep visitors at the site and attract repeat visitors. People spend more time at a **sticky** Web site and are thus exposed to more advertising.

The second issue is that very few Web sites have sufficiently large numbers of visitors to compete with mass media outlets such as radio or television. Although a few Web sites have succeeded in attracting the large general audience that major advertisers have traditionally wanted to reach, most successful advertising on the Web is targeted at specific groups. The set of characteristics that marketers use to group visitors is called **demographic information**, which includes things such as address, age, gender, income level, type of job held, hobbies, and religion. It can be difficult to determine whether a given Web site is attracting a specific market segment unless that site collects demographic information from its visitors—information that visitors often are reluctant to provide because of privacy concerns.

One solution to this second problem has been found by an increasing number of specialized information Web sites. These sites are successful in using an advertising-supported revenue model because they draw a specialized audience that certain advertisers want to reach. These sites do not need to gather demographic information from their visitors because anyone drawn to the site will have the specific set of interests that makes them a prized target for certain advertisers. In most cases, advertisers will pay high enough rates to support the operation of the site and in some cases, the advertising revenue is large enough to make these sites quite profitable.

Two examples of successful advertising-supported sites that appeal to audiences with specific interests are **The Huffington Post** and the **Drudge Report**. Each of these Web sites appeals to people who are interested in politics (liberal and conservative, respectively). Advertisers that want to target an audience with a specific political interest are willing to pay rates that are high enough to make these sites profitable enterprises. Online news sites that focus their coverage on a particular town or metropolitan area can use the advertising-supported revenue model successfully. Companies that want to reach potential customers in that area would find such sites to be useful for targeted marketing, since the Web sites would draw visitors with a specific interest in the geographic area.

Similarly, **HowStuffWorks** is a Web site that explains, as the name suggests, how things work. Each set of Web pages in the site attracts visitors with a highly focused interest. For example, a visitor looking for an explanation of how heating stoves work would be a good prospect for advertisers that sell heating stoves. HowStuffWorks does not need to obtain any specific information from its visitors; the fact that visitors are viewing the heating stoves information page is enough justification for charging heating stove companies a higher rate for ads on those pages. HowStuffWorks has a collection of pages that appeal to an array of visitors with highly focused interests. Thus, it is an attractive online advertising option for a wide variety of companies because the site has a collection of pages on a broad range of very specific products and processes that would be attractive to a variety of consumers, each of whom has a highly focused interest in one or more of them.

These three strategies—general interest, specific interest, and collection of specific interests—for implementing an advertising-supported revenue model are summarized in Figure 3-2.

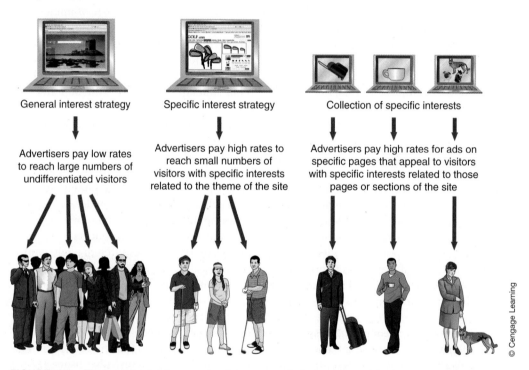

General interest strategy

Advertisers pay low rates to reach large numbers of undifferentiated visitors

Specific interest strategy

Advertisers pay high rates to reach small numbers of visitors with specific interests related to the theme of the site

Collection of specific interests

Advertisers pay high rates for ads on specific pages that appeal to visitors with specific interests related to those pages or sections of the site

© Cengage Learning

FIGURE 3-2   Three strategies for an advertising-supported revenue model

Some companies have been successful using the general interest strategy shown in Figure 3-2 by operating a Web portal. A **portal** or **Web portal** is a site that people use as a launching point to enter the Web (the word "portal" means "doorway"). A portal almost always includes a Web directory or search engine, but it also includes other features that help visitors find what they are looking for on the Web and thus make the Web more useful. Most portals include features such as shopping directories, white pages and yellow pages searchable databases, free e-mail, chat rooms, file storage services, games, and personal and group calendar tools.

One of the leading Web portal sites is **Yahoo!**, which was one of the first Web directories. A **Web directory** is a listing of hyperlinks to Web pages. Because the Yahoo!

portal's search engine presents visitors' search results on separate pages, it can include advertising on each results page that is triggered by the terms in the search. For example, when the Yahoo! search engine detects that a visitor has searched on the term *new car deals*, it can place a Ford ad at the top of the search results page. Ford is willing to pay more for this ad because it is directed only at visitors who have expressed interest in new cars. Besides Yahoo!, portal sites that use the general interest strategy today include Google and Bing. Smaller general interest sites, such as the Web directory refdesk.com, have had more difficulty attracting advertisers than the larger sites.

Not all portals use a general interest strategy, however. Some portals are designed to help visitors find information within a specific knowledge domain. The technology portal C-NET is one example of this type of site. C-NET uses the collection of specific interest strategy. The entire site is devoted to technology products and the site includes many reviews of specific technologies and related products. Advertisers pay more to have their ad appear near a discussion of a technology related to their product or on a page that reviews the product.

Travel portals such as Kayak have also been successful as advertising-supported online businesses. The Kayak site allows visitors to specify travel dates and destinations, and then searches multiple sites to find the best airfares, car rentals, and hotel rooms. It searches provider sites such as those of the airlines, hotels, and car rental companies, but it also searches sites that consolidate travel products and sell them at reduced prices. Kayak benefits its visitors by saving them the trouble of visiting multiple sites to find the best travel deals. And it sells targeted advertising space to companies that want to reach travelers with near-term travel plans.

Advertising-Supported Online Classified Ad Sites

Many newspapers and magazines publish all or part of their print content on the Web. They sell advertising to cover the costs of converting their print content to an online format and operating the Web site. Some publications, such as local shopping news and alternative press newspapers, have always been fully supported by advertising revenues and are distributed at retail locations and newsstands without charge. Many of these publications have made an easy transition to an advertising-supported revenue model. Most newspapers and magazines, however, have relied on subscription and newsstand revenue to supplement their advertising revenue. These publications have had a more difficult time in making their online editions generate sufficient revenue.

It remains unclear whether an online presence helps or hurts the business operations of these publishers. Although a Web site can provide greater exposure for the publication's name and a larger audience for advertising that it carries, an online edition also can divert sales from the print edition. Like retailers or distributors whose online sales lead to the loss of their brick-and-mortar sales, publishers also experience sales losses as a result of online distribution. Newspapers and other publishers worry about these sales losses because they are very difficult to measure.

In addition to the concern about lost sales of print editions, most newspaper and magazine publishers have found that the cost of operating their Web sites cannot be covered by the revenue they generate from selling advertising on the sites. Many publishers continue to experiment with various other ways of generating revenue from their Web sites. There is no consensus among media industry analysts regarding whether a pure advertising-supported revenue strategy can work for newspapers or magazines in the long run. As you will learn later in this chapter, many of these companies have experimented with combinations of revenue and subscription revenue sources and this experimentation will likely continue into the foreseeable future.

In the past, newspapers generated a significant percentage of their revenue from their classified advertising pages. You have already learned that targeted advertising can command higher rates than general advertising. Newspaper classified advertising was the original version of targeted advertising. Each ad is placed in a specific classification and only readers interested in that type of ad will read that classification. For example, a person looking for an apartment to rent would look in the Rentals classification. The growth of classified advertising Web sites has been very bad for newspapers. Sites such as **craigslist** now carry many free classified ads that would once have produced substantial classified advertising revenue for local newspapers. Craigslist and similar sites run most ads for free, only charging for a small proportion of the ads they carry (craigslist charges for job ads, brokered rental ads in New York City, and a few other categories). Craigslist generates enough revenue to continue operating, but many other classified advertising sites generate substantial revenue, replacing newspapers' historical role as the primary carrier of classified ads.

The most successful targeted classified advertising category has been Web employment sites. Companies such as **CareerBuilder.com** offer international distribution of employment ads. These sites offer advertisers access to targeted markets. When a visitor specifies an interest in, for example, engineering jobs in Dallas, the results page can include a targeted ad for which an advertiser will pay more because it is directed at a specific market segment. Other employment ad sites, such as **The Ladders**, charge both job seekers and employers for ads and access to those ads.

Employment ad sites such as **Monster.com** also target specific categories of job seekers by including short articles on topics of interest. These articles increase the site's stickiness and attract people who are not necessarily looking for a job. This is a good tactic because people who are not looking for a job are often the candidates most highly sought by employers.

Another type of online classified advertising business is the used vehicle site. Trader Publishing has printed advertising newspapers for many years and now operates the **AutoTrader.com** site. Similar sites accept paid advertising from individuals and companies that want to sell cars, motorcycles, and boats.

A product that is likely to be useful after the original buyer uses it is an appropriate item for inclusion in a classified advertising site. Classified advertising sites for used musical instruments, comic books, and used golf equipment are just a few examples.

Advertising-Subscription Mixed Revenue Models

In an **advertising-subscription mixed revenue model**, which has been used for many years by traditional print newspapers and magazines, subscribers pay a fee, but also accept some level of advertising. On Web sites that use the advertising-subscription mixed revenue model, subscribers are typically subjected to much less advertising than they are on sites supported completely by advertising. Firms have had varying levels of success in applying this mixed revenue model and a number of companies have moved to or from this model as they try to find the best way to generate revenue online.

Two of the world's most widely-circulated newspapers, *The New York Times* and *The Wall Street Journal*, have each used an advertising-subscription mixed model since they first took their publications online. *The Wall Street Journal's* mixed model is weighted toward subscription revenue. The site allows nonsubscriber visitors to view the classified ads and certain stories from the newspaper, but most of the content is reserved for subscribers who pay an annual fee for access to the site. Visitors who already subscribe to the print edition are offered a reduced rate on subscriptions to the online edition. As you will learn later in this chapter, *The New York Times* has gone through a number of changes to its revenue model, but for most of its online life it has made substantial portions of its content available at no cost and relied more heavily on advertising than *The Wall Street Journal*.

Most newspapers and magazines that use the advertising-subscription mixed revenue model for their online publications make most of their content available online, but a number of them do restrict the amount of free content as *The Wall Street Journal* does. Figure 3-3 shows the revenue models used by a number of newspapers and magazines, including those that use the advertising-supported model, the advertising-subscription mixed model with substantial content freely available, and the advertising-subscription mixed model with most content available only to subscribers.

Advertising-Supported	Advertising-Subscription Mixed Supported	
Most or All Content Free to All Visitors	Substantial Content Free to All Visitors	Most Content Available Only to Subscribers
The Boston Globe	BusinessWeek	The Economist
Cleveland Plain Dealer	Chronicle of Higher Education	Foreign Affairs
Financial Times	Forbes	Harvard Business Review
Newsweek	Inc. Magazine	National Geographic
InStyle	The Los Angeles Times	Nature
PC Magazine	The New York Times	Scientific American
San Francisco Chronicle	The Washington Post	Sports Illustrated
Smithsonian		Technology Review
Time		The Times
		The Wall Street Journal

FIGURE 3-3 Revenue models used by online editions of newspapers and magazines

Sports fans visit the **ESPN** site for all types of sports-related information. Leveraging its brand name from its cable television businesses, ESPN is one of the most-visited sports sites on the Web. It sells advertising and offers a vast amount of free information, but die-hard fans can subscribe to its Insider service to obtain access to even more sports information. Thus, ESPN uses a mixed model that includes advertising and subscription revenue, but it only collects the subscription revenue from Insider subscribers, who make up a small portion of site visitors.

Consumers Union, the publisher of product evaluations and ratings monthly magazine *Consumer Reports*, operates a Web site, **ConsumerReports.org**, that relies exclusively on subscriptions (that is, it is a purely subscription-supported site). Consumers Union is a not-for-profit organization that does not accept advertising as a matter of policy because it might appear to influence its research results. Thus, the site is supported by a combination of subscriptions and a small amount of charitable donations. The Web site does offer some free information as a way to attract subscribers and fulfill its organizational mission of encouraging improvements in product safety.

Fee-for-Transaction Revenue Models

In the **fee-for-transaction revenue model**, businesses offer services for which they charge a fee that is based on the number or size of transactions they process. Some of these services, including stock trading and online banking, lend themselves well to operating on the Web. To the extent that companies can offer Web site visitors the information they need about the transaction, companies can offer much of the personal service formerly

Selling on the Web

provided by human agents. If customers are willing to enter transaction information into Web site forms, these sites can provide options and execute transactions much less expensively than traditional transaction service providers. The removal of these traditional service providers is an example of **disintermediation**, which occurs when an intermediary, such as a human agent, is cut from a value chain. The introduction of a new intermediary, such as a fee-for-transaction Web site, into a value chain is called **reintermediation**.

Stock Brokerage Firms: Two Rounds of Disintermediation

Online stock brokerage firms use a fee-for-transaction model. They charge their customers a commission for each trade executed. In the past, stockbrokers offered investment advice and made specific buy and sell recommendations to customers in addition to their transaction execution services. They did not charge for this advice, but they did charge substantial commissions on the trades they executed. In the United States, these commission rates were set by a government agency and were the same for each stockbroker. Thus, because they could not compete on price, the best way for brokerage firms to compete was to offer more and better investment advice.

After the U.S. government deregulated the securities trading business in the early 1970s, a number of discount brokers opened, including the highly successful **Charles Schwab** firm. These discount brokers distinguished themselves by not offering any investment advice and charging very low commissions. They did not employ account executives (as the traditional brokerage firms did) because they did not need to offer the same level of personalized service; the attraction to customers was their low commission rates. Traditional brokers had provided free research to all of their customers, but many of those customers neither wanted nor valued the research. Those customers were very happy to move their business to the discount brokers who provided fast, inexpensive trade execution only. As this shift occurred, individual stockbrokers were disintermediated from the industry value chain.

A second round of disintermediation occurred in the 1990s as new online brokerage firms took business away from the discount brokers who had earlier taken business away from traditional brokers. The Web made it possible for firms such as **E*Trade Financial** to compete with both traditional and discount brokers by offering investment advice posted on their Web pages or sent in e-mailed newsletters. This advice was similar to that offered by a traditional broker, but could be provided without many of the costs of distributing the advice that traditional brokers had incurred (such as stockbroker salaries, overhead, and the costs of printing and mailing paper newsletters). These Web-based brokerage firms could also offer fast execution of trades by having customers enter data into Web page forms, thus competing with the discount brokers.

Of course, the full-line brokers found that they were simultaneously losing business to both the discount brokers and the online brokers. In response, both discount brokers and the few surviving traditional brokers opened stock trading and research information Web sites in attempts to take back some of their business from the online brokers. After two rounds of disintermediation and the financial crisis of 2008, the brokerage firms that remain today do most of their business online. **TD Ameritrade** is one example of a surviving firm that offers a combination of investment advice and advanced trading tools to a wide range of customers online.

Insurance Brokers

Other sales agency and brokerage businesses have moved substantial portions of their operations online. Although insurance companies themselves were slow to offer policies and investments for sale online, a number of intermediaries that sell insurance policies from a variety of companies have been online since the early days of the Web. Quotesmith, which began business in 1984 as a policy-quoting service for independent insurance brokers, decided in 1996 to sell its policy price quotes directly to the public over the Internet. By quoting policies and accepting applications directly, Quotesmith disintermediated the independent insurance agents with whom it formerly worked. Although Quotesmith is no longer in business, similar sites such as **InsWeb** and **Insurance.com** continue to provide quotes from multiple insurance carriers online directly to consumers.

As you learned in the case at the beginning of this chapter, **Progressive** provides quotes on its Web site for both its insurance products and for its competitors' products. **The General** (General Automobile Insurance Services) uses its Web site to reach auto insurance buyers who might have had trouble getting insurance from other companies. It advertises its online insurance quotes as being "fast and anonymous." By offering a comfortable environment to potential customers who have been rejected by other companies because of credit problems or traffic tickets, The General has been successful in this specific niche of the insurance market. Today, most major insurance companies offer information and policies for sale on their Web sites.

Event Tickets

Before the Web made online sales possible, obtaining tickets for concerts, shows, and sporting events could be a challenge. Some venues only offered tickets for sale at their own box offices, and others sold tickets through ticket agencies that were difficult for patrons to find or impossible to reach by telephone. The Web gave event promoters the ability to sell tickets from one virtual location to customers practically anywhere in the world. Established ticket agencies such as **Ticketmaster** were early participants in online ticket sales and earn a fee on every ticket they sell.

In addition to the original sale of tickets, the Web created opportunities for those who deal in secondary market tickets (tickets that have already been sold by the event's producer and that are being offered for resale to other persons). Companies such as **StubHub** and **TicketsNow** operate as brokers to connect owners of tickets with buyers in this market. These ticket resellers earn fees on tickets they resell for others, but they can also profit by buying blocks of tickets and reselling them at a higher price. Both ticket brokers and ticket resellers reduce transaction costs for both buyers and sellers of tickets by creating a central marketplace that is easy to find and that facilitates buyer-seller negotiation.

Online Banking and Financial Services

Because financial services do not involve a physical product, they are easy to offer on the Web. The greatest concerns that most people have when they consider moving their financial transactions to the Web are security and the reliability of the financial institution, which are the same concerns that exist in the physical world. However, on the Web, it is much more difficult for a firm to establish its reputation for security and trust than it is in the physical world, where massive buildings and clearly visible room-sized safes can help create the necessary image. Some people who are willing to buy products and services online are unwilling to trust a Web site for their banking services, but the number who do is growing. A 2010 comScore research study reported that 64 percent of U.S. banking customers pay their bills online. In Chapter 11, you will learn more about how online payments and other financial transactions are processed.

Most banks that entered the online banking business did so by offering some of their services on the Web. They generally began with sites that offered account balances and statements, then added bill pay, account transfers, loan applications, and other services. Some firms started completely new online banks that were not affiliated with any existing bank (such as the **First Internet Bank of Indiana**). Banks benefit from serving their customers online because it costs the bank less to provide services online than to provide those same services through personal interactions with bank employees in a branch office.

Although online banks let customers pay their bills electronically, many customers still receive their bills in the mail. Those who do receive their bills online must often visit a different Web site to view each online bill. A **bill presentment** service provides an electronic version of an invoice or billing statement (such as a credit card bill or a mobile phone services statement) with all of the details that would appear in the printed document. As online banks add bill presentment services that allow their customers to view all of their bills on the bank's Web site (and pay each of them with a single click), they are finding that more of their customers are willing to do their banking on the Web.

Another important feature that an increasing number of online banks now offer is **account aggregation**, which is the ability to obtain bank, investment, loan, and other financial account information from multiple Web sites and display it all in one location at the bank's Web site. Many of a bank's best customers have credit card, loan, investment, and brokerage accounts with several different financial institutions. Having all of this information collected in one place is very helpful to these customers. Some banks have created their own account aggregation and bill presentment software, but companies such as **Yodlee** sell these services to banks and other financial institutions. The number of banks that offer aggregation is expected to continue to grow.

Travel

In the past, travel agents earned substantial commissions on each airplane ticket, hotel reservation, auto rental, or vacation that they booked. These commissions were paid to the travel agent by the transportation or lodging provider. Thus, the traditional revenue model in the travel agency business was a fee-for-transaction, similar to the model of stock brokerage firms.

When the Internet became available to commercial users, a number of online travel agencies began doing business on the Web. Existing travel agencies did not, in general, rush to the new medium. They believed that the key value they added, personal customer service, could not be replaced with a Web site.

In recent years, most airlines and auto rental companies have reduced the amount of the commissions they pay travel agents. In some cases, they have stopped paying commissions at all. Most cruise lines and hotels continue to pay commissions. And many hotels sell blocks of rooms to travel agents who can then resell them as part of vacation packages. Some airlines also sell blocks of seats to travel agents. Online travel sites have much larger volume than traditional travel agencies and are thus able to buy larger blocks of hotel rooms and airline seats.

Online travel sites have evolved to make money in various ways. They all collect any commissions that are paid. And they buy and sell rooms and airline seats, but most of them, including **Travelocity**, which was based on the Sabre computer system that traditional travel agencies used to book flights and hotel rooms (Travelocity is owned by Sabre), and Microsoft's **Expedia** subsidiary, run advertising on their Web sites in a combined advertising-fee revenue model. In 2001, a consortium of five major U.S. airlines launched **Orbitz**, which became one of the most visited travel sites on the Web. The Orbitz home page appears in Figure 3-4.

FIGURE 3-4   Orbitz home page

The online travel sites were able to disintermediate many traditional travel agencies. By expanding rapidly online, they were able to negotiate better deals on hotel rooms and airline seats that they purchased for resale. With their scale of operations and low cost per transaction, they were able to continue operating profitably on the reduced airline ticket commissions. These factors combined to hasten the end of the traditional travel agency.

Some smaller travel agencies have survived; these agencies most often specialize in cruise vacations. Cruise lines still view travel agents as an important part of their selling strategy and continue to pay commissions to travel agents on the sales that they make. Web sites that make discounted cruise packages easy to search, such as VacationsToGo.com, or that provide detailed information about cruises have been successful in this travel industry niche.

Other small travel agencies have been successful by following a reintermediation strategy with a focus on specific groups of travelers. These travel agents identify a group of travelers with specific needs and sell travel packages designed for that group. For example, surf vacations have become increasingly popular. The stereotypical surfer of years gone by (a young unemployed male) has been replaced by a much broader demographic. Today's surfers often have significant financial resources and enjoy surfing in exotic locations. Web sites such as WaveHunters.com have followed a reintermediation strategy and cater to this specialized market. Travel agencies that specialize in unusual or exotic destinations, such as Antarctica, have also been successful as intermediaries if they have particular expertise, knowledge, or local contacts that help them create custom itineraries. These sites also include advertising as part of their online presences and revenue models.

Automobile Sales

Traditional auto dealers buy cars from the manufacturer and sell them to consumers. They provide showrooms and salespeople to help customers learn about product features, arrange financing, and make a purchase decision. Dealers make their profits by charging a markup on each vehicle sale in addition to charging fees for service, warranty extensions, and other add-ons. In the United States, most states have laws that prevent auto manufacturers from selling directly to consumers, which provides some protection from disintermediation for auto dealers. Almost all auto dealers negotiate the prices at which they sell their cars; thus, the salesperson's job includes extracting the highest possible price from the consumer. Many people do not like negotiating car prices, especially if they have taken the time to learn about car features, arrange financing, and are ready to purchase a car without further assistance from a salesperson.

Autobytel and similar firms, such as Edmunds.com, provide an information service to car buyers. They offer an independent source of information, reviews, and recommendations regarding auto makes and models. Some of these firms offer customers the ability to select a specific car (model, color, options) at a price the firm determines. The firm then finds a local dealer that has such a car and is willing to sell it for the determined price. An alternative approach is for the firm to locate dealers in the buyer's area that are willing to sell the car specified by the buyer (including make, model, options, and color) for a small premium over the dealer's nominal cost. After the firm introduces the buyer to the dealer, that buyer can purchase the car without negotiating with a salesperson. The firm charges participating dealers a fee for this service. In effect, these firms are disintermediating the individual salesperson. To the extent that the salesperson provides little value to the consumer, these firms are reducing the transaction costs in the process. The car salesperson is disintermediated and the Web site becomes the new intermediary in the transaction, which is an example of reintermediation. Some auto sales sites also sell advertising on their sites, which makes them, like the online travel agencies, examples of mixed fee-for-transaction and advertising-supported revenue models.

Real Estate and Mortgage Loans

Other fee-for-transaction businesses use Web sites to solicit business, including real estate brokers and mortgage loan brokers. Most real estate brokerage firms have a strong online presence, including information about properties they have for sale or rent, along with contact information for individual brokers affiliated with their offices. Many individual real estate brokers operate their own Web sites as well. The industry's trade association, the National Association of Realtors, sponsors a Web site, Realtor.com, that carries detailed descriptions and photos of houses listed for sale by its member firms. Although very few (if any) real estate transactions are completed online, these Web sites play an important role in bringing buyers and sellers together.

Although the financial crisis of 2008 dramatically reduced the number of mortgage brokers in business, a number of them continue to do business online. Both GMAC Mortgage and E-LOAN still provide information and take mortgage loan applications online.

The complexity and size of real estate transactions have made it difficult for online activities to displace completely the work done by individual real estate and mortgage brokers. Thus, this is one line of business that has been highly resistant to disintermediation caused by online technologies. The changes caused by online elements in the real estate and mortgage businesses have been minor.

Fee-for-Service Revenue Models

Companies are offering an increasing variety of services on the Web for which they charge a fee. These are neither broker services nor services for which the charge is based on the number or size of transactions processed. The fee is based on the value of the service provided. These **fee-for-service revenue models** range from games and entertainment to financial advice and the professional services of accountants, lawyers, and physicians.

Online Games

Computer and video games are a huge industry. In the United States alone, more than $14 billion per year is spent on these types of games. A substantial portion of that revenue is generated online. Although many sites that offer games relied on advertising revenue in the past (and some, such as **GSN.com**, still do), a growing number, including **MSN Games** and **Sony Online Entertainment**, include premium games in their offerings. Site visitors must pay to play these premium games, either by buying and downloading software to install on their computers, or by paying a subscription fee to enter the premium games area on the site. Almost all game sites include some elements of advertising in their revenue models.

Professional Services

State laws have been one of the main forces preventing U.S. professionals (such as physicians, lawyers, accountants, and engineers) from extending their practices to the Web. Since most professionals are licensed by individual states, state laws can prevent them from practicing their professions on the Web because online patients or clients would likely be located in other states. If they were to offer their services online to persons in other states, professionals could be charged with unlicensed practice in those other states. State laws concerning the imputed location of service delivery are vague; it can be difficult to determine exactly where a service provided online actually occurs. This uncertainty arises because most state professional practice laws were written long before the Internet existed.

Although some medical, legal, and other professional practices allow patients to make appointments online, and a few professionals do online consultations, most are reluctant to do any element of their practices on the Web. Many professionals are worried about protecting the privacy of their patients or clients online.

The **Law on the Web** site offers legal consultations on a variety of matters for residents of the United Kingdom. Accounting professionals in the United States can be located through the **CPA Directory**, and a number of legal referral sites can direct site visitors to local attorneys. The online version of the well-known Martindale-Hubbell lawyer directory is also available online at **Martindale.com**.

Although a large number of Web sites offer general health information, physicians and other health care professionals have been reluctant to sell specific advice to specific patients online. The difficulty of diagnosing medical problems without a physical examination of the patient is a significant barrier to providing most types of health care services online, but some physicians are beginning to offer online consultations to patients with whom they have an ongoing, established relationship.

The **Cope Today** Web site does offer online therapy to patients in the United States. The site connects potential patients with therapists licensed in that patient's jurisdiction, so the therapist providing the online consultation complies with state professional practice laws. Online consultations are done by text or video chat. The site's founder notes that some conditions, such as depression or anxiety, might be easier to treat online since the patient does not need to leave home to see a therapist.

Free for Many, Fee for a Few

Chris Anderson, the editor of *Wired Magazine*, proposed in 2004 that the economics of producing and selling digital products is substantially different from the economics of producing and selling physical products. In his books (see references to his work in the For Further Study and Research section at the end of this chapter), he explains that physical products benefit from the production of standardized versions that generate economies of scale. Because each product requires materials and labor, using the same materials allows large producers to buy those materials at lower costs by ordering in bulk. Labor costs can be reduced by training workers to do specific production tasks efficiently. Since most of the cost of a physical product is in the manufacture of each unit (as opposed to the design of the prototype), the key to making a profit is to reduce the cost of manufacturing. Digital products work differently. They tend to have large up-front costs. Once those costs are incurred, additional units can be made at very low additional cost. For example, a software program can cost thousands of dollars to create. It can take many hours of expensive programmer time to design, code, and test. But once it is in production, creating additional units (especially if those units are distributed in digital form, online) costs very little. Making minor changes in the program so that it works better for different types of customers can be inexpensive, too. Thus, the economics of digital products are quite different from the economics of physical products.

The result of Anderson's logic is that it can be profitable to offer a digital product to a large number of customers for free, and then charge a small number of customers for an enhanced, specialized, or otherwise differentiated version of the product. If you can charge the small number of customers enough to cover the cost of developing the digital product and yield a profit, you can give away many copies of the product, especially if those free copies entice more paying customers for the enhanced product. For example, Yahoo! offers free e-mail accounts to site visitors. This draws visitors to the Yahoo! site and allows the company to sell some advertising on the pages that display the e-mail service. But some e-mail users will want an enhanced version of the service. Perhaps they want pages with no advertising, the ability to send large attachments with their e-mails, or more storage space for their e-mails. Yahoo! charges for a premium version of its service that offers these features. It costs the company very little to offer this service, but it generates considerable revenue.

In the physical world, this free sample logic works in reverse. Companies selling physical products have often used a mixture of free and for-sale products. For example, a bakery might have a plate of cookies available for customers to taste. The bakery hopes that enough customers will be impressed with the taste of the free cookies that they will buy cookies or other baked goods. They give away a small number of physical products to boost sales.

CHANGING STRATEGIES: REVENUE MODELS IN TRANSITION

Many companies have gone through transitions in their revenue models as they learn how to do business successfully on the Web. As more people and businesses use the Web to buy goods and services, and as the behavior of those Web users changes, companies often find that they must change their revenue models to meet the needs of those new and changing Web users. Some companies created electronic commerce Web sites that needed many years to grow large enough to become profitable. This is not unusual; both CNN and ESPN took more than 10 years to become profitable and they had both created new businesses in television, which was an existing and well-established medium. Many Web companies found that their unprofitable growth phases were lasting longer than they had anticipated and were forced either to change their revenue models or go out of business.

This section describes the revenue model transitions undertaken by five different companies as they gained experience in the online world and faced the changes that occurred in that world. In the second wave of electronic commerce, these and other companies might well face the need to make further adjustments to their revenue models.

Subscription to Advertising-Supported Model

Microsoft founded its **Slate** magazine Web site as an upscale news and current events publication. Because *Slate* included experienced writers and editors on its staff, many people expected the online magazine to be a success. Microsoft believed that the magazine had a high value, too. At a time when most online magazines were using an advertising-supported revenue model, *Slate* began charging an annual subscription fee after a limited free introductory period.

Although *Slate* drew a wide readership and received acclaim for its incisive reporting and excellent writing, it was unable to draw a sufficient number of paid subscribers. At its peak, *Slate* had about 27,000 subscribers generating annual revenue of $500,000, which was far less than the cost of creating the content and maintaining the Web site. *Slate* is now operated as an advertising-supported site. Because it is a part of Microsoft, *Slate* does not report its own profit numbers. Microsoft maintains the *Slate* site as part of its Bing portal, so it is likely that the value of the publication to Microsoft is to increase the portal's stickiness.

Advertising-Supported to Advertising-Subscription Mixed Model

Another upscale online magazine, **Salon.com**, which has also received acclaim for its innovative content, has moved its revenue model in the direction opposite of *Slate*'s transition. After operating for several years as an advertising-supported site, *Salon.com* began offering an optional subscription version of its site called *Salon Premium*, which was free of advertising and could be downloaded for later offline reading on the subscriber's computer.

The subscription version offering was motivated by the company's inability to raise the additional money from investors that it needed to continue operations. The subscription version has gone through a number of changes over the year and now includes access to additional content such as downloadable music, e-books, and audio books. The premium version of the site, now called **Salon Core**, also includes subscriptions to various print magazines, access to sports content, music, and a preferential access to the site's writers and editors.

Advertising-Supported to Fee-for-Services Model

Xdrive Technologies opened its original advertising-supported Web site in 1999. Xdrive offered free disk storage space online to users. The users saw advertising on each page and had to provide personal information that allowed Xdrive to send targeted e-mail advertising to them. Its offering was very attractive to Web users who had begun to accumulate large files, such as MP3 music files, and wanted to access those files from several computers in different locations.

After two years of offering free disk storage space, Xdrive found that it was unable to pay the costs of providing the service with the advertising revenue it had been able to generate. After being bought by AOL in 2005, Xdrive switched to a subscription-supported model (AOL-registered users were eligible for a small free storage service) and began selling the service to business users as well as individuals. In recent years, disk drive costs have dropped and Xdrive frequently adjusted its monthly fee downward. AOL finally closed the service in 2009.

Companies that have successful online storage businesses today, such as **Carbonite** or **Dropbox**, generally charge a fee for their services that is based on the amount of storage used. Some companies use the "free for many, fee for a few" revenue model. Amazon.com and Google both offer consumer data storage services that are free up to a certain point, with additional storage available for a monthly or annual fee.

Advertising-Supported to Subscription Model

Northern Light was founded in August 1997 as a search engine, but a search engine that did more than search the Web. It also searched its own database of journal articles and other publications to which it had acquired reproduction rights. When a user ran a search, Northern Light returned a results page that included links to Web sites and abstracts of the items in its own database. Users could then follow the links to Web sites, which were free, or purchase access to the database items.

Thus, Northern Light's revenue model was a combination of the advertising-supported model used by most other Web search engines plus a fee-based information access service, similar to the subscription services offered by ProQuest and EBSCO that you learned about earlier in this chapter. The difference in the Northern Light model was that users could pay for just one or two articles (the cost was typically $1–$5 per article) instead of paying a large amount of money for unlimited access to its database on an annual subscription basis. Northern Light also offered subscription access to most of its database to companies, schools, and libraries.

In January 2002, Northern Light decided that the advertising revenue it was earning from the ads it sold on search results pages was insufficient to justify continuing to offer that service. It stopped offering public access to its search engine and converted to a new revenue model that was primarily subscription supported. Northern Light's new model generates revenue from annual subscriptions to large corporate clients. Its main products today include SinglePoint, a search engine that runs on corporate databases, and MI Analyst, a meaning extraction tool used in business research applications.

Multiple Changes to Revenue Models

Over its 240-year publishing history, Encyclopædia Britannica has developed one of the most respected brand names in research and education. Beginning in 1768 as a sort of precomputer-age frequently asked questions (FAQ) list, a group of academics developed the encyclopedia out of collected notes they had made while conducting research and decided to publish them as a series of articles.

The company has been through a number of revenue model transitions as it developed its current online business strategy. When Encyclopædia Britannica first moved online in 1994, it began with two Web-based offerings. The Britannica Internet Guide was a free Web navigation aid that classified and rated information-laden Web sites. It featured reviews written by Britannica editors who also selected and indexed the sites. The company's other Web site, Encyclopædia Britannica Online, contained the full text and pictures from the print encyclopedia. It was available for a subscription fee or as part of the Encyclopædia Britannica CD package. Britannica's intention was to use the free site to attract users to the paid subscription site.

In 1999, disappointed by low subscription sales of Encyclopædia Britannica Online, Britannica converted to a free, advertising-supported site. In terms of Web site traffic, the new revenue model was a huge success. The first day the new free site, Britannica.com, became available it had more than 15 million visitors, forcing Britannica to shut down for two weeks to upgrade its servers. The site offered full content of the encyclopedia's print edition in searchable form, plus access to the *Merriam-Webster's Collegiate Dictionary* and the *Britannica Book of the Year*. One of the most successful aspects of the site was the way it integrated the Britannica Internet Guide Web-rating service with its print content. The Britannica Store sold the CD version of the encyclopedia along with other educational and scientific products to help generate revenue.

Unfortunately, advertising sales were not what the company had hoped. After two years of trying to generate a profit using this advertising-supported model, Britannica returned to the mixed model it continues to use today. In this mixed model, the company offers free online access to summaries of encyclopedia articles and the *Merriam-Webster's Collegiate Dictionary*, but the full text of the encyclopedia is only available to visitors who pay an annual fee of about $70.

Britannica went from being a print publisher to a seller of information on the Web to an advertising-supported Web site to a mixed advertising subscription model—three major revenue model transitions—in just a few short years. The main value that Britannica has to sell is its reputation and the expertise of its editors, contributors, and advisors. After exploring these different revenue models, the company has decided that the best way to capitalize on its reputation and expertise is through a mixed revenue model of subscriptions and advertising support, with the bulk of its revenue coming from subscriptions. Britannica also generates revenue by selling books, CDs, DVDs, and software with an educational theme through its online products store.

The New York Times Web site has gone through several revenue model transitions since opening in the mid-1990s. Originally, the site was purely advertising supported and included most of the content in the print edition of the newspaper. It has always charged a subscription fee for its premium crossword puzzles and for time-limited access to its chess column to print subscribers. The first revenue model also included a fee for access to older articles stored in the newspaper's archives.

In 2005, *The New York Times* decided to limit access to much of its most desirable content to subscribers and began charging a fee for access to its Op Ed and news columns. The fee also allowed access to the crossword puzzles and the older articles in the archives. All of the limited-access content was also available to print edition subscribers. This program brought in about 227,000 subscribers, which at $44.95 per year generated about $10 million in revenue.

By 2007, the newspaper became convinced that it could earn more advertising revenue by providing free access to those pages than it was earning in subscription fees, so it went back to relying on an advertising-supported revenue stream. The newspaper charged only for access to the crossword puzzles and for older articles in the archives. With this change, the traffic to the Web site nearly doubled, reaching an average of 30 million unique visitors per month. However, the recession of 2008 caused advertising revenue to drop and the company began considering other alternatives.

In 2011, disappointed with the level of advertising revenue, the company adopted a rather complex program that gave the newspaper some flexibility in what it would put online (in case there was a major story it wanted to cover broadly) yet that would generate more revenue than the advertising-based revenue model it had been using for the previous four years. In the new plan, site visitors could read 20 articles a month at no charge. When a visitor attempts to view the 21st article, the site offers several subscription plans (currently priced between $15–$35 per month) that include unlimited access to the Web site and various levels of access through mobile phones. This type of barrier, which is triggered by a specific level of usage, has become known as a **pay wall**. Subscribers to the print edition are given unlimited access to the site.

The publishers of the newspaper are hopeful that this mixed revenue model will provide an acceptable balance between the editors' desire to have as many people as possible read the paper and the need to generate sufficient revenue to keep the newspaper operating. Their experience with this revenue model will doubtless be watched closely by the entire industry.

REVENUE STRATEGY ISSUES FOR ONLINE BUSINESSES

In the first part of this chapter, you learned about the revenue models that companies are using on the Web today. In this section, you will learn about some issues that arise when companies implement those models. You will also learn how companies deal with those issues.

Channel Conflict and Cannibalization

Companies that have existing sales outlets and distribution networks often worry that their Web sites will take away sales from those outlets and networks. For example, Levi Strauss & Company sells its Levi's jeans and other clothing products through department stores and other retail outlets. The company began selling jeans to consumers on its Web site in mid-1998. Many of the department stores and retail outlets that had been selling Levi's products for many years complained to the company that the Web site was now competing with them. In January 2000, Levi Strauss announced it would stop selling its clothing products on its own Web site. Such a **channel conflict** can occur whenever sales activities on a company's Web site interfere with its existing sales outlets. The problem is also called **cannibalization** because the Web site's sales consume sales that would be made in the company's other sales channels. In recent years, the Levi's Web site resumed selling products directly to consumers, but it includes a Store Locator link that helps customers find a nearby store if they want to buy in person. Both Levi Strauss and the retail stores it sells through have agreed that the sales through the Web site are insignificant. Over time, many Levi's retailers have opened online stores themselves, so they see the Levi's site as less of a threat than they did in 2000.

Maytag, the manufacturer of home appliances, found itself in the same position as Levi Strauss. It created a Web site that allowed customers to order directly from Maytag. After less than two years of making direct online sales and receiving many complaints from its authorized distributors and resellers, Maytag decided to incorporate online partners into its Web site store design. Now, after searching and gathering information about specific products from the Maytag Web site, a customer can click a Where to Buy link and be directed to a nearby Maytag retailer.

Both Levi's and Maytag faced channel conflict and cannibalization issues with their retail distribution partners. Their established retailers sold many times the dollar volume than either company could ever hope to sell on their own Web sites. Thus, to avoid angering their retailers, who could always sell competing products, both Levi's and Maytag decided that it would be best to work with their retail partners. Similar issues can also arise within a company if that company has established sales channels that would compete with direct sales on the company's own Web site.

Eddie Bauer, a retailer of clothing and outdoor gear, was selling through a catalog and retail stores located primarily in major shopping malls when it decided to begin selling products on its Web site. The company believed that it could make online sales more attractive by allowing customers to return unwanted products that they had purchased online at the retail store locations. The managers of these stores were concerned about the time it would take for their sales associates to process these returns and about having to add the items to their stores' inventories. In a retail store operation, managing labor costs and inventory are very important in achieving store profitability. The managers at the company's catalog division were also worried. They feared that sales through the Web site would cannibalize sales through the catalog.

By making adjustments in the managers' compensation and bonus plans, Eddie Bauer was able to convince all of the managers to support the Web site. The retail store managers were credited with an inventory and labor cost allowance for each Web site return they handled. The catalog division managers were given a credit for existing catalog customers who purchased goods from the Web site. By giving their customers access to the company's products through a coordinated presence in all three distribution channels, Eddie Bauer was able to increase overall sales to those customers. This type of solution is called **channel cooperation**.

Strategic Alliances

As you learned in Chapter 1, when two or more companies join forces to undertake an activity over a long period of time, they are said to create a strategic alliance. When companies form a strategic alliance, they are operating in the network form of organization that you learned about in Chapter 1. Companies form strategic alliances for many purposes. An increasing number of businesses are forming strategic alliances to sell on the Web. For example, the relationships that Levi's created with its retail partners by giving them space on the Levi's Web site to sell Levi's products is an example of a strategic alliance.

Earlier in this chapter you learned about Yodlee, the account aggregation services provider, and the online bank sites that offer these services to consumers. The relationship between Yodlee and its bank clients is another example of a strategic alliance. Yodlee can concentrate on developing the technology and services while the banks provide the customers. Account aggregation services decrease the likelihood that customers will consider moving to another bank, which helps the bank hold on to its customers. Thus, both parties benefit from the strategic alliance.

Amazon.com has forged a number of strategic alliances with existing firms. As you learned in Chapter 1, Amazon joined with Target to sell that discount retailer's products on a Target-branded Web site. Amazon.com has also formed strategic alliances with many smaller companies to offer their products for sale on the Amazon.com Web site.

Luxury Goods Strategies

Some types of products can be difficult to sell online. This is particularly true for expensive luxury goods and high-fashion clothing items that customers generally want to see in person or touch. Many luxury brands hesitated to offer their products online for fear of alienating the upscale physical stores that sold their products. For example, clothier Lilly Pulitzer launched its Web site in 2000, but did not sell on the site until 2008, fearing that it would lose some of the luxury cachet it derived from limiting its sales outlets.

Some upscale brands overcome this obstacle by limiting the range of their online offerings. For example, luxury brand Chanel, which launched its retail site in 2010, and Calvin Klein do not offer all of their products online. Chanel sells fragrance and skincare products online but not its clothing lines. Calvin Klein does not sell its couture line online, but it does sell its ready-to-wear lines on its Web site.

In large part, luxury retailers limit their sales online out of concern that some or all of their products' features must be experienced in person and cannot be adequately represented online. One industry that has overcome this obstacle, however, is the retail jewelry business. After years of slow online sales, jewelry sales have grown rapidly in recent years. Retailers such as Blue Nile and Ice.com operate highly successful online jewelry stores. Even general retailers such as Costco offer $50,000 diamond rings online. Helping these stores overcome resistance is the general availability of independent appraisal certificates for diamonds and other high-priced jewelry items. Another important factor is the stores' well-advertised "no questions asked" return policies.

Overstock Sales Strategies

In the fast-changing clothing business, retailers have always had to deal with the problem of overstocks—products that did not sell as well as hoped. Many retailers use outlet stores to sell their overstocks. Lands' End found that its overstocks Web page worked so well that it has closed some of its physical outlet stores. Many other retailer Web sites include a link to separate sections for overstocks or clearance sales of end-of-season merchandise.

An online overstocks store works well because it reaches more people than a physical store and it can be updated more frequently than a printed overstocks catalog. Overstocks and clearance sale pages have become a standard element of clothing retailers' Web sites.

CREATING AN EFFECTIVE BUSINESS PRESENCE ONLINE

Businesses have always created a presence in the physical world by building stores, factories, warehouses, and office buildings. An organization's **presence** is the public image it conveys to its stakeholders. The **stakeholders** of a firm include its customers, suppliers, employees, stockholders, neighbors, and the general public. Most companies tend not to worry much about the image they project until they grow to a significant size—until then, they are too focused on just surviving to spare the effort. On the Web, presence can be much more important. Many customers and other stakeholders of a Web business know the company only through its Web presence. Creating an effective Web presence can be critical even for the smallest and newest firms operating on the Web.

Identifying Web Presence Goals

When a business creates a physical space in which to conduct its activities, its managers focus on very specific objectives. Few of these objectives are image driven. The new company must find a location that will be convenient for its customers, with sufficient floor space and features to allow the selling activity to occur. A new business must balance its needs for inventory storage space and employee work space with the costs of obtaining that space. The presence of a physical business location results from satisfying these many other objectives and is rarely a main goal of designing the space.

A firm's physical location must satisfy so many other business needs that it often runs out of the resources it would need to convey a good presence. On the Web, businesses and other organizations have the luxury of building their Web sites with the main goal of creating a distinctive presence. A good Web site design can provide many image-creation and image-enhancing features very effectively—it can serve as a sales brochure, a product showroom, a financial report, an employment ad, and a customer contact point. Each entity that establishes a Web presence should decide which features the Web site can provide and which of those features are the most important to include. An effective site is one that creates an attractive presence that meets the objectives of the business or organization. A list of these objectives, along with some examples of Web site design strategies that can help accomplish them, appears in Figure 3-5.

Objectives	Strategies
Attracting visitors to the Web site	Include links to the Web site (or specific pages) in marketing e-mails
Making the site interesting enough that visitors stay and explore	Product reviews, comparison features, advice on how to use a product or service
Convincing visitors to follow the site's links to obtain information	Clearly labeled links that include a hint of the information to be obtained by following them
Creating an impression consistent with the organization's desired image	Using established branding elements such as logos, characters used in other advertising media, slogans, or catchphrases
Building a trusting relationship with visitors	Ensuring the validity and objectivity of information presented on the site
Reinforcing positive images that the visitor might already have about the organization	Presenting testimonials, information about awards, links to external reviews or articles about the organization or its products and services
Encouraging visitors to return to the site	Featuring current information about the organization or its products and services that is regularly updated

FIGURE 3-5   Web presence objectives and strategies

Making Web Presence Consistent with Brand Image

Different firms, even those in the same industry, might establish different Web presence goals. For example, Coca Cola and Pepsi are two companies that have established powerful brand images in the same business, but they have developed significantly different Web presences. These two companies frequently change their Web pages, but the Coca Cola page usually includes a trusted corporate image such as the Coke bottle. Alternatively, the Pepsi page is usually filled with links to a variety of activities and product-related promotions.

These Web presences convey the images each company wants to project. Each presence is consistent with other elements of the marketing efforts of these companies—Coca Cola's traditional position as a trusted classic, and Pepsi's position as the upstart product favored by a younger generation.

Most auto manufacturers' Web sites convey a consistent brand image. They usually include links to detailed information about each model, a dealer locator page, information about the company, and a set of shopping tools such as configuration pages for each model.

Not-for-Profit Organizations

Auto makers enhance their images by providing useful information to customers on their Web sites. The main function of their Web sites, however, is to promote their products and get customers in touch with a dealer who can sell them a car. For other organizations, the image-enhancement capability is a key goal of their Web presence efforts. Not-for-profit organizations are an excellent example of this. They can use their Web sites as a central resource for communications with their varied and often geographically dispersed constituencies.

A key goal for the Web sites of many not-for-profit organizations is information dissemination. The Web allows these groups to integrate information dissemination with fund-raising in one location. Visitors who become engaged in the issues presented are usually just one or two clicks away from a page offering memberships or other opportunities to donate using a credit card. Web pages also provide a two-way contact channel for people who are engaged in the organization's efforts but who do not work directly for the organization—for example, many not-for-profits rely on volunteers and coordination with other organizations to accomplish their goals. This combination of information dissemination and a two-way contact channel is a key element on any successful electronic commerce Web site. For example, the **American Civil Liberties Union (ACLU)**, which is devoted to the advocacy of individual rights in the United States, includes many communication opportunities on its Web site.

The ACLU home page, shown in Figure 3-6, gives visitors an opportunity to learn about the organization and contribute money or join if their interests are piqued by what they see. The ACLU home page includes links to information about each major issue on which the ACLU has taken a position. The ACLU's Web site is especially valuable to it because the organization serves many different constituencies, not all of whom agree with the ACLU or with each other on all issues. If the ACLU were to create a print newsletter that contained interesting information for some of its supporters, that same information might offend other supporters. The Web site allows visitors to select the issues in which they are interested—and only those issues.

FIGURE 3-6 ACLU home page

Not-for-profit organizations can use the Web to stay in touch with existing stakeholders and identify new opportunities for serving them. Political parties want to offer information about party positions on issues, recruit members, keep existing members informed, and provide communication links to visitors who have questions about the party. All the major U.S. political parties have Web sites, and each year candidates running for public office set up their own Web sites. In addition, political organizations that are not affiliated with a specific party, such as the nonpartisan **Center for Responsive Politics**, also accomplish similar goals with their Web presences.

WEB SITE USABILITY

Research indicates that few businesses accomplish all of their goals for their Web sites in their current Web presences. Even sites that succeed in achieving most of these goals often fail to provide sufficient interactive contact opportunities for site visitors.

In this section, you will learn how the Web is different from other ways in which companies have communicated with their customers, suppliers, and employees in the past. You will learn how companies can improve their Web presences by making their sites accessible to more people and easier to use, and by making sure that their sites encourage visitors to trust and even develop feelings of loyalty toward the organization behind the Web site.

How the Web Is Different

Through years of trial, error, and research, firms have come to realize that doing business online differs greatly from doing business in the physical world. When firms first started creating Web sites in the mid-1990s, they often built simple sites that conveyed basic information about their businesses. Few firms conducted any market research to see what kinds of things potential visitors might want to obtain from these Web sites, and even fewer considered what business infrastructure adjustments would be needed to service the site. For example, few firms had e-mail address links on their sites. Those firms that did include an e-mail link often understaffed the department responsible for answering visitors' e-mail messages. Thus, many site visitors sent e-mail messages that were never answered.

This failure to understand how the Web is different from other presence-building media continues to be an important reason that so many businesses do not achieve their Web objectives. To learn more about this issue, see Jakob Nielsen's classic **Failure of Corporate Websites** page in the Web Links; the article was written in 1998, but still accurately describes far too many Web sites. In revisiting the issue in 2009 (see **Top 10 Information Architecture Mistakes**), Nielsen found that a surprising number of Web sites still contained the same kinds of architectural and navigational flaws that impair site visitors' ability to find information.

Most Web sites that are designed to create an organization's presence in the Web medium include links to a fairly standard information set. The site should give the visitor easy access to the organization's history, a statement of objectives or mission statement, information about products or services, financial information, and a way to communicate with the organization. Sites achieve varying levels of success based largely on how they offer this information. Presentation is important, but so is realizing that the Web is an interactive medium. The Web gives even large companies the ability to engage in two-way, meaningful communication with their customers. Companies that do not make effective use of this ability will lose customers to competitors that do.

Meeting the Needs of Web Site Visitors

Businesses that are successful on the Web realize that every visitor to their Web site is a potential customer or partner. Thus, an important concern for businesses crafting Web presences is the variation in visitor characteristics. People who visit a Web site seldom arrive by accident; they are there for a reason.

would contain the same financial information in different formats; visitors can then choose the format that best suits their immediate needs. Visitors looking for a specific financial fact might choose the HTML file so that the information appears in their Web browsers. Other visitors who want a copy of the entire annual report as it was printed would select the PDF file and either view it in their browsers or download and print the file. Visitors who want to conduct analyses on the financial data would download the spreadsheet file and perform calculations using the data in their own spreadsheet software.

To be successful in conveying an integrated image and offering information to potential customers, businesses should try to meet the accessibility goals shown in Figure 3-8 when constructing their Web sites.

Business Web sites need to:

- Offer easily accessible facts about the organization
- Allow visitors to experience the site in different ways and at different levels
- Provide visitors with a meaningful, two-way (interactive) communication link with the organization
- Sustain visitor attention and encourage return visits
- Offer easily accessible information about products and services and how to use them

FIGURE 3-8 Accessibility goals for business Web sites

Trust and Loyalty

When companies first started selling on the Web, many of them believed that their customers would use the abundance of information to find the best prices and disregard other aspects of the buying experience. For some products, this may be true; however, most products include an element of service. When customers buy a product, they are also buying that service element. A seller can create value in a relationship with a customer by nurturing the customer's trust and developing it into loyalty. Business researchers have found that a 5 percent increase in customer loyalty (measured as the proportion of returning customers) can yield profit increases of 25 to 80 percent.

Even when products are commodity items, the service element can be a powerful differentiating factor for which customers will pay extra. These services include such things as delivery, order handling, help with selecting a product, and after-sale support. Because many of these services are things that a potential customer cannot evaluate before purchasing a product, the customer must trust the seller to provide an acceptable level of service.

When a customer has a positive service experience with a seller, that customer begins to trust the seller. When a customer has multiple good experiences with a seller, that customer feels loyal to the seller. Thus, the repetition of satisfactory service can build customer loyalty, which can prevent a customer from seeking alternative sellers who offer lower prices.

Many companies doing business on the Web spend large amounts of money to obtain customers. If they do not provide levels of customer service that lead customers to develop trust in and loyalty to the firm, the companies are unlikely to recover the money they spend to attract the customers in the first place, much less earn a profit.

Customer service is a problem for many electronic commerce sites. Recent research indicates that customers rate most retail electronic commerce sites to be average or low in customer service. A common weak spot for many sites is the lack of integration between the companies' call centers and their Web sites. As a result, when a customer calls with a complaint or problem with a Web purchase, the customer service representative does not have information about Web transactions and is unable to resolve the caller's problem.

Even in the second wave of electronic commerce, e-mail responsiveness of electronic commerce sites is disappointing. Many major companies are slow to respond to e-mail inquiries about product information, order status, or after-sale problems. A significant number of companies in these studies never acknowledged or responded to the e-mail queries.

Usability Testing

An increasing number of companies are realizing the importance of usability testing, however, most companies do not perform any usability testing on their Web sites. As its name suggests, **usability testing** is the testing and evaluation of a site by its owner to ensure ease of use for site visitors. As the practice of usability testing becomes more common, more Web sites will meet the goals outlined previously in this chapter.

Experts estimate that average electronic commerce Web sites frustrate as many as half of their potential customers to the point that they leave without buying anything. Even the best sites lose many customers because the sites are confusing or difficult to use. Simple changes in site usability can increase customer satisfaction and sales. For example, some Web sites do not include telephone contact information in the belief that not staffing a call center will save the business money. However, if your customers cannot reach you, they will not continue to do business with you. Most customers will give up when they cannot communicate with you when they need to, using the medium they prefer for that communication.

Companies that have done usability tests, such as Eastman Kodak, T. Rowe Price, and Maytag, have found that they can learn a great deal about meeting visitor needs by conducting focus groups and watching how different customers navigate through a series of Web site test designs. Industry analysts agree that the cost of usability testing is so low compared to the total cost of a Web site design or overhaul that it should almost always be included in such projects. Two pioneers of usability testing are Ben Shneiderman and Jakob Nielsen. Dr. Shneiderman founded the University of Maryland Human-Computer Interaction Lab and has published a number of books on interface design. Dr. Nielsen's Alertbox Web site includes information about how to conduct usability testing and how to use its results to improve Web site design and operation. In 2011, he published an excellent summary of usability issues on the Alertbox site titled E-Commerce Usability.

Because usability testing is fairly inexpensive, many companies run usability tests periodically on their Web sites. Although user behavior is quite stable over time, Web sites evolve and are changed almost constantly. Many times these changes can affect Web site structure and navigation in unexpected and unintended ways. A regular program of usability testing can identify these issues and allow companies to resolve them before they cause user frustration and lost sales.

Customer-Centric Web Site Design

An important part of a successful electronic business operation is a Web site that meets the needs of potential customers. In the list of goals for constructing Web sites that you learned about earlier in the chapter, the focus was on meeting the needs of all site visitors (which might include customers, potential customers, investors, potential contributors for charitable organizations, business partners, suppliers, potential employees, and the general public). Putting the customer at the center of all site designs is called a **customer-centric** approach to Web site design. A customer-centric approach leads to some guidelines that Web designers can follow when creating a Web site that is intended to meet the specific needs of *customers*, as opposed to all Web site visitors. These guidelines include the following:

- Design the site around how visitors will navigate the links, not around the company's organizational structure.
- Allow visitors to access information quickly.
- Avoid using inflated marketing statements in product or service descriptions.
- Avoid using business jargon and terms that visitors might not understand.
- Build the site to work for visitors who are using the oldest browser software on the oldest computer connected through the lowest bandwidth connection—even if this means creating multiple versions of Web pages.
- Be consistent in use of design features and colors.
- Make sure that navigation controls are clearly labeled or otherwise recognizable.
- Test text visibility on a range of monitor sizes; text can become too small to read on a small monitor and so large it shows jagged edges on a large monitor.
- Check to make sure that color combinations do not impair viewing clarity for color-blind visitors.

Web sites that are designed for mobile device users should follow a few additional guidelines. These rules help accommodate the use of devices with very small screens (compared to laptop or desktop computer users) and the tendency of mobile device users to be even less patient than other Web users.

- Text should be extremely concise; there is no space for excess verbiage on a mobile device screen.
- Navigation must be clear, intuitive, and easy to see.
- The set of available functions should be limited to those likely to be used by site visitors in a mobile setting (the page can include links to the more complete, non-mobile version of the site).
- Creating a dedicated Web site for mobile users is almost always essential because the needs of mobile users are so different from those of other users.
- Conduct usability tests by having potential site users navigate several versions of the site.

Web marketing consultant Kristin Zhivago of **Zhivago Marketing Partners** has a number of recommendations for Web sites that are designed specifically to meet the needs of online customers. She encourages Web designers to create sites focused on the customer's buying process rather than the company's perspective and organization. For example, she suggests that companies examine how much information their Web sites provide and how useful that information is for customers. If the site does not provide substantial "content for your click" to visitors, they will not become customers.

Using these guidelines when you create your site can help make visitors' Web experiences more efficient, effective, and memorable. Usability is an important element of creating an effective Web presence.

USING THE WEB TO CONNECT WITH CUSTOMERS

An important element of a corporate Web presence is communicating with site visitors who are customers or potential customers. In this section, you will learn how Web sites can help firms identify and reach out to customers.

The Nature of Communication on the Web

Most businesses are familiar with two general ways of identifying and reaching customers: personal contact and mass media. These two approaches are often called **communication modes** because they each involve a characteristic way (or mode) of conveying information from one person to another (or communicating). In the **personal contact** model, the firm's employees individually search for, qualify, and contact potential customers. This personal contact approach to identifying and reaching customers is sometimes called **prospecting**. In the **mass media** approach, firms prepare advertising and promotional materials about the firm and its products or services. They then deliver these messages to potential customers by broadcasting them on television or radio, printing them in newspapers or magazines, posting them on highway billboards, or mailing them.

Some experts distinguish between broadcast media and addressable media. **Addressable media** are advertising efforts directed to a known addressee and include direct mail, telephone calls, and e-mail. Since few users of addressable media actually use address information in their advertising strategies, in this book, we consider addressable media to be mass media. Many businesses use a combination of mass media and personal contact to identify and reach customers. For example, Prudential uses mass media to create and maintain the public's general awareness of its insurance products and reputation, whereas its salespeople use prospecting techniques to identify potential customers. Once an individual becomes a customer, Prudential maintains contact through a combination of personal contact and mailings.

The Internet is a medium with unique qualities. It occupies a central space in the continuum of media choices. It is not a mass medium, even though a large number of people now use it and many companies seem to view their Web sites as billboards or broadcasts. Nor is the Internet a personal contact tool, although it can provide individuals the convenience of making personal contacts through e-mail and newsgroups. Jeff Bezos, founder of Amazon.com, described the Web as the ideal tool for reaching what he calls "the hard middle"—markets that are too small to justify a mass media campaign, yet too large to cover using personal contact. Figure 3-9 illustrates the position of the Web as a customer contact medium, located between the large markets addressed by mass media and the highly focused markets addressed by personal contact selling and promotion techniques.

Mass media
One-to-many

Seller

Sends a few carefully crafted messages to all

Thousands or millions of viewers, listeners, or readers

The Web
Many-to-one
and
many-to-many

Seller

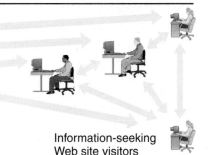

Information-seeking Web site visitors

Personal contact
One-to-one

Salesperson

Customer or prospect

FIGURE 3-9   Business communication modes

To help you better understand the differences shown in Figure 3-9, read the following scenario. The scenario assumes that you have heard about a new book, but would like to learn more about it before buying it. Consider how your information acquisition process would vary, depending on the medium you used to gather the information.

- *Mass media*: You might have been exposed to general promotional messages from book publishers that have created impressions about quality associated with particular book brands. If your existing knowledge includes a brand identity for the book's publisher, these messages might influence your perceptions of the book. You might have been exposed to an ad for the title on television, radio, or in print. You might have heard the book's author interviewed on a radio program or read a review of the book in a publication such as *The New York Times Book Review* or *Booklist* magazine. Notice that most of these process elements involve you as a passive recipient of information. This communication channel is labeled "Mass media" and appears at the top of Figure 3-9. Communication in this model flows from one advertiser to many potential buyers and thus is called a **one-to-many communication model**. The defining characteristic of the mass media promotion process is that the seller is active and the buyer is passive.

- *Personal contact*: Small-value items are not frequently sold through this medium because the costs of devoting a salesperson's efforts to a small sale are prohibitive. However, in the case of books, local bookshop owners and employees often devote considerable time and resources to developing close relationships with their customers. Although each individual book sale is a small-value transaction, people who frequent local bookshops tend to buy large numbers of books over time. Thus, the bookseller's investment in developing personal contacts is often rewarded. In this scenario, you may visit your local bookshop and strike up a conversation with a knowledgeable bookseller. In the personal contact model, this would most likely be a bookseller with whom you have already established a relationship. The bookseller would offer an opinion on the book based on having read that book, books by the same author, or reviews of the book. This opinion would be expressed as part of a two-way conversational interchange. This interchange usually includes a number of conversational elements (small talk, such as discussions about the weather, local sports, or politics) that are not directly related to the transaction you are considering. These other interchanges are part of the trust-building and trust-maintaining activities that businesses undertake to develop the relationship element of the personal contact model. The underlying **one-to-one communication model** appears at the bottom of Figure 3-9 and is labeled "Personal contact." The defining characteristic of information gathering in the personal contact model is the wide-ranging interchange that occurs within the framework of an existing trust relationship. Both the buyer and the seller (or the seller's representative) actively participate in this exchange of information.

- *The Web*: To obtain information about a book on the Web, you could search for Web site references to the book, the author, or the subject of the book. You would likely identify a number of Web sites that offer such information. These sites might include those of the book's publisher, firms that sell books on the Web, independent book reviews, or discussion groups focused on the book's author or genre. *The New York Review of Books* and *Booklist* magazine, both staples of mass media book promotion, are available online. Book review sites that did not originate in a print edition, such as *BookBrowse*, also appear on the Web. Most online booksellers maintain searchable space on their sites for readers to post reviews and comments about specific titles. If the author of the book is famous, there might even be independent Web fan sites devoted to him or her. If the book is about a notable person, incident, or time period, you might find Web sites devoted to those notable topics that include reviews of books related to the topic. You could examine any number of these resources to any extent you desired. You might encounter some advertising material created by the publisher while searching the Web. However, if you choose not to view the publisher's ads, you will find it as easy to click the Back button on your Web browser as it is to surf television channels with your remote control. The Web affords you many communication channels. Figure 3-9 shows only one of the communication models that can occur when using the Web to search for product information. The model labeled "The Web" in Figure 3-9 is the **many-to-one communication model**. The Web gives you the flexibility to use a one-to-one model (as in the personal contact model) in which you communicate over the Web with an individual working for the seller, or engage in **many-to-many communications** with other potential buyers. The defining characteristic of a product information search on the Web is that the buyer actively participates in the search and controls the length, depth, and scope of the search.

Summary

In this chapter, you learned that businesses are using six main approaches to generate revenue on the Web: the Web catalog, digital content sales, advertising-supported, advertising-subscription mixed, fee-for-transaction, and fee-for-service models. You learned how these models work and what kinds of businesses use which models. You also learned that some companies have changed models as they learned more about their customers and the business environment in which their Web sites operate.

Companies sometimes face the challenges of channel conflict and cannibalization either within their own organizations or with the companies that have traditionally provided sales distribution to consumers for them. In accordance with the network model of organization that you learned about in Chapter 1, companies undertaking electronic commerce initiatives sometimes form strategic alliances with other companies to obtain their skills in Web site operation.

By understanding how the Web differs from other media and by designing a Web site to capitalize on those differences, companies can create an effective Web presence that delivers value to visitors. Every organization must anticipate that visitors to its Web site arrive with a variety of expectations, prior knowledge, and skill levels, and are connected to the Internet through a range of technologies. Knowing how these factors can affect the visitor's ability to navigate the site and extract information from the site can help organizations design better, more usable Web sites. Enlisting the help of users when building test versions of the Web site is also a good way to create a Web site that represents the organization well.

Firms must understand the nature of communication on the Web so they can use it to identify and reach the largest possible number of customers and qualified prospects. Using a many-to-one communication model enables Web sites to effectively reach potential customers.

Key Terms

Account aggregation

Addressable media

Advertising-subscription mixed revenue model

Advertising-supported revenue model

Bill presentment

Cannibalization

Catalog model

Channel conflict

Channel cooperation

Communication modes

Customer-centric

Demographic information

Digital content revenue model

Digital Rights Management (DRM)

Disintermediation

Fee-for-service revenue model

Fee-for-transaction revenue model

Mail-order

Many-to-many communications

Many-to-one communication model

Marketing channel

Mass media

One-to-many communication model

One-to-one communication model

Pay wall

Personal contact

Personal shopper

Portal

Presence

Prospecting

Reintermediation

Stakeholders

Stickiness

Sticky

Usability testing

Virtual model

Web catalog revenue model

Web directory

Web portal

Review Questions

1. Write a paragraph in which you outline the problems that a Web site with a narrow focus that is directed at a small audience could face if it were to rely exclusively on advertising revenue.

2. Define the term "stickiness" as used in electronic commerce. In one or two paragraphs, explain why it is important for advertising-supported Web sites to have this characteristic.

3. Define "channel conflict" and describe in one or two paragraphs how a company might deal with this issue.

4. In two paragraphs, explain why a customer-centric Web site design is so important, yet is so difficult to accomplish.

5. Define the term "presence," and then write about 100 words in which you explain why companies that do business on the Web must be more concerned about presence than firms that operate only in the physical world. Be sure to include a discussion of the characteristics that differ between online and offline presences.

6. Many businesses offer free samples of their products or services to potential customers to induce them to become customers. Write a paragraph in which you describe how this strategy can be implemented online. Be sure to note how the amount of sampling that is likely to be beneficial differs in the online environment and the physical world.

Exercises

1. Assume you are in the market for an Android-based tablet device to use at school. List five features or characteristics that would be important for you to have in such a device. Using your list, write about 100 words on whether you would be most likely to find a product that meets your specific needs by shopping online or in a physical store. If you believe that shopping both online and in person would be a good idea for you, explain which characteristics or features you would evaluate as you shopped online versus in person.

2. You have been hired as a consultant by your local town newspaper, the *Midland Clarion*. Midland is a small town of about 10,000 residents. The *Clarion* has seen its paid subscription list decline steadily over the past 10 years, but the number of visitors to the newspaper's Web site, which includes all of the stories that run in the paper, has steadily increased over the same time period. The *Clarion's* publisher is considering charging an annual subscription fee for access to the editorials and columns on the Web site. The classified and display advertising, along with short summaries of news stories, would continue to be available at no cost. Prepare a report of about 300 words in which you outline the advantages and disadvantages of this advertising-subscription mixed revenue model for the *Clarion*.

3. High-end jewelry retailers such as Cartier, Harry Winston, and Tiffany often use Adobe's Flash software to create their Web sites. In about 200 words, present three arguments for and three arguments against the use of Flash animations in sites such as these. Consider the retailers' objectives, the characteristics of the products being sold, and the type of customers who visit these sites.

4. Prepare a report in which you evaluate the usability of two Web sites that sell GPS navigation devices. A list of links to companies that sell this type of product is included in the Web Links for this exercise, but you may use other sites if you wish. In your report, evaluate how easy it is to learn about and purchase the product. Your report should include a section of about 100 words in which you describe the criteria you used in your evaluation (such as a feature comparison tool, ease of determining shipping costs, clarity and usefulness of product photographs or illustrations, and so on), a section of about 200 words that

summarizes your findings, and a section of about 100 words in which you present your conclusion.

5. Visit the Web sites of two museums. The Web Links for this exercise include a list of links to museums, but you may use other sites if you wish. Write a report of 100 words in which you describe the process on each site for making a donation to the museum. Evaluate how the site's design encourages visitors to donate; for example, is it easy to find a link to a page where you can make a donation? Does the site encourage donations throughout the site's various pages? Provide at least one recommendation for improving each of the two sites you selected.

6. Many real estate agents today have Web sites that list the properties they have for sale. These agents also advertise the properties on Realtor.com and sometimes in television ads. However, most real estate agents would tell you that personal contact provides their most important connections with clients, potential clients, and client referral sources. Write three paragraphs in which you briefly describe the things that real estate agents can best accomplish through (1) their Web sites, (2) mass media advertising, and (3) personal contact.

Cases

C1. Lonely Planet

In 1972, Tony and Maureen Wheeler were newlyweds who decided to have one last adventurous travel experience before settling down. Their trip was an overland trek from London to Australia through Asia. So many other travelers asked them about their experiences that they sat down at their kitchen table and wrote a book titled *Across Asia on the Cheap*. They published the book themselves and were surprised by how many copies they sold. More than three decades and 60 million books later, their publishing enterprise has turned out to be one of the most successful in history.

The Wheelers' publishing company, Lonely Planet, has grown rapidly, with typical annual sales increases of 15 percent or more. In 2007, BBC Worldwide purchased a 75 percent ownership interest in the company and purchased the rest of the company's stock in 2011. Lonely Planet TV now produces a variety of travel and documentary programs that appear on cable networks throughout the world. As a BBC subsidiary, the company does not release sales figures, but industry analysts estimate current annual revenues to be about $110 million. Lonely Planet publishes more than 600 titles and holds a 20 percent share of the travel guide market. The company has more than 450 employees in its U.K., U.S., French, and Australian offices performing editorial, production, graphic design, and marketing tasks. Travel guide content is written by a network of more than 200 contract authors in more than 20 countries. These authors are knowledgeable about everything from visa regulations to hotel prices to the names of the hottest new entertainment spots. The combined expertise of the in-house staff and the in-country authors has kept Lonely Planet ahead of its competitors for many years.

Lonely Planet also offers travel services that include a phone card, hotel and hostel room-booking, airplane tickets, European rail travel reservations and tickets, package tours, and travel insurance. These services are sold by telephone and on the Lonely Planet Web site.

The Web site has won numerous awards, including the Society of American Travel Writers Silver Award and a spot on *Time* magazine's "Fifty Best Web Sites" list. The site was launched in 1994 and includes an online store in which Lonely Planet publications are sold. However, the site's main draws are its comprehensive collection of information about travel destinations and its online discussion area, the Thorn Tree, which has nearly a half million registered users. The company has had trouble turning any of this information into a source of revenue generation.

Despite its excellent Web site and its use of new technologies, most of Lonely Planet's revenues are still generated by book sales. The typical production cycle of a travel guide is about eight months long. This is the time it takes to commission authors, conduct research, work through several drafts of writing and editing, select photos, create the physical book, and print it. This production cycle causes new books to be almost a year out of date by the time they are published. Only the most popular titles are revised annually. Other titles are on two-, three-, or four-year revision cycles. The time delay in publication means that many details in the guides are outdated or wrong; restaurants and hotels close (or move), exchange rates and visa regulations change, and once-hot night spots are abandoned by fickle clientele.

Lonely Planet publications are well researched and of high quality, but the writers do not work continually because the books are not published continually. The Web site often has information that is more current than the published travel guides.

The site's online shop does offer some custom guides, which are parts of its existing travel guides packaged in different ways, and it does let customers buy specific chapters from its books, but it still is largely focused on selling books, although the site does offer PDF files that can be downloaded to mobile devices. Lonely Planet has adopted some new technologies, but has not used them to change its revenue model in any major way or to make basic changes in the production of its main product, the travel guides.

Required:

1. Review the company's offerings for Apple iPhone and iPad products and for Android smart phones (Trippy). Evaluate those products and identify opportunities for other products or services that the company could offer for mobile devices that would take advantage of Internet technologies (including wireless technologies for mobile devices) and address customers' concerns about the timeliness and currency of information in the printed travel guides.

2. Prepare a report in which you analyze the marketing channel conflicts and cannibalization issues that Lonely Planet faces as it is currently operating. Suggest solutions that might reduce the revenue losses or operational frictions that result from these issues.

3. Many loyal Lonely Planet customers carry their travel guides (which can be several hundred pages thick) with them as they travel around the world. In many cases, these customers do not use large portions of the travel guides. Also, Internet access can be a problem for many of these customers while they are traveling. Describe a digital product (or products) other than the PDFs of book chapters it currently offers that might address this customer concern and also yield additional revenue for Lonely Planet. Your answer here could build on ideas that you developed in your solution to Requirement 1.

Note: Your instructor might assign you to a group to complete this case and might ask you to prepare a formal presentation of your results to your class.

C2. Association for the Study of International Business

The Association for the Study of International Business (ASIB) is an organization of researchers, professors, and business executives interested in the study, analysis, and promotion of business activities beyond domestic borders. Mario DiPonetti, ASIB's executive director, has hired you as a consultant to help him map out a future Web revenue strategy for the association.

The ASIB has about 3000 members located in countries throughout the world; however, about half of its members are in the United States. Each member pays an annual membership

Crawford, W. 2004. "Keeping the Faith: Playing Fair with Your Visitors," *EContent*, 27(4), September, 42–43.

Cyr, D., M. Head, H. Larios, and B. Pan. 2009. "Exploring Human Images in Website Design: A Multi-method Approach," *MIS Quarterly*, 33(3), 539–575.

Demery, P. 2011. "Training, Technology, and Teamwork Help E-retailers Derive More Sales and Profits from Live Chat," *Internet Retailer*, November, 14–16.

Doonar, J. 2004. "It's Not Such a Lonely Planet," *Brand Strategy*, January, 24–25.

The Economist. 2010. "Charging for Content: Media's Two Tribes," 396(8689), July 3–9, 63.

Egol, M., H. Hawkes, and G. Springs. 2009. "Reinventing Print Media," *strategy+business*, 56, Autumn, 80–83.

Enright, A. 2011. "Classy Examples: Luxury Brands Show How to Sell High-ticket Items Online and Build Trust," *Internet Retailer*, May 31. (http://www.internetretailer.com/2011/05/31/classy-examples)

Greenstein, S. and M. Devereux. 2006. *The Crisis at Encyclopaedia Britannica. Kellogg School of Management Case 5-306-504*. Evanston, IL: Northwestern University.

Gupta, S. and C. Mela. 2009. "What Is a Free Customer Worth?" *Harvard Business Review*, 86(11), 102–109.

Holmes, E. 2009. "CBS's TV.com Boosts Offerings in Bid to Secure Foothold," *The Wall Street Journal*, January 12, B3.

Jones, K., L. Leonard, and C. Riemenschneider. 2009. "Trust Influencers on the Web," *Journal of Organizational Computing & Electronic Commerce*, 19(3), 196–213.

Kemp, T. 2000. "Wal-Mart No Web Mart," *InternetWeek*, October 9, 1–2.

Leski, M. 2011. "Reading: From Paper to Pixels," *IEEE Security & Privacy*, 9(4), July-August, 76–79.

McCoy, A. 2008. "Reel Estate: Downloads Are Changing the Movie Rental Landscape," *Pittsburgh Post-Gazette*, February 6. (http://www.post-gazette.com/pg/08037/854979-42.stm)

Medical Economics. 2009. "Website to Offer Online Visits Nationwide," August 7, 18.

Miller, C. and J. Bosman. 2011. "E-books Outsell Print Books at Amazon," *The New York Times*, May 19. (http://www.nytimes.com/2011/05/20/technology/20amazon.html)

Netherby, J. 2009. "Zucker has Hulu Profit in Sight," *Video Business*, June 1, 3, 21.

Nicholls, J. 2011. "Perusing Google eBookstore," *Collection Management*, 36(2), March, 131–136.

Nielsen, J. 1999. *Designing Websites With Authority: Secrets of an Information Architect*. Indianapolis, IN: New Riders.

Nielsen, J. 2000. "Flash: 99% Bad," *Alertbox*, October 29. (http://www.useit.com/alertbox/20001029.html)

Nielsen, J. 2001. "Usability Metrics," *Alertbox*, January 21. (http://www.useit.com/alertbox/20010121.html)

Nielsen, J. 2011. "E-commerce Usability," *Alertbox*, October 24. (http://www.useit.com/alertbox/ecommerce.html)

Nielsen, J., K. Coyne, and M. Tahir. 2001. "Make It Usable," *PC Magazine*, 20(3), February 6, IPO1–IPO6.

Nielsen, J. and M. Tahir. 2002. *Homepage Usability: 50 Websites Deconstructed*. Indianapolis, IN: New Riders.

Nielsen Norman Group. 2011. *Non-profit and Charity Website Usability: 116 Design Guidelines*. Fremont, CA: Nielsen Norman Group.

Palvia, P. 2009. "The Role of Trust in E-commerce Relational Exchange: A Unified Model," *Information & Management*, 46(4), 213–220.

Pegoraro, R. 2005. "Priorities for the Store-Shopping List," *The Washington Post*, August 28, F1.

Pérez-Peña, R. 2007. "Times to End Charges on Web Site," *The New York Times*, September 18. (http://www.nytimes.com/2007/09/18/business/media/18times.html)

Peters, J. 2011. "Times' Online Pay Model Was Years in the Making," *The New York Times*, March 20. (http://www.nytimes.com/2011/03/21/business/media/21times.html)

Rayport, J. and J. Sviokla. 1995. "Exploiting the Virtual Value Chain," *Harvard Business Review*, 73(6), November–December, 75–85.

Rueter, T. 2011. "Home Depot Enables Online Shoppers to Pick Up Purchases Inside Stores," *Internet Retailer*, September 2. (http://www.internetretailer.com/2011/09/02/home-depot-enables-online-shoppers-pick-items-stores)

Sanderfoot, A. and C. Jenkins. 2001. "Content Sites Pursue Fee-Based Model," *Folio: The Magazine for Magazine Management*, 30(6), 15–16.

Schiller, K. 2011. "Google Opens eBookstore," *Information Today*, 28(1), January, 8.

Schwartz, E. 1997. *Webonomics*. New York: Broadway Books.

Schwartz, E. 1999. *Digital Darwinism*. New York: Broadway Books.

Seelye, K. 2005. "Why Newspapers Are Betting on Audience Participation," *The New York Times*, July 4, C2.

Shneiderman, B. 1997. *Designing the User Interface: Strategies for Effective Human-Computer Interaction.* Reading, MA: Addison-Wesley.

Sklar, J. 2009. *Principles of Web Design, Fourth Edition*. Boston, MA: Course Technology.

Smith, E. 2008. "Napster to Sell Downloads for Most Music Players," *The Wall Street Journal*, January 7, B2.

Spira, J. 2011. "Internet TV: Almost Ready for Prime Time," *IEEE Spectrum*, 48(7), July, 24–26.

Stambor, Z. 2011. "Customer Service: Video and Chat Help E-retailers Get Personal With Customers," *Internet Retailer*, June 30. (http://www.internetretailer.com/2011/06/30/customer-service)

Steel, E. 2007. "Job-Search Sites Face a Nimble Threat; Online Boards Become Specialized, Challenging Web-Print Partnerships," *The Wall Street Journal*, October 9, B10.

Stone, B. 2008. "Netflix Partners With LG to Bring Movies Straight to TV," *The New York Times*, January 3. (http://www.nytimes.com/2008/01/03/technology/03netflix.html)

Stross, R. 2011. "The Therapist Will See You Now, Via the Web," *The New York Times*, July 9. (http://www.nytimes.com/2011/07/10/technology/bringing-therapists-to-patients-via-the-web.html)

Tedeschi, B. 2005. "New Era of Ticket Resales: Online and Aboveboard," *The New York Times*, August 29, C4.

Tian, X. and B. Martin. 2011. "Impacting Forces on eBook Business Models Development," *Publishing Research Quarterly*, 27(3), 230–246.

Trachtenberg, J. 2007. "Borders Business Plan Gets a Rewrite," *The Wall Street Journal*, March 22, B1–B2.

Weingarten, M. 2001. "Flash Backlash," *The Industry Standard*, March 5. (http://www.thestandard.com/article/0,1902,22330,00.html)

Weiss, T. 2000. "Walmart.com Back Online After Four-Week Overhaul," *Computerworld*, 34(45), November 6, 24.

Williams, T. 2005. "NYTimes.com to Offer Subscription Service," *The New York Times*, May 17, C5.

Wu, J. 2011. *Global Recorded Music Market Forecast*. Boston: Strategy Analytics.

Zeitchik, S. 2003. "New Worlds at Lonely Planet," *Publishers Weekly*, 250(25), June 23, 12.

Zimmerman, A. 2000. "Wal-Mart Launches Web Site for a Third Time, This Time Emphasizing Speed and Ease," *The Wall Street Journal*, October 31, B12.

CHAPTER 4

MARKETING ON THE WEB

INTRODUCTION

In September 1997, a new gift shop opened for business on the Web. There were already many gift shops on the Web at that time; however, this store, named 911Gifts.com, carried items that were chosen specifically to meet the needs of last-minute gift shoppers. Including 911—the emergency telephone number used in most parts of the United States—in the store's name was intended to convey the impression of crisis-solving urgency. The company's two major strengths were its promise of next-day delivery on all items and its site layout, in which gift selections were organized by holiday rather than by product type. A harried shopper could simply click the Mother's Day gifts link and view a set of gift choices appropriate for that holiday that were ready for immediate delivery. The site also included a reminder service to help its customers avoid another emergency gift situation on the next holiday.

By 1999, the company was doing about $1 million in annual sales. It carried about 500 products, and each of the products was chosen to yield a gross margin of at least 40 percent. 911Gifts.com was successful, but the company's founders wanted to build wider awareness of their brand. They realized that building a brand would require a substantial investment of funds and skills beyond what they had in the company at that time. Thus, they hired Hilary Billings, who had built the Pottery Barn catalog business at Williams-Sonoma, to create a brand-building strategy and obtain financing to implement that strategy. Billings completely overhauled the 911Gifts.com marketing plan. She used the new marketing plan to bring in more than $30 million from investors for a rebranding and total redesign of the company's Web site. In October 1999, the new brand was launched as **RedEnvelope**, named after an Asian tradition of enclosing gifts of money in a simple red envelope. The new brand was intended to convey a sense of elegant simplicity rather than the feeling of panic and emergency solutions conveyed by the old brand name.

The product line was updated to fit the new image. About 300 products were dropped and replaced with different products that focus groups had judged to be more appealing. The new product line had a higher average gross margin than the old line. Billings launched a massive brand-awareness campaign that included online advertising, buses in seven major cities painted red and festooned with large red bows, and print advertising in upscale publications. The most important change in advertising strategy was the launch of a print catalog. RedEnvelope catalogs are mailed to customers to coincide with major gift-giving holidays and serve as additional reminders. Because RedEnvelope sells a small set of products that are chosen for their visual appeal and for the status they are intended to convey, the full-color, lushly illustrated print catalogs are a powerful selling tool.

One year later, the results of this extensive makeover and substantial monetary investment were clear. RedEnvelope had tripled its number of customers and had increased sales by more than

400 percent. The company chose a specific part of the gifts market and targeted its offerings to meet the needs and desires of those customers. The company created a brand, a marketing plan, and a set of advertising and promotion strategies that would expose the company to the largest portion of that market it could afford to reach. The most important point is that RedEnvelope matched its inventory selection, delivery methods, and marketing efforts to each other and to the needs of its customers.

Since 2008, RedEnvelope has been part of Provide Commerce, a company that operates online gift businesses such as ProFlowers and Shari's Berries. The company continues to use print catalogs and a focus on upscale product lines to keep its sales increasing each year. Marketing an online business often requires the use of a combination of marketing techniques that sometimes include traditional approaches such as print catalogs.

WEB MARKETING STRATEGIES

In this chapter, you will learn how companies are using the Web in their marketing strategies to advertise their products and services and promote their reputations. Increasingly, companies are classifying customers into groups and creating targeted messages for each group. The sizes of these targeted groups can be smaller when companies are using the Web—in some cases, just one customer at a time can be targeted. New research into the behavior of Web site visitors has even suggested ways in which Web sites can respond to visitors who arrive at a site with different needs at different times. This chapter will also introduce you to some of the ways companies are making money by selling advertising on their Web sites.

Most companies use the term **marketing mix** to describe the combination of elements that they use to achieve their goals for selling and promoting their products and services. When a company decides which elements it will use, it calls that particular marketing mix its **marketing strategy**. As you learned in Chapter 3, companies—even those in the same industry—try to create unique brand presences in their markets. A company's marketing strategy is an important tool for conveying its branding and advertising messages to current and prospective customers. A company's Web presence is an element of that marketing strategy.

The Four Ps of Marketing

Most marketing classes organize the essential issues of marketing into the **four Ps of marketing**: product, price, promotion, and place. **Product** is the physical item or service that a company is selling. Elements such as quality, design, features, characteristics, and even the packaging make up the product. These intrinsic characteristics of the product are important, but customers' perceptions of the product, called the product's **brand**, can be as important as the actual characteristics of the product.

The **price** element of the marketing mix is the amount the customer pays for the product. In recent years, marketing experts have argued that companies should think of price in a broader sense, that is, the total of all financial costs that the customer pays (including transaction costs) to obtain the product. This total cost is subtracted from the benefits that a customer derives from the product to yield an estimate of the **customer value** obtained in the transaction. Later in this book, you will learn how the Web can create new opportunities for creative pricing and price negotiations through online auctions, reverse auctions, and group buying strategies. These Web-based opportunities are helping companies find new ways to create increased customer value.

Promotion includes any means of spreading the word about the product. It requires decisions about advertising, public relations, personal selling, and overall promotion of the product. On the Internet, possibilities abound for communicating with existing and potential customers. In this chapter, you will learn how organizations use their Web sites, e-mail strategies, and social media as communication tools for promoting their products and services.

For years, marketing managers dreamed of a world in which instant deliveries would give all customers exactly what they wanted when they wanted it. The issue of **place** (also called **distribution**) is the need to have products or services available in many different locations. The problem of getting the right products to the right places at the best time to sell them has plagued companies since commerce began. Although the Internet does not solve all of these logistics and distribution problems, it can certainly help. For example, digital products (such as information, news, software, music, video, and e-books) can be delivered almost instantly through the Internet. Companies that sell products that must be shipped have found that the Internet gives them much better shipment tracking and inventory control tools than they have ever had before. Figure 4-1 depicts the components of the four Ps of marketing and shows their contributions to overall marketing strategy.

Product
Quality
Design
Features
Branding
Packaging
Customer perception

Price
Value to customer
Price of competing products
Customer price sensitivity
Discounts
Differential pricing

Marketing Strategy

Promotion
Advertising
Public relations
Personal selling
Online communications

Place
Distribution channels
Market coverage
Logistics
Inventory management

© Cengage Learning

FIGURE 4-1 The four Ps of marketing contribute to marketing strategy

Product-Based Marketing Strategies

In Chapter 3, you learned about the importance of a company's Web presence and how this presence must integrate with the brand or other established images the company uses in its promotional activities. Most companies offer a variety of products that appeal to different groups. When creating a marketing strategy, managers must consider both the nature of their products and the nature of their potential customers.

Managers at many companies think of their businesses in terms of the products and services they sell. This **product-based structure** is a logical way to think of a business because companies spend a great deal of effort, time, and money to design and create those products and services. If you ask managers to describe what their companies are selling, they usually provide you with a detailed list of the physical objects they sell or use to create a service. When customers are likely to buy items from particular product categories, or are likely to think of their needs in terms of product categories, this type of product-based organization makes sense. Most office supplies stores on the Web believe their customers think of their needs using a product category structure. For example, both the **Home Depot** and **Staples** use product categories as a very strong organizing theme in the design of their Web sites.

Many other online businesses use a similar product-based marketing strategy. **Sears**, a company that sold its products through catalogs and later in physical stores for many years before adding online sales, uses a product-based structure on its Web site. Most companies that used print catalogs in the past organized them by product category, and this design theme has carried over into many of their Web sites.

Many retailers that began as catalog-based businesses organize their Web sites from an internal viewpoint, that is, according to the way that they arranged their product design and manufacturing processes. If customers arrive at these Web sites looking for a specific type of product, this approach works well. Alternatively, customers who are looking to fulfill a specific need, such as outfitting a new office or choosing a graduation gift, rather than find a specific product, might not find these Web sites as useful.

Many marketing researchers and consultants advise companies to think as if they were their own customers and to design their Web sites so that customers find them to be enabling experiences that can help customers meet their individual needs. Sometimes this requires the Web site to offer alternative shopping paths. For example, an online florist's Web site could allow customers to specify an arrangement that includes specific flowers or colors (satisfying customers with a desire for a specific product), yet provide a separate shopping path for customers who want to buy an arrangement for a specific occasion (birthday, anniversary, Mother's Day, and so on). Similarly, toy sites provide users with filtering options so they can select price range, type of toy, recipient age range, cost, and so on.

Customer-Based Marketing Strategies

In Chapter 3, you learned that the Web creates an environment that allows buyers and sellers to engage in complex communications modes. The communication structures on the Web can become much more complex than those in traditional mass media outlets such as broadcast and print advertising. When a company takes its business to the Web, it can create a Web site that is flexible enough to meet the needs of many different users. Instead of thinking of their Web sites as collections of products, companies can build their sites to meet the differing needs of various types of customers.

A good first step in building a customer-based marketing strategy is to identify groups of customers who share common characteristics. Creating a Web site that acknowledges those groups and treats each differently can make the site more accessible and useful to each group. This is difficult to do because most organizations think about their Web sites as models of their activities, which they view from an internal perspective. For example, early university Web sites were often organized around the internal elements of the school (such as departments, colleges, and programs) in an implementation of their own internal perspective. Today, most university home pages include links to separate sections of the Web site designed for specific stakeholders, such as current students, prospective students, parents of students, potential donors, and faculty. This construction reflects the external perspective of each different user group that might use the Web site.

COMMUNICATING WITH DIFFERENT MARKET SEGMENTS

Identifying groups of potential customers is just the first step in selling to those customers. An equally important component of any marketing strategy is the selection of communication media to carry the marketing message.

In the physical world, companies can convey large parts of their messages by the way they construct buildings and design their floor spaces. For example, banks have traditionally been housed in large, solid-looking buildings that provide passersby an ample view of the main safe and its thick, sturdy door. Banks use these physical manifestations of reliability and strength to communicate an important part of their service offerings—that a customer's money is safe and secure with the bank.

Media selection, or choosing where to market and advertise a company, can be critical for an online firm because it does not have a physical presence. The only contact a potential customer might have with an online firm could be the image it projects through the media and through its Web site. The challenge for online businesses, especially new online businesses, is to convince customers to trust them even though they do not have a physical presence.

Trust, Complexity, and Media Choice

As you learned in Chapter 3 (see Figure 3-9), the Web provides a communication mode that is an intermediate step between mass media and personal contact, but it is a very broad step. Using the Web to communicate with potential customers offers many of the advantages of personal contact selling and many of the cost savings of mass media. Figure 4-2 shows how these three information dissemination modes compare on the important dimensions of trust and product (or service) complexity.

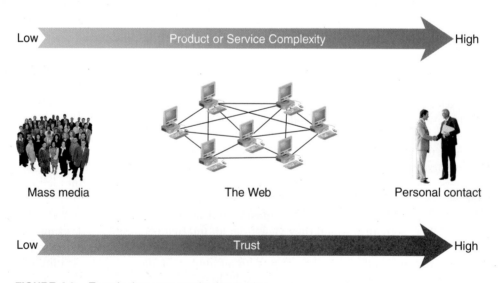

FIGURE 4-2   Trust in three communication modes

Although mass media offers the lowest level of trust, many companies continue to use it successfully. The cost of mass media advertising can be spread over the many people in its large audiences. For example, the cost of creating and running a television ad can be millions of dollars, but that ad will be viewed by millions of people. Thus, the cost of advertising per viewer is very low. Its low cost per viewer makes mass media advertising attractive to many companies.

In 2009, Ford Motor Company shifted a significant portion of its advertising budget from traditional car ads that focused on new vehicle features to ads that told stories about how Ford managers are running the company more intelligently than its competitors. Ford's two major U.S.-based competitors, Chrysler and General Motors, both needed taxpayer bailout money to survive the global economic recession. Ford was able to use mass media advertising to draw a sharp contrast between itself and its competitors in the minds of millions of potential customers. The message was straightforward and could be delivered to all of Ford's customers (and potential customers) using the same language and images. Thus, mass media advertising was an ideal choice for delivering this message.

After years of being barraged by television and radio commercials, many people have developed a resistance to the messages conveyed in mass media. The impact on an audience of the shouted expression "New and improved!" is very low. The overuse of superlatives has caused many people to distrust or ignore much mass media. Television remote controls have mute buttons and make channel surfing easy for a reason. Attempts to re-create mass media advertising on the Web are likely to fail for the same reasons—many people ignore or resist messages that lack content of any specific personal interest to them.

Mass media advertising campaigns that are successful often rely on the passive nature of the media consumption experience. People watching television or listening to radio are usually in a passive and receptive state of mind. Thus, advertisers can include messages in mass media advertising that recipients might not consider valid or convincing if they were actively evaluating those statements. The messages are accepted by recipients because they are in a nonquestioning and passive state of mind. In contrast, Web users are actively engaged in the medium, with hands on the keyboard and mouse, as they view Web pages. This active state of mind makes Web users far more likely to evaluate critically the advertising messages they see and less likely to accept the content of those messages in the same passive way that television viewers accept the content of television commercials.

The level of complexity inherent in the product or service is also an important factor in media choice. Products that have few characteristics or that are easy to understand can be promoted well using mass media. Because mass media is expensive to produce, most companies use it to deliver short messages (although there are exceptions, such as infomercials). Highly complex products and services are best promoted through personal contact, which allows the potential customer to ask clarifying questions during the promotional presentation.

The Web occupies a wide middle ground and can be used for delivering short but focused messages that promote, but it can also be used to deliver longer and more complex messages. The Web can even be used to engage the potential customer in a back-and-forth dialog similar to that used in personal contact selling. Most important, a properly designed Web site can give potential customers the ability to choose their level of interaction. A company can present a mass media type of message that a site visitor can click to access a more detailed message. If the visitor still wants more information, the site can offer the opportunity for interactive communication (such as an online chat) with a customer service representative. Thus, the Web can offer elements of mass media messaging, personal contact interaction, and anything in between.

Companies can use the Web to capture some of the benefits of personal contact, yet avoid some of the costs inherent in that approach. One way to do this is to use some of the new communication tools that the Internet provides. For example, people can post their thoughts on a Web site and invite others to add commentary. Individuals have used this type of Web site, known as a **Web log** or **blog**, as an outlet for expressing their political, religious, and other strongly felt beliefs. Today, many companies use blogs as a communication device. For example, retailers use blogs to give their online stores a personality and provide customers with a reason to visit their Web sites even if they are not shopping.

Another way that companies can develop an involvement among their customers is to use social media to create discussions about new products, promotions, and even advertising campaigns. **Social media** is a general term for Web sites such as Facebook or Google+ and online communication technologies such as Twitter that allow participants to exchange ideas and report news and information updates to each other. You will learn more about social media and mobile device (including smart phones and tablets) applications that many people use with social media in Chapter 6.

Blogs and social media provide ways for companies to engage in two-way online communications that more closely resemble the high-trust personal contact mode of communication than the low-trust mass media mode. They also allow companies to achieve these benefits without incurring the high cost of traditional personal contact techniques.

Market Segmentation

Companies' responses to the decrease in advertising effectiveness were to identify specific portions of their markets and target them with specific advertising messages. This practice, called **market segmentation**, divides the pool of potential customers into groups, or **segments**. Segments are usually defined in terms of customer characteristics such as age, gender, marital status, income level, and geographic location. Thus, for example, unmarried men between the ages of 19 and 25 might be one market segment.

In the early 1990s, firms began identifying smaller and smaller market segments for specific advertising and promotion efforts. This practice of targeting very small market segments is called **micromarketing**. However, the low cost per viewer of traditional mass media advertising campaigns becomes much higher when mass media methods are used to target very small market segments. This cost increase hampered the success of micromarketing strategies. Even though micromarketing was an improvement over mass media advertising, it still used the same basic approach and suffered from the weaknesses of that approach.

Marketers have traditionally used three categories of variables to identify market segments. One variable is location. Firms divide their customers into groups by where they live or work. In this type of segmentation, called **geographic segmentation**, companies create different combinations of marketing efforts for each geographical group of customers. The grouping can be by nation, state (or province), city, or even by neighborhood. Alternatively, companies can develop one marketing strategy for urban customers, another for suburban customers, and yet a third for rural customers.

The second category uses information about age, gender, family size, income, education, religion, or ethnicity to group customers. This type of segmentation is called **demographic segmentation**. Demographic variables are frequently used by traditional marketers because research has shown that customers' need for and usage of products are strongly related to these types of variables. Demographic segmentation also exists on the Web. For example, a number of sites are devoted to women's issues or directed at specific age groups (such as teenagers) whose members tend to download music and purchase trendy clothing or video games. Often, demographic and geographic segmenting strategies are combined. For example, an airline might target middle-income families living in Wisconsin and Michigan with midwinter advertising for vacation trips to Florida.

In **psychographic segmentation**, marketers try to group customers by variables such as social class, personality, or their approach to life. For example, an auto company might direct advertising for a sports car to customers who are gregarious and have a high need for achievement. The use of psychographic segmentation has increased dramatically in recent years as marketers attempt to identify characteristic lifestyles and then design advertising to reach people who see themselves as having a particular lifestyle.

Companies that advertise on television often create messages designed to reach the likely audiences of various types of programs. These audiences represent one or more market segments. The market segments can be geographic, demographic, psychographic, or a combination of these. Figure 4-3 presents some examples from the television medium that show how companies do this.

Type of television program	Type of advertising
Children's cartoons	Children's toys and games
Daytime dramas	Household and laundry goods, pet foods
Late-night talk shows	Snack foods and nonprescription sleep aids
Golf tournaments	Golf equipment, investment services, and life insurance
Baseball and football games	Snack foods, beer, autos
Documentary films	Books, CDs, educational DVDs

FIGURE 4-3   Television advertising messages tailored to program audience

Children's television shows are likely to feature advertising for products that appeal to children. Ads on daytime dramas are directed at people who are home during the day and who thus might be interested in household and laundry care products. These people are more likely than others to own pets, so they also will see ads for pet foods. Advertisers on late-night talk shows often direct their ads at people who might have trouble falling asleep. Advertisers also believe that this late-night audience is receptive to promotions for snack foods to eat while watching these programs or for nonprescription medications for ailments that might be keeping them up so late.

Advertisers use sports programming as a vehicle for two different market segments. Some sports shows, such as golf tournaments or tennis matches, appeal to higher-income viewers. Other sports shows, such as baseball or football game broadcasts, appeal to viewers with more moderate incomes. As a result, programs that cover golf or tennis are more likely to include ads for investment and insurance products and luxury automobiles than are baseball or football programs. Also, because viewers of golf tournaments and tennis matches are likely to play the sport, these programs often include ads for game equipment. Baseball or football games rarely include ads for game equipment because few viewers of these games are participants in the sport themselves.

Programs that feature documentaries (such as those on the History Channel or the Discovery Channel) often carry ads for books, book clubs, CDs, and educational DVDs. Advertisers have found that these types of products appeal to the intellectual, arts-loving audiences of these programs.

Companies do much more than just match advertising messages to market segments. They also build a sales environment for their product or service that corresponds to the market segment they are trying to reach. In the physical world, store design and layout are often directed at specific market segments. If you walk through a shopping mall, you can observe that colors, displays, lighting, background music, and even the clothes worn by sales clerks vary with the targeted segment. For example, a clothing store for teenagers presents a completely different experience to its customers than a clothing store that sells expensive, conservative attire targeted toward more mature women with larger incomes.

Market Segmentation on the Web

The Web gives companies an opportunity to present different store environments online. For example, if you visit the home pages of Juicy Couture and Talbots, you will find that both pages are well designed and functional. However, they are each directed to different market segments. The Juicy Couture site is targeted at young, fashion-conscious buyers. The site uses a wide variety of typefaces, bold graphics, and photos of brightly colored products to convey its tone. The emphasis is to make a bold fashion statement and, presumably, become the envy of your friends. In contrast, the Talbots site is rendered in a more subtle, conservative style. The site is designed for older, more established buyers. The messages emphasized are stability and timeless elegance. These images appeal to a market segment of people looking for classics instead of the latest trends.

In the physical world, retail stores have limited floor and display space. These limitations often force physical stores to decide on one particular message to convey. Exceptions do exist, such as a music store that has a separate room for classical recordings (with background music that differs from the rest of the store) or a large department store that can use lighting and display space differently in each department; however, smaller retail stores usually choose the one image that appeals to most of their customers. On the Web, retailers can provide separate virtual spaces for different market segments. For example, Dell's home page includes links to separate sections of its site for home users, small and medium businesses, public sector organizations, and large enterprises. Some Web retailers provide the ultimate in targeted marketing—they allow their customers to create their own stores, as you will learn in the next section.

Offering Customers a Choice on the Web

Dell has done many things well in its online business. Its Web site offers customers a number of different ways to do business with the company. Its U.S. home page includes links for each major group of customers it has identified, including home, small business, medium and large business, government, education, and health care. Once the site visitor has selected a customer category, specific products and product categories are available as links.

Dell Premier accounts give users a high level of customer-based market segmentation. In these accounts, Dell offers each customer its own Dell Web site. Dell can customize a company's Premier account pages to show product selections for which price and terms have already been negotiated. Dell even allows individual employees of its customers to create their own personalized pages within their companies' Premier pages. This highly customized approach to offering products and services that match the needs of a particular customer is called **one-to-one marketing**. The Internet gives marketers the best opportunity for highly customized interactions with customers that they have had since the heyday of the door-to-door salesperson in the 1940s and 1950s.

BEYOND MARKET SEGMENTATION: CUSTOMER BEHAVIOR AND RELATIONSHIP INTENSITY

In the previous sections, you learned how companies can target as market segments groups of customers that share common characteristics. You also learned how one-to-one marketing gives companies a chance to create Web experiences that are unique to each individual customer. The next step—beyond market segmentation, even beyond one-to-one marketing—is when companies use the Web to target specific customers in different ways at different times.

Segmentation Using Customer Behavior

In the physical world, businesses can sometimes create different experiences for customers in response to their needs. For example, a company might decide that its mission is to sell prepared meals to hungry customers. A given potential customer responds to hunger in different ways at different times. If a person is hungry in the morning, but late for work, that person might drive through a fast-food restaurant or grab a quick cup of coffee at the train station. Lunch might be a sandwich ordered and delivered to the office, or it could require a nice restaurant if a client needs to be entertained. Dinner could be at a restaurant with friends, take-out food from a neighborhood Chinese restaurant, or a delivered pizza.

The point is that the same person requires different combinations of products and services depending on the occasion. In general, the creation of separate experiences for customers based on their behavior is called **behavioral segmentation**. When based on things that happen at a specific time or occasion, behavioral segmentation is sometimes called **occasion segmentation**.

Usually, businesses that operate in the physical world can meet only one or a few of a customer's differing behavioral needs. For example, the Chinese restaurant mentioned earlier might offer dining room service and take-out service, but it probably would not offer a drive-through window or a morning coffee kiosk. Very few restaurants are able to offer everything from fast food through a five-course dinner. In the online world, it is much easier to design a single Web site that meets the needs of visitors who arrive in different behavioral modes. Thus, a Web site design can include elements that appeal to different behavioral segments.

Marketing researchers study how and why people prefer different combinations of products, services, and Web site features and how these preferences are affected by their modes of interaction with the site. Market researchers know that people want Web sites that offer a range of interaction possibilities. Remember that a particular person might visit a particular Web site at different times with different needs and will want an interaction that meets those needs on each visit. Customizing visitor experiences to match the site usage behavior patterns of each visitor or type of visitor is called **usage-based market segmentation**. Researchers have identified common patterns of online behavior and grouped patterns into categories. One set of categories that marketers use today includes browsers, buyers, and shoppers.

Browsers

Some visitors to a company's Web site are just surfing or browsing. Web sites intended to appeal to potential customers in this mode must offer them something that piques their interest. The site should include words that are likely to jog the memories of visitors and remind them of something they want to buy on the site.

These keywords are often called **trigger words** because they prompt a visitor to stay and investigate the products or services offered on the site. Links to explanations about the site or instructions for using the site can be particularly helpful to this type of customer. A site should include extra content related to the product or service the site sells. For example, a Web site that sells camping gear might offer reviews of popular camping destinations with photos and online maps. Such content can keep a visitor who is in browser mode interested long enough to stay at the site and develop a favorable impression of the company. Once visitors have developed this favorable impression, they are more likely to buy on this visit or bookmark the site for a return visit.

Buyers

Visitors who arrive in buyer mode are ready to make a purchase right away. The best thing a site can offer a buyer is a direct route into the purchase transaction. For visitors who first choose a product from a printed catalog, many Web sites include a text box on their home pages that allows visitors to enter the catalog item number. This places that item in the site's shopping cart and takes the buyer directly to the shopping cart page. A **shopping cart** is the part of a Web site that keeps track of selected items for purchase and automates the purchasing process.

Perhaps the ultimate in shopping cart convenience is the 1-Click feature offered by **Amazon.com**, which allows customers to purchase an item with a single click. Any items that a customer purchases using the 1-Click feature within a 90-minute time period are aggregated into one shipment. Amazon.com has a patent on the 1-Click feature. You will learn more about such business process patents and other legal issues in Chapter 7.

Shoppers

Some customers arrive at a Web site knowing that it offers items they are interested in buying. These visitors are motivated to buy, but they are looking for more information before they make a purchase decision. For the visitor who is in shopper mode, a site should offer comparison tools, product reviews, and lists of features. Sites such as **Crutchfield** and **Best Buy** allow customers to specify the level of detail presented for each product, sort products by brand, or price, and compare products with each other side by side.

Remember that a person might visit a Web site one day as a browser and then return later as a shopper or a buyer. People do not retain behavioral categories from one visit to the next—even for the same Web site.

Alternative Models

Although many companies work with these three visitor categories, other researchers are exploring alternative models. Much of Web site visitor behavior is not yet well understood. One study conducted by major consulting firm McKinsey & Company examined the online behavior of 50,000 active Internet users and identified six different groups. Following are the six behavior-based categories and their characteristic traits:

- *Simplifiers* are users who like convenience. They are attracted by sites that make doing business easier, faster, or otherwise more efficient than is possible in the physical world.
- *Surfers* use the Web to find information, explore new ideas, and shop. They like to be entertained, and they spend far more time on the Web than other people. To attract surfers, sites must offer a wide variety of content that is attractive, well displayed, and constantly updated.
- *Bargainers* are in search of a good deal. Although they make up less than 10 percent of the online population, they make up more than half of all visitors to the eBay auction site. They enjoy searching for the best price or shipping terms and are willing to visit many sites to do that.
- *Connectors* use the Web to stay in touch with other people. They are intensive users of chat rooms, instant messaging services, social networking sites, electronic greeting card sites, and Web-based e-mail. Connectors tend to be new to the Web, less likely than other people to purchase on the Web, and actively trying to learn what the Web has to offer them.
- *Routiners* return to the same sites over and over again. They use the Web to obtain news, stock quotes, and other financial information. Routiners like the comfort of working with a user interface that they know well.
- *Sportsters* are similar to routiners, but they tend to spend time on sports and entertainment sites rather than news and financial information sites. Because they view the Web as an entertainment vehicle, sportsters are attracted by sites that are interactive and attractive.

Other research studies have identified similar sets of characteristics and categories. Companies in different industries or lines of business identify somewhat different sets of characteristics and group their Web site visitors using different names. The challenge for Web businesses is to identify which groups are visiting their sites and formulate ways of generating revenue from each segment. For example, some of these groups (such as simplifiers and bargainers) are ready to buy and would be interested in seeing specific product or service offerings. Other groups (such as surfers, routiners, and sportsters) would be good targets for specific types of advertising messages. As more researchers study Web site visitor behavior, perhaps the industry will learn how to recognize the various modes in which visitors arrive and then channel them into the appropriate sections of the site. Until then, many Web sites use Dell's approach, in which visitors are asked to identify themselves as belonging to a particular category of customer when they enter the sites.

Customer Relationship Intensity and Life-Cycle Segmentation

One goal of marketing is to create strong relationships between a company and its customers. The reason that one-to-one marketing and usage-based segmentation are so valuable is that they help to strengthen companies' relationships with their customers. Good customer experiences can help create an intense feeling of loyalty toward the company and its products or services. Researchers have identified several stages of loyalty as customer relationships develop over time. A five-stage model of customer loyalty that is typical of these models appears in Figure 4-4.

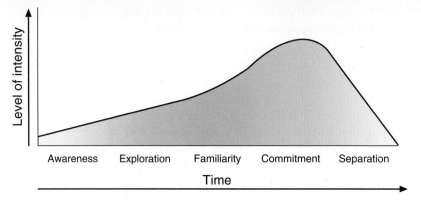

© Cengage Learning

FIGURE 4-4   Five stages of customer loyalty

This model shows the increase in intensity of the relationship as the customer moves through the first four stages: awareness, exploration, familiarity, and commitment. In the fifth stage, separation, a decline occurs and the relationship terminates. Not all customers go through the full five stages; some stop at a stage and continue the relationship at that level of intensity or terminate the relationship at that point. Some customers in a particular stage might have contact with the company online while other customers in the same stage encounter the company offline. Companies should strive for a consistent customer experience at a particular life-cycle stage. That is, customers should experience the same level and quality of service whether they encounter the company online or offline. Online and offline customer contact points are often called **touchpoints**, and the goal of providing similar levels and quality of service at all touchpoints is called **touchpoint consistency**.

As the figure shows, changes in the nature of the relationship do not occur suddenly as a customer moves from one stage to the next. Within each stage, the level of intensity changes gradually as the customer moves through that stage. The characteristics of the five stages are outlined in the next sections.

Awareness

Customers who recognize the name of the company or one of its products are in the awareness stage of customer loyalty. They know that the company or product exists, but have not had any interaction with the company. Advertising a brand or a company name is a common way for companies to achieve this level of relationship with potential customers.

Exploration

In the exploration stage, potential customers learn more about the company or its products. The potential customer might visit the company's Web site to learn more, and the two parties will often communicate by telephone or e-mail. A large amount of information interchange can occur between the parties at this stage.

Familiarity

Customers who have completed several transactions and are aware of the company's policies regarding returns, credits, and pricing flexibility are in the familiarity stage of their relationship with the company. In this stage, they are as likely to shop and buy from competitors as they are from the company.

Commitment

After experiencing a considerable number of highly satisfactory encounters with a company, some customers develop a fierce loyalty or strong preference for the products or brands of that company. These customers have reached the commitment stage and are often willing to tell others about how happy they are with their interactions. To lure customers from the familiarity stage to the commitment stage, companies sometimes make concessions on prices or terms. Usually, the value of the strong relationship is worth more to the company than the costs of these concessions.

Separation

Over time, the conditions that made the relationship valuable might change. The customer might be severely disappointed by changes in the level of service (either as provided by the company or as perceived by the customer) or product quality. The company can also evaluate the relationship and conclude that the loyal, committed customer is costing too much to maintain. As the intensity of the relationship fades, the parties enter a separation stage.

An important goal of any marketing strategy should be to move customers into the commitment stage as rapidly as possible and keep them there as long as possible. Companies want to see customers move into the separation stage only if they are costing more to serve than they are worth.

Life-Cycle Segmentation

Analyzing how customers' behavior changes as they move through the five stages can yield information about how they interact with the company and its products in each stage. The five stages are sometimes called the **customer life cycle**, and using these stages to create groups of customers that are in each stage is called **life-cycle segmentation**. Two companies that undertake continuing research into market segmentation and how companies can use segment information to develop better relationships with their customers are Claritas and Donnelley Marketing.

Claritas created one of the first segment marketing databases, named PRIZM, in the early 1970s. Claritas built PRIZM to take advantage of people's tendency to live near other people with similar tastes and preferences. Thus, PRIZM identifies the demographic characteristics of people by neighborhood. Claritas developed a number of other products that offer marketers databases with specific demographic, financial, and psychographic characteristics. Donnelley Marketing offers similar products, such as its Buyer Behavior Indicator and Affluence Models databases. Both Donnelley and Claritas extended their research from traditional direct marketing to help firms sell online. You can learn more about these companies and their products by following their links in the Web Links for this chapter.

Acquisition, Conversion, and Retention of Customers

One goal of the strategies and tactics you will learn about in the rest of this chapter is to attract new visitors to a Web site. The benefits of acquiring new visitors are different for Web businesses with different revenue models. For example, an advertising-supported site is interested in attracting as many visitors as possible to the site and then keeping those visitors at the site as long as possible. That way, the site can display more advertising messages to more visitors, which is how the site earns a profit. For sites that operate a Web catalog, charge a fee for services, or are supported by subscriptions, attracting visitors to the site is only the first step in the process of turning those visitors into customers. The total amount of money that a site spends, on average, to draw one visitor to the site is called the **acquisition cost**.

The second step that a Web business wants to take is to convert the first-time visitor into a customer. This is called a **conversion**. For advertising-supported sites, the conversion is usually considered to happen when the visitor registers at the site, or, in some cases, when a registered visitor returns to a site several times. For sites with other revenue models, the conversion occurs when the site visitor buys a good or service or subscribes to the site's content. The total amount of money that a site spends, on average, to induce one visitor to make a purchase, sign up for a subscription, or (on an advertising-supported site) register, is called the **conversion cost**. Most managers use a cumulative definition for conversion cost; that is, conversion cost includes acquisition cost.

For many Web businesses, the conversion cost is greater than the profit earned on the average sale (or the average first sale). In such cases, the Web business must induce the customer to return to the site and buy again (or renew the subscription, or view more advertising). Customers who return to the site one or more times after making their first purchases are called **retained customers**. Different businesses use different measures for determining when a customer is a retained customer. Some companies consider a customer retained if he or she returns just once and purchases again. Others use some number of subsequent purchases or some number of subsequent purchases within a specific time frame. The costs of inducing customers to return to a Web site and buy again are called **retention costs**.

Companies have found that measuring acquisition, conversion, and retention costs is important because it gives them an idea of which advertising and promotion strategies are successful. These measurements are more precise than classifying customers into the five stages of loyalty in the customer life-cycle model. It is much easier to determine, for example, whether a customer has been converted or retained than it is to determine whether that customer is in the familiarity stage or the commitment stage. A company that is evaluating its promotion campaign can measure the conversion costs and compare them to the profit generated by the average first-time sale. Most companies are very interested in retaining customers, because the cost of acquiring a new customer is between 3 and 15 times (depending on the type of business) the cost of retaining an existing customer.

In the rest of this chapter, you will learn some specific techniques that can be elements of successful Web marketing strategies. Remember that each of these techniques makes sense only when used in concert with another. Not all techniques work well in all situations. For example, in the chapter's opening case, RedEnvelope found that a print catalog could be an integral part of promoting its online sales. RedEnvelope's success does not mean that printing catalogs is a good idea for all Web businesses (see the Kozmo Learning from Failures feature). It is only a good idea if it provides customers with recognizable value and augments the rest of the company's marketing strategy.

LEARNING FROM FAILURES

Kozmo

Throughout New York City, people in their homes late at night crave entertainment and snack foods. Kozmo was launched in 1998 to meet the needs of those New Yorkers. With its orange-jacketed delivery people riding bicycles or motor scooters, Kozmo promised delivery of most items within an hour of ordering. Kozmo did not offer as wide a range of items as most convenience stores, so its main competitive advantage was its delivery service. Kozmo attempted to become profitable by adding high-margin items, such as DVD players and Sony PlayStations, and expanding its delivery areas to include higher-income neighborhoods. In addition to Manhattan, Kozmo operated for a short time in Houston and San Diego. In these cities, the higher average distances between deliveries made it even more difficult to cover costs.

Despite its best efforts, Kozmo was unable to create an image that was much different from that of a convenience store on wheels. Kozmo found it difficult to convince customers that delivered snack food items and videos were significantly more valuable than snack food items and videos on the shelves of nearby convenience stores. Most of Kozmo's product line consisted of items for which most people were accustomed to paying low prices.

In March 2001, just one month before closing operations, Kozmo announced a marketing plan that included spending $2.5 million to print and circulate 400,000 catalogs. The plan was a last-ditch attempt to increase brand awareness, gain new customers, and convince people who did not have an Internet connection to use Kozmo's phone order service. Unlike RedEnvelope, however, the Kozmo catalog was not a part of an integrated business plan and did not provide the same kind of added value that RedEnvelope's catalog provides—a bag of potato chips does not gain much appeal by appearing in a full-color catalog photo.

The lesson from Kozmo's experience is that using one element from a marketing strategy that worked for one company is no guarantee that it will work for every company. Marketing techniques are effective only when implemented as part of an integrated strategy that fits the company's products and gives customers a compelling reason to buy.

Customer Acquisition, Conversion, and Retention: The Funnel Model

Marketing managers need to have a good sense of how their companies acquire and retain customers. They often must evaluate competing marketing strategies to determine which of the strategies is the most effective at attracting and retaining customers. The **funnel model of customer acquisition, conversion, and retention** is used as a conceptual tool to understand the overall nature of a marketing strategy, but it also provides a clear structure for evaluating specific strategy elements.

The funnel model is very similar to the customer life-cycle model you learned about earlier in this chapter; however, the funnel model is less abstract and does a better job of showing the effectiveness of two or more specific strategies. The funnel is a good analogy for the operation of a marketing strategy because almost every marketing strategy starts with a large number of prospects and converts fewer and fewer of those prospects into serious prospects, customers, and finally, loyal customers. One example of a funnel model appears in Figure 4-5.

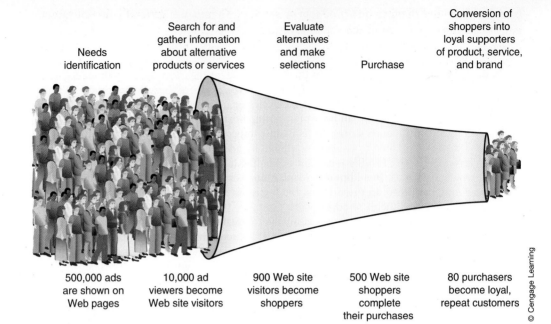

Needs identification	Search for and gather information about alternative products or services	Evaluate alternatives and make selections	Purchase	Conversion of shoppers into loyal supporters of product, service, and brand
500,000 ads are shown on Web pages	10,000 ad viewers become Web site visitors	900 Web site visitors become shoppers	500 Web site shoppers complete their purchases	80 purchasers become loyal, repeat customers

© Cengage Learning

FIGURE 4-5   Funnel model of customer acquisition, conversion, and retention

In this funnel model of the steps that potential customers take as they become loyal, the repeat customers are on the right side of the figure. The top of the figure explains the increasing level of commitment that occurs in each step. Using market research and past history as a guide, the marketing manager develops the numbers that show the effectiveness of the planned strategy. The wider the right end of the funnel, the better the strategy; that is, the more prospects are converted into loyal customers. The funnel model can be used in planning marketing strategies by comparing the projected results shown in the diagram with the results for alternative strategies shown in separate diagrams. The funnel model can also be used to show results that can then be compared with the costs of running the marketing campaign. Either way, the model gives marketing managers a tool for conceptualizing and evaluating alternative strategies.

ADVERTISING ON THE WEB

Advertising is all about communication. The communication might be between a company and its current customers, potential customers, or even former customers that the company would like to regain. To be effective, firms should send different messages to each of these audiences.

The five-stage customer loyalty model shown in Figure 4-4 (in the previous section) can be helpful in creating the messages to convey to each of these audiences. In the awareness stage, the advertising message should inform. The message could describe a new product, suggest new uses for existing products, or describe specific improvements to a product. Audiences in the exploration stage should receive messages that explain how a product or service works and encourage switching to that brand. In the familiarity stage, the advertising message should be persuasive—convincing customers to purchase specific products or request that a salesperson call. Customers in the commitment stage should be sent reminder messages. These ads should reinforce customers' good feelings about the brand and remind them to buy products or services. Companies generally do not target ads at customers who are in the separation stage.

Most companies that launch electronic commerce initiatives already have advertising programs in place. Online advertising should always be coordinated with existing advertising efforts. For example, print ads should include the company's URL. Banner ads are the dominant advertising format in use on the Web. Other online ad formats include pop-up ads, pop-behind ads, interstitial ads, and active ads.

Banner Ads

Most advertising on the Web uses banner ads. A **banner ad** is a small rectangular object on a Web page that displays a stationary or moving graphic and includes a hyperlink to the advertiser's Web site. Banner ads are versatile advertising vehicles—their graphic images can help increase awareness, and users can click them to open the advertiser's Web site and learn more about the product. Thus, banner ads can serve both informative and persuasive functions.

Early banner ads used a simple graphic, usually in GIF format, that loaded with the Web page and remained on the page until the user moved to another page or closed the browser. Today, a variety of **animated GIFs** and **rich media objects** created using Shockwave, Java, or Flash are used to make attention-grabbing banner ads. These ads can be rotated so that each time the Web page is loaded into a browser, the ad changes.

Although Web sites can create banner ads in any dimensions, advertisers decided early in the life of electronic commerce that it would be easier to standardize the sizes. The standard banner sizes that most Web sites have voluntarily agreed to use are called **interactive marketing unit (IMU) ad formats**. The Interactive Advertising Bureau (IAB) is a not-for-profit organization that promotes the use of Internet advertising and encourages effective Internet advertising. The IAB has established voluntary standards for IMUs. As the Web grew, so did the creativity of Web advertisers. They were using an increasing number of IMU ad formats including pop-up ads, buttons, and ads that filled entire page borders. By 2003, advertisers were using more than 15 different IMU ad formats and the IAB decided to encourage its members to agree to use only four standard formats.

These formats are now called the **universal ad package (UAP)** and are the most common formats used on the Web today. Many advertisers use these four standard formats because they know that almost every Web site will be able to display their ads in those formats properly. The UAP formats (and their IAB specifications) include the following:

1. Medium rectangle (300 × 250 pixels)
2. Rectangle (180 × 150 pixels)
3. Leaderboard (728 × 90 pixels)
4. Wide skyscraper (160 × 600 pixels)

A **leaderboard ad** is a banner ad that is designed to span the top or bottom of a Web page. A **skyscraper ad** is a banner ad that is designed to be placed on the side of a Web page and remain visible as the user scrolls down through the page. You can learn more about banner ads, including examples of the latest IAB-approved sizes, by following the Web Links to the IAB Web site.

Most advertising agencies that work with online clients can create banner ads as part of their services. Web site design firms can also create banner ads. Charges for creating banner ads range from about $100 to more than $5000, depending on the complexity of the ad. Companies can make their own banner ads by using a graphics program or the tools provided by some Web sites. AdDesigner.com is an advertising-supported Web site that lets visitors design their own banner ads and download them for free. **AdReady** offers free "do-it-yourself" ad-creation service alongside its professional creative services.

Banner Ad Placement

Companies have three different ways to arrange for other Web sites to display their banner ads. The first is to use a banner exchange network. A **banner exchange network** coordinates ad sharing so that other sites run one company's ad while that company's site runs other exchange members' ads. Usually, the exchange requires each member site to accept two ads on its site for every one of its ads that appears on another member's site. The exchange then makes its profit by selling the extra ad space to other businesses. Companies in the banner exchange business include **HitExchange** and **Voltrank**.

Because banner exchanges are free, many smaller online businesses use them; however, it is often difficult to find a group of other Web sites that have formed an exchange or that belong to an exchange that are not direct competitors. This limitation prevents many businesses from using banner exchange networks.

The second way that businesses can place their banner advertising is to find Web sites that appeal to one of the company's market segments and then pay those sites to carry the ads. This can take considerable time and effort. Smaller sites might not have an established pricing policy for advertising. Larger sites usually have high standard rates that they discount for larger customers. Smaller customers generally pay the standard rates. A company can hire an advertising agency to negotiate lower rates and help with ad placement. A full-service advertising agency can help design the ads, create the banners, and identify appropriate Web sites on which to display them. Agencies that do a lot of Internet work can often negotiate lower advertising rates with sites because the agencies can consolidate their clients' budgets and buy large blocks of advertising space at one time.

A third way to place banner advertising is to use a banner advertising network. A **banner advertising network** acts as a broker between advertisers and Web sites that carry ads. The larger banner advertising networks, such as **DoubleClick** and **ValueClick**, offer many of the same services as comprehensive ad agencies and often broker space primarily on larger Web sites (such as Yahoo!) that have high traffic rates and are, thus, more expensive. The smaller firms, on the other hand, often sell only leftover discounted space.

New Strategies for Banner Ads

When banner ads first appeared on the Web in the mid-1990s, they were a novelty for Web surfers. As users saw more ads, however, the ads lost their ability to attract attention. Click-through rates, which had been as high as 2 percent when banner ads were first introduced, have steadily dropped and now range from .3 percent to .5 percent, depending on the site's content. Although some recent research suggests that Web site visitors see and are influenced by banner ads that they do not click, advertisers are reluctant to pay for ads that do not produce directly measurable results.

To battle the decrease in click-through rates, banner ad designers first introduced animated GIFs with moving elements in the hopes that they might be more attractive to the user's eye than stationary graphics. When animated GIFs failed to halt the decline, designers created ads that included rich media effects, such as movie clips. They also added interactive effects by writing Java programs that could respond to a user's click with some action (other than simply loading the advertiser's page into the browser). Some of these interactive ads even act like miniature video games.

Some designers created banner ads that appear to be dialog boxes in the hope that confused users would click them. Several examples of this type of banner ad are shown in Figure 4-6. These ads are designed to induce users to click a button in the ad to fix the "error," but the banners actually link to Web sites or begin installing a program on the user's computer.

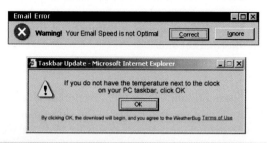

FIGURE 4-6   Disguised banner ads

© Cengage Learning

Text Ads

An ad format that is deceptively simple but very effective is the text ad. A **text ad** is a short promotional message that does not use any graphic elements and is usually placed along the top or right side of a Web page. Google was the first company to use text ads successfully on the Web. Google places text ads on its search results pages. When you visit Google and use it to search for information, the page that provides the links relevant to your search query includes short text ads for products or services related to your search query. Google found that these ads were less obtrusive than banner ads and that they were very effective because they reached people who were interested in learning more about something (as reflected in the search query they had entered) related to the advertisers' products or services.

Text ads were so unobtrusive that Google was criticized when it first included them on its pages. Observers noted that site visitors might not be able to distinguish the paid ads from the search results. In response to this criticism, Google and most other search sites that use text ads now clearly label the ads to prevent users from being confused.

The use of text ads was one of the innovations that helped Google become one of the leading search sites on the Web. It gave Google an effective way to earn money while providing users with a useful search experience.

A number of sites that provide information use text ads in another way by turning some of the text in the stories they display into hyperlinks that lead to advertisers' sites. This type of advertising is called an **inline text ad**. Newspaper, magazine, and other sites that users visit to learn more about a topic can use this technique. For example, a newspaper site might have a story about local banks. Banks that are mentioned in the story could have their names presented in the story as links to ads for the banks' services. The newspaper would charge the advertising banks a fee for placing the link in the story. Another way information sites use text ads is to include them in the middle of the running text of a story as a separate, blocked-off paragraph. These paragraphs are often labeled "sponsored links" or something similar so that readers understand that they are looking at a link to an ad. This use of inline text ads is common in online magazines devoted to specific industries and in general information sites.

Other Web Ad Formats

The steady decline in the effectiveness of banner ads has prompted advertisers to explore other formats for Web ads. One of these formats is the pop-up ad. A **pop-up ad** is an ad that appears in its own window when the user opens or closes a Web page. The window in which the ad appears does not include the usual browser controls. The only way to dismiss the ad is to click the small close button in the upper-right corner of the window's frame. Many users find pop-up ads extremely annoying. A particularly irritating variation on the pop-up ad technique occurs at Web sites that open more than one pop-up ad when a user leaves the site or closes the browser. If the user does not act quickly enough, the browser spawns multiple windows and can even crash the computer.

Another type of pop-up ad is called the pop-behind ad. A **pop-behind ad** is a pop-up ad that is followed very quickly by a command that returns the focus to the original browser window. The result is an ad that is parked behind the user's browser, waiting to appear when the browser is closed.

Despite user objections to pop-up ads (in all their variations), an increasing number of Web sites are using them as a way of delivering a larger advertising image in a more forceful way. Some users have responded by using **ad-blocking software** that prevents banner ads and pop-up ads from loading. Most Web browsers can be configured not to display many of these ads; however, any site that uses methods for navigation that are similar to those used to deliver ads (such as pop-up information windows) cannot operate as intended in the reconfigured browser. Some researchers have found that pop-up ads not only annoy users, they actually create lasting bad will among users toward the company whose products are depicted in the ads. Despite these findings, many advertisers find pop-up ads to be effective tools for drawing customers to their sites and continue to use them.

Another intrusive ad format is the **interstitial ad**. When a user clicks a link to load a page, the interstitial ad opens in its own browser window, instead of the page that the user intended to load (the general meaning of the word "interstitial" is something that comes between two other things). Many interstitial ads close automatically, allowing the intended page to open in the existing browser window. Other interstitials require the user to click a button before they close. Because they open in a full-size browser window, interstitial ads offer the advertiser even more space than the pop-up ad format. These ads also completely cover the Web page that the user was trying to see. Many users find interstitials even more annoying than pop-up ads because they are larger and a more forceful interruption of the Web-browsing experience.

Rich media ads, also called **active ads**, are another ad format. These ads generate graphical activity that "floats" over the Web page itself instead of opening in a separate window. These ads always contain moving graphics and usually include audio and video elements. One of the first rich media ads featured the figure of a little man who walked into the displayed Web page, unrolled a movie poster, and then pasted the poster onto the Web page (covering up part of the Web page content—content that a user might have been reading!). After about 10 seconds, the figure walked off the page and the poster disappeared. While it was open on the page, the poster was an active link to the movie's Web site.

Another early rich media ad showed a Ford Explorer driving into the Web page. The Web page appeared to shake with the vibrations of the Explorer as it drove through. Rich media ads are certainly attention grabbers and are even more intrusive than pop-ups or interstitials because they occur on the Web page itself and offer users no obvious way to dismiss them.

Rich media ads are also used on Web sites that deliver video. For example, a Web site that provides television shows or video news updates will often include a rich media ad at the beginning of the video clip. A visitor opens the video and must view a 15- or 30-second ad before the content begins to play.

Mobile Device Advertising

In recent years, the use of mobile devices that are connected to the Internet, such as smart phones and tablets, has grown tremendously. The programs that run on these devices, called **mobile apps** (which is a short form of the term "mobile software applications") perform a variety of functions such as calendar, contact management, Web browsing, e-mail, and entertainment. A number of mobile apps provide connectivity to specific Web sites or groups of Web sites. You will learn more about the business of selling mobile apps in Chapter 6.

Some of the sellers of mobile apps include an advertising element in their revenue models. These apps include **mobile ads** that display messages from advertisers (other than the seller of the app). For example, the mobile app of *The New York Times* has a small bar at the bottom of the screen that displays ads. Some productivity and game software also includes advertising that appears on a part of the screen or as a separate screen that must be clicked through to get to the productivity tool or game. The advertising space on mobile apps is sold in the same way that banner advertising on Web sites is sold.

Site Sponsorships

Some Web sites offer advertisers the opportunity to sponsor all or parts of their sites. These **site sponsorships** give advertisers a chance to promote their products, services, or brands in a more subtle way than by placing banner or pop-up ads on the sites (although some sponsorship packages include a certain number of banner and pop-up ads).

Companies that buy Web site sponsorships have goals that are similar to those of sporting event sponsors or television program sponsors; that is, they want to tie the company or product name to an event or a set of information. The idea is that the quality of the event or information set will carry over to the company's products, services, or brands. In general, sponsorships are used to build brand images and develop reputations rather than to generate immediate sales. A site sponsorship can be exclusive, which prevents any other companies from sponsoring the site, or it can be shared, which means that other companies can be co-sponsors of the site. In general, an exclusive site sponsorship will cost more than a shared site sponsorship.

In some cases, the sponsor is given the right to create content for the site or to weave its advertising message into the site's content. This practice can raise ethical concerns if not done carefully. Sites that offer content spots to sponsors should always identify the content as an advertisement or as provided by the sponsor. Unfortunately, many sites do not use clear labels for sponsored content. This can confuse site visitors who are unable to distinguish between editorial content and advertising. Sites that offer medical information, for example, should be especially careful to distinguish between information that is generated by the site's reporters or editorial staff and information that is provided by pharmaceutical companies or medical device manufacturers.

Online Advertising Cost and Effectiveness

As more companies rely on their Web sites to make a favorable impression on potential customers, the issue of measuring Web site effectiveness has become important. Mass media efforts are measured by estimates of audience size, circulation, or number of addressees. When a company purchases mass media advertising, it pays a dollar amount for every thousand people in the estimated audience. This pricing metric is called **cost per thousand** (**CPM**; the "M" is from the Roman numeral for "thousand").

Measuring Web audiences is more complicated because of the Web's interactivity and because the value of a visitor to an advertiser depends on how much information the site gathers from the visitor (for example, name, address, e-mail address, telephone number, and other demographic data). Because each visitor voluntarily chooses whether to provide these bits of information, all visitors are not of equal value. Internet advertisers have developed some Web-specific metrics for site activity, but these are not generally accepted and are currently the subject of considerable debate.

A **visit** occurs when a visitor requests a page from the Web site. Further page loads from the same site are counted as part of the visit for a specified period of time. This period of time is chosen by the administrators of the site and depends on the type of site. A site that features stock quotes might use a short time period because visitors may load the page to check the price of one stock and reload the page 15 minutes later to check another stock's price. A museum site would expect a visitor to load multiple pages over a longer time period during a visit and would use a longer visit time window. The first time that a particular visitor loads a Web site page is called a **trial visit**; subsequent page loads are called **repeat visits**. Each page loaded by a visitor counts as a **page view**. If the page contains an ad, the page load is called an **ad view**.

Some Web pages have banner ads that continue to load and reload as long as the page is open in the visitor's Web browser. Each time the banner ad loads is an **impression**. If the visitor clicks the banner ad to open the advertiser's page, that action is called a **click** or **click-through**. Banner ads are often sold on a CPM basis where the "thousand" is 1000 impressions. Rates vary greatly and depend on how much demographic information the Web site obtains about its visitors and what kinds of visitors the site attracts, but most rates range between $1 and $50 CPM. Exclusive site sponsorships can be more expensive, sometimes hitting $100 CPM. And context-related text ads on sites with demographics that are very good for the particular targeted text ad can reach $200 CPM.

Rates have varied throughout the history of the Web. As the online advertising market grew, rates slowly climbed, peaking in the late 1990s, when they ranged from $5 to $100. After that time, they gradually drifted downward to their current levels. Figure 4-7 shows a comparison of CPM rates for banner ads and other Web advertising media to CPM rates for advertising placed in traditional media outlets.

Medium	Description	Audience size	Cost per thousand (CPM)
Network television	30-second commercial	10 million–50 million	$5–$30
Local television station	30-second commercial	50,000–2 million	$3–$25
Cable television	30-second commercial	100,000–500,000	$8–$20
Radio	60-second commercial	50,000–2 million	$1–$18
Major metro newspaper	Full-page ad	100,000–600,000	$80–$130
Regional edition of a national magazine	Full-page ad	50,000–900,000	$40–$80
Local magazine	Full-page ad	3000–80,000	$100–$140
Direct mail coupon pack	Mailed in letter-sized envelope	10,000–200,000	$15–$20
Billboard	Highway billboard	100,000–3 million	$2–$5
World Wide Web	Banner ad	10,000–50 million	$1–$50
World Wide Web	Rich media ad	10,000–50 million	$18–$50
World Wide Web	Text ad	10,000–50 million	$1–$200
World Wide Web	Site sponsorship (exclusive)	10,000–50 million	$60–$100
World Wide Web	Site sponsorship (shared)	10,000–50 million	$20–$50
Targeted e-mail	Single mailing	10,000–10 million	$5–$15
Mobile ads	App-embedded	10,000–5 million	$10–$15

FIGURE 4-7   CPM rates for advertising in various media

One of the most difficult things for companies to do as they move onto the Web is gauge the costs and benefits of advertising on the Web. Many companies have developed new metrics to evaluate the number of desired outcomes their advertising yields. For example, instead of comparing the number of click-throughs that companies obtain per dollar of advertising, they measure the number of new visitors to their site who buy for the first time after arriving at the site by way of a click-through. They can then calculate the advertising cost of acquiring one customer on the Web and compare that to how much it costs them to acquire one customer through traditional channels.

Effectiveness of Online Advertising

After years of experimenting with a variety of online advertising formats, the effectiveness of online advertising remains difficult to measure. One major problem has been the lack of a single industry standard measuring service, such as the service that the Nielsen ratings provide for television broadcasting or the Audit Bureau of Circulations procedures provide for the print media. In 2004, a joint task force of the Interactive Advertising Bureau (IAB) and the Institute of Practitioners in Advertising (IPA) created a set of media measurement guidelines that all online advertisers can use to produce comparable ad view numbers.

Although the task force guidelines have helped to establish measures of ad views, difficulties remain in assessing the effectiveness of online advertising because site visitors change their Web surfing behaviors and habits as they gain experience using the Web. For example, an experienced Web user is far less likely than a new Web user to click a banner ad. Declining click-through rates might not be a good indicator of the success of online advertising, however. Many companies are finding that online advertising can be an important element in a comprehensive marketing strategy that uses several different media to deliver messages to potential customers. Recent survey results show that more potential car purchasers would be influenced by an online ad than by a television ad. Very few people would buy a car based solely on information contained in an online ad, but online ads might prove to be an effective way of building brand recognition and conveying information about cars to potential buyers. You can learn more about current developments in online advertising effectiveness by visiting the **AdAge.com**, **eMarketer**, and **Online Publishers Association** Web sites.

Most marketing analysts do agree that online advertising is much more effective if it is properly targeted. Online ads that reach site visitors who are looking for something specific that is related to the ad's message are much more successful than ads viewed by a general population. Thus, market segmentation is an important element in online advertising success. One useful marketing tool that uses market segmentation successfully is e-mail marketing, the subject of the next section.

E-MAIL MARKETING

Sociologists and cultural anthropologists have proclaimed e-mail to be one of the greatest tools for human communication to be developed in the 20th century. Because advertising is a process of communication, it is easy to see that e-mail can be a very powerful element in any company's advertising strategy. Many businesses would like to send e-mail messages to their customers and potential customers to announce new products, new product features, or sales on existing products. However, industry analysts have severely criticized some companies for sending e-mail messages to customers or potential customers. Some companies have even faced legal action after sending out mass e-mailings.

Unsolicited Commercial E-Mail (UCE, Spam)

Spam, also known as **unsolicited commercial e-mail (UCE)** or **bulk mail**, is electronic junk mail and can include solicitations, advertisements, or e-mail chain letters. The origin of the term spam is generally believed to have come from a song performed by the British comedy troupe, Monty Python, about Hormel's canned meat product, SPAM. In the song, an increasing number of people join in repeating the song's chorus: "Spam spam spam spam, spam spam spam spam, lovely spam, wonderful spam…" Just as in the song, e-mail spam is a tiresome repetition of meaningless text that eventually drowns out any other attempt at communication.

Besides wasting people's time and their computer disk space, spam can consume large amounts of Internet capacity. If one person sends a useless e-mail to a million people, that unsolicited mail consumes Internet resources for a few moments that would otherwise be available to other users. Once merely an annoyance, spam has become a major problem for companies. In addition to consuming bandwidth on company networks and space on e-mail servers, spam distracts employees who are trying to do their jobs and requires them to spend time deleting the unwanted messages. A considerable number of spam messages include content that can be offensive to recipients. Some employers worry that their employees might sue them, arguing that the offensive spam they receive while working contributes to a hostile work environment, which can be grounds for harassment allegations. Industry analysts estimate that spam costs businesses more than $30 billion per year in the direct costs of dealing with it and in lost productivity of employees who are subjected to it. You will learn about the legal issues surrounding spam in Chapter 7, and you will learn about the technical issues related to spam and some strategies for battling it in Chapter 8.

Sending e-mail messages to Web site visitors who expressly request the e-mail messages is a completely different story. A key element in any e-mail marketing strategy is to obtain customers' approvals before sending them any e-mail that includes a marketing or promotional message. By obtaining these approvals, as you will learn in the next section, companies can avoid being accused of engaging in spam.

Permission Marketing

Many businesses are finding that they can maintain an effective dialog with their customers by using automated e-mail communications. Sending one e-mail message to a customer can cost less than 1 cent if the company already has the customer's e-mail address. Purchasing the e-mail addresses of people who ask to receive specific kinds of e-mail messages adds between a few cents and a dollar to the cost of each message sent. Another factor to consider is the conversion rate. The **conversion rate** of an advertising method is the percentage of recipients who respond to an ad or promotion. Conversion rates on requested e-mail messages range from 10 percent to more than 30 percent. These are much higher than the click-through rates on banner ads, which are currently under .5 percent and decreasing.

The practice of sending e-mail messages to people who request information on a particular topic or about a specific product is called **opt-in e-mail** and is part of a marketing strategy called **permission marketing**. Seth Godin, the founder of YoYoDyne and later the vice president for direct marketing at Yahoo!, developed this marketing strategy and publicized it in a book he wrote with Don Peppers titled *Permission Marketing*. Godin argues that, as the pace of modern life quickens, time becomes a valuable commodity. Most marketing efforts that traditional businesses use to promote their products or services depend on potential customers having enough time to listen to sales pitches and pay attention to the best ones. As time becomes more precious to everyone, people no longer wish to hear and evaluate advertising and promotional appeals for products and services in which they have no interest. ConstantContact and Yesmail are two companies that offer permission-based e-mail and related services.

Thus, a marketing strategy that sends specific information only to people who have indicated an interest in receiving information about the product or service being promoted should be more successful than a marketing strategy that sends general promotional messages through the mass media. Companies such as Return Path offer opt-in e-mail services. These services provide the e-mail addresses to advertisers at rates that vary depending on the type and price of the product being promoted, but range from a minimum of about $1 to a maximum of 25–30 percent of the selling price of the product.

Combining Content and Advertising

One strategy for getting e-mail accepted by customers and prospects that many companies have found successful is to combine useful content with an advertising e-mail message. Articles and news stories that would interest specific market segments are good ways to increase acceptance of e-mail.

E-mail messages that include large articles or large attachments (such as graphics, audio, or video files) can fill up recipients' in boxes very quickly, so many advertisers send content by inserting hyperlinks into e-mail messages. The hyperlinks should take customers to the content, which is stored on the company's Web site. Once customers are viewing pages on the Web site, it is easier to induce them to stay on the site and consider making purchases. Using hyperlinks that lead to a Web page instead of embedding content in e-mail messages is especially important if the content requires a browser plug-in to play (as many audio and video files do). The Web page can provide a link to the needed plug-in software.

An important element in any marketing strategy is coordination across media outlets. If a company is using e-mail to promote its products or services, it should make sure that any other marketing efforts it is undertaking at the same time, such as press releases, print media ads, or broadcast media ads, are delivering a message that is consistent with the e-mail campaign's message.

Outsourcing E-Mail Processing

Many companies find that the number of customers who opt-in to information-laden e-mails can grow rapidly. The job of handling e-mail lists and mass mailing software can quickly outgrow the capacity of the company's information technology staff. A number of companies offer e-mail management services, and most small to midsized companies outsource their e-mail processing operations to an e-mail processing service provider.

The Additional Information section of the Web Links for this chapter includes links to several companies that offer e-mail processing and management services. These companies will manage an e-mail campaign for a cost of between 1 and 5 cents per valid e-mail address. Many of these companies will also help their clients purchase lists of e-mail addresses from companies that compile such lists.

TECHNOLOGY-ENABLED CUSTOMER RELATIONSHIP MANAGEMENT

The nature of the Web, with its two-way communication features and traceable connection technology, allows firms to gather much more information about customer behavior and preferences than they can gather using micromarketing approaches. Now, companies can measure a large number of things that are happening as customers and potential customers gather information and make purchasing decisions. The information that a Web site can gather about its visitors (which pages were viewed, how long each page was viewed, the sequence, and similar data) is called a **clickstream**.

Technology-enabled relationship management is important when promoting and selling on the Web. **Technology-enabled relationship management** occurs when a firm obtains detailed information about a customer's behavior, preferences, needs, and buying patterns, *and* uses that information to set prices, negotiate terms, tailor promotions, add product features, and otherwise customize its entire relationship with that customer.

Although companies can use technology-enabled relationship management concepts to help manage relationships with vendors, employees, and other stakeholders, most companies currently use these concepts to manage customer relationships. Thus, technology-enabled relationship management is often called **customer relationship management (CRM), technology-enabled customer relationship management,** or **electronic customer relationship management (eCRM).** Figure 4-8 lists seven dimensions of the customer interaction experience and shows how technology-enabled customer relationship management differs from traditional seller–customer interactions in each of those dimensions.

Dimensions	Technology-enabled customer relationship management	Traditional relationships with customers
Advertising	Provide information in response to specific customer inquiries	"Push and sell" a uniform message to all customers
Targeting	Identify and respond to specific customer behaviors and preferences	Market segmentation
Promotions and discounts offered	Individually tailor to customer	Same for all customers
Distribution channels	Direct or through intermediaries; customer's choice	Through intermediaries chosen by the seller
Pricing of products or services	Negotiated with each customer	Set by the seller for all customers
New product features	Created in response to customer demands	Determined by the seller based on research and development
Measurements used to manage the customer relationship	Customer retention; total value of the individual customer relationship	Market share; profit

FIGURE 4-8   Technology-enabled relationship management and traditional customer relationships

CRM as a Source of Value in the Marketspace

Harvard Business School researchers Jeffrey Rayport and John Sviokla observed that firms today do business in both a physical world and a virtual, information world. Rayport and Sviokla distinguish between commerce in the physical world, or marketplace, and commerce in the information world, which they term the **marketspace.** In the information world's marketspace, digital products and services can be delivered through electronic communication channels, such as the Internet.

In Chapter 1, you learned that the value chain model described the primary and support activities that firms use to create value. This value chain model is valid for activities in the physical world and in the marketspace. However, value creation requires different processes in the marketspace. By understanding that value creation in the marketspace is different, firms can identify value opportunities effectively in both the physical and information worlds.

For years, businesses have viewed information as a part of the value chain's supporting activities, but they have not considered how information itself might be a source of value. In the marketspace, firms can use information to create new value for customers. Many electronic commerce Web sites today offer customers the convenience of an online order history, make recommendations based on previous purchases, and show current information about products in which the customer might be interested.

Successful Web-marketing approaches all involve enabling the potential customer to find information easily and customizing the depth and nature of that information; such approaches should encourage the customer to buy. Firms should track and examine the behaviors of their Web site visitors, and then use that information to provide customized, value-added digital products and services in the marketspace. Companies that use these technology-enabled relationship management tools to improve their contact with customers are more successful on the Web than firms that adapt advertising and promotion strategies that were successful in the physical world, but are less effective in the virtual world.

In the early days of the Web, many companies attempted to create comprehensive CRM systems that captured every bit of information about every customer. Many of these systems failed because they were overly complex and required company staff to spend too much time entering data. In recent years, companies have had more success with CRM systems that are less ambitious in scope. By limiting data collection to key facts that matter to salespeople and customers, these systems provide valuable information, yet they do not overly burden sales and administrative staff with data entry work. More companies are getting better at automating the collection of data, which also increases the likelihood that a CRM implementation will be successful.

Today's CRM systems use information gathered from customer interactions on the company's Web site and combine them with information gathered from other customer interactions, such as calls to customer service departments. As you learned earlier in this chapter, the occurrence of contact between the customer and any part of the company is called a customer touchpoint. A good CRM system will gather information from every customer touchpoint and combine it with information from other sources about industry trends, general economic conditions, and market research about changes in general preference levels that might affect demand for the company's products or services.

In a CRM system, the multiple sources of information about customers, their preferences, and their behavior is entered into a large database called a **data warehouse**. On a regular basis, analysts query the data warehouse using sophisticated software tools to perform data mining and statistical modeling. **Data mining** (also called **analytical processing**) is a technique that examines stored information and looks for patterns in the data that are not yet known or suspected. In CRM, analysts might apply data mining techniques to the data warehouse and find that customers often buy two specific products at the same time. By offering both products together at a reduced price whenever a customer views either product page, the company could increase sales of both products. **Statistical modeling** is a technique that tests theories that CRM analysts have about relationships among elements of customer and sales data. For example, a statistical model could be used to test whether free shipping increases sales enough to cover the cost of offering the free shipping. Figure 4-9 shows the elements in a typical CRM system.

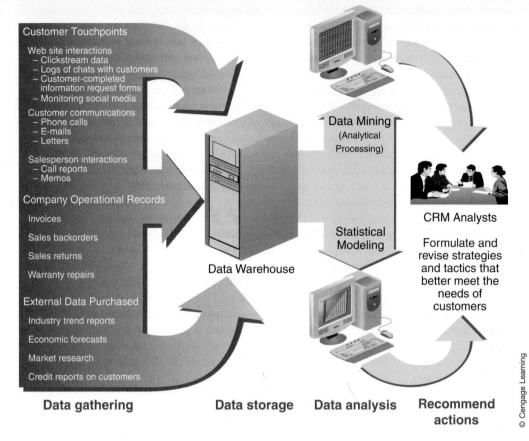

Customer Touchpoints

Web site interactions
- Clickstream data
- Logs of chats with customers
- Customer-completed information request forms
- Monitoring social media

Customer communications
- Phone calls
- E-mails
- Letters

Salesperson interactions
- Call reports
- Memos

Company Operational Records

Invoices

Sales backorders

Sales returns

Warranty repairs

External Data Purchased

Industry trend reports

Economic forecasts

Market research

Credit reports on customers

Data Warehouse

Data Mining (Analytical Processing)

Statistical Modeling

CRM Analysts

Formulate and revise strategies and tactics that better meet the needs of customers

Data gathering   **Data storage**   **Data analysis**   **Recommend actions**

© Cengage Learning

FIGURE 4-9   Elements of a typical CRM system

You can obtain updates on current developments in CRM at the **destinationCRM.com** Web site. You can learn more about data warehousing at the **Data Warehousing Information Center** and about data mining at **The Data Mine**. In Chapter 9, you will learn more about software tools and other technologies that companies are using to implement CRM.

CREATING AND MAINTAINING BRANDS ON THE WEB

A known and respected brand name can present to potential customers a powerful statement of quality, value, and other desirable characteristics in one recognizable element. Branded products are easier to advertise and promote because each product carries the reputation of the brand name. Companies have developed and nurtured their branding programs in the physical marketplace for many years. Consumer brands such as Ivory soap, Walt Disney entertainment, Maytag appliances, and Ford automobiles have been developed over many years with the expenditure of tremendous amounts of money. However, the value of these and other trusted major brands far exceeds the cost of creating them.

Elements of Branding

The key elements of a brand, according to researchers at the advertising agency Young & Rubicam, are differentiation, relevance, and perceived value. Product differentiation is the first condition that must be met to create a product or service brand. The company must clearly distinguish its product from all others in the market. This makes branding difficult for commodity products such as salt, nails, or plywood—difficult, but not impossible.

A classic example of branding a near-commodity product is Procter & Gamble's creation of the Ivory brand more than 100 years ago. The company was experimenting with manufacturing processes and had accidentally created a bar soap that contained a high percentage of air. When one of the workers noted that the soap floated in water, the company decided to sell the soap using this differentiating characteristic in packaging and advertising by claiming "it floats." Thus was the Ivory soap brand born. **Procter & Gamble** maintains this brand differentiation on its Web site even today by maintaining a separate **Ivory Soap** site.

The second element of branding—relevance—is the degree to which the product offers utility to a potential customer. The brand only has meaning to customers if they can visualize its place in their lives. Many people understand that **Tiffany & Co.** creates a highly differentiated line of jewelry and gift products, but very few people can see themselves purchasing and using such goods.

The third branding component—perceived value—is a key element in creating a brand that has value. Even if your product is different from others on the market and potential customers can see themselves using this product, they will not buy it unless they perceive value. Some large fast-food outlets have well-established brands that actually work against them. People recognize these brands and avoid eating at these restaurants because of negative associations—such as low overall quality and high-fat-content menu items. Figure 4-10 summarizes the elements of a brand.

Element	Meaning to customer
Differentiation	In what significant ways is this product or service unlike its competitors?
Relevance	How does this product or service fit into my life?
Perceived value	Is this product or service good?

FIGURE 4-10   Elements of a brand

If a brand has established that it is different from competing brands and that it is relevant and inspires a perception of value to potential purchasers, those purchasers will buy the product and become familiar with how it provides value. Brands become established only when they reach this level of purchaser understanding and acceptance.

Unfortunately, brands can lose their value if the environment in which they have become successful changes. A dramatic example is Digital Equipment Corporation (DEC). For years, DEC was a leading manufacturer of midrange computers. When the market for computing shifted to personal computers, DEC found that its branding did not transfer to the personal computers that it produced. The consumers in that market did not see the same perceived value or differentiation in DEC's personal computers that the buyers of midrange systems had seen for years. This is an important element of branding for Web-based firms to remember, because the Web is still evolving and changing at a rapid pace.

Emotional Branding vs. Rational Branding

Companies have traditionally used emotional appeals in their advertising and promotion efforts to establish and maintain brands. Branding experts Ted Leonhardt and Bill Faust have described "brand" as "an emotional shortcut between a company and its customer." These emotional appeals work well on television, radio, billboards, and in print media because the ad targets are in a passive mode of information acceptance. However, emotional appeals are difficult to convey on the Web because it is an active medium controlled to a great extent by the customer. Many Web users are actively engaged in such activities as finding information, buying airline tickets, making hotel reservations, and obtaining weather forecasts. These users are busy people who will rapidly click away from emotional appeals.

Marketers are attempting to create and maintain brands on the Web by using **rational branding**. Companies that use rational branding offer to help Web users in some way in exchange for their viewing an ad. Rational branding relies on the cognitive appeal of the specific help offered, not on a broad emotional appeal. For example, Web e-mail services give users a valuable service—an e-mail account and storage space for messages. In exchange for this service, users see an ad on each page that provides this e-mail service.

Affiliate Marketing Strategies

Of course, this leveraging approach works only for firms that already have Web sites that dominate a particular market. As the Web matures, it will be increasingly difficult for new entrants to identify unserved market segments and attain dominance. A tool that many new, low-budget Web sites are using to generate revenue is affiliate marketing. In **affiliate marketing**, one firm's Web site—the affiliate firm's—includes descriptions, reviews, ratings, or other information about a product that is linked to another firm's site that offers the item for sale. For every visitor who follows a link from the affiliate's site to the seller's site, the affiliate site receives a commission. The affiliate site also obtains the benefit of the selling site's brand in exchange for the referral.

The affiliate saves the expense of handling inventory, advertising and promoting the product, and processing the transaction. In fact, the affiliate risks no funds whatsoever. Amazon.com was one of the first companies to create a successful affiliate program on the Web. Most of Amazon.com's affiliate sites are devoted to a specific issue, hobby, or other interest. Affiliate sites choose books or other items that are related to their visitors' interests and include links to the seller's site on their Web pages. Books, music, and video products are naturals for this type of shared promotional activity, but sellers of other products and services also use affiliate marketing programs to attract new customers to their Web sites.

One of the more interesting marketing tactics made possible by the Web is **cause marketing**, which is an affiliate marketing program that benefits a charitable organization (and, thus, supports a "cause"). In cause marketing, the affiliate site is created to benefit the charitable organization. When visitors click a link on the affiliate's Web page, a donation is made by a sponsoring company. The page that loads after the visitor clicks the donation link carries advertising for the sponsoring companies. Many companies have found that the click-through rates on these ads are much higher than the typical banner ad click-through rates.

Affiliate Commissions

Affiliate commissions can be based on several variables. In the **pay-per-click model**, the affiliate earns a commission each time a site visitor clicks the link and loads the seller's page. This is similar to the click-through model of charging for banner advertising, and the rates paid per thousand click-throughs are similar to those paid for banner ads.

In the **pay-per-conversion model**, the affiliate earns a commission each time a site visitor is converted from a visitor into either a qualified prospect or a customer. An example of a seller that might use the qualified prospect definition is a credit card-issuing bank. The bank might decide that its best strategy is to pay affiliates only when the visitor turns out to be a good credit risk. Alternatively, the bank might decide it wants to pay the affiliate only if the visitor is approved for the card and then accepts the card (completes the sale). A site that pays its affiliates on completed sales usually pays a percentage of the sale amount rather than a fixed amount per conversion. Some sites use a combination of these methods to pay their affiliates. Commissions on completed sales range from 5 to 20 percent of the sale amount, depending on variables such as the type of product, the strength of the product's brand, how profitable the product is, and the size of an average order.

You can learn more about affiliate programs by visiting an affiliate program broker site that offers affiliate program opportunities for a number of Web sites. An **affiliate program broker** is a company that serves as a clearinghouse or marketplace for sites that run affiliate programs and sites that want to become affiliates. These brokers also often provide software, management consulting, and brokerage services to affiliate program operators. **LinkShare** and **Commission Junction** are two popular affiliate program brokers. Other companies offer affiliate program brokering along with other marketing services.

Viral Marketing Strategies and Social Media

Traditional marketing strategies have always been developed with an assumption that the company would communicate with potential customers directly or through an intermediary acting on behalf of the company, such as a distributor, retailer, or independent sales organization. Because the Web expands the types of communication channels available, including customer-to-customer communication, another marketing approach, viral marketing, has become popular on the Web. **Viral marketing** relies on existing customers to tell other people—the company's prospective customers—about the products or services they have enjoyed using. Much as affiliate marketing uses Web sites to spread the word about a company, viral marketing approaches use word of mouth through individual customers to do the same thing. The number of customers increases the way a virus multiplies, thus the name.

BlueMountainArts, an electronic greeting card company, purchased very little advertising but grew rapidly. Electronic greeting cards are e-mail messages that include a link to the greeting card site. When people received Blue Mountain Arts electronic greeting cards in their e-mail, they clicked a link in the e-mail message that opened the Blue Mountain Arts Web site in their browser. Once at the Blue Mountain Arts site, they were likely to search for cards that they might like to send to other friends. A greeting card recipient might send electronic greeting cards to several friends, who could then send greetings to their friends. Each new visitor to the site could spread the "virus," which in this case was the knowledge of Blue Mountain Arts. By late 1999, when the company was sold to At Home Corporation for $780 million, Blue Mountain had more than 10 million people visiting its site each month. Blue Mountain Arts built a large following using its approach to viral marketing. Today, the site requires visitors to pay for a subscription before they can send electronic greeting cards. However, the site's original strategy of offering free greetings combined with a viral marketing strategy helped it build a large customer base very quickly.

Today, many viral marketing campaigns involve use of social media sites such as **Facebook** or **Google+** and social communication media such as **Twitter**. A key element to understand when doing promotional activities in these social environments is that people do not use social media to shop; they use social media to socialize. This means that marketing with social media is best done using an indirect approach. Instead of informing the community that it has something to sell, a company is more likely to generate viral activity by encouraging members of the community who use their products to discuss how desirable the product or service is. Getting the community to discuss a product or service in a positive way is the goal, rather than simply delivering a promotional message to the community. Direct advertising communications, whether they are postings on sites like

Facebook or Google+, or are tweets (as communications in Twitter are known), are likely to be ignored by the community.

Some companies make the mistake of posting a large number of information items in the social media environment. Because most people active in social media have a large number of friends, sites such as Facebook include mechanisms for filtering out information periodically. If you post too often, your posts can be filtered out by these mechanisms before very many people see them. The key to viral marketing in this environment is to post frequently enough that your presence appears to be active, but not so often that your posts or tweets get lost in the clutter or filtered out of the environment.

In Facebook, tags are a method of linking to someone else. If your company has a Facebook page, you can post information on that page and everyone who is your Facebook "friend" will see it. If you include the name of another company (or person) who has a Facebook page in your information posting, you can include that name as a tag, which will cause your information to appear on their Facebook wall as well. And everyone who is their friend will see your posting even if those people or companies are not a friend of your Facebook page. This can expand the reach of your posting and can start the viral flow of information. Figure 4-11 illustrates the viral nature of social media marketing.

1. Emily posts a status report that includes a tag for her favorite brand of boots, Fuzzter.

2. Emilly's friends all see the post about the Fuzzter boots and several of them share the post.

3. Friends of Emily's friends are now aware of the Fuzzter brand of boots.

FIGURE 4-11   Viral marketing through social media

© Cengage Learning

Marketing on the Web

The number of individuals who associate with your social media site is a good metric for organizations to track as they assess the success of their viral marketing activities. On social media Web sites, followers of a particular company's discussion activity are called **fans**.

In absolute numbers, these metrics can be hard to interpret; however, monitoring changes in the metrics can provide a readily available measure of the success of specific initiatives. For example, Extreme Pizza distributed a wave of coupon promotions through a combined Facebook/Twitter campaign. Their combined number of associated individuals (sometimes called collectively a **fan base**) increased by almost 60 percent in 10 days. The company interpreted this as a major success. Using multiple social media outlets (such as Extreme Pizza did in this example) is a good strategy in a viral marketing campaign because different customers will favor different social media sites and technologies.

SEARCH ENGINE POSITIONING AND DOMAIN NAMES

Potential customers find Web sites in many different ways. Some site visitors are referred by a friend or click a link on a referring Web site. Others are referred by an affiliate marketing partner of the site. Some see the site's URL in a print advertisement or on television. Others arrive unintentionally after typing a URL that is similar to the company's name. But many site visitors are directed to the site by a search engine or directory Web site.

Search Engines and Web Directories

A **search engine** is a Web site that helps people find things on the Web. Search engines contain three major parts. The first part, called a **spider**, a **crawler**, or a **robot** (or simply **bot**), is a program that automatically searches the Web to find Web pages that might be interesting to people. When the spider finds Web pages that might interest search engine site visitors, it collects the URL of the page and information contained on the page. This information might include the page's title, keywords included in the page's text, and information about other pages on that Web site. In addition to words that appear on the Web page, Web site designers can specify additional keywords in the page that are hidden from the view of Web site visitors, but that are visible to spiders. These keywords are enclosed in an HTML tag set called meta tags. The word "meta" is used for this tag set to indicate that the keywords describe the content of a Web page and are not themselves part of the content.

The spider returns this information to the second part of the search engine to be stored. The storage element of a search engine is called its **index** or **database**. The index checks to see if information about the Web page is already stored. If it is, it compares the stored information to the new information and determines whether to update the page information. The index is designed to allow fast searches of its very large amount of stored information.

The third part of the search engine is the search utility. Visitors to the search engine site provide search terms, and the **search utility** takes those terms and finds entries for Web pages in its index that match those search terms. The search utility is a program that creates a Web page that is a list of links to URLs that the search engine has found in its index that match the site visitor's search terms. The visitor can then click the links to visit those sites. You will learn more about the technologies used in search engines in later chapters of this book.

Some search engine sites also provide classified hierarchical lists of categories into which they have organized commonly searched URLs. Although these sites are technically called Web directories, most people refer to them as search engines. The most popular of these sites, such as Yahoo!, include a Web directory and a search engine. They give users the option of using the search engine to find categories of URLs as well as the URLs themselves. This combination of Web directory and search engine can be a powerful tool for finding things on the Web. **Nielsen//NetRatings**, the online audience measurement and analytics consulting firm, issues press releases that list the most frequently visited Web sites. Search engine and Web directory sites regularly appear on these lists.

Marketers want to make sure that when a potential customer enters search terms that relate to their products or services, their companies' Web site URLs appear among the first 10 returned listings. The weighting of the factors that search engines use to decide which URLs appear first on searches for a particular search term is called a **search engine ranking**. For example, if a site is near the top of the list of links returned for the search term "auto," that site is said to have a high search engine ranking for "auto." The combined art and science of having a particular URL listed near the top of search engine results is called **search engine positioning**, **search engine optimization**, or **search engine placement**. For sites that obtain most of their visitors from search engines, a high ranking that places their URL near the top of the list of links returned by the search engine is extremely important.

Paid Search Engine Inclusion and Placement

Today, a number of search engine sites make it easier to obtain good ad placement on search results pages—but for a price. These search engine sites offer companies a **paid placement**, which is the option of purchasing a top listing on results pages for a particular set of search terms. A paid placement also is called a **sponsorship** or a **search term sponsorship**; however, these search term sponsorships are not the same thing as the general site sponsorships you learned about earlier in this chapter. The rates for paid placements vary tremendously depending on the desirability of the search terms to potential sponsors. For example, a search term such as "rental car" would likely be more expensive than a search term such as "frictionless ball bearing" because the potential audience for rental car advertising is much larger than the number of people interested in a specialized industrial product like ball bearings.

Another option for companies is to buy banner ad space at the top of search results pages that include certain terms. For example, Chevrolet might want to buy banner ad space at the top of all search results pages that are generated by queries containing the words "new" and "car." Most search engine sites sell banner ad space on this basis. An increasing number sell space on results pages for the most desirable terms only to companies that agree to package deals that include paid placement and banner ad purchases.

Search engine positioning is a complex subject. A number of consulting firms do nothing but advise companies on positioning strategy. Entire books have been written on the subject and several major conferences are devoted to the subject each year.

Spending on online advertising grew rapidly in the early days of the Web. The amount spent in the United States went from virtually zero in 1995 to about $8 billion in 2000. The Internet slump of 2001–2002 did result in a drop to about $6 billion, but since then, the growth was remarkable through the 2008–2009 recession. After a small drop in 2009, growth has resumed and is expected to continue. Figure 4-12 shows the amount of online advertising sold and projected to be sold in the United States from 2006 through 2015.

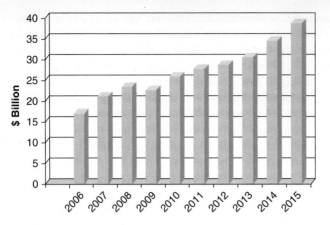

Source: Adapted from reports by ClickZ, eMarketer, Forrester Research, Nielsen//NetRatings, and Internet Retailer.

FIGURE 4-12   U.S. online advertising expenditures, actual and projected

Online advertising is growing much faster than any other type of advertising or advertising spending in general. Thus, online advertising is becoming a larger proportion of all advertising. Figure 4-13 shows how online advertising compares to other U.S. advertising. Online advertising in the rest of the world is expanding rapidly as well, but outside the United States online advertising is a smaller proportion of total advertising.

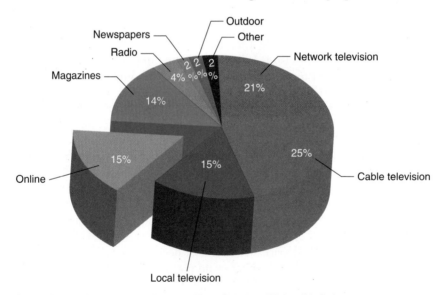

Source: Adapted from reports by eMarketer, MagnaGlobal, and Nielsen//NetRatings.

FIGURE 4-13   U.S. advertising expenditures by medium, 2012 estimates

The business of selling search engine inclusions and placements is complex because many search engines do not sell inclusion and placement rights on their pages directly to advertisers. They use **search engine placement brokers**, which are companies that aggregate inclusion and placement rights on multiple search engines and then sell those combination packages to advertisers. **LookSmart** is an example of a large search engine placement broker. Another reason for the complexity in this business is that recent years have brought a flurry of mergers and acquisitions. For example, in 2003, Yahoo!

purchased Overture, a search engine placement broker. This put Yahoo! in the business of selling advertising for several of its major competitors (who had been using Overture as their search engine placement broker). The most popular search engine, Google, does not use a placement broker to sell search term inclusion and placement for its site. Google sells these services directly through its **Google AdWords** program.

An excellent resource for keeping up with the rapid changes in this business is Danny Sullivan's **Search Engine Land** Web site. Although some of the content on the site is limited to paid subscribers, the site does include many free resources and explanations that are useful for learning about search engines, placement brokers, and search engine optimization in general.

Web sites that offer content can also participate in paid placement. Google offers its **AdSense** program to sites that want to carry ads that match the content offered on the site. Other companies, such as **Kanoodle** and Yahoo!'s **Overture** division, offer similar ad brokerage services, but Google is the leader in this market, reporting more than $2 billion in AdSense advertising sales in 2010. The content site receives a placement fee from the broker in exchange for the ad placement and the broker sells the placement slots to interested advertisers. These techniques in which ads are placed in proximity to related content are sometimes called **contextual advertising**.

Of course, this approach is not without its flaws. In 2003, the *New York Post* ran a sensational story that described a gruesome murder. The murder victim's body had been cut into pieces, which the murderer hid in a suitcase. When the newspaper's Web site ran the story, it appeared with a paid placement ad for luggage. The ad broker's software had noted the word "suitcase" in the story and decided that it would be the perfect place for a luggage ad. Today, ad brokers use more sophisticated software and human reviewers to prevent this type of error; however, some industry analysts believe that contextual advertising on content sites will never be as successful as paid placement on search engine pages. They argue that search engine pages are provided to site visitors looking for something specific, often as part of a purchasing process. Content sites are used to explore and learn about more general things. Thus, an ad on a search engine results page will always be more effective than an ad on a content site page.

Another variation of paid placement ads uses search engine results pages that are generated in response to a search for products or services in a specific geographical area. This technique, called **localized advertising**, places ads related to the location on the search results page. Localized advertising came about as a result of local search services. In 2004, Google launched a local search service that lets users search by ZIP code or local address. All of the other major search engine and Web directory sites followed Google's lead and now offer some form of localized search, either as part of their main search page or as a separate service. The local advertising market (in outlets such as the Yellow Pages) is estimated to be more than $25 billion, a very attractive market for online advertisers.

Web Site Naming Issues

Companies that have a well-established brand name or reputation in a particular line of business usually want the URLs for their Web sites to reflect that name or reputation. Obtaining identifiable names to use on the Web can be an important part of establishing a Web presence that is consistent with the company's existing image in the physical world.

Two airlines that started their online businesses with troublesome domain names have both purchased more suitable domain names. Southwest Airlines' domain name was www.iflyswa.com until it purchased www.southwest.com. Delta Air Lines' original domain name was www.delta-air.com. After several years of complaints from confused customers who could never remember to include the hyphen, the company purchased the domain name www.delta.com.

Companies often buy more than one domain name. Some companies buy additional domain names to ensure that potential site visitors who misspell the URL will still be redirected (through the misspelled URL) to the intended site. For example, Yahoo! owns the name Yahow.com. Other companies own many URLs because they have many different names or forms of names associated with them. For example, General Motors' main URL is GM.com, but the company also owns GeneralMotors.com, Chevrolet.com, Chevy.com, GMC.com, and many others. In 1995, Procter & Gamble purchased hundreds of domain names that included the names of its products, such as Crisco.com, Folgers.com, Jif.com, and Pampers.com. It also bought names related to its products such as Flu.com, BadBreath.com, Disinfect.com, and Stains.com. Procter & Gamble hoped that people searching the Web for information about stains, for example, would find the Stains.com site, which featured links to the company's cleaning products. Procter & Gamble even purchased Pimples.com and Underarms.com.

Buying, Selling, and Leasing Domain Names

In 1998, a poster art and framing company named Artuframe opened for business on the Web. With quality products and an appealing site design, the company was doing well, but it was concerned about its domain name, which was www.artuframe.com. After searching for a more appropriate domain name, the company's president found the Web site of Advanced Rotocraft Technology, an aerospace firm, at the URL www.art.com. After finding out that Advanced Rotocraft Technology's site was drawing 150,000 visitors each month who were looking for something art related, Artuframe offered to buy the URL. The aerospace firm agreed to sell the URL to Artuframe for $450,000. Artuframe immediately relaunched as **Art.com** and experienced a 30 percent increase in site traffic the day after implementing the name change.

The newly named site did not rely on the name change alone, however. It entered a joint marketing agreement with Yahoo! that placed ads for Art.com on art-related search results pages. Art.com also created an affiliate program with businesses that sell art-related products and not-for-profit art organizations. Although Art.com was ultimately unsuccessful in building a profitable business on the Web and liquidated in mid-2001, the domain name was snapped up immediately by already profitable Allwall.com for an undisclosed amount. The new Allwall.com site, relaunched with the Art.com domain name, experienced a 100 percent increase in site visitors within the first month.

The market for domain names continues to be active, with names that include general topic terms (especially those that are sensational) often bringing high prices. Although eCompanies' 1999 purchase of Business.com for $7.5 million was the record holder for many years, more recent sales have exceeded that number. For example, Insure.com sold in 2009 for $16 million. Many domain name sales details are kept private, but some of the highest prices paid that have been reported in the media appear in Figure 4-14.

Domain name	Price
Insure.com	$16.0 million
Fund.com	$10.0 million
Business.com	$7.5 million
Diamond.com	$7.5 million
Beer.com	$7.0 million
Israel.com	$5.9 million
Casino.com	$5.5 million
Toys.com	$5.1 million
Slots.com	$5.0 million
asSeenonTV.com	$5.0 million
Korea.com	$5.0 million
Property.com and Properties.com	$4.0 million
Altavista.com	$3.3 million
Candy.com	$3.0 million
Loans.com	$3.0 million
Wine.com	$3.0 million
Gambling.com	$2.5 million
Autos.com	$2.2 million
Mortgages.com	$2.2 million

FIGURE 4-14   Domain names that sold for more than $2 million

Although most domains that have high value are in the .com TLD, the name engineering.org sold at auction to the American Society of Mechanical Engineers, a not-for-profit organization, for just under $200,000.

Some companies and individuals invested their money in the purchase of highly desirable domain names. Instead of selling these names to the highest bidder, some of these domain name owners decided to retain ownership of the domain names and lease the rights to the names to companies for a fixed time period. Usually, these domain name lessors rent their domain names through URL brokers.

URL Brokers and Registrars

Several legitimate online businesses, known as **URL brokers**, are in the business of selling, leasing, or auctioning domain names that they believe others will find valuable. Companies selling "good" (short and easily remembered) domain names include BuyDomains.com and GreatDomains.

Companies can also obtain domain names that have never been issued, or that are currently unused, from a domain name registrar. The Internet Corporation for Assigned Names and Numbers (ICANN; about which you learned in Chapter 2) maintains a list of accredited registrars. Many of these registrars offer domain name search tools on their Web sites. A company can use these tools to search for available domain names that might meet their needs. Another service offered by domain name registrars is domain name parking. **Domain name parking**, also called **domain name hosting**, is a service that permits the purchaser of a domain name to maintain a simple Web site (usually one page) so that the domain name remains in use. The fees charged for this service are usually much lower than those for hosting an active Web site.

Summary

In this chapter, you learned how companies can use the principles of marketing strategy and the four Ps of marketing to achieve their goals for selling products and offering services on the Web. Some companies use a product-based marketing strategy and some use a customer-based strategy. The Web enables companies to mix these strategies and give customers a choice about which approach they prefer.

Market segmentation using geographic, demographic, and psychographic information can work as well on the Web as it does in the physical world. The Web gives companies the powerful added ability to segment markets by customer behavior and life-cycle stage, even when the same customer exhibits different behavior during different visits to the company's site.

Online advertising has become more intrusive since it was introduced in the mid-1990s, even though research has shown that users find such ads to be irritating. You learned how companies are using various types of online ads, including banners, pop-ups, pop-behinds, text, inline text, and interstitials to promote their sites to potential customers. Permission marketing and opt-in e-mail offer alternatives that can be used with or instead of Web page ads. Context-sensitive text ads are a rapidly growing form of online advertising that users find less intrusive than other online advertising media.

Many companies are using the Web to manage their relationships with customers in new and interesting ways. By understanding the nature of communication on the Web, companies can use it to identify and reach the largest possible number of qualified customers. Technology-enabled customer relationship management can provide better returns for businesses on the Web than the traditional unaided approaches of market segmentation and micromarketing.

Firms on the Web can use rational branding instead of the emotional branding techniques that work well in mass media advertising. Some businesses on the Web are sharing and transferring brand benefits through affiliate marketing and cooperative efforts among brand owners. Others are using viral marketing strategies in online social media to increase awareness of their brands and the size of their customer bases.

Successful search engine positioning and domain name selection can be critical for many businesses in their quests for new online customers. A growing number of advertisers are paying for inclusion and placement services to guarantee that their sites' URLs appear among the top results provided to potential customers by search engines. They are also paying for placement of advertising messages in those pages and on other sites, such as content sites and local information sites. The most important theme in this chapter is that companies must integrate the Web marketing tools they use into a cohesive and customer-sensitive overall marketing strategy.

Key Terms

Acquisition cost	Banner exchange network
Active ads	Behavioral segmentation
Ad-blocking software	Blog
Ad view	Bot
Affiliate marketing	Brand
Affiliate program broker	Brand leveraging
Analytical processing	Bulk mail
Animated GIFs	Cause marketing
Banner ad	Click
Banner advertising network	Clickstream

Click-through

Contextual advertising

Conversion

Conversion cost

Conversion rate

Cost per thousand (CPM)

Crawler

Customer life cycle

Customer relationship management (CRM)

Customer value

Database

Data mining

Data warehouse

Demographic segmentation

Distribution

Domain name hosting

Domain name parking

Electronic customer relationship management (eCRM)

Fan

Fan base

Four Ps of marketing

Funnel model of customer acquisition, conversion, and retention

Geographic segmentation

Impression

Index

Inline text ad

Interactive marketing unit (IMU) ad formats

Interstitial ad

Leaderboard ad

Life-cycle segmentation

Localized advertising

Marketing mix

Marketing strategy

Market segmentation

Marketspace

Micromarketing

Mobile ads

Mobile apps

Occasion segmentation

One-to-one marketing

Opt-in e-mail

Page view

Paid placement

Pay-per-click model

Pay-per-conversion model

Permission marketing

Place

Pop-behind ad

Pop-up ad

Price

Product

Product-based structure

Promotion

Psychographic segmentation

Rational branding

Repeat visit

Retained customers

Retention costs

Rich media ads

Rich media objects

Robot (bot)

Search engine

Search engine optimization

Search engine placement

Search engine placement brokers

Search engine positioning

Search engine ranking

Search term sponsorship

Search utility

Segments

Shopping cart

Site sponsorships

Skyscraper ad

Social media

Spam

Spider

Sponsorship

Statistical modeling

Technology-enabled customer relationship management

Technology-enabled relationship management

Text ad

Touchpoints

Touchpoint consistency

Trial visit

Trigger words

Universal ad package (UAP)

Unsolicited commercial e-mail (UCE)

URL brokers

Usage-based market segmentation

Viral marketing

Visit

Web log (blog)

Review Questions

1. Briefly define the term "customer value" and explain how it is calculated.

2. In about 100 words, describe at least two different online communication tools that a company could use to promote its products or services.

3. Some online stores use a product-based marketing strategy as the basis for their Web site design. Others use a customer-based strategy. In about two paragraphs, briefly define each strategy. Then find two specific Web sites, one that uses each strategy. Write an additional paragraph for each Web site in which you evaluate whether the strategy each has chosen is effective for the particular product or service they are selling.

4. In about 100 words, explain how the level of complexity of a product can affect a company's choice of communication modes it might use to disseminate information about that product.

5. In about 200 words, explain how the achieved trust level of a company's communications using blogs and social media compare with similar communication efforts conducted using mass media and personal contact.

6. In about 100 words, briefly describe micromarketing and explain how the availability of online marketing channels made micromarketing easier to accomplish. Be sure to include a discussion of usage-based market segmentation in your answer.

7. In about 100 words, define and distinguish between the concepts of customer acquisition cost, conversion cost, and retention cost.

8. In about 100 words, describe the benefits of using inline text ads rather than banners and other display ads online.

9. Briefly state the three elements of a brand. Then, assume you are the marketing director for PerfectSeasons, a new line of cookware sponsored by a famous celebrity chef. In about 300 words, describe how you would promote each of the three brand elements for this new product line on the celebrity chef's Web site.

Exercises

1. Visit FTD.com to examine how that company implements occasion segmentation. Write a report of approximately 200 words in which you describe two clear examples of occasion segmentation on the site and explain why an online florist would mix occasion segmentation with product segmentation rather than use one or the other separately.

2. Assume you are a consultant to TopSpin, a tennis equipment manufacturer that sells its products directly to customers on the Web. Review Figure 4-3, which describes what types of television programs would be good hosts for various types of advertising. Applying the logic presented in Figure 4-3, create a list of four Web sites (other than Web sites devoted specifically to the sport of tennis or tennis equipment) in which TopSpin should consider placing advertising to support its online sales effort. For each Web site you identify, write one paragraph in which you explain why that site would be a good advertising outlet to reach TopSpin's potential customers.

10 percent clicked-through). Ninety percent of those who opened the e-mail watched the video. The third e-mail continued the slightly increasing trends for opening and attention (34 percent opened, and 94 percent listened to the audio), but the click-through rate was much higher than the previous two e-mails (14 percent). Also, the dollar amount of donations increased with each subsequent e-mailing. The e-mail campaign raised more than $450,000 in its six-week period.

Oxfam coordinated this e-mail effort with other awareness activities it was conducting in the same time period. The organization sent letters to supporters who had not provided e-mail addresses and ran ads in two newspapers (*The Independent* and *The Guardian*) that carried messages similar to those in the e-mails.

Required:

1. Oxfam used its existing opt-in e-mail list only for this campaign; it did not purchase (or borrow from other charitable organizations) any additional e-mail addresses. Evaluate this decision. In about 200 words, explain the advantages and disadvantages of acquiring other e-mail addresses for a campaign of this nature.

2. For this campaign, Oxfam chose to use e-mails that contained HTML, audio, and video elements rather than using plain-text e-mails. In about 100 words, describe the advantages and disadvantages of using formats other than plain-text in this type of e-mail campaign. Be sure to identify any specific trade-offs that Oxfam faced in deciding not to use plain-text e-mail.

3. Oxfam used HTML in the first e-mail, video in the second, and audio in the third. Evaluate the use of different e-mail formats for this type of message and consider the sequencing of the formats that Oxfam used in this campaign. In about 300 words, summarize the considerations that would affect a decision to use a particular sequence of e-mail formats in a campaign such as this and evaluate the sequence that Oxfam used.

4. A manager at Oxfam might be tempted to conclude that the sequence of formats used in the e-mail messages was related to the increase in donations over the six weeks of the campaign. In about 100 words, present at least two reasons why this would be an incorrect conclusion.

5. If Oxfam were to undertake a similar emergency fund-raising effort today, it might use social media. In about 300 words, describe how Oxfam could use Facebook, Google+, and Twitter in combination with its existing online resources to enhance or replace the e-mail campaign described in the case.

Your instructor might assign you to a group to complete this case and might ask you to prepare a formal presentation of your results to your class.

C2. Montana Mountain Biking

Jerry Singleton founded Montana Mountain Biking (MMB) 18 years ago. MMB offers one-week guided mountain biking expeditions based in four Montana locations. Most of MMB's new customers hear about the company and its tours from existing customers. Many of MMB's customers come back every year for a mountain biking expedition; about 80 percent of the riders on any given expedition are repeat customers.

Jerry is happy with this high repeat percentage, but he is worried that MMB is missing a large potential market. He has been reluctant to spend a lot of money on advertising. About 10 years ago, he spent $80,000 on a print advertising campaign that included ads in several outdoor interest and sports magazines, but the ads did not generate enough additional customers to cover the cost of the advertising. Five years ago, a marketing consultant advised Jerry that the ads had not been placed well. The magazines did not reach the serious mountain bike enthusiast, which is MMB's true target market. After all, a casual mountain bike rider would probably not be drawn to a week-long expedition.

Another concern of Jerry's is that more than 90 percent of MMB's customers come from neighboring states. Jerry has always thought that MMB was not reaching the sizable market of serious mountain bike enthusiasts in California. He talked to the marketing consultant about buying an address list and sending out a promotional mailing, but producing and mailing the letters seemed too expensive. The cost of renting the list was $0.10 per name, but the printing and mailing were $4 per letter. There were 60,000 addresses on the list, and the consultant told him to expect a conversion rate of between 1 and 3 percent. At best, the mailing would yield 1800 new customers and MMB's profit on the one-week expedition was only about $100 per customer. It looked like the conversion cost would be about $246,000 (60,000 × $4.10) to obtain a profit of $180,000 (1800 × $100). The consultant explained that it was an investment; because MMB had such a high customer retention rate, the profit from the new customers in the second or third years would exceed the one-time cost of the mailing in the first year. Jerry was not convinced.

Nine years ago, MMB launched its first Web site. It included information about the company and its tours, but Jerry did not see any need to include an expedition-booking function on the site. He did think about selling caps and jackets with the MMB logo, but that idea never was implemented. The MMB logo is well known in the mountain biking community in the upper Midwest.

The MMB Web site includes an e-mail address so that visitors to the site can send an e-mail requesting more information about the expeditions. Robin Davis, one of MMB's expedition leaders, is an amateur photographer who has taken many photos while on the trails over the years. Last year, she had those photos digitized and put them on the MMB Web site. The number of e-mail inquiries increased dramatically within a month. Many of the inquiries were about MMB's expeditions, but a surprising number asked for permission to use the photos, or asked if MMB had more photos like those for sale. Jerry is not quite sure what to make of the popularity of those photos. He is, after all, in the mountain bike expedition business.

Required:

1. Review the five stages of customer loyalty shown in Figure 4-4 and prepare a report of about 200 words in which you classify MMB's customers. Estimate the percentage of MMB customers who fall into each of the five categories. Support your classification with logic and evidence from the case narrative.

2. In a report of about 200 words, recommend an e-mail marketing strategy for MMB. In your recommendation, consider the results of MMB's earlier print mail advertising campaign, your answer to the first requirement, and the potential offered by permission marketing.

3. In about 300 words, explain how MMB could use social media-based viral marketing tactics to gain new customers and cement its relationships with existing customers. In your answer, be sure to discuss features that MMB should include on its Web site and its Facebook page to support the viral marketing strategy.

4. Prepare a report of about 500 words in which you outline an affiliate marketing strategy for MMB. Include a description of the types of Web sites that MMB should attempt to recruit as affiliates, and present at least five examples of specific sites that would be good referral sources.

Note: Your instructor might assign you to a group to complete this case and might ask you to prepare a formal presentation of your results to your class.

For Further Study and Research

Agarwal, A., D. Harding, and J. Schumacher. 2004. "Organizing for CRM," *The McKinsey Quarterly*, June, 80–91.

Andrews, R. and I. Currim. 2004. "Behavioral Differences Between Consumers Attracted to Shopping Online Vs. Traditional Supermarkets: Implications for Enterprise Design and Marketing," *International Journal of Internet Marketing and Advertising*, 1(1), January–March, 38–61.

Armitt, C. 2004. "Case Study: Crisis in Sudan E-mail Campaign," *New Media Age*, September 2, 22.

Bayer, J. and E. Servan-Schreiber. 2011. "Gaining Competitive Advantage Through the Analysis of Customers' Social Networks," *Direct, Data and Digital Marketing Practice*, 13(2), October, 106–118.

Beck, K. 2011. "Pizza Chain Goes Extreme on Facebook," *CRM Magazine*, 15(6), June, 38–39.

Blair, J. 2001. "Behind Kozmo's Demise: Thin Profit Margins," *The New York Times*, April 13. (http://www.nytimes.com/2001/04/13/technology/13KOZM.html)

Bruton, C. and G. Schneider. 2003. "Multiple Channels for Online Branding," *Academy of Marketing Studies Journal*, 7(1) 109–114.

Case, C. and D. King. 2011. "Twitter Usage in the Fortune 50: A Marketing Opportunity?" *Journal of Marketing Development and Competitiveness*, 5(3), 94–101.

Cashier Live. 2011. "Top Five Retail Facebook Strategies," *Small Biz Bee*, August 2. (http://smallbizbee.com/index/2011/08/02/top-5-retail-facebook-strategies/)

Chan, A., J. Dodd, and R. Stevens. 2004. *The Efficacy of Pop-ups and the Resulting Effect on Brands*. Oxfordshire, UK: Bunnyfoot Universality.

Chan, Y. 2009. "Effects Beyond Click-through: Incidental Exposure to Web Advertising." *Journal of Marketing Communications*, 15(4), September, 227–246.

Chen, Y., S. Fay and Q. Wang. 2011. "The Role of Marketing in Social Media: How Online Consumer Reviews Evolve," *SSRN Working Paper*, January 11. (http://ssrn.com/abstract=1710357)

Clifford, S. 2009. "Put Ad on Web. Count Clicks. Revise." *The New York Times*, May 31, BU1, BU5.

Coyle, P. 2010. "What Are Average CPM Rates for Online Sports Ads in 2010?" *Coyle Media*, June 18. (http://www.coylemedia.com/2010/06/18/what-are-average-cpm-rates-for-online-sports-ads-in-2010/)

Delio, M. 2001. "Kozmo Kills the Messenger," *Wired News*, April 13. (http://www.wired.com/news/business/0,1367,43025,00.html)

Dover, D. 2011. *Search Engine Optimization Secrets*. Indianapolis: Wiley.

Gardner, E. 1999. "Art.com," *Internet World*, March 15, 13. (http://www.iw.com/print/1999/03/15/)

Godin, S. 2005. *All Marketers Are Liars: The Power of Telling Authentic Stories in a Low-Trust World*. New York: Portfolio.

Godin, S. and D. Peppers. 1999. *Permission Marketing: Turning Strangers into Friends, and Friends into Customers*. New York: Simon & Schuster.

Hanlon, P. and J. Hawkins. 2008. "Expand Your Brand Community Online," *Advertising Age*, January 7, 14–15.

Harvard Business Review. 2003. "How to Measure the Profitability of Your Customers," 81(6), June, 74.

Heffernan, V. 2011. "Google's War on Nonsense," *The New York Times*, June 26. (http://opinionator.blogs.nytimes.com/2011/06/26/googles-war-on-nonsense/)

Hinz, O., B. Skiera, C. Barrot, and J. Becker. 2012. "Seeding Strategies for Viral Marketing: An Empirical Comparison," *Journal of Marketing*, January, forthcoming.

Hoffman, D. and T. Novak. 2000. "How to Acquire Customers on the Web," *Harvard Business Review*, 78(3), May–June, 179–188.

Interactive Advertising Bureau (IAB). 2008. *IAB Ad Campaign Measurement Process Guidelines*. New York: IAB. (http://www.iab.net/media/file/ad_campaign_measurement_2008.pdf)

Interactive Advertising Bureau (IAB). 2009. *IAB Audience Reach Measurement Guidelines*. New York: IAB. (http://www.iab.net/media/file/audience_reach_022009.pdf)

Ives, N. 2007. "Forecast for '08 is OK, But Only Online Shines," *Advertising Age*, December 3, 3–4.

Jiang, T. and A. Tuzhilin. 2009. "Improving Personalization Solutions through Optimal Segmentation of Customer Bases," *IEEE Transactions on Knowledge & Data Engineering*, 21(3), March, 305–320.

Jones, K. 2008. *Search Engine Optimization: Your Visual Blueprint for Effective Internet Marketing*. Indianapolis: Wiley.

Jothi, P., M. Neelamalar, and R. Prasad. 2011. "Analysis of Social Networking Sites: A Study on Effective Communication Strategy in Developing Brand Communication," *Journal of Media and Communication Studies*, 3(7), July, 234–242.

Jukic, B., D. Dravitz, N. Jukic, A. Tekleab, L. Meamber, and L. Dashnaw. 2009. "Multilevel Information Presentation Strategy and Customer Reaction: An Empirical Investigation in an Online Setting," *Journal of Organizational Computing & Electronic Commerce*, 19(3), July–September, 173–195.

Kaplan, A. and M. Haenlein. 2011. "The Early Bird Catches the News: Nine Things You Should Know About Micro-blogging," *Business Horizons*, 54(2), March–April, 105–113.

Kaye, K. 2011. "Online Ad Industry Rebounded in 2010," *ClickZ*, April 13. (http://www.clickz.com/clickz/news/2043354/online-industry-rebounded-2010)

Kennedy, A. and K. Hauksson. 2012. *Global Search Engine Marketing*. Indianapolis: Que.

Kilby, N. 2007. "Doubling Your Search Efforts," *Marketing Week*, May 17, 31–34.

Kiley, D. and B. Helm. 2009. "The Great Trust Offensive," *BusinessWeek*, September 28, 38–42.

King, D. 2008. "Waiting for the Day that Search Becomes Four-Dimensional," *New Media Age*, January 17, 13.

Koprowski, G. 1998. "The (New) Hidden Persuaders: What Marketers Have Learned About How Consumers Buy on the Web," *The Wall Street Journal*, December 7, R10.

Kunz, M., B. Hackworth, P. Osborne, and J. High. 2011. "Fans, Friends, and Followers: Social Media in the Retailers' Marketing Mix," *Journal of Applied Business and Economics*, 12(3), 61–68.

Leonhardt, T. and B. Faust. 2001. "Brand Power: Using Design and Strategy to Create the Future," *Design Management Journal*, 12(1), Winter, 10–13.

Maddox, K. 2004. "The Return of the Boom," *B to B*, 89(7), 23.

MagnaGlobal. 2011. *2011 Advertising Forecast*. New York: MagnaGlobal.

Marckini, F. 2001. *Search Engine Positioning*. San Antonio, TX: Republic of Texas Press.

Masters, D. 2007. "Inline Text Ads," *Success on the Web*, September 5. (http://successonthe-web.blogspot.com/2007/09/inline-text-ads.html)

McKay, L. 2009. "Microsites to Serve Microsegments," *CRM Magazine*, 13(8), August, 21–22.

McWilliams, B. 2002. "Dot-Com Noir: When Internet Marketing Goes Sour," *Salon.com*, July 1. (http://www.salon.com/tech/feature/2002/07/01/spyware_inc/index.html)

Meyer, M. and L. Kolbe. 2005. "Integration of Customer Relationship Management: Status Quo and Implications for Research and Practice," *Journal of Strategic Marketing*, 13(3), September, 175–198.

New Media Age. 2004. "Has Branding Got Lost Amid Search?" September 2, 21–22.

Overholt, A. 2004. "Search for Tomorrow," *Fast Company*, August, 69–71.

Oxfam. 2011. *Oxfam Annual Report 2009–10*. Oxford, UK: Oxfam.

Payne, A. and P. Frow. 2005. "A Strategic Framework for Customer Relationship Management," *Journal of Marketing*, 69(4), October, 167–176.

Plosker, G. 2004. "What Does Paid Search Mean to You?" *Online*, 28(5), September–October, 49–51.

PricewaterhouseCoopers. 2011. *IAB Internet Advertising Revenue Report: 2010 Full Year Results*. New York: Interactive Advertising Bureau. (http://www.iab.net/media/file/IAB_Full_year_2010_0413_Final.pdf)

PricewaterhouseCoopers. 2011. *IAB Internet Advertising Revenue Report: 2011 First Six Months Results*. New York: Interactive Advertising Bureau. (http://www.iab.net/media/file/IAB-HY-2011-Report-Final.pdf)

Ralphs, M. 2011. "Built In or Bolt On: Why Social Currency Is Essential to Social Media Marketing," *Direct, Data and Digital Marketing Practice*, 12(3), January, 211–215.

Rapoza, J. 2004. "Annoying Web Ads Redux," *eWeek*, 21(15), April 12, 70.

Rayport, J. and J. Sviokla. 1994. "Managing in the Marketspace," *Harvard Business Review*, 72(6), November–December, 141–150.

Rayport, J. and J. Sviokla. 1995. "Exploiting the Virtual Value Chain," *Harvard Business Review*, 73(6), November–December, 75–85.

Rigby, D. and D. Ledingham. 2004. "CRM Done Right," *Harvard Business Review*, 82(11), November, 118–127.

Ryals, L. 2005. "Making Customer Relationship Management Work: The Measurement and Profitable Management of Customer Relationships," *Journal of Marketing*, 69(4), October, 252–261.

Sandoval, G. 2001. "Kozmo to Shut Down, Lay Off 1,100," *News.com*, April 11. (http://www.zdnet.com/ecommerce/stories/main/0,10475,5081050,00.html)

Schneider, G. and C. Bruton. 2003. "Communication Modalities for Commercial Speech on the Internet," *Journal of Organizational Culture, Communication, & Conflict*, 7(2) 89–94.

Schwarz, E. 2010. "Snapshots From the Digital Media Marketsphere," *Technology Review: Business Impact*, October, 20–22.

Seda, C. 2004. *Search Engine Advertising*. Indianapolis, IN: New Riders.

Simonite, T. 2010. "Why Can't Internet Ads Be Sold Like TV Commercials?" *Technology Review: Business Impact*, October, 26–27.

Tedeschi, B. 2005. "Blogging While Browsing, But Not Buying," *The New York Times*, July 4. (http://www.nytimes.com/2005/07/04/technology/04ecom.html)

Vega, T. 2011. "Online Ad Revenue Continues to Rise," *The New York Times*, April 13. (http://mediadecoder.blogs.nytimes.com/2011/04/13/online-ad-revenue-continues-to-rise/)

Weber, T. 2001. "Can You Say 'Cheese'? Intrusive Web Ads Could Drive Us Nuts," *The Wall Street Journal*, May 21, B1.

CHAPTER 5

BUSINESS-TO-BUSINESS ACTIVITIES: IMPROVING EFFICIENCY AND REDUCING COSTS

LEARNING OBJECTIVES

In this chapter, you will learn about:

- How businesses use the Internet to improve purchasing, logistics, and other business process activities
- Electronic data interchange and how it works
- How businesses have moved some of their electronic data interchange operations to the Internet
- Supply chain management and how businesses are using Internet technologies to improve it
- Electronic marketplaces and portals that make purchase–sale negotiations easier and more efficient

INTRODUCTION

Since the first large companies evolved during the Industrial Revolution, they have tried to find ways to cut costs. The first major efforts were directed at finding ways to manufacture products more efficiently. Results included standardized processes, use of machinery, and the assembly line. Later, these companies looked to cut waste in their purchasing, logistics, and management operations. After years of improving their internal processes, businesses began to look outside their own organizations for opportunities to reduce costs.

Beginning with manufacturing operations, then following with transportation services, logistics, advertising, market research, accounting, and human resources; going to outside providers of these specialized business functions became common. As transportation and shipping improved, outside providers of manufacturing operations could be located in other countries. However, language barriers, differing business customs and practices, and high data transfer costs prevented many business services from being obtained overseas.

More than a billion of the world's 7 billion people live on less than two dollars a day. Charitable organizations devote substantial resources to providing the basic necessities of life to them every year. A longer-term solution is to help those living in these conditions to find ways to grow food, start businesses, and eventually build their own industries. The lack of infrastructure (water, electricity, and roads) in poor countries has, limited the kinds of business activities that can be started in these countries. But the Internet has started to change that.

When California high school student Leila Janah won a scholarship at age 16, she decided to use the scholarship to fund a year in Ghana where she taught English and creative writing. She was impressed with the eagerness and talent of her students. When she returned to the United States, she completed a degree at Harvard and went to work in international development. In 2008, after thinking about how she could empower young people in poor rural areas, Janah realized that the Internet could offer a pathway out of poverty for them. She started **Samasource**, a not-for-profit organization that facilitates connections between these potential workers and work that large high-tech companies need to have done.

Samasource enters into contracts with large companies that have specific business process tasks they need accomplished, such as data entry, transcriptions, creating captions for images, error-checking information in databases, translating text, and so on. Samasource then breaks down these projects into small tasks that workers can perform anywhere in the world, as long as they have an Internet connection. Samasource has outfitted work centers in rural areas of Africa, South Asia, and

Haiti with inexpensive computers or smart phones, generators that provide electricity, and satellite dishes that provide Internet connectivity. In its first three years of operation, Samasource has provided more than 1600 workers with more than $1 million in payments for their work. These workers are not highly skilled, but can handle specific work if the tasks are broken down and organized for them. Samasource serves as the intermediary between organizations that have complex needs and the workers who can do pieces of the work if it is coordinated for them. Many of these workers were unemployed or, if employed, were earning less than two dollars per day. With Samasource, they can earn two dollars an hour in many cases.

Samasource joins other charitable organizations in this activity. Jeremy Hockenstein was a management consultant working in Cambodia where he met young people in Internet cafes eagerly learning English, inspired by the prospects of business globalization. Despite their enthusiasm, good jobs were scarce. In 2001, Hockenstein co-founded Digital Divide Data, a not-for-profit organization that partners with local schools in Cambodia, Laos, and Kenya to offer students a half-day job training and internship experience. Its workers typically spend four years in the program while going to school. On the job, they learn English, computer skills, and perform data entry, digital content conversion, optical scanning clean-up, and similar tasks.

Organizations such as Samasource and Digital Divide Data help businesses in the developed world get tasks accomplished more cost effectively. At the same time, they help build worker knowledge and skills in less developed countries that can help industries grow there. Global industries see this development of trained workforces that can eventually support manufacturing industries as a good long-term strategy. Until computers become perfect and the world's records are completely digitized, there will be a need for people to help with transcriptions, captioning, tagging, error-correction, and data verification. The Internet helps deliver this work in a way that does a great deal of good for people in need around the world.

PURCHASING, LOGISTICS, AND BUSINESS SUPPORT PROCESSES

In the previous two chapters, you learned about strategy issues that arise when businesses and other organizations provide information to potential customers. In terms of the value chain model described in Chapter 1, you learned about the primary activities: identify customers, market and sell, and deliver. You also became familiar with a number of business models for selling on the Web. Although many of these business models are used in business-to-business electronic commerce, the emphasis in Chapters 3 and 4 was on business-to-consumer advertising, promotion, and sales activities.

In this chapter, you will learn how companies use electronic commerce to improve their business processes, including purchasing and logistics primary activities and all of the processes relating to their support activities (which include finance and administration, human resources, and technology development). You can refer to Figure 1-9 in Chapter 1 for a review of primary activities and support activities. Although the work might not be as glamorous as designing a Web site or creating an advertising campaign, the potential for cost reductions and business process improvements in purchasing, logistics, and support activities is tremendous.

An important characteristic of purchasing, logistics, and support activities is flexibility. A purchasing or logistics strategy that works this year may not work next year. Fortunately, economic organizations are evolving from the hierarchical structures used since the Industrial Revolution to new, more flexible network structures. These network structures are, in many cases, made possible by the transaction cost reductions that companies realize when they use Internet and Web technologies to carry out business processes. For example, the use of other organizations to perform specific activities is called **outsourcing**. U.S.-based companies such as **Paychex** and **TriNet** handle payroll, human resources, health insurance, and other employee benefit plans for thousands of companies that have decided to outsource those business processes.

When the outsourcing is done by organizations in other countries, it is often called **offshoring**. Outsourcing and offshoring have existed for decades, but the activities outsourced were typically manufacturing activities. For example, Apple or Motorola would offshore the manufacture of their U.S.-designed mobile phones by having them manufactured and assembled in less-developed Asian countries. The Internet has enabled companies to offshore many nonmanufacturing activities such as purchasing, research and development, record keeping, and information management. This type of offshoring is often called **business process offshoring**. Offshoring that is done by or through not-for-profit organizations who use the business activity to support training or charitable activities in less developed parts of the world (such as the organizations described in the opening case for this chapter) is sometimes called **impact sourcing** or **smart sourcing**. It can be done in countries that do not yet have the infrastructure to support manufacturing activities.

Purchasing Activities

Purchasing activities include identifying and evaluating vendors, selecting specific products, placing orders, and resolving any issues that arise after receiving the ordered goods or services. These issues might include late deliveries, incorrect quantities, incorrect items, and defective items. By monitoring all relevant elements of purchase transactions, purchasing managers can play an important role in maintaining and improving product quality and reducing costs. In Chapter 1, you learned how companies can organize their strategic business unit activities using an industry value chain. The part of an industry value chain that precedes a particular strategic business unit called that business unit's **supply chain**. A company's supply chain for a particular product or service includes all the activities undertaken by every predecessor in the value chain to design, produce, promote, market, deliver, and support each individual component of that product or service. For example, the supply chain of an automobile manufacturer includes every activity undertaken by each individual component supplier, including engine manufacturers, steel fabricators, glass manufacturers, wiring harness assemblers, and thousands of others.

The Purchasing Department within most companies traditionally has been charged with buying all of these components at the lowest price possible. Usually, Purchasing staff did this by identifying qualified vendors and asking them to prepare bids that described what they would supply and how much they would charge. The Purchasing staff would then select the lowest bid that still met the quality standards for the component. This bidding process led to a very competitive environment with a large number of suppliers; this process focused excessively on the cost of individual components and ignored the total supply chain costs, including the cost to the manufacturing organization of dealing with such a large number of suppliers. As you learned in Chapter 1, many managers call this function "procurement" instead of "purchasing" to distinguish the broader range of responsibilities. Procurement generally includes all purchasing activities, plus the monitoring of all elements of purchase transactions. It also includes managing and developing relationships with key suppliers. Another term that is used to describe procurement activities is supply management. In many companies, procurement staff must have high levels of product knowledge to identify and evaluate appropriate suppliers. The part of procurement activity devoted to identifying suppliers and determining the qualifications of those suppliers is called **sourcing**. In Chapter 1, you learned that the use of Internet technologies in procurement activities is called e-procurement. Similarly, the use of Internet technologies in sourcing activities is called **e-sourcing**. Specialized Web-purchasing sites can be particularly useful to procurement professionals responsible for sourcing. The business purchasing process is usually much more complex than most consumer purchasing processes. Figure 5-1 shows the steps in a typical business purchasing process.

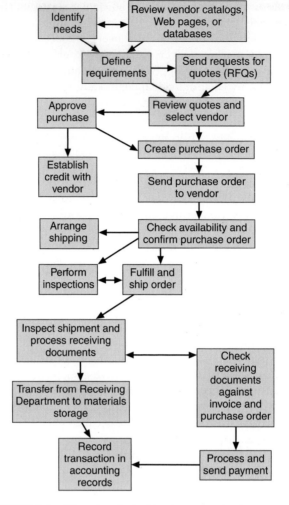

FIGURE 5-1   Steps in a typical business purchasing process

As you can see, the business purchasing process includes many steps. The business purchasing process also requires a number of people to coordinate their individual activities as part of the process. In large companies, the Procurement Department that supervises the purchasing process might include hundreds of employees who supervise the purchasing of materials, inventory for resale, supplies, and all of the other items that the company needs to buy. The total dollar amount of the goods and services that a company buys during a year is called its **spend**. In large companies, the spend can be many billions of dollars. Managing the spend in those companies is an important function and can be a key element in a company's overall profitability. Major international manufacturing companies have spends that exceed $50 billion and can process millions of purchase orders each year. By using Internet technologies in their purchasing, logistics, and support business processes, such companies can save billions of dollars each year.

For many years, the National Association of Purchasing Management has been the main organization for procurement professionals. In 2002, the association changed its name to the **Institute for Supply Management (ISM)**. ISM runs conferences, publishes a monthly journal (*Inside Supply Management*), and offers helpful information on its Web site. Many of the articles in the journal discuss implementations of Internet technologies in purchasing and logistics. Full-time students who want to learn more about supply management can join ISM at no cost.

Direct vs. Indirect Materials Purchasing

Businesses make a distinction between direct and indirect materials. **Direct materials** are those materials that become part of the finished product in a manufacturing process. Steel manufacturers, for example, consider the iron ore that they buy to be a direct material. The procurement process for direct materials is an important part of any manufacturing business because the cost of direct materials is usually a very large part of the cost of the finished product. Large manufacturing companies, such as auto manufacturers, engage in two types of direct materials purchasing. In the first type, called **replenishment purchasing** (or **contract purchasing**), the company negotiates long-term contracts for most of the materials that it will need. For example, an auto manufacturer estimates how many cars it will make during a year and contracts with two or three steel mills to supply most of the steel it will need to build those cars. By negotiating the contracts in advance and guaranteeing the purchase, the auto manufacturer obtains low prices and good delivery terms. Of course, actual demand never matches expected demand perfectly. If demand is higher than the auto company's estimate, it must buy additional steel during the year. These purchases are made in a loosely organized market that includes steel mills, warehouses, speculators (who buy and sell contracts for future delivery of steel), and companies that have excess steel that they purchased on contract (demand for their products was lower than they had anticipated). This market is called a **spot market**, and buying in this market, the second type of direct materials purchasing, is called **spot purchasing**. **Indirect materials** are all other materials that the company purchases, including factory supplies such as sandpaper, hand tools, and replacement parts for manufacturing machinery.

Large companies usually assign responsibility for purchasing direct and indirect materials to separate departments. Most companies include the purchase of nonmanufacturing goods and services—such as office supplies, computer hardware and software, and travel expenses—in the responsibilities of the indirect materials Procurement Department. Many vendors that manufacture general industrial merchandise and standard machine tools for a variety of industries have created Web sites through which their customers can purchase materials. A number of customers buy these indirect material products on a recurring basis, and many of them are commodities, that is, standard items that buyers usually select using price as their main criterion. These indirect materials items are often called **maintenance, repair, and operating (MRO)** supplies. Procurement professionals generally use the terms "indirect materials" and "MRO supplies" interchangeably. Most companies have a difficult time controlling MRO spending from a centralized procurement office because many MRO purchases are numerous and small in dollar value. One way that Procurement Departments control MRO spending is by issuing **purchasing cards** (usually called **p-cards**). These cards, which resemble credit cards, give individual managers the ability to make multiple small purchases at their discretion while providing cost-tracking information to the procurement office.

By using a Web site to process orders, the vendors in this market can save the costs of printing and shipping catalogs and handling telephone orders. They can also keep price and quantity information continually updated, which would be impossible to do in a printed catalog. Some industry analysts estimate that the cost to process an MRO order through a Web site can be less than one-tenth of the cost of handling the same order by telephone.

Two of the largest MRO suppliers in the world are **McMaster-Carr** and **W.W. Grainger**. The Grainger Web site offers more than 900,000 different products for sale. Grainger's online store, which appears in Figure 5-2, offers visitors a variety of ways to access information about and order Grainger products.

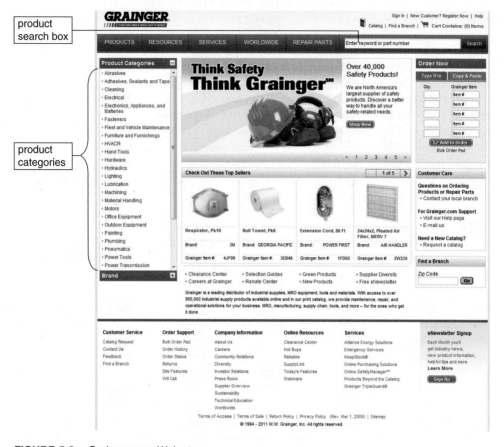

FIGURE 5-2   Grainger.com Web store

Office equipment and supplies are items that are used by a wide variety of businesses. Market leaders **Office Depot** and **Staples** each have well-designed Web sites devoted to helping business Purchasing Departments buy these routine items as easily as possible. **Digi-Key** and **Newark.com** are leading online sellers of electronic parts.

Logistics Activities

The classic objective of logistics is to provide the right goods in the right quantities in the right place at the right time. Logistics management is an important support activity for both the sales and the purchasing activities in a company. Businesses need to ensure that the products they sell to customers are delivered on time and that the raw materials they buy from vendors and use to create their products arrive when needed. The management of materials as they go from the raw materials storage area through production processes to become finished goods is also an important part of logistics.

Logistics activities include managing the inbound movements of materials and supplies and the outbound movements of finished goods and services. Thus, receiving, warehousing, controlling inventory, scheduling and controlling vehicles, and distributing finished goods are all logistics activities. The Web and the Internet are providing an increasing number of opportunities to manage these activities better as they lower transaction costs and provide constant connectivity between firms engaged in logistics management. Web-enabled automated warehousing operations are saving companies millions of dollars each year. Major transportation companies such as Schneider National, Ryder Supply Chain, and J.B. Hunt now want to be seen by their customers as information management firms as well as freight carriers.

For example, the Schneider Track and Trace system delivers real-time shipment information to Web browsers on its customers' computers. This system shows the customer which freight carrier is transporting a shipment, where the shipment is, and when it should arrive at its destination. J.B. Hunt, which operates more than 100,000 trucks, trailers, and containers, implemented a Web site that lets its customers track their shipments themselves. With customers doing their own tracking, J.B. Hunt needs far fewer customer service representatives. Also, J.B. Hunt found that its customers could monitor their own shipments more effectively than the company, saving J. B. Hunt more than $12,000 per week in labor and lost shipment costs. When transportation and freight companies engage in the business of operating all or a large portion of a customer's materials movement activities, the company is called a **third-party logistics (3PL) provider**. For example, Ryder has a multiyear contract to design, manage, and operate all of Whirlpool's inbound freight activities and is considered a 3PL provider to Whirlpool.

Both FedEx and UPS have freight-tracking Web pages available to their customers. Firms that run their own trucking operations have implemented tracking systems that use global positioning satellite (GPS) technology to monitor vehicle movements. Many of these freight-handling companies also provide 3PL services to other businesses as a way to generate additional revenue from their investments in tracking technologies. The marriage of GPS and portable computers with the Internet was an excellent example of second-wave electronic commerce. The addition of smart phone technologies to the mix is an example of the third-wave electronic commerce.

Business Process Support Activities

Activities that support all of a business' processes include finance and administration tasks, the operation of human resources, and technology development activities. Finance and administration business processes include activities such as making payments, processing payments received from customers, planning capital expenditures, and budgeting and planning to ensure that sufficient funds will be available to meet the organization's obligations as they come due. The operation of the computing infrastructure and database management functions of the organization is also an administration activity. Human resource processes include activities such as hiring, training, and evaluating employees; administering benefits; and complying with government record-keeping regulations. Technology development includes networking research scientists into virtual collaborative workgroups, sharing research results, publishing research papers online, and providing connections to outside sources of research and development services. Figure 5-3 summarizes these categories of support activities.

Finance and Administration	Human Resources	Technology Development
Making payments to suppliers Processing payments from customers Planning capital expenditures Budgeting Planning operations Operating computing infrastructure	Hiring employees Training employees Evaluating employees Administering benefit programs Compliance with government record-keeping regulations	Creating and maintaining virtual collaborative research workgroups Posting research results Publishing research reports online Connecting researchers to outside sources of research and development services

FIGURE 5-3   Categories of support activities

Human resources, payroll, and retirement plan services are all areas in which small and midsized companies often look for outside help. These business processes are subject to many detailed rules and regulations that often require an expert to decipher. A wide range of companies offer human resource management services online. Firms such as **CheckPointHR** offer a full range of services online; others, such as **CompuPay**, specialize in payroll processing services, which are also available online. These business process outsourcing providers duplicate their clients' human resources and/or payroll functions on a password-protected Web site that is accessible to clients' employees. The employees can then access their employers' benefits information, find the answers to frequently asked questions, and even perform benefit option calculations. Larger firms build these types of functions into their own internal systems.

One common support activity that underlies multiple primary activities is training. In many companies, the Human Resources Department handles training. Other companies may decentralize this function and have individual departments administer it. For example, insurance firms expend large amounts of resources on sales training. In most insurance companies, the Sales and Marketing Department administers this training. By putting training materials on the company intranet, insurance companies can distribute the training materials to many different sales offices, yet coordinate the use of those materials in the corporate headquarters sales office.

The Swedish telecommunications giant Ericsson runs an extranet for current and former employees, families of those employees, and employees of approved business partners. Ericsson has more than 120,000 employees scattered across the globe. One part of this extranet includes a Web site that enables current employees, retirees, and other recipients of payments from the company's medical and retirement plans to efficiently track their benefits. Another part of the extranet includes a Web site designed to facilitate knowledge management. **Knowledge management** is the intentional collection, classification, and dissemination of information about a company, its products, and its processes. This type of knowledge is developed over time by individuals working for or with a company and is often difficult to gather and distill. You will learn more about knowledge management and the software tools used to facilitate it in Chapter 9.

Ericsson managers hope that their knowledge network will generate new ideas, help solve problems, and improve business processes throughout the international organization. Designers of the system have identified their biggest challenge: to direct the information they collect in the extranet to projects and product development activities that will benefit from that information. You can learn more about knowledge management in general at the **KMWorld** Web site. In Chapter 9, you will learn about software that companies can use to build knowledge management systems.

E-Government

Although governments do not typically sell products or services to customers, they perform many functions for the individual citizens, businesses, and other organizations that they serve. Many of these functions can be enhanced by the use of electronic commerce. Governments also operate business-like activities; for example, they employ people, buy supplies from vendors, and distribute benefit payments of many kinds. They also collect a variety of taxes and fees from their constituents (you will learn more about how governments use the Web in administering their tax laws in Chapter 7). The use of electronic commerce by governments and government agencies to perform these functions is often called **e-government**.

In 2000, the U.S. government's Financial Management Service (FMS) opened its **Pay.gov** Web site. The FMS is the agency responsible for receiving the government's trillions of dollars of tax, license, and other fee revenue. It is also responsible for paying out trillions of dollars in Social Security benefits, veterans' benefits, tax refunds, and other disbursements. Federal agencies can link their Web sites to Pay.gov, which lets site visitors pay taxes and fees they owe to these agencies using their credit cards, debit cards, or various forms of electronic funds transfer. The U.S. government's Bureau of Public Debt operates the **TreasuryDirect** site, which allows individuals to buy savings bonds and financial institutions to buy treasury bills, bonds, and notes.

Following the terrorist attacks of September 11, 2001, the U.S. government became aware of a lack of activity coordination and information sharing among several of its agencies, including the Federal Bureau of Investigation (FBI), the Central Intelligence Agency (CIA), and the Bureau of Customs and Border Protection. A number of initiatives that use Internet technologies are under way to increase the availability of information within and among these agencies under the auspices of the **Department of Homeland Security (DHS)**.

Other countries' national governments use e-government to reduce administrative costs and provide better service to stakeholders. In the United Kingdom, the **Department for Work and Pensions** Web site provides information on unemployment, pension, and social security benefits. Smaller countries also have portal Web sites, such as **Singapore Government Online**, that provide information and enable citizens to interact with their governments online.

State governments also have Web sites for conducting business and interacting with their citizenry. In 2001, the state of California opened its one-stop portal site, **CA.GOV**, a recent version of which appears in Figure 5-4.

link to business laws, regulations, and information about doing business with California

FIGURE 5-4 State of California portal site

This site gives visitors access to every California government agency and state operation. Site visitors can transact a wide array of business with the state—from renewing a driver's license to reserving a campsite. The site gives Californians one site through which they can conduct virtually all of their business with their state. For businesses, the site offers the full text of all California business laws and regulations. It also provides information about how to sell to and buy from the state and its agencies.

Most other U.S. state governments (and, in other countries, provincial or regional governments) have similar Web sites. States can reduce the cost of providing services while providing those services more efficiently by using Web technologies to serve their stakeholders. The most common services offered by states and similar regional governments are the following: access to the text of state laws and regulations, renewal of licenses, promotion of the state to businesses considering new locations, job listings, promotion of tourism in the state, tax forms and filing information, and information for companies that want to do business with the state.

Most local governments now have Web sites that offer residents a variety of information. The Web sites of larger cities (such as Minneapolis or New Orleans) include transcripts of city council meetings, local laws and regulations, business license and tax administration functions, and promotional information about the city for new residents or businesses seeking new locations. Smaller cities, towns, and villages are also using the Web to communicate with residents (see the Cheviot, Ohio, Web site for one example). These local government Web sites have proven to be useful general communication tools in the aftermath of natural disasters.

Network Model of Economic Organization in Purchasing: Supply Webs

In Chapter 1, you learned about the three different forms of economic organization: markets, hierarchies, and networks. One trend that is becoming clear in purchasing, logistics, and support activities is the shift away from hierarchical structures toward network structures. The traditional purchasing model had one hierarchically structured firm negotiating purchase terms with several similarly structured supplier firms, playing each supplier against the others. As is typical in a network organization, more businesses are now giving their Procurement Departments new tools to negotiate with suppliers, including the possibility of forming strategic alliances. For example, a buying firm might enter into an alliance with a supplier to develop new technology that will reduce overall product costs. The technology development might be done by a third firm using research conducted by a fourth firm. Such alliances and outsourcing contracts are examples of the move toward network economic structures that you learned about in Chapter 1.

While reading the previous sections in this chapter, you might have noticed that companies can have other firms perform various support activities for them. These outsourcing and offshoring arrangements are examples of firms moving toward a network model of economic organization. Consider a business that uses one supplier to manage its payroll, another to administer its employee benefits plans, and a third to handle its document storage needs. The document storage service supplier might store the documents of the payroll service supplier and the benefits administration firm. The payroll service supplier might handle the payroll for the benefits administration firm. A fourth firm might provide online backup storage for the files of the other three companies. Of course, the payroll firm and the employee benefits firm might form a marketing partnership to sell both of their services to particular market segments. The document storage firm and the online backup storage firm might form a similar strategic alliance. Some researchers who study the interaction of firms within an industry value chain are beginning to use the term **supply web** instead of "supply chain" because many industry value chains no longer consist of a single sequence of companies linked in a single line, but include many parallel lines that are interconnected in a web or network configuration.

Highly specialized firms can now exist and trade services very efficiently on the Web. The Web is enabling this shift from hierarchical to network forms of economic organization. These emerging networks of firms are more flexible and can respond to changes in the economic environment much more quickly than hierarchically structured businesses. You can learn more about the economics of networked organizations at the Network Economics Web site maintained by the University of California, Berkeley. The roots of Web technology for business-to-business transactions, however, lie in a hierarchically structured approach to inter-firm information transfer: electronic data interchange.

ELECTRONIC DATA INTERCHANGE

In Chapter 1, you learned that electronic data interchange (EDI) is a computer-to-computer transfer of business information between two businesses that uses a standard format of some kind. The two businesses that are exchanging information are trading partners. Firms that exchange data in specific standard formats are said to be **EDI compatible**. The business information exchanged is often transaction data; however, it can also include other information related to transactions, such as price quotes and order status inquiries. Transaction data in business-to-business (B2B) transactions includes the information traditionally included on paper documents. The data from invoices, purchase orders, requests for quotations, bills of lading, and receiving reports accounts for more than 75 percent of all information exchanged by U.S. trading partners. EDI was the first form of electronic commerce to be widely used in business—beginning some 20 years before anyone used the term "electronic commerce."

Understanding EDI is important because most B2B electronic commerce is based on EDI or adapted from EDI. It is also important because EDI is still the single most commonly used technology in online B2B transactions. The dollar amount of EDI transactions today is about equal to that of all other B2B transaction technologies combined. This section provides a brief history of EDI and explains how it works. It also explains why EDI is better than processing mountains of paper transactions.

Early Business Information Interchange Efforts

The emergence of large business organizations in the late 1800s and early 1900s brought with it the need to create formal records of business transactions. By the 1950s, companies were using computers to keep records of internal transactions, but information flows between businesses required paper documents (purchase orders, invoices, bills of lading, checks, remittance advices, and so on) because one company's computers could not communicate with other companies' computers. Generating these paper forms (by hand or as printed computer output), mailing them, and then having recipients enter the data from them into their computer systems was slow, inefficient, expensive, redundant, and unreliable. By the 1960s, businesses with large transaction volumes had begun exchanging information with each other by shipping punched cards or reels of magnetic tape. During the 1960s and 1970s, data communications technologies improved, allowing businesses to transfer much of this intercompany information over telephone lines instead.

Although these information transfer agreements between trading partners increased efficiency and reduced errors, they were not an ideal solution. Because the data translation programs that one business wrote would frequently not work on other businesses' computers, each company participating in these information exchanges had to spend considerable money to write their own programs. Only large companies could afford this investment. Smaller or lower-volume businesses could not afford to participate.

In 1968, a number of freight and shipping companies joined together to attack their collective paperwork burden. They created a standardized information set that included all the data elements that shippers commonly included on bills of lading, freight invoices, shipping manifests, and other paper forms. Instead of printing a paper form, shippers could convert information about shipments into a computer file that they could send to any freight company that had adopted the standard. The freight company could then transfer the standardized data into its own information systems. The costs saved by not printing or handling forms, not re-entering data, and avoiding errors were significant, even for smaller shippers and freight carriers.

Although these industry-specific data interchange standards were helpful, their benefits were limited to members of the standard-setting groups in those specific industries. Most businesses that are in one industry buy goods and services from businesses that are in other industries. For example, a machinery manufacturer might buy materials from steel mills, paint distributors, electrical assembly contractors, and container manufacturers. Almost every business needs to buy office supplies and the services of freight and transportation companies. Thus, full realization of economies and efficiencies required standards that could be used by companies in all industries.

Emergence of Broader Standards: The Birth of EDI

Toward the end of the 1970s, standard-setting groups and several large companies that were frustrated by fragmented industry standards decided to mount a major effort to create a set of cross-industry standards for electronic components, mechanical equipment, and other widely used items. The **American National Standards Institute (ANSI)** has been the coordinating body for standards in the United States since 1918. ANSI does not set standards, but it maintains procedures for the development of national standards and accredits committees that follow those procedures. In 1979, ANSI chartered a new committee to develop uniform EDI standards. This committee is called the **Accredited Standards Committee X12 (ASC X12)**. The ASC X12 committee and its subcommittees include information systems professionals from hundreds of businesses. The administrative body that coordinates ASC X12 activities is the Data Interchange Standards Association (DISA). The ASC X12 standard currently includes specifications for several hundred **transaction sets**, which are the names of the formats for specific business data interchanges.

The X12 standards were quickly adopted by major firms in the United States, but businesses in other countries continued to use their own national standards. In the mid-1980s, the United Nations Economic Commission for Europe invited North American and European EDI experts to work together on designing a common set of EDI standards based on the successful experiences of U.S. firms in using the ASC X12 standards. In 1987, the United Nations published its first standards under the title **EDI for Administration, Commerce, and Transport (EDIFACT, or UN/EDIFACT)**. The DISA and the UN/EDIFACT group have attempted to develop a single common set of international standards several times since 2000; however, these attempts have never succeeded. Today, both standards continue to exist. Companies that do business worldwide must either make their EDI software work with both standards or use a software product that does conversions between the standards. Figure 5-5 lists some of the more commonly used transaction sets, showing the paper document from which the transaction set was devised along with the identifiers of the ASC X12 and the UN/EDIFACT versions of the transaction set.

Transaction Description	Transaction Set Identifiers	
	ASC X12	**UN/EDIFACT**
Ordering Transactions		
Purchase Order	850	ORDERS
Purchase Order Acknowledgement	855	ORDRSP
Purchase Order Change	860	ORDCHG
Request for Quotation	840	REQOTE
Response to Request for Quotation	843	QUOTES
Shipping Transactions		
Ship Notice/Manifest (Advance Shipping Notice)	856	DESADV
Bill of Lading (Shipment Information)	858	IFTMCS
Receiving Advice	861	RECADV
Sales and Payment Transactions		
Invoice	810	INVOIC
Freight Invoice	859	IFTFCC
Payment Order/Remittance Advice	820	REMADV

FIGURE 5-5   Commonly used EDI transaction sets

How EDI Works

Although the basic idea behind EDI is straightforward, its implementation can be complicated, even in fairly simple business situations. For example, consider a company that needs a replacement for one of its metal-cutting machines. This section describes the steps involved in making this purchase using a paper-based system, and then explains how the process would change using EDI. In both of these examples, we assume that the vendor uses its own vehicles instead of a common carrier to deliver the purchased machine.

Paper-Based Purchasing Process

The buyer and the vendor in this example are not using any integrated software for business processes internally; thus, each information-processing step results in the production of a paper document that must be delivered to the department handling the next step. Information transfer between the buyer and vendor is also paper based and can be delivered by mail, courier, or fax. The information flows that occur in the paper-based version of the purchasing process example are shown in Figure 5-6.

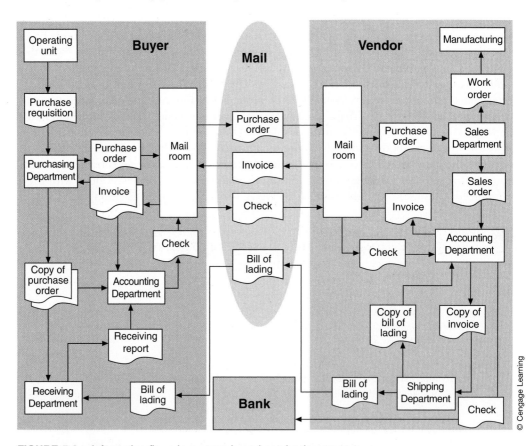

FIGURE 5-6   Information flows in a paper-based purchasing process

© Cengage Learning

Once the production manager in the operating unit decides that the metal-cutting machine needs to be replaced, the following process begins:

1. The production manager completes a purchase requisition form and sends it to Purchasing. This requisition describes the machine that is needed to perform the metal-cutting operation.

2. Purchasing contacts vendors to negotiate price and terms of delivery. When Purchasing has selected a vendor, it prepares a purchase order and forwards it to the mail room.

3. Purchasing also sends one copy of the purchase order to the Receiving Department so that Receiving can plan to accept delivery when scheduled; Purchasing sends another copy to Accounting to advise it of the financial implications of the order.

4. The mail room sends the purchase order it received from Purchasing to the selected vendor by mail or courier.

5. The vendor's mail room receives the purchase order and forwards it to its Sales Department.

6. The vendor's Sales Department prepares a sales order that it sends to its Accounting Department and a work order that it sends to Manufacturing. The work order describes the machine's specifications and authorizes Manufacturing to begin work on it.

7. When the machine is completed, Manufacturing notifies Accounting and sends the machine to shipping.

8. The Accounting Department sends the original invoice to the mail room and a copy of the invoice to the Shipping Department.

9. The mail room sends the invoice to the buyer by mail or courier.

10. The vendor's Shipping Department uses its copy of the invoice to create a bill of lading and sends it with the machine to the buyer.

11. The buyer's mail room receives the invoice at about the same time as its Receiving Department receives the machine with its bill of lading.

12. The buyer's mail room sends one copy of the invoice to Purchasing so the Purchasing Department knows that the machine was received, and sends the original invoice to Accounting.

13. The buyer's Receiving Department checks the machine against the bill of lading and its copy of the purchase order. If the machine is in good condition and matches the specifications on the bill of lading and the purchase order, Receiving completes a receiving report and delivers the machine to the operating unit.

14. Receiving sends a completed receiving report to Accounting.

15. Accounting makes sure that all details on its copy of the purchase order, the receiving report, and the original invoice match. If they do, Accounting issues a check and forwards it to the mail room.

16. The buyer's mail room sends the check by mail or courier to the vendor.

17. The vendor's mail room receives the check and sends it to Accounting.

18. Accounting compares the check to its copies of the invoice, bill of lading, and sales order. If all details match, Accounting deposits the check in the vendor's bank and records the payment received.

EDI Purchasing Process

The information flows that occur in the EDI version of this sample purchasing process are shown in Figure 5-7. The mail service has been replaced with the data communications of an EDI network, and the flows of paper within the buyer's and vendor's organizations have been replaced with computers running EDI translation software.

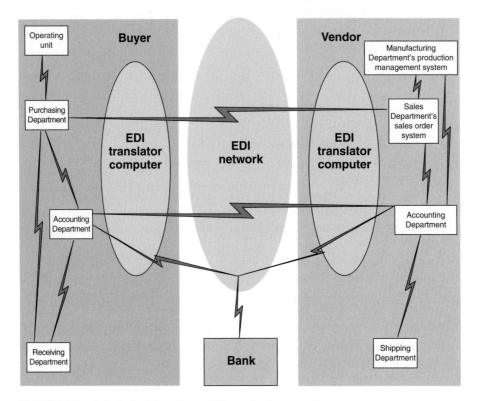

FIGURE 5-7 Information flows in an EDI purchasing process

In the EDI purchasing process, when the operating unit manager decides that the metal-cutting machine needs to be replaced, the following process begins:

1. The operating unit manager sends an electronic message to its Purchasing Department. This message describes the machine that is needed to perform the metal-cutting operation.

2. Purchasing contacts vendors by telephone, e-mail, or through their Web sites to negotiate price and terms of delivery. After selecting a vendor, Purchasing sends a message that the buyer's EDI translator computer converts to a standard format purchase order transaction set that goes through an EDI network to the vendor where the message is routed through its EDI translator and sent to the Sales Department. At that point, the message is automatically entered into the vendor's Manufacturing Department production management system (where the machine's specifications are provided so Manufacturing can begin work on building it) and the vendor's accounting system (in their Accounting Department).

3. Purchasing also sends electronic messages to the buyer's Receiving Department (so it can plan to accept delivery when it is expected) and to the

buyer's Accounting Department with details such as the agreed purchase price.

4. When the machine is completed, Manufacturing notifies Accounting and sends the machine to the vendor's Shipping Department.

5. The vendor's Shipping Department sends an electronic message to its Accounting Department indicating that the machine is ready to ship. It also sends an electronic message to its EDI translator computer that indicates the machine is ready to ship. The EDI translator computer converts the message into a standard 856 transaction set (Advance Ship Notification) and forwards it through the EDI network to the buyer.

6. The vendor's Accounting Department sends a message to its EDI translator computer, which converts the message to the standard invoice transaction set and forwards it through the EDI network to the buyer's EDI translator computer before the buyer's Receiving Department receives the machine. The computer then converts the invoice data to a format that the buyer's information systems can use. The invoice data becomes immediately available to both the buyer's Accounting and Receiving Departments.

7. When the machine arrives, the buyer's Receiving Department checks the machine against the invoice information on its computer system. If the machine is in good condition and matches the specifications shown in the buyer's system, Receiving sends a message to Accounting confirming that the machine has been received in good order. It then delivers the machine to the operating unit.

8. The buyer's Accounting Department system compares all details in the purchase order data, receiving data, and decoded invoice transaction set from the vendor. If all the details match, the accounting system notifies its bank to reduce the buyer's account and increase the vendor's account by the amount of the invoice. The EDI network may provide services that perform this task.

As you can see by comparing the paper-based purchasing process in Figure 5-6 to the EDI purchasing process in Figure 5-7, the departments are exchanging the same messages among themselves, but EDI reduces paper flow and streamlines the interchange of information among departments within a company and between companies. The paper-based system has 18 individual steps compared to the eight steps required to complete this transaction using EDI. The three key elements (shown in Figure 5-7) that alter the process so dramatically are the EDI network (instead of the mail service) that connects the two companies and the two EDI translator computers that handle the conversion of data from the formats used internally by the buyer and the vendor to standard EDI transaction sets.

Value-Added Networks

Trading partners can implement the EDI network and EDI translation processes in several ways. Each of these ways uses one of two basic approaches: direct connection or indirect connection. The first approach, called **direct connection EDI**, requires each business in the network to operate its own on-site EDI translator computer (as shown in Figure 5-7). These EDI translator computers are then connected directly to each other using leased telephone lines. Because dedicated leased-lines are expensive, only a few very large companies still use direct connection EDI, which is illustrated in Figure 5-8.

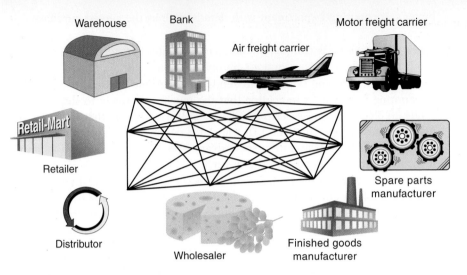

FIGURE 5-8   Direct connection EDI

Instead of connecting directly to each of its trading partners, a company might decide to use the services of a value-added network. As you learned in Chapter 1, a value-added network (VAN) is a company that provides communications equipment, software, and skills needed to receive, store, and forward electronic messages that contain EDI transaction sets. To use the services of a VAN, a company must install EDI translator software that is compatible with the VAN. Often, the VAN supplies this software as part of its operating agreement.

To send an EDI transaction set to a trading partner, the VAN customer connects to the VAN using a dedicated or dial-up telephone line and then forwards the EDI-formatted message to the VAN. The VAN logs the message and delivers it to the trading partner's mailbox on the VAN computer. The trading partner then dials in to the VAN and retrieves its EDI-formatted messages from that mailbox. This approach is called **indirect connection EDI** because the trading partners pass messages through the VAN instead of connecting their computers directly to each other. Figure 5-9 illustrates indirect connection EDI using a VAN.

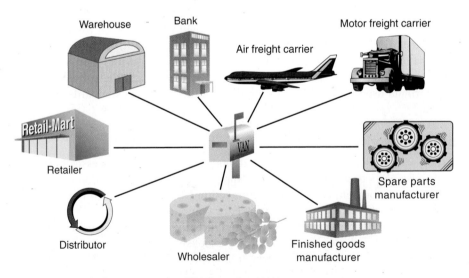

FIGURE 5-9   Indirect connection EDI through a VAN

Companies that provide VAN services include **CovalentWorks**, **EasyLink Services**, **GXS**, **Kleinschmidt**, **Promethean Software Services**, and **SPS Commerce**. Advantages of using a VAN are as follows:

1. Users need to support only the VAN's one communications protocol instead of many possible protocols used by trading partners.
2. The VAN can provide translation between different transaction sets used by trading partners (for example, the VAN can translate an ASC X12 set into a UN/EDIFACT set).
3. The VAN can perform automatic compliance checking to ensure that the transaction set is in the specified EDI format.
4. The VAN records message activity in an audit log. This VAN audit log becomes an independent record of transactions; this record can be helpful in resolving disputes between trading partners.

Because EDI transactions are business contracts and often involve large amounts of money, the existence of an independent audit log helps establish nonrepudiation. **Nonrepudiation** is the ability to establish that a particular transaction actually occurred. It prevents either party from repudiating, or denying, the transaction's validity or existence.

In the past, VANs had one serious disadvantage: cost. Most VANs required an enrollment fee, a monthly maintenance fee, and a transaction fee ranging from a few cents to a dollar that was levied on each transaction. The up-front cost of implementing indirect connection EDI, including software, VAN enrollment fee, and hardware, could easily exceed $20,000.

Today, VAN costs are much lower because VANs use the Internet instead of leased telephone lines to connect to their customers. Costs to begin EDI are less than $5000, with monthly fees under $100 that include a generous transaction allowance. Even small companies find that they can engage indirect connection EDI and sell to large industrial and retail companies that require their vendors to use EDI.

Companies that provide VAN services today all use the Internet as their main data communication technology. EDI on the Internet is called **Internet EDI** or **Web EDI**. It is also called **open EDI** because the Internet is an open architecture network, as you learned in Chapter 2. The **EDIINT (Electronic Data Interchange-Internet Integration**, also abbreviated **EDI-INT**) set of protocols is the most common set used for the exchange of EDI transaction sets over the Internet.

Most EDIINT exchanges today are encoded using the **AS2 (Applicability Statement 2)** specification, which is based on the HTTP rules for Web page transfers, although some companies are using a more secure specification, **AS3 (Applicability Statement 3)**. Wal-Mart, for example, requires all of its vendors to use the EDIINT protocol transmitted using AS2. Both AS2 and AS3 transmissions return secure electronic receipts to the senders for every transaction, which helps establish nonrepudiation.

EDI Payments

Some EDI transaction sets provide instructions to a trading partner's bank. These transaction sets are negotiable instruments; that is, they are the electronic equivalent of checks. All banks have the ability to perform electronic funds transfers (EFTs), which are the movement of money from one bank account to another. You learned about EFTs in Chapter 1. The bank accounts involved in EFTs may be customer accounts or the accounts that banks keep on their own behalf with each other. When EFTs involve two banks, they are executed using an **automated clearing house (ACH)** system, which is a service that banks use to manage their accounts with each other. In the United States, banks can use the ACH operated by the U.S. Federal Reserve Banks or one of the private ACHs operated by a group of banks or a separate company. You will learn more about how banks process ACH payments in Chapter 11.

SUPPLY CHAIN MANAGEMENT USING INTERNET TECHNOLOGIES

You learned earlier in this chapter that the part of an industry value chain that precedes a particular strategic business unit is called a supply chain. Many companies use strategic alliances, partnerships, and long-term contracts to create relationships with other companies in the supply chains for the products that they manufacture or sell. These relationships can be quite complex, with suppliers helping their customers develop new products, specify product features, refine product specifications, and identify needed product improvements. In many cases, companies are able to reduce costs by developing close relationships with a few suppliers rather than negotiating with a large number of suppliers each time they need to buy materials or supplies. When companies integrate their supply management and logistics activities across multiple participants in a particular product's supply chain, the job of managing that integration is called **supply chain management**. The ultimate goal of supply chain management is to achieve a higher-quality or lower-cost product at the end of the chain.

Value Creation in the Supply Chain

In recent years, businesses have realized that they can save money and increase product quality by taking a more active role in negotiations with suppliers. By engaging suppliers in cooperative, long-term relationships, companies have found that they can work together with these suppliers to identify new ways to provide their own customers with faster, cheaper, and better service. By coordinating the efforts of supply chain participants, firms that engage in supply chain management are reaching beyond the limits of their own organization's hierarchical structure and creating a new network form of organization among the members of the supply chain.

Supply chain management was originally developed as a way to reduce costs. It focused on very specific elements in the supply chain and tried to identify opportunities for process efficiency. Today, supply chain management is used to add value in the form of benefits to the ultimate consumer at the end of the supply chain. This requires a more holistic view of the entire supply chain than had been common in the early days of supply chain management.

Businesses that engage in supply chain management work to establish long-term relationships with a small number of very capable suppliers. These suppliers, called **tier-one suppliers**, in turn develop long-term relationships with a larger number of suppliers that provide components and raw materials to them. These **tier-two suppliers** manage relationships with the next level of suppliers, called **tier-three suppliers**, that provide them with components and raw materials. A key element of these relationships is trust between the parties. The long-term relationships created among participants in the supply chain are called **supply alliances**. The level of information sharing that must take place among the supply chain participants can be a major barrier to entering into these alliances. Firms are not accustomed to disclosing detailed operating information and often perceive that information disclosure might hurt the firm by placing it at a competitive disadvantage.

For example, Dell Computer has been able to reduce supply chain costs by sharing information with its suppliers. The moment Dell receives an order from a customer, it makes that information available to its tier-one suppliers, who can then better plan their production based on Dell's exact demand trends. For example, a supplier of disk drives can change its production plans immediately when it sees a shift in Dell's customer orders from computers with one size disk drive to another, usually larger, size disk drive. This prevents the supplier from overproducing the smaller drive, which reduces the supplier's costs (for unsold drives) and costs in the supply chain overall (the supplier does not need to charge more for the disk drives it does sell to Dell to recover the cost of the unsold drives).

In exchange for the stability of the closer, long-term relationships, buyers expect annual price reductions and quality improvements from suppliers at each stage of the supply chain. However, all supply chain participants share information and work together to create value. Ideally, the supply chain coordination creates enough value that each level of supplier can share the benefits of reduced cost and more efficient operations. Supply chain management has been gaining momentum during the past decade and is supported by major purchasing groups such as the Supply Chain Council. By working together, supply chain members can reduce costs and increase the value of the product or service to the ultimate consumer.

One area in which differences in organizational goals often arise is described by Marshall Fisher in his 1997 *Harvard Business Review* article. He explains that firms often organize themselves to achieve either efficient process goals or market-responsive flexibility goals. Some companies structure themselves to be efficient producers, whereas others structure themselves to be flexible producers. The kinds of things that allow a firm to be an efficient, low-cost producer are exactly the things that prevent a firm from being flexible enough to respond to market changes. For example, the efficient producer invests in expensive machines that can stamp out large numbers of low-cost items. This investment drives down the cost of production, but makes it difficult for the producer to be flexible. A large investment in specialized machinery prevents that producer from reconfiguring the plant layout. If even one member of the supply chain for a product that requires flexible production operates as an efficient producer (instead of as a flexible producer), every other firm in the supply chain suffers. The efficient producer creates bottlenecks that hamper the best efforts of all other supply chain members. Clear communication up and down the supply chain can keep each participant informed of what the ultimate consumer demands. The participants can then plot a strategy to meet those demands.

Clear communications, and quick responses to those communications, are key elements of successful supply chain management. Technologies, and especially the technologies of the Internet and the Web, can be very effective communications enhancers. For the first time, firms can effectively manage the details of their own internal processes and the processes of other members of their supply chains. Software that uses the Internet can help all members of the supply chain review past performance, monitor current performance, and predict when and how much of certain products need to be produced. Figure 5-10 lists the advantages of using Internet technologies in supply chain management. The only major disadvantage of using Internet technologies in supply chain management is the cost of the technologies. In most cases, however, the advantages provide value that greatly exceeds the cost of implementing and maintaining the technologies.

Suppliers can:

- Share information about changes in customer demand
- Receive rapid notification of product design changes and adjustments
- Provide specifications and drawings more efficiently
- Increase the speed of processing transactions
- Reduce the cost of handling transactions
- Reduce errors in entering transaction data
- Share information about defect rates and types

FIGURE 5-10   Advantages of using Internet technologies in supply chain management

Increasing Supply Chain Efficiencies

Many companies are using Internet and Web technologies to manage supply chains in ways that yield increases in efficiency throughout the chain. These companies have found ways to increase process speed, reduce costs, and increase manufacturing flexibility so that they can respond to changes in the quantity and nature of ultimate consumer demand.

For example, Boeing, the largest producer of commercial aircraft in the world, faces a huge task in keeping its production on schedule. Each airplane requires more than 1 million individual parts and assemblies, and each airplane is custom configured to meet the purchasing airline's exact specifications. These parts and assemblies must be completed and delivered on schedule or the production process comes to a halt.

Using EDI and Internet links, Boeing works with suppliers so that they can provide exactly the right part or assembly at exactly the right time. Even before an airplane enters into production, Boeing makes the engineering specifications and drawings available to its suppliers through secure Internet connections. As work on the airplane progresses, Boeing keeps every member of the supply chain continually informed of completion milestones achieved and necessary schedule changes. Instead of waiting 36 months for delivery, customers can now have their new airplanes in 10 months or less.

Although Dell Computer is famous for its use of the Web to sell custom-configured computers to individuals and businesses, it has also used technology-enabled supply chain management to give customers exactly what they want. Dell's tier-one suppliers have access to a secure Web site that shows them Dell's latest sales forecasts, along with other information about planned product changes, defect rates, and warranty claims. In addition, the Web site tells suppliers who Dell's customers are and what they are buying. All of this information helps these tier-one suppliers plan their production much better than they could otherwise. The information sharing goes in both directions in Dell's supply chain: tier-one suppliers are required to provide Dell with current information on their defect rates and production problems. As a result, all members of the supply chain work together to reduce inventories, increase quality, and provide high value to the ultimate consumer. The improved coordination between Dell and its tier-one suppliers has reduced the amount of inventory Dell must keep on hand from three weeks' sales to two hours' sales. Ultimately, Dell wants to see inventory levels measured in minutes. By increasing the amount of information it has about its customers, Dell has been able to dramatically reduce the amount of inventory it must hold. Dell has also shared this information with members of its supply chain. This kind of cooperative work requires a high level of trust. To enhance this trust and develop a sense of community, Dell maintains discussion boards as an open forum in which its supply chain members can share their experiences in dealing with Dell and with each other.

For Boeing, Dell, and other firms, the use of Internet and Web technologies in managing supply chains has yielded significantly increased process speed, reduced costs, and increased flexibility. All of these attributes combine to allow a coordinated supply chain to produce products and services that better meet the needs of the ultimate consumer.

Materials-Tracking Technologies

Tracking materials as they move from one company to another and as they move within the company has always been challenging. Companies have been using optical scanners and bar codes for many years to help track the movement of materials. In many industries, the integration of bar coding and EDI has become prevalent. Figure 5-11 shows a typical bar-coded shipping label that is used in the auto industry. Each bar-coded element is a representation of an element of the ASC X12 transaction set number 856, Advance Shipping Notice. If you examine the figure carefully, you can see that five of the 856 transaction set's elements have been bar coded (including Part Number, Quantity Shipped, Purchase Order Number, Serial Number, and Packing List Number).

FIGURE 5-11 Shipping label with bar-coded elements from EDI transaction set 856, Advance Shipping Notice

These bar codes allow companies to scan materials as they are received and to track them as they move from the materials warehouse into production. Companies can use this bar-coded information along with information from their EDI systems to manage inventory flows and forecast materials needs across their supply chains.

Large online retailers such as Amazon.com, Target, and Kohl's maintain fulfillment centers from which they ship products that customers have ordered online. Tracking systems, called **real-time location systems (RTLS)**, in these fulfillment centers use bar codes to monitor inventory movements and ensure that goods are shipped as quickly as possible.

In the second wave of electronic commerce, companies are integrating new types of tracking into their Internet-based materials-tracking systems. The most promising technology now being used is **Radio Frequency Identification Devices (RFIDs)**, which are small chips that use radio transmissions to track inventory. RFID technology has existed for many years, but until recently, it required each RFID to have its own power supply (usually a battery). RFIDs can be read much more quickly and with a higher degree of accuracy than bar codes. Bar codes must be visible to be scanned. RFID tags can be placed anywhere on or in most items and are readable even when covered with packing materials, dirt, or plastic bands. A bar-code scanner must be placed within a few inches of the bar code. Most RFID readers have a range of about six feet.

An important development in RFID technology is the passive RFID tag, which can be made cheaply and in very small sizes. A passive RFID tag does not need a power source. It receives a radio signal from a nearby transmitter and extracts a tiny amount of power from that signal. It uses the power it extracts to send a signal back to the transmitter. That signal includes information about the inventory item to which the RFID tag has been affixed. RFID tags are small enough to be installed on the face of credit cards or sewn into clothing items.

In 2003, Wal-Mart began testing the use of RFID tags on its merchandise for inventory tracking and control. Wal-Mart initiated a plan to have all of its suppliers install RFID tags in the goods they shipped to the retailer. Wal-Mart wanted suppliers to do this within three years. Having all incoming inventory RFID tagged would allow Wal-Mart to manage its inventory better and reduce the incidence of stockouts. A **stockout** occurs when a retailer loses sales because it does not have specific goods on its shelves that customers want to buy. Many of Wal-Mart's suppliers found the RFID tags, readers, and the computer systems needed to manage tagged inventory to be quite expensive. These suppliers pushed Wal-Mart to slow down the implementation of its plan. Wal-Mart responded by encouraging suppliers to use RFID tags, but focused its energies on developing pilot projects within Wal-Mart to test RFID-based inventory management systems.

Many industry observers have concluded that general acceptance of RFID tagging will not occur in most industries until 2015. Although the cost of a passive RFID tag is now below eight cents, even that small cost can be prohibitive for companies that ship large volumes of low-priced goods. The cost of RFID tags is expected to continue dropping, and as it does, more and more companies will find them to be useful in an increasingly wide range of situations. You can learn more about current developments in this technology by visiting the **RFID Journal** online. Figure 5-12 shows a typical passive RFID tag.

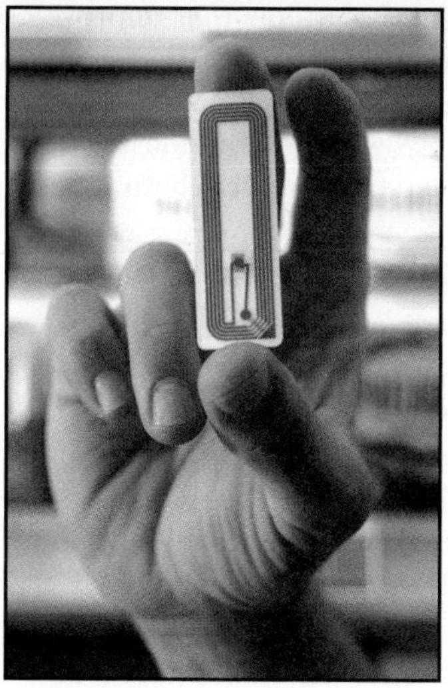

FIGURE 5-12 Passive RFID tag

Courtesy, Moeller-Horcher. Source: Metro

Creating an Ultimate Consumer Orientation in the Supply Chain

One of the main goals of supply chain management is to help each company in the chain focus on meeting the needs of the consumer at the end of the supply chain. Companies in industries with long supply chains have, in the past, often found it difficult to maintain this customer focus, which is sometimes called an **ultimate consumer orientation**. Instead, companies have directed their efforts toward meeting the needs of the next member in the supply chain. This short-sighted approach can cause companies to miss opportunities to add value in subsequent steps of the chain.

One company that pioneered the use of Internet technology to go beyond the next step in its value chain is Michelin North America. Michelin has a highly respected brand name and reputation in the tire business. However, most consumers rely on local tire dealers to make specific recommendations when they need replacement tires for their vehicles. Michelin spends a great deal of money on direct advertising to its ultimate consumers. This advertising is directed at maintaining Michelin's powerful brand and convincing the consumer of the value of Michelin tires. The advertising and brand building effort can be wasted, however, if the consumer goes to a local tire dealer who recommends another brand.

Michelin launched an online business initiative in 1995 called BIB NET (after the company's famous Michelin Man mascot, whose name is Bibendum). The goal of this initiative was to sell more Michelin tires to consumers, but the initiative was directed at Michelin's tire dealers, not the ultimate consumers. BIB NET was an extranet that allowed tire dealers to access tire specifications, inventory status, and promotional information about Michelin products through a simple-to-use Web browser interface. Before BIB NET, dealers calling Michelin for product information were sometimes placed on hold. A dealer who is talking to a customer cannot afford to wait on hold. By giving dealers the power to access Michelin product information directly and immediately, Michelin saved money (maintaining a Web page is much less expensive than answering thousands of phone calls) and gave dealers better service. Dealers using BIB NET are much less likely to recommend a competitor's tires to their customers.

Because Internet technologies are tools that improve communications at a very low cost, they are ideal aids for enhancing the creation of a highly coordinated and effective supply chain. A number of polls and studies confirm that most information technology and purchasing managers believe that information technology is helping to improve their firms' relationships with suppliers and supply chain management initiatives.

Building and Maintaining Trust in the Supply Chain

The major issue that most companies must deal with in forming supply chain alliances is developing trust. Continual communication and information sharing are key elements in building trust. Because the Internet and the Web provide excellent ways to communicate and share information, they offer new avenues for building trust. Most procurement professionals have built trust on years of doing business with the same vendors. In many industries, vendors send sales representatives to call on buyers regularly. Vendors also participate actively in trade shows and conferences. By giving buyers frequent opportunities to interact with vendor representatives, vendors help build trust.

Vendors are finding that the Web gives them an opportunity to stay in contact with their customers more easily and less expensively. Although most buyers still see vendor sales representatives regularly, e-mail and the Web give them nearly instant access to their sales representative and other vendor personnel. By providing comprehensive information at a moment's notice, vendors can build buyers' trust in the vendor's ability to deliver products and provide the personalized service that buyers need.

Many supply chain management researchers are working on new ways to accumulate information about supplier performance and report that information to supply chain partners. This type of monitoring and reporting could help companies establish trust more quickly. Many issues, such as the objectivity and validity of performance measurements, must still be resolved before these information networks become generally accepted and used by the supply chain community. The task of developing information exchange resources that can provide supplier performance summaries is one of the great challenges that B2B electronic commerce faces in its second wave.

ELECTRONIC MARKETPLACES AND PORTALS

In the late 1990s, a number of industry-focused hubs opened and began offering marketplaces and auctions in which companies in the industry could contact each other and transact business. The idea was that these hubs would offer a doorway (or portal) to the Internet for industry members. Because these hubs were vertically integrated (that is, each hub would offer services to just one industry), they were called **vertical portals**, or **vortals**. In this section, you will learn how these B2B electronic marketplaces were conceived, developed, and operated as this sector of electronic commerce matured from 1997 through the present.

Independent Industry Marketplaces

The first vertical portals were trading exchanges focused on a particular industry. These vertical portals became known by various names that highlighted different elements of their collective nature, including **industry marketplaces** (focused on a single industry), **independent exchanges** (not controlled by a company that was an established buyer or seller in the industry), or **public marketplaces** (open to new buyers and sellers just entering the industry). These portals are also known collectively as **independent industry marketplaces**. Ventro opened its first industry marketplace, Chemdex, in early 1997 to trade in bulk chemicals. To leverage the high investment it had made in trading exchange technology, Ventro followed Chemdex with other Web marketplaces, including Promedix in specialty medical supplies, Amphire Solutions in food service, MarketMile in general business products and services, and a number of others. Other companies were quick to follow in Ventro's chosen markets as well as many others.

The number of new entrants into these businesses grew rapidly. By mid-2000, there were more than 2200 independent exchanges in a wide variety of industries. As venture capital funding became scarce for companies that were not earning profits—and virtually none of these marketplaces were earning profits—many of them closed. By 2010, there were fewer than 50 industry marketplaces still operating. Ventro, for example, has closed all of the dozens of marketplaces it had opened during the boom years. It turned out that no more than one or two independent marketplaces in any particular industry could survive.

Some of the industry pioneers who closed their industry marketplace operations, such as Ventro, were able to build successful businesses selling the software and technology that they developed to run their marketplaces. Today, leading software vendors such as IBM, Microsoft, Oracle, and SAP also offer products that can be used to build B2B marketplaces. In the mid-2000s, B2B marketplace models gradually replaced independent marketplaces as the dominant form of operation in this type of electronic commerce. You will learn about four of these B2B marketplace models—private stores, customer portals, private company marketplaces, and industry consortia-sponsored marketplaces—in the remainder of this section.

LEARNING FROM FAILURES

MetalSite

Although a number of small steel manufacturing plants (called minimills) have opened in the past 20 years, most of the world's steel is still produced in very large steel mills. In these steel mills, it is economical to produce steel only in large batches. Because of the high cost of reconfiguring machinery, a steel mill set up to create one type of steel (for example, rolled sheets) requires significant time and money to change over to produce another type of steel (for example, bar steel). To minimize these changeover costs, steel mills produce steel products in large batches to meet estimated demand rather than actual orders. Because production quantities are designed to meet estimated demand instead of actual demand, steel mills often have overproduction of some items.

Companies such as Bethlehem Steel, with annual revenues of more than $4 billion and 14,000 employees, solved this problem in the past by sending faxes to potential buyers of their excess production. Buyers would respond with a bid on the product in which they were interested, and Bethlehem would negotiate with them to determine price and delivery terms.

In 1998, MetalSite was one of the first metal trading exchanges to begin doing business on the Web. These exchanges offered manufacturers such as Bethlehem an efficient way to reach a larger market for their excess production. By mid-2000, there were more than 200 metal exchanges operating on the Web. These exchanges were following a reintermediation strategy; that is, they were entering the supply chain of the steel industry to provide some added value that had not existed in the supply chain before. However, most industry analysts agreed that there was no need for more than one or two exchanges in the steel industry. In 2001, metal trading exchange sites began to fail.

Continued

MetalSite had grown rapidly. With more than $35 million of investors' money, MetalSite was able to sign up 24,000 registered users and by mid-2001, was trading about $30 million worth of steel each month. However, its commissions of between 1 percent and 2 percent on each trade did not yield enough money to cover operating costs. The steel business was in a downturn along with the rest of the U.S. economy, and the downward pressure on commissions from competing exchanges was increasing rapidly. The major steel companies were discussing ways to form alliances to operate their own exchanges. After three years of operation and a desperate last-minute search for new investors, MetalSite closed in August 2001.

MetalSite had entered a business that could not support more than a few companies, and it was unable to become one of the survivors. The lesson from MetalSite's experience is that a reintermediation strategy must add significant value to the supply chain, and the company pursuing that strategy must be able to construct significant barriers that competitors must overcome to enter the business. MetalSite was unable to do either and thus failed. Many other B2B exchange sites that found themselves in similar competitive situations have also failed.

Private Stores and Customer Portals

As established companies in various industries watched new businesses open marketplaces, they became concerned that these independent operators would take control of transactions from them in supply chains—control that the established companies had spent years developing. Large companies that sell to many relatively small customers can exert great power in negotiating price, quality, and delivery terms with those customers. These sellers feared that industry marketplaces would dilute that power.

Many of these large sellers had already invested heavily in Web sites that they believed would meet the needs of their customers better than any industry marketplace. For example, Cisco and Dell offer private stores for each of their major customers within their selling Web sites. A **private store** has a password-protected entrance and offers negotiated price reductions on a limited selection of products—usually those that the customer has agreed to purchase in certain minimum quantities. Other companies, such as Grainger, provide additional services for customers on their selling Web sites. These **customer portal** sites offer private stores along with services such as part number cross referencing, product usage guidelines, safety information, and other services that would be needlessly duplicated if the sellers were to participate in an industry marketplace.

Private Company Marketplaces

Similarly, large companies that purchase from relatively small vendors can exert comparable power over those vendors in purchasing negotiations. The Procurement Departments of these companies can install procurement software (you will learn more about all types of electronic commerce software in Chapter 9), generally referred to as **e-procurement software**, that allows a company to manage its purchasing function through a Web interface. It automates many of the authorizations and other steps (see Figure 5-1) that are part of business procurement operations.

Although e-procurement software was originally designed to help manage the MRO procurement process, today it includes other marketplace functions, such as requests for quote posting areas, auctions, and integrated support for purchasing direct materials. E-procurement software for large companies can cost millions of dollars for licensing fees, installation, and customization; however, a growing number of companies are offering e-procurement software for smaller businesses.

Companies that implement e-procurement software usually require their suppliers to bid on their business. For example, an office supplies provider would create a schedule of prices at which it would sell to the company. The company would then compare that pricing to bids from other suppliers. The selected supplier would provide product price and description information to the company, which would insert that information into its e-procurement software. This permits authorized employees to order office supplies at the negotiated prices through a Web interface.

When industry marketplaces opened for business, these larger companies were reluctant to abandon their investments in e-procurement software or to make the software work with industry marketplaces' software—especially in the early years of industry marketplaces when there were many of them in each industry. These companies use their power in the supply chain to force suppliers to deal with them on their own terms rather than negotiate with suppliers in an industry marketplace.

As marketplace software became more reliable, many of these companies purchased software and technology consulting services from companies, such as Ventro and e-Steel, that had abandoned their industry marketplace businesses and were offering the software they had developed to companies that wanted to develop private marketplaces. A **private company marketplace** is a marketplace that provides auctions, request for quote postings, and other features (many of which are similar to those of e-procurement software) to companies that want to operate their own marketplaces.

Industry Consortia-Sponsored Marketplaces

Some companies had relatively strong negotiating positions in their industry supply chains, but did not have enough power to force suppliers to deal with them through a private company marketplace. These companies began to form consortia to sponsor marketplaces. An **industry consortia-sponsored marketplace** is a marketplace formed by several large buyers in a particular industry.

Figure 5-13 summarizes the characteristics of five general forms of marketplaces that exist in B2B electronic commerce today. The information in the figure comes from several sources, but the structure of the figure is adapted from one presented by Warren Raisch, a Web marketplace consultant, in his book *The eMarketplace*.

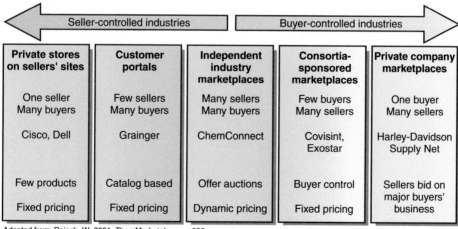

Seller-controlled industries			Buyer-controlled industries	
Private stores on sellers' sites	**Customer portals**	**Independent industry marketplaces**	**Consortia-sponsored marketplaces**	**Private company marketplaces**
One seller Many buyers	Few sellers Many buyers	Many sellers Many buyers	Few buyers Many sellers	One buyer Many sellers
Cisco, Dell	Grainger	ChemConnect	Covisint, Exostar	Harley-Davidson Supply Net
Few products	Catalog based	Offer auctions	Buyer control	Sellers bid on major buyers' business
Fixed pricing	Fixed pricing	Dynamic pricing	Fixed pricing	

Adapted from: Raisch, W. 2001. *The eMarketplace*, p. 225.

FIGURE 5-13   Characteristics of B2B marketplaces

Although the figure shows five distinct B2B marketplace categories, the lines between them are not always clear. For example, Dell has from time to time sold other companies' products on its private store site, which would make it more like a customer portal than a private store. As the B2B marketplace industry matures, it is unlikely that one type of marketplace will become dominant. Most B2B experts believe that a variety of marketplaces with the characteristics of these five general categories will continue to exist for some time.

Summary

In this chapter, you learned how companies are using Internet technologies in a variety of ways to improve their business processes for purchasing, logistics, and support activities. Companies and other large organizations, such as government agencies, are finding it more important than ever to extend the reach of their enterprise planning and control activities beyond their organizations' legal definitions to include parts of other organizations. This emerging network model of organization was introduced in Chapter 1 and is used in this chapter to describe the growth in interorganizational communications and coordination. In many cases, organizations outsource some of their business processes to companies that specialize in those processes. Some of those business process service providers are located in other countries and can perform the work at a much lower cost.

EDI, the first example of electronic commerce, was first developed by freight companies to reduce the paperwork burden of processing repetitive transactions. The spread of EDI to virtually all large companies has led smaller businesses to seek an affordable way to participate in EDI. The Internet is now providing the inexpensive communications channel that EDI lacked for so many years and is allowing smaller companies to participate in Internet EDI.

The increase in communications capabilities offered by the Internet and the Web is, and will continue to be, an important force driving the adoption of supply chain management techniques in a variety of industries. Supply chain management can be implemented and enhanced through the use of online technologies. Increasingly, firms are connecting with their supply chain alliance partners and other companies, such as 3PL providers, to become more efficient and provide more value to the ultimate consumer of their value chains' products and services.

The emergence of industry electronic marketplaces in the mid-1990s gave way to the development of several different models for B2B electronic commerce, including private stores, customer portals, private marketplaces, and industry consortia-sponsored marketplaces.

Key Terms

Accredited Standards Committee X12 (ASC X12)

American National Standards Institute (ANSI)

AS2 (Applicability Statement 2), AS3 (Applicability Statement 3)

Automated clearing house (ACH)

Business process offshoring

Contract purchasing

Customer portal

Direct connection EDI

Direct materials

EDI compatible

EDI for Administration, Commerce, and Transport (EDIFACT, or UN/EDIFACT)

EDIINT (Electronic Data Interchange-Internet Integration, EDI-NET)

E-government

E-procurement software

E-sourcing

Impact sourcing

Independent exchange

Independent industry marketplace

Indirect connection EDI

Indirect materials

Industry consortia-sponsored marketplace

Industry marketplace

Internet EDI

Knowledge management

Maintenance, repair, and operating (MRO)

Nonrepudiation

Offshoring

Open EDI

Outsourcing

Private company marketplace

Private store

Public marketplace

Purchasing card (p-card)

Radio Frequency Identification Devices (RFIDs)

Real-time location systems (RTLS)

Replenishment purchasing

Smart sourcing

Sourcing

Spend

Spot market

Spot purchasing

Stockout

Supply alliances

Supply chain

Supply chain management

Supply web

Third-party logistics (3PL) provider

Tier-one suppliers

Tier-three suppliers

Tier-two suppliers

Transaction sets

Ultimate consumer orientation

Vertical portal (vortal)

Web EDI

Review Questions

1. In about two paragraphs, distinguish between outsourcing and offshoring as they relate to business processes.

2. Define "direct materials" and "indirect materials." List three or more reasons that a large company would have two separate departments to manage the purchasing activities for each.

3. In about 100 words, explain the differences between contract purchasing and spot purchasing.

4. Define "logistics." In one paragraph, explain what services a third-party logistics provider might offer.

5. In 100 words, define "e-government" and describe three services or functions that should be offered on a small town's Web site.

6. In about 100 words, describe how and why the economies and efficiencies realized by early business information interchange efforts were limited.

7. In two paragraphs, explain what changes allowed smaller companies to participate in electronic data interchange (EDI).

8. In about 100 words, explain why trust is an important element in supply chain alliances.

9. In one or two paragraphs, describe the key elements of an independent industry marketplace.

Exercises

1. Using your library or your favorite search engine, identify the main reason a medium-sized manufacturing company might want to use p-cards for its MRO spending. Summarize your findings in two or three paragraphs.

2. You have just started work as an intern in the Purchasing Department of Westridge Systems, a manufacturer of electronic control systems for manufacturing assembly lines. You do not know much about electronic equipment, but your supervisor has given you the task of identifying vendors who sell oscilloscopes that interface with personal computers. Use the ThomasNet Web site to locate at least three vendors who offer such a product. For

each company, determine whether it offers products for sale on a Web site that discloses prices and details about the products' specifications. Summarize what you have learned from your research about how each vendor sells its oscilloscopes online in a report of approximately 150 words.

Note: Your instructor might ask you to prepare a formal presentation of your findings in class.

3. You work for Cobalt Milling, a manufacturer of steel fittings and cutting tools for the metal-working industry. Susan, the company's purchasing manager, wants you to look into improving vendor relationships through Internet EDI. Select two of the EDI service providers from the list provided in the Web Links that would be suitable for a small manufacturing firm, and examine their Web sites. In a memo to Susan of approximately 100 words, explain why the two providers you chose would be suitable for Cobalt Milling.

4. A number of standard-setting organizations offer memberships to business firms. You are working for Grace Henry, chief information officer (CIO) of Flex-Electric, a midsize company that manufactures components for electronic medical and laboratory instruments. Flex-Electric uses EDI to process transactions with both its vendors (purchases) and customers (sales). The company is also exploring ways in which it might use RFID tags to track its inventories. Grace asks you to learn more about the international supply chain standard-setting organization, **GS1**. Prepare a memo to Grace of about 300 words in which you outline the purposes of the organization and identify areas in which it is active that might be useful for Flex-Electric to know about.

5. Companies in a particular supply chain can work together to eliminate costs from the supply chain. In many cases, these cost savings are not shared evenly among the companies in the supply chain. Using research resources on the Web or in your library, identify an industry in which savings are not shared equally. In two or three paragraphs, explain why some supply chain participants in your chosen industry can obtain more benefit than others from cost reductions in the supply chain.

6. Some business and political leaders argue that offshoring is dangerous because it can move jobs from developed countries to less developed countries. Others argue that although offshoring might displace workers in the short run, in the longer term, everyone benefits by having developing economies create new industries, products, and markets for products and services provided by companies in the developed world. Using resources in your library or online, present two arguments for and two arguments against a U.S. company offshoring the maintenance of its customer database (error checking, removing duplicate entries, and so on).

Cases

C1. Harley-Davidson

Harley-Davidson manufactures high-end motorcycles and sells them worldwide. The company sells more than $4 billion in motorcycles and related products each year, and has one of the most recognized brands in the world. However, business was not always so good for the company. In the 1980s, the company was on the brink of bankruptcy. Facing increasing competition from Japanese and German manufacturers, Harley-Davidson had allowed its quality standards and cost controls to slip. In a legendary business turnaround, the company rebuilt itself. Harley-Davidson completely changed its supply chain to fulfill the expectations of its brand-aware customers.

Over a period of several years, Harley-Davidson reduced its number of suppliers from 4000 to fewer than 350. More important, it began to work with those suppliers to reduce costs throughout the supply chain. Each supplier is expected to find ways (with the help and cooperation of Harley-Davidson) to reduce manufacturing costs and improve quality every year. This was the only way Harley-Davidson believed it could avoid moving its factories to lower-cost locations in other countries. The efforts paid off and the company still manufactures its motorcycles only in the United States.

In 2000, the company decided to focus its cost reduction and quality improvement efforts on its information technology infrastructure. Because it had been so successful in working with its suppliers to reduce manufacturing costs and improve quality, Harley-Davidson wanted to do the same thing with information technology. By using Internet technologies to share information throughout the supply chain, the company hoped to find opportunities for efficiencies and cost reductions at all stages of the process of creating motorcycles.

When the company first talked with its suppliers about its information technology initiative, those suppliers noted that each of Harley-Davidson's main factories used different invoices, production schedules, and purchasing procedures. The suppliers explained that this created difficulties for them when they dealt with more than one factory and increased their cost of doing business with Harley-Davidson. Thus, one of the first things the company did was to standardize forms and procedures. Then it moved to require all suppliers to use EDI. For smaller suppliers, the company set up a Web site that had Internet EDI capabilities. The smaller suppliers could simply log in to the Web site and conduct EDI transactions through their Web browsers.

This Web browser interface grew to become a complete extranet portal called **Harley-Davidson Supply Net**. All suppliers now use the portal to consolidate orders, track production schedule changes, obtain inventory forecasts in real time, and obtain payments for materials shipped. The portal also allows suppliers to obtain product testing information, part specifications, and product design drawings.

Key elements in both EDI and the Web portal systems have been bar codes and scanners. Most of Harley-Davidson's individual parts and all shipments are bar coded. The bar-code information is integrated with the materials tracking, invoicing, and payment information in the systems and is made available, as appropriate, to suppliers. Harley-Davidson uses bar-code standards developed by the **Automotive Industry Action Group**.

Required:

1. Become familiar with RFID technology and its potential uses in Harley-Davidson's supply chain using the information presented in this chapter and information you obtain through the Web Links, your favorite search engine, and your library. In about 400 words, evaluate the advantages and disadvantages for Harley-Davidson of replacing its bar codes and scanners technology with RFID.

2. In about 100 words, compare and contrast the issues that Wal-Mart and other large retailers faced when they tried to implement RFID in their supply chains with those that Harley-Davidson will likely face as it moves into RFID implementations with its suppliers.

3. When Harley-Davidson implements RFID, it will likely use the technology to help manage its relationships with its main customers, which are the local dealerships that sell motorcycles and use replacement parts in their repair shops. In about 200 words, outline the issues that will likely arise when Harley-Davidson begins requiring RFID tracking of spare parts inventories at its dealers.

4. In a report of about 300 words, develop and present an approximate timetable for the adoption of RFID technology with specific recommendations about where Harley-Davidson should first implement it. For example, RFID tags could be installed in

motorcycles as they leave final assembly, in various parts before they are shipped from suppliers, or in subassemblies as they are created at various Harley-Davidson manufacturing operations. Justify the time delays you propose in the adoption of RFID at each stage of the supply chain.

Note: Your instructor might assign you to a group to complete this case, and might ask you to prepare a formal presentation of your results to your class.

C2. American Packaging Machinery

American Packaging Machinery (APM) is a company that provides repair and maintenance services to companies that operate large packaging systems. Packaging systems are arrangements of machinery that place items in containers such as boxes or bags and apply plastic shrink wrap to the containers. These machines must be adjusted regularly, and they have hundreds of parts that can wear out or fail. APM offers service contracts on most major packaging systems. A typical service contract provides for an APM technician to make regular visits to the customer site to perform preventive maintenance. The service contract also includes a certain number of emergency repair visits per year. APM also sends technicians to perform repairs for companies that do not have service contracts.

APM technicians are paid by the hour, with additional pay for overtime hours and time they work outside of standard working hours, such as weekends and holidays. APM technicians are members of a labor union, the International Brotherhood of Electrical Workers (IBEW), which negotiates pay rates and working conditions for the technicians. APM subtracts union dues from each technician's weekly paycheck and submits the total dues collected each week to the IBEW regional office. The union contract currently provides that APM technicians are covered by a medical insurance plan underwritten by the Prudential Trust Insurance Company. Although APM pays most of the insurance premium, technicians do pay a part of the premium cost. This contribution to the premium is withheld from their paychecks each week.

You are the director of electronic commerce for APM and you report to Laura Adams, APM's CIO. Laura asks for your help in outlining a new automated system she wants to install, which would use EDI and EFTs to handle APM's technician payroll and related transactions. She has provided the following narrative that describes how the system will work:

1. Technicians will record their time worked by entering the start and stop times for each job into a program that runs on their handheld computers (the technicians already use these handheld computers to look up wiring and mechanical diagrams for the machinery on which they work and to receive their job assignments). The time-worked information will be transmitted from the handheld computer to APM's Payroll Department.

2. The Payroll Department will summarize the time-worked information and send it to supervisors' desktop computers. Each supervisor will indicate an authorization for each technician's time-worked, overtime, and holiday/weekend hours. That authorization will be returned by the system each day to the Payroll Department.

3. The Payroll Department will summarize the time-worked information each week and calculate gross pay, deductions, and net pay for each employee. The deductions include the federal and state taxes that must be withheld by law, the contribution to the medical insurance premium, and the union dues that are withheld under the IBEW union contract.

4. The Payroll Department will send an electronic summary of the payroll information, including deductions, to the Accounting Department, which will prepare payroll tax returns and make the necessary entries in the APM accounting system to record payroll and the related tax expenses.

5. The Payroll Department will send electronic authorizations to APM's bank to make the necessary EFTs to deposit: the amount of each technician's net pay to that technician's bank account; the amount of each tax withheld to the account of the appropriate government agency; the amount of the total contributions to the medical insurance premium to the insurance company's account; and the amount of the union dues withheld to the IBEW's account. Most of these accounts are at other banks.

6. The Payroll Department will send electronic notifications to Prudential Trust and the IBEW regional office, notifying them of the transferred amounts each week.

7. The Payroll Department will send an electronic summary of the hours worked by each technician and the amount of gross pay, including overtime and holiday/weekend pay, to the APM union steward's desktop computer. The union steward is an APM technician who is elected by the technicians to monitor the terms of the union contract and handle any grievances that arise between the technicians and APM management.

Required:

1. Draw a diagram of the proposed payroll EDI and EFT system (you can use Figure 5-7 as a guide).

2. List and briefly describe any problems or issues that you think might arise in the implementation of this system.

3. Provide a rationale and recommendation as to which elements of this system—if any— you think APM should hire an outside company to implement.

Note: Your instructor might assign you to a group to complete this case, and might ask you to prepare a formal presentation of your results to your class.

For Further Study and Research

Abid-Ali, A. 2009. "Driving Efficiency with RFID," *Electronics Weekly*, May 13, 11–12.

Albrecht, C., D. Dean, and J. Hansen. 2005. "Marketplace and Technology Standards for B2B E-commerce: Progress, Challenges, and the State of the Art," *Information & Management*, 42(6), September, 865–875.

Asher, A. 2007. "Developing a B2B E-Commerce Implementation Framework: A Study of EDI Implementation for Procurement," *Information Systems Management*, 24(4), Fall, 373–390.

Benton, E. 2010. "Leila Janah, Founder of Samasource," *Fast Company*, March 23. (http://www.fastcompany.com/article/leila-janah-samasource)

Bills, S. 2009. "Fed EDI Service for Small-Bank Clients," *American Banker*, June 15, 10.

Binns, S. 2004. "Businesses Miss Benefits of High-Tech Radio Tagging," *Supply Management*, 9(2), January 22, 13.

Bornstein, D. 2011. "Workers of the World, Employed," *The New York Times*, November 3. (http://opinionator.blogs.nytimes.com/2011/11/03/workers-of-the-world-employed/)

Bovel, D. and M. Joseph. 2000. "From Supply Chain to Value Net," *Journal of Business Strategy*, 21(4), July–August, 24–28.

Boye, J. 2008. "Enterprise Portal Market Overview 2008," *KM World*, 17(5), May, 8–10.

Bunyaratavej, K., J. Doh, E. Hahn, A. Lewing, and S. Massini. 2011. "Conceptual Issues in Services Offshoring Research: A Multidisciplinary Review," *Group & Organization Management*, 36(1), February, 70–102.

Clark, P. 2001. "MetalSite Kills Exchange, Seeks Funding," *B to B*, 86(13), June 25, 3.

Cleary, M. 2001. "Metal Meltdown Doesn't Deter New Ventures," *Interactive Week*, 8(27), July 9, 29.

Commercial Carrier Journal. 2008. "EDI's a Habit Hard to Break," January, 58–59.

Demery, P. 2010. "How Wine Country Gift Baskets Saves With Drop-Shipping," *Internet Retailer*, August 26. (http://www.internetretailer.com/2010/08/26/how-wine-country-gift-baskets-saves-drop-shipping)

DiSera, M. 2009. "How to Improve ROI with RFID," *Control Engineering*, 56(4), April, 48–51.

Dobbs, J. 1999. *Competition's New Battleground: The Integrated Value Chain*. Cambridge, MA: Cambridge Technology Partners.

Drickhamer, D. 2003. "EDI is Dead! Long Live EDI!" *Industry Week/IW*, 252(4), April, 31–35.

Duvall, M. 2007. "Wal-Mart Changes its Faltering RFID Strategy to Lure More Suppliers, But Insists it's not Turning Back," *Baseline*, October, 43–55.

Financial Executive. 2008. "E-procurement," 24(1), February, 61.

Fisher, M. 1997. "What Is the Right Supply Chain for Your Product?" *Harvard Business Review*, 75(2), March-April, 105–116.

Fraser, J. 2007. "Commercial Tools Boost Partner Connection in the Value Network," *Manufacturing Business Technology*, 25(11), November, 43.

Friedman, T. 2006. *The World Is Flat: A Brief History of the Twenty-first Century*. New York: Farrar, Straus and Giroux.

Fries, J., A. Turri, D. Bello, and R. Smith. 2010. "Factors That Influence the Implementation of Collaborative RFID Programs," *Journal of Business & Industrial Marketing*, 25(8), 590–595.

Fulcher, J. 2007. "Internet-based EDI May be Reliable and Less Expensive, but not Necessarily Easier," *Manufacturing Business Technology*, 25(6), June, 40–42.

Huang, Z., B. Janz, and M. Frolick. 2008. "A Comprehensive Examination of Internet-EDI Adoption," *Information Systems Management*, 25(3), Summer, 273–286.

Karpinski, R. 2002. "Wal-Mart Mandates Secure, Internet-Based EDI for Suppliers," *Internet-Week*, September 12. (http://www.internetwk.com/security02/INW20020912S0011)

Kay, R. 2009. "QuickStudy: Extensible Business Reporting Language (XBRL): The SEC Mandates It, How Does It Work?" *Computerworld*, October 5. (http://www.computerworld.com/s/article/342881/XBRL_Extensible_Business_Reporting_Language)

Kenney, M., S. Massini, and T. Murtha. 2009. "Offshoring Administrative and Technical Work: New Fields for Understanding the Global Enterprise. *Journal of International Business Studies*, 40, 887–900.

Lekakos, G. 2007. "Exploiting RFID Digital Information in Enterprise Collaboration," *Industrial Management & Data Systems*, 107(8), 1110–1122.

Lewin, A. and H. Volberda. 2011. "Co-evolution of Global Sourcing: The Need to Understand the Underlying Mechanisms of Firm Decisions to Offshore," *International Business Review*, 20(3), June, 241–251.

Massini, S., N. Perm-Ajchariyawong, and A. Lewin. 2010. "The Role of Corporate-wide Offshoring Strategy in Directing Organizational Attention to Offshoring Drivers, Risks, and Performance. *Industry and Innovation*, 17(4), 337–371.

McCartney, L. and A. Virzi. 2007. "GlobalSpec: The Little Engine that Could," *Baseline*, October, 57–58.

Messmer, E. 2007. "Dot-com Survivor Stays the Course: Covisint Remains a Valuable Player in Auto Industry E-commerce," *Network World*, 24(43), November 5, 18.

Morgan, J. and R. Monczka. 2003. "Why Supply Chains Must Be Strategic," *Purchasing*, April 17, 42–45.

Noormohammadi, M. 2011. "Samasource Provides Jobs for Poor Via the Internet," *Voice of America*, December 10. (http://www.voanews.com/english/news/Samasource-Provides-Jobs-for-Poor-Via-the-Internet-135376738.html)

Ngai, E. and F. Riggins. 2008. "RFID: Technology, Applications, and Impact on Business Operations," *International Journal of Production Economics*, 112(2), April, 507–509.

Purchasing. 2001. "MetalSite Shuts Operations While Seeking New Owner," July 5, 32.

Purchasing. 2004. "Easing into E-procurement with Indirect Spend," February 19, 35–36.

Raisch, W. 2001. *The eMarketplace: Strategies for Success in B2B Ecommerce*. New York: McGraw-Hill.

RFID Journal. 2011. "How Much Does an RFID Tag Cost Today?" (http://www.rfidjournal.com/faq/20/85)

Rueter, T. 2011. "Faster Fulfillment," *Internet Retailer*, May 17. (http://www.internetretailer.com/2011/05/17/faster-fulfillment)

Ryder, K. 2011. "Five Useful iPad Apps," *The Wall Street Journal Asia Scene Blog*, February 24. (http://blogs.wsj.com/scene/2011/02/24/five-useful-ipad-apps/)

Silver, B. 2005. "Content in the Age of XML," *Intelligent Enterprise*, June 1, 24–26.

Songini, M. 2004. "Supply Chain System Failures Hampered Army Units in Iraq," *Computerworld*, 38(30), July 26, 1–2.

Stockdale, R. and C. Standing. 2002. "A Framework for the Selection of Electronic Marketplaces: A Content Analysis Approach," *Internet Research: Electronic Networking Applications and Policy*, 12(3), 221–234.

Sullivan, L. 2004. "Ready to Roll," *Information Week*, March 8, 45–47.

Sullivan, M. 2001. "High-Octane Hog," *Forbes*, 168(6), September 10, 8–10.

Supplier Selection & Management Report. 2003. "How Harley-Davidson Teamed With 16 Major Suppliers To Cut Costs," 3(1), January, 1–3.

Tanner, C., R. Wölfle, P. Schubert, and M. Quade. 2008. "Current Trends and Challenges in Electronic Procurement: An Empirical Study," *Electronic Markets*, 18(1), January, 6–18.

Taylor, D. 2004. "No Time to Spare: A Guide to Supply Chain Performance Management," *Intelligent Enterprise*, 7(10), June 12, 20–24.

Taylor, D. and A. Terhune. 2001. *Doing E-Business: Strategies for Thriving in an Electronic Marketplace*. New York: John Wiley & Sons.

Ufelder, S. 2004. "B2B Survivors: Why Did Some Online Exchanges Survive While Many Others Failed?" *Computerworld*, February 2, 27–29.

Ustundag, A. and M. Tanyas. 2009. "The Impact of Radio Frequency Identification (RFID) Technology on Supply Chain Costs," *Transportation Research*, 45(1), January, 29–38.

Waugh, R. and S. Elliff. 1998. "Using the Internet to Achieve Purchasing Improvements at General Electric," *Hospital Material Management Quarterly*, 20(2), November, 81–83.

Whang, S. 2010. "Timing of RFID Adoption in a Supply Chain," *Management Science*, 56(2), February, 343–355.

Yao, Y., M. Dresner, and J. Palmer. 2009. "Private Network EDI vs. Internet Electronic Markets: A Direct Comparison of Fulfillment Performance," *Management Science*, 55(5), May, 843–852.

Zang, Y. and L Wu. 2010. "Application of RFID and RTLS Technology in Supply Chain Enterprise," *Proceedings of the 2010 Sixth International Wireless Communications Networking and Mobile Computing Conference*, September 23–25, 1–4.

CHAPTER 6

SOCIAL NETWORKING, MOBILE COMMERCE, AND ONLINE AUCTIONS

LEARNING OBJECTIVES

In this chapter, you will learn:

- How social networking emerged from virtual communities
- How social networking tools such as blogs are used in online business activities
- About mobile technologies that are now used to do business online
- How online auctions and auction-related businesses have become a major new commercial activity introduced as part of electronic commerce

INTRODUCTION

In 2003, Mark Zuckerberg and several other students at Harvard University were working independently on ways to create online information spaces that would network Harvard students with each other. Zuckerberg's Web site rapidly became successful, attracting more than half of Harvard's undergraduate student body as participants. These students posted photos and information about themselves and their activities.

Although Zuckerberg ran into some resistance from the school's administration and was forced to take down his site, he believed the concept had merit. So he continued working on the idea and, after dropping out of school and moving to California, he and two fellow students launched

TheFacebook.com, a networking site for college and university students, in 2004. One of PayPal's founders, Peter Theil, invested $500,000 in the fledgling enterprise and helped the company raise an additional $38 million over the next two years.

By 2006, the company had purchased the domain name Facebook.com for $200,000 and had signed major advertising deals, including a three-year agreement with Microsoft. Facebook had gradually expanded the range of users it allowed to set up pages on the site, and by 2006, it was open to everyone. As other Web sites that offered similar functions became less popular, **Facebook** continued to grow.

In 2011, Facebook reported having more than 750 million regular users (60 percent of whom were located outside the United States) and was valued at $50 billion in a round of share offerings to Goldman Sachs and overseas private investors. Analysts estimate that the privately held company earns about $500 million per year on revenue of about $2 billion. In this chapter, you will learn about Facebook and other Web sites that earn profits by facilitating visitors' connections to each other.

FROM VIRTUAL COMMUNITIES TO SOCIAL NETWORKS

In Chapters 3 and 4, you learned how businesses use the Web to create online identities, reach customers, and sell products and services to those customers. In Chapter 5, you learned how businesses are using the Web to purchase goods and work with their suppliers more effectively. In all three of these chapters, the focus was on how companies are using the Web to improve the things that they have been doing for years; primarily buying and selling. In this chapter, you will learn how companies are using the Web to do things that they have never done before. The Web makes it possible for people to form online communities that are not limited by geography. Individuals and companies with common interests can meet online and discuss issues, share information, generate ideas, and develop valuable relationships.

As you learned in earlier chapters, the Internet reduces transaction costs in value chains and offers an efficient means of communication to anyone with an Internet connection. Combining the Internet's transaction cost-reduction potential with its role as a facilitator of communication among people has led companies to develop new ways of making money on the Web by serving as relationship facilitators.

This section begins with a brief history of online communities, and then outlines how companies today operate Web sites that promote relationships among site visitors and businesses that advertise or otherwise participate on the sites.

Virtual Communities

A **virtual community**, also called a **Web community** or an **online community**, is a gathering place for people and businesses that does not have a physical existence. Howard Rheingold described the characteristics of these communities in his 1993 book, *The Virtual Community*, which is widely recognized as the definitive book on the subject. Virtual communities began online even before the Internet was in general use. **Bulletin board systems (BBSs)** were computers that allowed users to connect through modems (using dial-up connections through telephone lines) to read and post messages in a common area, or electronic bulletin board. BBSs often hosted discussions on specific topics or issues related to specific geographic regions. Many BBSs were free, but some charged a monthly membership fee. Other discussion board services followed, provided by commercial enterprises such as Compuserv, Prodigy, and GEnie. These companies generated revenue by charging a monthly fee and selling advertising. Usenet newsgroups were another early form of virtual community. Started at Duke University in 1979, **Usenet** was a set of interconnected computers devoted to storing information on specific topics. **Usenet newsgroups** were message posting areas on those computers in which interested persons (primarily from the education and research communities) could discuss those topics.

Today, Web chat rooms and sites devoted to specific topics or the general exchange of information, photos, or videos can constitute virtual communities. These communities offer people a way to connect with each other and discuss common issues and interests. The social interaction in these communities can be considerable and many sociologists believe that the communication and relationship-forming activities that occur online are similar to those that occur in physical communities. The rest of this section describes the development of these communities into the Web sites that people use today to form and maintain relationships online.

Early Web Communities

One of the first Web communities was the **WELL**. The WELL, which is an acronym for "whole earth 'lectronic link," predates the Web. It began in 1985 as a series of BBS dialogs among the authors and readers of the *Whole Earth Review*. Members of the WELL pay a monthly fee to participate in its forums and conferences. The WELL was home to many of the researchers who created the Internet and the Web along with a number of noted writers and artists. In 1999, Salon.com bought the WELL and continues to operate it as a monthly subscription service.

As the Web emerged in the mid-1990s, its potential for creating new virtual communities was quickly exploited. In 1995, Beverly Hills Internet opened a virtual community site that featured two Webcams aimed down Hollywood streets; the site also had links to entertainment information Web sites. Members were given free space on the site to create their own Web pages. The Webcams never did attract much traffic, but the offer of free Web space did. As the site grew, it changed its name to GeoCities and earned revenue by selling advertising that appeared on members' Web pages and pop-up pages that opened whenever a visitor accessed a member's site. GeoCities grew rapidly and was purchased in 1999 by Yahoo! for $5 billion. Yahoo! operated the site for ten years before closing it in 2009.

Other similar sites became virtual communities. Tripod was founded in 1995 in Massachusetts and offered its participants free Web page space, chat rooms, news and weather updates, and health information pages. Like GeoCities, Tripod sold advertising on its main pages and on participants' Web pages. Theglobe.com also began in 1995 as a class project at Cornell University. The site included bulletin boards, chat rooms, discussion areas, and personal ads. Theglobe.com sold advertising to support the site's operation.

Later additions included news feeds, an online art gallery, and shopping pages. The company fell victim to the Web slowdown of 2000 and closed in 2001 after suffering declines in its advertising revenue.

The idea behind these early Web community sites continues to inspire online business endeavors. Virtual communities evolved into the social networking sites that emerged in the second wave of electronic commerce, as you will learn in the next section.

Social Networking Emerges

Virtual communities provided an important service to the small number of people who regularly used the Internet in its early days. As the Internet and Web grew, many of these communities found that their original purpose as a place for sharing the new experiences of online communication began to fade. In the second wave of electronic commerce, a new phenomenon in online communication began. People who were using the Internet no longer found a single common bond in the very fact that they were using the Internet. Instead, they were finding that a variety of common interests—for example, gardening, specific medical issues, or parenting—created the basis for online interaction. Later Internet communities were formed in which the Internet itself was no longer the focal point of the community, but was simply a tool that enabled communication among community members. These Web sites, designed to facilitate interactions among people, are called social networking sites. A **social networking site** is a Web site that allows individuals to create and publish a profile, create a list of other users with whom they share a connection (or connections), control that list, and monitor similar lists made by other users. In this section, you will learn about the evolution of social networking sites.

One of the first sites, Six Degrees, started in 1997. Six Degrees was based on the idea that no more than six persons separated anyone in the world from any other person. The site was unable to generate sufficient revenue to continue operations and closed in 2000.

More successful social networking sites followed several years later. **Friendster** was founded by Jonathan Abrams in 2002. Friendster was the first Web site to include most of the features found today in all social networking sites. After growing rapidly in the United States and in Asia, the company's membership outstripped its technological ability to handle their activities. Further, the company's management team was unable to agree on strategy for dealing with competition from new U.S. social networking sites such as **MySpace**, **Tribe.net**, and **Facebook**. In Asia, local language social networking sites such as **GREE** and **mixi** in Japan and **QQ.com** and **Renren** in China eroded Friendster's early successes. **Orkut** (named for the Google employee who developed the site in 2004) never really caught on in the United States, but became the top social networking site in both Brazil and India.

LinkedIn, a site devoted to facilitating business contacts, was founded in 2003 and allows users to create a list of trusted business contacts. Users then invite others to participate in several forms of relationships on the site, each of which is designed to help them either find jobs, find employees, or develop connections to business opportunities. LinkedIn has become the most popular business-focused social networking site in the world.

Other social networking sites have met with varying degrees of success. Some sites have developed a following by offering specific features; for example, **YouTube** (now owned by Google) popularized the inclusion of videos in social networking sites, and has become a popular social networking site for younger Web users. **Twitter** offers users a way to send short messages to other uses who sign up to follow their messages (called **tweets**). In 2011, Google introduced **Google+**, a new social networking site to compete with Facebook, which it identified as its primary competition. Figure 6-1 shows the launch year for some of the more successful social networking sites.

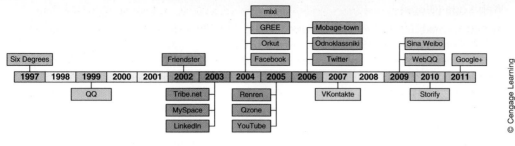

FIGURE 6-1　Social networking Web sites

The general idea behind all of these sites is that people are invited to join by existing members who think they would be valuable additions to the community. The site provides a directory that lists members' locations, interests, and qualities; however, the directory does not disclose the name or contact information of members. A member can offer to communicate with any other member, but the communication does not occur until the intended recipient approves the contact (usually after reviewing the sender's directory information).

In addition to searching the directory of the community, members can make connections with new contacts through friends they have established in the community (perhaps starting with the person who invited them to join). By gradually building up a set of connections, members can develop contacts within the community that might prove valuable later.

The expansion of social networking sites into all corners of the world continues as we move into the third wave of electronic commerce. In 2008, Google moved Orkut's headquarters to Brazil to acknowledge the location of its primary audience. In addition to the Chinese and Japanese sites mentioned earlier, successful social networking sites in local languages have emerged in Germany (**Xing**), the Netherlands (**Hyves**), Russia (**VKontakte** and **Odnoklassniki**), and Spain (**Tuenti**). Figure 6-2 shows the leading social networking sites in several areas of the world.

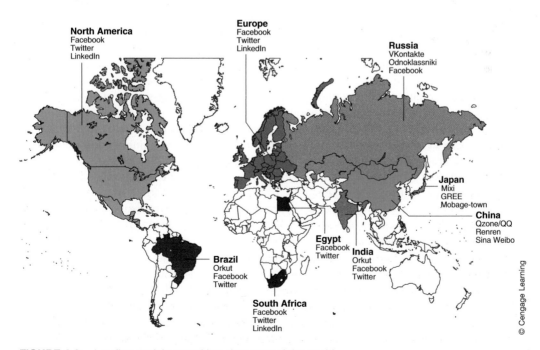

FIGURE 6-2　Leading social networking sites around the world

Web Logs (Blogs)

As you learned in Chapter 4, Web logs, or blogs, are Web sites that contain commentary on current events or specific issues written by individuals. Many blogs invite visitors to add comments, which the blog owner may or may not edit. The result is a continuing discussion of the topic with the possibility that many interested persons will contribute to that discussion. Because blog sites encourage interaction among people interested in a particular topic, they are a form of a social networking site. Sites such as **Twitter** are considered to be **microblogs** because they function as a very informal blog site with entries (messages, or tweets) that are limited to 140 characters in length.

Most of the early blogs were focused on technology topics or on topics about which people have strong beliefs (for example, political or religious issues). The 2004 U.S. elections saw the first major use of blogs as a political networking tool. In previous elections, candidates had Web sites and political parties send out e-mail messages to supporters and potential donors, but in the 2004 elections, these activities were coordinated in a new way. Individuals working alone or with established political organizations set up Web sites that provided a place for people interested in a candidate or an issue to communicate with each other. These sites allowed people to discuss issues, plan strategies, and even arrange in-person meetings called **meetups**. By the 2008 U.S. elections, all of the major candidates were using blogs, microblogs, and social networking activities as essential tools for communicating their messages, organizing volunteers, and raising money. As we enter the third wave of electronic commerce, social networking tools are being used to organize all sorts of charitable fund-raising and support activities as well.

After seeing the success of blogs, microblogs, and social networks as political networking tools, many retailers embraced these tools as a way to engage Web site visitors who were not ready to buy from the site, but who were interested in the products or services offered. Marketing and supply chain managers also saw the benefits of these social networking activities in enhancing their B2B relationships. Many companies include blogs as part of their online offerings. These blogs discuss uses and technical specification issues regarding the products or services offered for sale.

CNN was a pioneer in including information from blogs and microblogs in its television newscasts. Other broadcasters and newspapers now incorporate blogs and social networking features in their Web sites. Small-town newspapers often depend on readers to contribute information about community issues and events. Newspapers of all sizes would rather run a blog with reader contributions tied into microblogs and social networking sites than pay reporters to write stories about events or issues that would interest only a small segment of their readership. By inviting information and opinion contributions, newspapers are finding they can reach younger readers who did not grow up reading print newspapers. This trend toward having readers help write online news is called **participatory journalism**.

In addition to running a blog that is part of an existing activity (such as a political campaign, retail business, or newspaper), blogs can become a business in themselves if they can generate financial support through fees or advertising. Jake Dobkins writes about New York City on the blog site **Gothamist**. Instead of drawing a salary from a newspaper as a food and entertainment reporter, he blogs about the latest in New York nightlife. Advertising revenue has been sufficient to support Dobkins and the site's cofounder, Jen Chung. Now with a staff of bloggers, editors, and ad salespeople, these entrepreneurs are expanding into other cities. Michael Arrington began blogging in 2005 about new online business startups. Again, instead of writing a column for a business magazine, he decided to put his research and reporting talents into his own business, which today is operating as **TechCrunch**, a successful advertising-supported Web site.

Social Networking Web Sites for Shoppers

The practice of bringing buyers and sellers together in a social network to facilitate retail sales is called **social shopping**. One of the first of these was **craigslist**, an information resource for San Francisco area residents that was created in 1995 by WELL member Craig Newmark. That community has grown to include information for most major cities in the United States and in several other countries. The site is operated by a not-for-profit foundation, and all postings other than help-wanted ads are free.

The **Etsy** Web site provides a marketplace for people who want to sell handmade items. The social network here includes buyers and sellers interested in crafts of all types. In fact, the sense of community is so strong that a separate site, **We Love Etsy**, exists to provide a place for Etsy buyers and sellers to share information.

Idea-Based Social Networking

Social networking sites form communities based on connections among people. Other Web sites create communities based on the connections between ideas. These more abstract communities are called **idea-based virtual communities** and the people who participate in them are said to be engaging in **idea-based networking**. The **del.icio.us** site calls itself a "social bookmarks manager." Individuals place Web page bookmarks with one-word tags that describe the Web page in a community-accessible location on the site. The bookmark–tag combinations are focused on ideas and the contributions of all community members build a shared base of knowledge about those ideas. Among the most active tag names on the site are words such as design, reference, tools, music, news, how to, and photography. Another idea-based virtual community that uses shared tags is **43 Things**.

Virtual Learning Networks

One form of social network you might have used is the **virtual learning network**. Many colleges and universities now offer courses that use distance learning platforms such as **Blackboard** for student-instructor interaction. These distance learning platforms include tools such as bulletin boards, chat rooms, and drawing boards that allow students to interact with their instructors and each other in ways that are similar to the interactions that might occur in a physical classroom setting. Some open-source software projects are devoted to the development of virtual learning communities, including **Moodle** and uPortal (maintained by the not-for-profit open source software development organization, **Jasig**). **Open source software** is developed by a community of programmers who make the software available for download at no cost. Other programmers then use the software, work with it, and improve it. Those programmers can submit their improved versions of the software back to the community. Open source software is an early and successful example of a virtual community. You can learn more about open-source software at the **Open Source Initiative** Web site.

Revenue Models for Social Networking Sites

By the late 1990s, virtual communities were selling advertising to generate revenue. Search engine sites and Web directories were also selling advertising to generate revenue. Beginning in 1998, a wave of purchases and mergers occurred among these sites. The new sites that emerged still used an advertising-only revenue-generation model and included all the features offered by virtual community sites, search engine sites, Web directories, and other information-providing and entertainment sites. These Web portals, which you first learned about in Chapter 3, are so named because their goal is to be every Web user's doorway to the Web.

Advertising-Supported Social Networking Sites

Visitors spend a greater amount of time at portal sites than they do at most other types of Web sites, which is attractive to advertisers. Other types of social networking sites can also draw large numbers of visitors who spend considerable time on the sites. This section describes how these characteristics make social networking sites appealing to advertisers.

Smaller social networking sites that have a more specialized appeal can draw enough visitors to generate significant amounts of advertising revenue, especially compared to the costs of running such a site. For example, software developer Eric Nakagawa posted a picture of a grinning fat cat on his Web site in 2007 with the caption "I can has cheezburger?" as a joke. He followed that with several more cat photos and funny captions over the next few weeks and added a blog so that people could post comments about the pictures. Within a few months, the site was getting more than 100,000 visitors a day. Nakagawa found that a site with that kind of traffic could charge between $100 and $600 per day for a single ad. Today, he spends his time fine-tuning the site to make it more attractive to visitors, who now submit their own photos and captions. **I Can Has Cheezburger** now generates a respectable income. Nakagawa has no illusions about expanding the site, hiring thousands of people, or selling stock to the public, but he is earning a comfortable living generated by a highly specialized social networking site.

As you learned in Chapter 3, sites that have higher numbers of visitors can charge more for advertising on the site. You also learned that stickiness (a Web site's ability to keep visitors on the site and attract repeat visitors) is also an important element of a site's attractiveness to advertisers. One rough measure of stickiness is how long each user spends at the site. Figure 6-3 lists the most popular Web sites in the world based on the number of users who accessed the sites during the month of August 2011.

Owner	Millions of unique visitors	Average time per unique visitor per month (H:MM)
Google	379	3:43
Microsoft	316	2:20
Facebook	301	6:21
Yahoo!	235	2:30
Wikimedia Foundation	156	0:13
eBay	134	1:16
InteractiveCorp	132	0:11
Amazon.com	129	:27
Apple Computer	117	1:58
AOL, Inc.	103	3:31

Adapted from reports for August 2011 published by The Nielsen Company at
http://www.nielsen.com/us/en/insights/top10s/internet.html

FIGURE 6-3   Popularity and stickiness of leading Web sites

The leading sites often have more than 200 million unique visitors per month. The figure also shows the average amount of time each visitor spends on the site each month (an estimate of stickiness). The information in both figures is adapted from **Nielsen** reports and shows sites grouped by owner (for example, the Apple Computer listing includes its iTunes store; the Google listing includes YouTube; the Microsoft listing includes Microsoft software support sites, MSN, and the Bing search engine; and InteractiveCorp includes Ask.com, Citysearch, Match.com, and Newsweek). Web sites that are social networking sites (such as AOL and Facebook) or that include social networking elements (such as eBay, Google, Microsoft, and Yahoo!) regularly appear on these Nielsen lists.

Because social networking sites often ask their members to provide demographic information about themselves, the potential for targeted marketing on these types of sites is very high. High visitor counts can yield high advertising rates for these sites. In the boom years of the first wave of electronic commerce, Web sites with high degrees of stickiness (which were usually Web portals) could obtain up-front cash payments from advertisers, which is very unusual for any kind of advertising sale. In recent years, all types of social networking sites have negotiated advertising deals that include a percentage of sales generated from sales leads on their sites. Second-wave advertising fees are based less on up-front site sponsorship payments and more on the generation of revenues from continuing relationships with people who use the social networking sites.

Mixed-Revenue and Fee-for-Service Social Networking Sites

Although most social networking sites use advertising to support their operations, some do charge a fee for some services. For example, the Yahoo! Web portal offers most of its services free (supported by advertising), but it does sell some of its social networking features, such as its All-Star Games package. Yahoo! also sells other features, such as more space to store messages and attached files, as part of its premium e-mail service. These fees help support the operation of the social networking elements of the site.

Some advertising-supported social networking sites have followed the lead of Yahoo! in a strategy called monetizing eyeballs or monetizing visitors. **Monetizing** refers to the conversion of existing regular site visitors seeking free information or services into fee-paying subscribers or purchasers of services. Sites that monetize visitors by charging them always worry about visitor backlash. They can never be sure how many existing visitors will pay for services that have been offered in some form at no cost.

Other social networking sites that use a mixed-revenue model are the financial information sites **The Motley Fool** and **TheStreet.com**. These sites offer investment advice, stock quotes, and financial planning help. Some of the information is provided at no cost, additional information is available to subscribers who pay no fee but who are required to provide personal information, and even more information is available to subscribers who agree to pay a fee.

Fee-Based Social Networking

An early attempt to monetize social networking by charging a fee for a specific service was the **Google Answers** site. Google Answers gave people a place to ask questions that were then answered by an expert (called a Google Answers Researcher) for a fee. Google administered a test to determine which members of the community were qualified to become Google Answers Researchers. Google operated this service from 2002 to 2006 (questions and answers posted during that time period are still available on the Web site). Similar services operated by Yahoo! (**Yahoo! Answers**) and Amazon (**Askville**) allow volunteers to answer questions, but provide no opportunity for researchers to earn fees. These services do generate advertising revenue for the sites, however.

After Google closed its service, a number of the people who had been Google Researchers joined together and started a similar service on the site **Uclue**. Researchers earn 75 percent of the total fee paid to Uclue. Advocates of using paid researchers argue that the quality of the answers is higher than on free sites and that the questions tend to be more serious and better formulated. Both approaches are examples of how Web sites can generate revenue by providing a place in which virtual communities can interact.

Microlending Sites

One of the most interesting uses of social networking on the Web has been the emergence of sites that function as clearinghouses for microlending activity. **Microlending** is the practice of lending very small amounts of money to people who are starting or operating small businesses, especially in developing countries. Microlending became famous in 2006 when Muhammad Yunus and the Grameen Bank won the Nobel Peace Prize for their work in developing microlending initiatives in Bangladesh.

A key element of microlending is working within a social network of borrowers. The borrowers provide support for each other and an element of pressure to ensure the loans are repaid by each member of the group. **Kiva** and **MicroPlace** are examples of social networking sites that bring together many small investors who lend money to groups and individuals all over the world who need loans to start or continue their small business ventures. Kiva partners with microfinance institutions that are knowledgeable about business conditions in their parts of the world. These institutions select local individuals they believe are good credit risks and help them post a loan request on the Kiva site. Lenders can review the loan requests and agree to fund part (or all) of the loan amount using the Kiva Web site. The loans, which typically range from a few hundred to a few thousand dollars, are scheduled to be repaid within short time periods ranging from a few months to a year.

Internal Social Networking

A growing number of large organizations have built internal Web sites that provide opportunities for online interaction among their employees. These sites also include important information for employees. These sites run on the intranets you learned about in Chapter 2. Organizations have saved significant amounts of money by replacing the printing and distribution of paper memos, newsletters, and other correspondence with a Web site. Internal social networking pages also provide easy access to employee handbooks, newsletters, and employee benefits information.

These organizations are also finding that an internal social networking Web site can become a good way of fostering working relationships among employees who are dispersed over a wide geographic area. For example, a global company could create a question and answer page for all of its equipment maintenance technicians. Such a page would provide mentoring and informal help functions for all the equipment maintenance technicians in the company.

Many companies are adding wireless connectivity to their internal community sites and are using this technology to extend the reach of the site to employees who are traveling, meeting with customers or suppliers, or telecommuting. These extended community sites are yet another example of a second-wave combination of technology (wireless communications) with a business strategy from the first wave (internal Web portals).

The use of mobile technology is becoming an important part of almost every social networking business strategy as people use their mobile phones to do everything from take photos they will post on Facebook to send tweets to their followers on Twitter. In the next section, you will learn about the potential for combining mobile phone technologies with the idea of social networking to create new online business opportunities.

Virtually all phones sold today include **short messaging service (SMS)**, which allows mobile phone users to send short text messages to each other. For years, mobile phones such as the BlackBerry have had the ability to send and receive e-mail, but until recently, many owners of these phones used them only for phone calls. However, two developments coincided in the United States in 2008 that made these phones more viable as devices for browsing the Web. First, high-speed mobile telephone networks grew dramatically in availability, and second, manufacturers began offering a wide variety of smart phones that include a Web browser (and a screen large enough to make it usable), an operating system, and the ability to run applications on that operating system. In this section, you will learn about the impact of this confluence of technologies on the potential for online business using these devices, called **mobile commerce (m-commerce)**.

Mobile Operating Systems

In Japan and parts of Southeast Asia, mobile commerce has been a much larger part of online business activity than it has elsewhere in the world (including in the United States). One reason is that these countries introduced high-capacity mobile phone networks long before U.S. network providers did. NTT DoCoMo, which is the largest phone company in Japan, pioneered mobile commerce in 2000 with its i-mode service. Starting with the sale of games and other programs that run on the phones, NTT DoCoMo has been a leader in expanding mobile commerce, including online shopping and payments.

In the United States, the introduction of smart phones and the high-capacity networks that make them able to support mobile commerce did not begin until 2008. These smart phones, such as the Apple iPhone, the Palm Pre, several BlackBerry models, and phones that use the Android operating system, opened the door for serious U.S. mobile commerce for the first time. Figure 6-4 shows several examples of designs used in smart phones today.

PixAchi/Shutterstock.com

Oleksiy Mark/Shutterstock.com

mkabakov/Shutterstock.com

Christian Delbert/Shutterstock.com

FIGURE 6-4   Smart phones come in a range of different styles

Some smart phones and wireless PDAs display Web pages using the **Wireless Application Protocol (WAP)**. WAP allows Web pages formatted in HTML to be displayed on devices with small screens, such as mobile phones. Another approach, made possible by increased screen resolution, is to display a normal Web page on the device. The Apple iPhone was one of the first devices to include touch screen controls that make viewing and navigating a normal Web page easy to do on a small handheld device. A third approach is to design Web sites to match specific smart phones. This is much more difficult to accomplish because there can be many different phones that use the same operating system, and each phone has a different interface (the buttons, touches, or gestures that perform specific functions often vary).

Apple, BlackBerry, and Palm each use their own proprietary operating systems. Some phone makers (including HTC, Motorola, and Nokia) that created their own operating systems and software applications for common functions such as calendar, contacts, and e-mail now use a standard operating system provided by a third party.

The most common third-party operating systems are Android, Windows Phone, and Symbian. Windows Phone (formerly Windows Mobile) is a proprietary operating system sold by Microsoft. Symbian started as a proprietary system but became open source in 2008. Nokia, the main user of Symbian, donated the software to the Symbian Foundation, which manages continued open source development of the operating system. However, in 2011, Nokia began building smart phones that use the Windows Phone operating system. Many industry analysts believe that Symbian, which had been a leader in early Internet-capable phones, is no longer competitive as an operating system for full-featured smart phones.

The most popular and fastest growing third-party operating system is Android, which was developed by Google. Android is open source, which allows smart phone manufacturers to use it at no cost. Most smart phone manufacturers that use Android add some customized features to the software's interface. Figure 6-5 shows the change in U.S. market shares for leading smart phone operating systems during recent years.

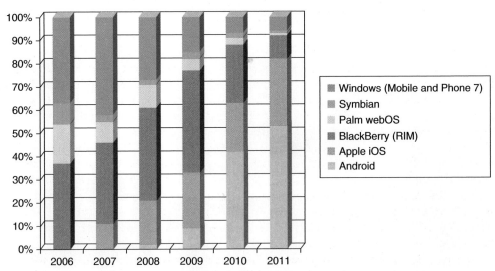

Source: The NPG Group, Consumer Tracking Service, Mobile Phone Track at http://www.bgr.com/2011/12/13/apple-and-google-dominate-smartphone-space-while-other-vendors-scramble/

FIGURE 6-5 Smart phone operating systems: U.S. market shares

Once a manufacturer chooses an operating system for its phones, the user cannot delete it and switch to a different operating system. Unlike computers, the operating system is integrated into the software the carrier uses to make the phone operate on its network. Most carriers will void the warranty on a phone if the user has modified the operating system in any way, although some users with technical skills do so. Modifying an Apple iPhone's operating system is called **jailbreaking** the phone. Modifying an Android operating system is called **rooting** the phone.

Mobile Apps

The emergence of common operating systems (instead of each phone manufacturer using its own operating system) occurred because the way software applications are developed and sold has changed. In the past, U.S. mobile phone companies generated revenue by controlling the application software (usually called **apps**) that could run on their phones. Companies would license the apps from software developers and then charge subscribers a monthly usage fee for each app. Apple turned this revenue strategy on its head when AT&T agreed to be Apple's sole carrier for the iPhone (that is, iPhones would only operate on the AT&T network) and agreed to allow Apple to sell apps for the phone directly.

The **Apple Apps for iPhone** online store was launched at the same time as the iPhone itself and became an instant success, making a wide variety of software available for the phone. Because Apple allowed independent developers to create apps and sell them (on a revenue-sharing basis) through their Apps for iPhone store, a number of software developers made hundreds of thousands of dollars for their creations. **Zynga**, a company that develops games for mobile phones, generates more than $1 billion in revenues each year selling its game apps for phones. Other firms, such as **Mutual Mobile**, provide software design and development services for companies that want apps to use in their own organizations.

BlackBerry and Palm have followed Apple's lead and now have apps stores of their own (**BlackBerry App World** and **Palm Pre Applications**), and the open source Android and Symbian phones also have software developers creating apps for them (see **Android Market** and **SymbianGear**). Many companies now develop apps for multiple platforms.

A number of apps do nothing more than provide a quick gateway to a company's Web site. Many online shopping destinations offer free apps that are optimized to provide users the best possible shopping experience on the small screen of a smart phone. Other apps are sold for a fee. Games, puzzles, productivity tools (such as contact managers, calendars, and task organizers), and reference works generally fall into this category. Most apps sell for $1 to $5, although prices can vary widely. Some newspaper and media sites offer free access to their online content through apps; others, such as *The New York Times*, offer subscriptions that can be accessed through their apps.

Some mobile app sellers include an advertising element in their revenue models. These apps include mobile ads that display messages from advertisers (other than the seller of the app). One common way to include ads is to display them in a small bar at the bottom of the app screen. Some apps include advertising that appears on a part of the screen or as a separate screen that must be clicked through to get to the app. The advertising space on mobile apps is sold in the same way that banner advertising on Web sites is sold (which you learned about in Chapter 4).

Companies that want to participate in mobile commerce should first review their Web sites to determine how well the site works when viewed on a mobile device. Many companies that are serious about connecting with mobile users are creating separate Web sites for mobile users. For example, the **Scottrade Mobile** Web site is optimized for display on the small screens used in smart phones. The site gives Scottrade customers quick access to financial markets information and to their trading accounts. **Yahoo! Mobile** provides a version of its Web portal that is similarly optimized for the small screen and limited controls of most smart phones.

Although the use of mobile phones for online banking is still in its early stages in the United States, forward-looking financial institutions such as **Wescorp** are working on ways to draw in younger customers by offering complete banking services through a Web site that is optimized for specific smart phones.

The Veterans Affairs Medical Center in Washington, D.C., has issued smart phones to its physicians. They use the phones to send and receive messages, but in addition, they can use the phones to read detailed information needed for treating patients. For example, cardiologists can read electrocardiograms (EKGs) on their smart phones, saving time and often a trip to the hospital. Other hospitals are using smart phones in equally creative ways. For example, diabetic patients can track what they eat, insulin injections, blood sugar readings, and their level of physical activities on their phones. Doctors treating these patients can access the data using their own smart phones and can better help patients manage their diabetes. The University of Louisville Medical School provides a suite of smart phone apps to their students, including Epocrates (a drug information database), and apps that let them look up information about diseases and access reference works.

Most smart phones have global positioning satellite (GPS) service capabilities, which means that apps that combine the phone user's location with the availability of retail stores and services can be interwoven into creative mobile business opportunities. For example, some apps can direct the user to specific business locations (such as restaurants, movie theaters, or auto repair facilities) based on the user's current location.

Most app development is done by trained programmers; however, there are tools such as **Swebapps** and **App Inventor** that provide a point-and-click interface for building simple apps. And sites such as **TaskCity** can connect a person or company that needs an app created with a programmer who can do the job.

Tablet Devices

In 2010, Apple introduced the iPad, a tablet device that is smaller than a laptop computer, yet larger than a smart phone. **Tablet devices** can be connected to the Internet through a wireless phone carrier's service or through a local wireless network. Most tablet devices can use both access modes and can switch between them. Within a year, many other manufacturers had introduced tablet devices to compete with the iPad.

Apple's iPad tablet devices run the company's proprietary iOS operating system. Most other manufacturers' tablet devices run the Android operating system. Some of Amazon.com's electronic book products, such as the Kindle Fire, have the ability to be used as online tablet devices. Because tablet devices' screens are larger, they are more likely to be used than smart phones to buy consumer products (most purchases completed on smart phones are for digital products such as music, videos, or apps).

Mobile Payment Apps

Since 2004, NTT DoCoMo has been selling mobile phones, called **mobile wallets** (*osaifu-ketai*, in Japanese), that function as credit cards. Although the individual applications on DoCoMo phones are not overwhelming (for example, one application lets you use a mobile phone to pay for a vending machine purchase in Japan), their combined capabilities generate a significant amount of business. Other countries that have a tradition of using cash for retail transactions have seen significant adoptions of mobile phone apps that allow them to be used to make payments. Very few people have credit cards in these countries and the convenience of using a mobile phone for payments has been very attractive.

In the United States, where the use of credit cards is widespread, the use of a mobile phone for payments has not been as compelling. However, in 2011, a number of companies began to offer retail store technologies that allow the use of smart phones as payment devices. American Express, Visa, and MasterCard have all made phone readers available to retailers. Google introduced its Google Wallet for Android phones. You will learn more about payment systems in Chapter 11.

In the next section, you will learn how online business pioneers adapted auctions, a very old business practice, to a new online business opportunity.

ONLINE AUCTIONS

In many ways, online auctions provide a business opportunity that is perfect for the Web. An auction site can charge both buyers and sellers to participate, and it can sell advertising on its pages. People interested in trading specific items can form a market segment that advertisers will pay extra to reach. Thus, the same kind of targeted advertising opportunities that search engine sites generate with their results pages are available to advertisers on auction sites. This combination of revenue-generating characteristics makes it relatively easy to develop online auctions that yield profits early in the life of the project.

One of the Internet's strengths is that it can bring together people who share narrow interests but are geographically dispersed. Online auctions can capitalize on that ability by either catering to a narrow interest or providing a general auction site that has sections devoted to specific interests. Before you learn more about online auctions, the next section introduces some basic auction terminology and principles.

Auction Basics

The earliest written records of auctions are from Babylon and date from 500 BC. In those auctions, men bid against each other for the women they wished to marry. Roman soldiers used auctions to liquidate the property they took from their vanquished foes. In AD 193, the Praetorian Guard auctioned off the entire Roman Empire after killing the Emperor Pertinax. In later years, Buddhist temples held auctions to sell off the possessions of deceased monks. Auctions became common activities in 17th-century England, where taverns held regular auctions of art and furniture. The 18th century saw the birth of two British auction houses—Sotheby's in 1744 and Christie's in 1766—that continue to be major auction firms today. The British settlers of the colonies that would become the United States brought auctions with them. Colonial auctions were used to sell farm equipment, animals, tobacco, and, sad to say, human beings.

In an auction, a seller offers an item or items for sale, but does not establish a price. This is called "putting an item up for bid" or "putting an item on the (auction) block." Potential buyers are given information about the item or some opportunity to examine it; they then offer **bids**, which are the prices they are willing to pay for the item. The potential buyers, or **bidders**, each have developed **private valuations**, or amounts they are willing to pay for the item. The whole auction process is managed by an **auctioneer**. In some auctions, people employed by the seller or the auctioneer can make bids on behalf of the seller. These people are called **shill bidders**. Shill bidders can artificially inflate the price of an item and may be prohibited from bidding by the rules of a particular auction.

English Auctions

Many different kinds of auctions exist. Most people who have attended or seen an auction on television have experienced only one type of auction, the **English auction**, in which bidders publicly announce their successive higher bids until no higher bid is forthcoming. At that point, the auctioneer pronounces the item sold to the highest bidder at that bidder's price. This type of auction is also called an **ascending-price auction**. An English auction is sometimes called an **open auction** (or **open-outcry auction**) because the bids are publicly announced; however, there are other types of auctions that use publicly announced bids that are also called open auctions.

In some cases, an English auction has a minimum bid, or reserve price. A **minimum bid** is the price at which an auction begins. If no bidders are willing to pay that price, the item is removed from the auction and not sold. In some auctions, a minimum bid is not announced, but sellers can establish a minimum acceptable price, called a **reserve price**, or simply **reserve**. If the reserve price is not exceeded, the item is withdrawn from the auction and not sold.

English auctions that offer multiple units of an item for sale and allow bidders to specify the quantity they want to buy are called **Yankee auctions**. When the bidding concludes in a Yankee auction, the highest bidder is allotted the quantity he or she bid. If items remain after satisfying the highest bidder, those remaining items are allocated to successive lower (next highest) bidders until all items are distributed. Although all successful bidders (except possibly the lowest successful bidder) receive the quantity of items on which they bid, they only pay the price bid by the *lowest* successful bidder.

To understand Yankee auctions better, consider this example. A seller places nine items up for bid. When the bidders stop increasing their bids, the successful bidders include the following: the highest bidder, who bid $85, quantity five; the second-highest bidder, who bid $83, quantity three; and the third-highest bidder, who bid $81, quantity four. All three of the successful bidders pay $81 per item, but the highest bidder receives five items, the second-highest bidder receives three items, and the third-highest bidder receives the one remaining item, despite having bid for a quantity of four, because only one is left after satisfying the quantity bids of the higher bidders.

English auctions have drawbacks for both sellers and bidders. Because the winning bidder is only required to bid a small amount more than the next-highest bidder, winning bidders tend not to bid their full private valuations, which prevents sellers from obtaining the maximum possible price. Bidders risk becoming caught up in the excitement of competitive bidding and then bidding more than their private valuations. This psychological phenomenon, called the **winner's curse**, has been extensively documented by William Thaler (see the Thaler reference in the "For Further Study and Research" section at the end of this chapter) and other behavioral economists.

Dutch Auctions

The **Dutch auction** is a form of open auction in which bidding starts at a high price and drops until a bidder accepts the price. Because the price drops until a bidder claims the item, Dutch auctions are also called **descending-price auctions**. Farmers' cooperatives in the Netherlands use this type of auction to sell perishable goods such as produce and flowers, which is how it came to be known as a "Dutch" auction. In most Dutch auctions, the seller offers a number of similar items for sale. One common implementation of a Dutch auction uses a clock that drops the price with each tick. The first bidder to call out "stop," which stops the clock, becomes the winning bidder. The winning bidder can take all or any part of the auctioned items at that price. If any items remain, the clock is restarted and continues to run until all the items are taken by successive lower bidders. A Dutch auction is often better for the seller because the bidder with the highest private valuation will not let the bid drop much below that valuation for fear of losing the item to another bidder. Dutch auctions are particularly good for moving large numbers of commodity items quickly. A few online stores have offered Dutch auctions from time to time. For several years, women's clothing retailer Coldwater Creek used Dutch auctions to sell closeout items on its site.

In 2004, Google used a Dutch auction to sell its stock to investors in its initial public offering. The financial community considered this use of a Dutch auction to be highly innovative and very successful. Google used the Dutch auction to obtain the highest price possible for its shares. In a similar financial transaction, online advertising and technology company LookSmart used a Dutch auction to buy back some of its stock. Usually, when a company announces a share buyback, the price of the stock moves upward and the company must pay an increasing price as it buys the shares on the open market. LookSmart announced a price range and let shareholders place bids that specified the number of shares and the price within that range at which they would be willing to sell. When the auction was over, LookSmart had repurchased exactly the number of shares it had wanted to buy at the lowest price it had specified, which meant that the Dutch auction worked very well for it.

First-Price Sealed-Bid Auctions

In **sealed-bid auctions**, bidders submit their bids independently and are usually prohibited from sharing information with each other. In a **first-price sealed-bid auction**, the highest bidder wins. If multiple items are auctioned, successive lower (next highest) bidders are awarded the remaining items at the prices they bid.

Second-Price Sealed-Bid Auctions

The **second-price sealed-bid auction** is the same as the first-price sealed-bid auction except that the highest bidder is awarded the item at the price bid by the *second*-highest bidder. At first glance, one might wonder why a seller would even consider such an auction because it gives the item to the winning bidder at a lower price. William Vickrey won the **1996 Nobel Prize in Economics** for his studies of the properties of this auction type. He concluded that it yields higher returns for the seller, encourages all bidders to bid the amounts of their private valuations, and reduces the tendency for bidders to collude. Because the winning bidder is protected from an erroneously high bid, all bidders tend to bid higher than they would in a first-price sealed-bid auction. Second-price sealed-bid auctions are commonly called **Vickrey auctions**.

Open-Outcry Double Auctions

The Chicago Board Options Exchange conducts **open-outcry double auctions** of commodity futures and stock options. The buy and sell offers are shouted by traders standing in a small area on the exchange floor called a trading pit. Each commodity or stock option is traded in its own pit. The action in a trading pit can become quite frenzied as 20 or 30 traders shout offers aloud. Double auctions, either sealed bid or open outcry, work well only for items of known quality, such as securities or graded agricultural products, which are regularly traded in large quantities. Such items can be auctioned without bidders inspecting the items before placing their bids.

Double Auctions

In a **double auction**, buyers and sellers each submit combined price–quantity bids to an auctioneer. The auctioneer matches the sellers' offers (starting with the lowest price and then going up) to the buyers' offers (starting with the highest price and then going down) until all the quantities offered for sale are sold to buyers. Double auctions can be operated in either sealed-bid or open-outcry formats. The **New York Stock Exchange** conducts sealed-bid double auctions of stocks and bonds in which the auctioneer, called a specialist, manages the market for a particular stock or bond issue. The specialist company must use its own funds, when necessary, to maintain a stable market in the specific security it manages. Although the specialist system has been in use for more than a century, critics have charged that specialists can and do use their knowledge to enrich themselves at the expense of investors. In 2007, the New York Stock Exchange added an electronic trading system that automatically matches buyer and seller offers. Today the automated system, which bypasses specialists, handles most of the trading volume on the exchange.

Reverse (Seller-Bid) Auctions

In a **reverse auction** (also called a **seller-bid auction**), multiple sellers submit price bids to an auctioneer who represents a single buyer. The bids are for a given amount of a specific item that the buyer wants to purchase. The prices go down as the bidding continues until no seller is willing to bid lower. Reverse auctions have been operated for consumers from time to time, but most reverse auctions involve businesses as buyers and sellers. In many business reverse auctions, the buyer acts as auctioneer and screens sellers before they can participate. You will learn more about specific implementations of reverse auctions later in this chapter.

The seven auction types described in this section are the most commonly used in business today. Figure 6-6 summarizes the key characteristics of each of these seven major auction types.

Auction type	Key characteristics
English auction	Starting from a low price, bidding increases until no bidder is willing to bid higher.
Dutch auction	Starting from a high price, bidding automatically decreases until the bidder accepts the price.
First-price sealed-bid auction	Secret bidding process; the highest bidder pays the amount of the highest bid.
Second-price sealed-bid auction (Vickrey auction)	Secret bidding process; the highest bidder pays the amount of the *second*-highest bid.
Double auction (open-outcry)	Buyers and sellers declare combined price–quantity bids. The auctioneer matches seller offers (lowest to highest) with buyer offers (highest to lowest). Buyers and sellers can modify bids based on knowledge gained from other bids.
Double auction (sealed-bid)	Buyers and sellers declare combined price–quantity bids. The auctioneer (specialist) matches seller offers (lowest to highest) with buyer offers (highest to lowest). Buyers and sellers cannot modify their bids.
Reverse auction (seller-bid)	Multiple sellers submit price bids to an auctioneer that represents a single buyer. The bids are for a given amount of a specific item that the buyer wants to purchase. Prices go down as the bidding continues until no seller is willing to bid lower.

FIGURE 6-6 Key characteristics of seven major auction types

Online Auctions and Related Businesses

Millions of people buy and sell all types of goods on consumer auction sites each year. Although the online auction business is changing rapidly as it grows, three broad categories of auction Web sites have emerged: general consumer auctions, specialty consumer auctions, and business-to-business auctions. Some industry analysts consider the two types of consumer auctions to be business-to-consumer electronic commerce. Other analysts believe that a more appropriate term for the electronic commerce that occurs in general consumer auctions is consumer-to-consumer or even **consumer-to-business** (because the bidders at a general consumer auction might be businesses). Their argument is that many sellers who participate in general consumer auctions are not really businesses; they are ordinary people who use these auctions to sell personal items instead of holding a garage sale, for example. Whether you prefer to think of online auctions as business-to-consumer, consumer-to-consumer, or consumer-to-business, the largest number of auction transactions occurs on general consumer auction sites.

General Consumer Auctions

The most successful consumer auction Web site today is eBay. Sellers and buyers must register with eBay and agree to the site's basic terms of doing business. Sellers pay eBay a listing fee and a sliding percentage of the final selling price. Buyers pay nothing to eBay. In addition to paying the basic fees, sellers can choose from a variety of enhanced and extra-cost services, including having their auctions listed in boldface type and featured in lists of preferred auctions.

In an attempt to address buyer concerns about seller reliability, eBay instituted a rating system. Buyers can submit ratings of sellers after doing business with them. These ratings are converted into graphics that appear with the seller's nickname in each auction in which the seller participates. Although this system is not perfect, many eBay bidders feel that it affords them some protection from unscrupulous sellers. eBay also uses buyer ratings of sellers to place restrictions on sellers (such as withholding funds for three weeks) or, if the ratings are low enough, prohibit them from selling on eBay at all. The converse is true also; sellers rate buyers, which provides sellers some protection from unscrupulous buyers.

Although eBay does not release any statistics about buyer and seller frauds, most industry observers agree that sellers face larger potential losses than buyers. Sellers' greatest risks are from buyers who use stolen credit card numbers or who place the winning bid but never contact the seller to conclude the transaction. Buyers' risks include sellers who never deliver or who misrepresent their merchandise. You will learn about ways that sellers and buyers can protect themselves later in this chapter.

The most common format used on eBay is a computerized version of the English auction. The eBay English auction allows the seller to set a reserve price. In eBay English auctions, the bidders are listed, but the bid amounts are not disclosed until after the auction is over. This is a slight variation on the in-person English auction, but because eBay always shows a continually updated high bid amount, a bidder who monitors the auction can see the bidding pattern as it occurs. The main difference between eBay and a live English auction is that bidders do not see the details of the bidding history (which bidders placed which bids when) until the auction is over. The eBay English auction also allows sellers to specify that an auction be made private. In an eBay private auction, the site never discloses bidders' identities and the prices they bid. At the conclusion of the auction, eBay notifies only the seller and the highest bidder. Another auction type offered by eBay is an increasing-price format for multiple-item auctions that eBay calls a Dutch auction. However, these auctions are actually the Yankee auction variant of an English auction.

In either type of eBay auction, bidders must constantly monitor the bidding activity if they intend to win the auction. All eBay auctions have a **minimum bid increment**, the amount by which one bid must exceed the previous bid, which is about 3 percent of the bid amount. To make bidding easier, eBay allows bidders to make a proxy bid. In a **proxy bid**, the bidder specifies a maximum bid. If that maximum bid exceeds the current bid, the eBay site automatically enters a bid that is one minimum bid increment higher than the current bid. As new bidders enter the auction, the eBay site software continually enters higher bids for all bidders who placed proxy bids. Although this feature is designed to make bidding require less bidder attention, if a number of bidders enter proxy bids on one item, the bidding rises rapidly to the highest proxy bid offered. This rapid rise in the current bid often occurs in the closing hours of an eBay auction, usually as the result of bidders raising their proxy bid levels.

To attract sellers who frequently offer items or who continually offer large numbers of items, eBay offers a platform called eBay stores within its auction site. At a very low cost, sellers can establish eBay stores that show items for sale as well as items being auctioned. This can help sellers generate additional profits from sales of items related to those offered in their auctions. These eBay stores are integrated into the auction site; that is, when a bidder searches for an item, the results page includes auctions and listings from sellers' eBay stores.

Competition in General Consumer Auctions

eBay has been so successful because it was the first major Web auction site for consumers that did not cater to a specific audience and because it advertises widely. eBay spends about $1 billion each year to market and promote its Web site. A significant portion of this promotional budget is devoted to traditional mass media outlets, such as television advertising. For eBay, such advertising has proven to be the best way to reach its main market: people who have a hobby or a very specific interest in items that are not locally available. Whether those items are jewelry, antique furniture, coins, first-edition books, or stuffed animals, eBay has created a place where people can become collectors, dispose of their collections, or trade out of their collections.

Because one of the major determinants of Web auction site success is attracting enough buyers and sellers to create markets in many different items, some Web sites that already have a large number of visitors entered the general consumer auction business. Yahoo! created an auction site patterned after eBay. Yahoo! believed that it could leverage its brand name and capitalize on its large number of site visitors to compete with eBay.

Yahoo! had some early success in attracting large numbers of auction participants, in part because it offered its auction service to sellers at no charge. Yahoo! was less successful in attracting buyers, resulting in less bidding action in each auction than generally occurs on eBay. In January 2001, Yahoo! began charging sellers in the face of dropping ad revenues in its other Web operations. Within one month, Yahoo! lost about 80 percent of its auction listings; however, the percentage of listed items that ended in a sale increased six-fold, and the dollar amount of completed auctions remained constant. Because Yahoo! draws a large number of visitors every month, the company hoped that it would be able to further increase participation in its auctions and attract some of the sellers who left in reaction to the fees. However, in 2005, Yahoo! reverted to its original policy of not charging fees to sellers. Despite its efforts, Yahoo! was unable to draw enough buyers or sellers to its U.S. auction site and closed the operation in 2007.

Amazon.com also added a general consumer auction to its list of products and services. Unlike eBay, which was profitable from the start, Amazon took seven years to earn its first small profits from all of its businesses. One way that Amazon attempted to compete with eBay was through its "Auctions Guarantee." This guarantee directly addressed concerns raised in the media by eBay customers about being cheated by sellers. When Amazon opened its Auctions site, it agreed to reimburse any buyer for merchandise purchased in an auction that was not delivered or that was "materially different" from the seller's representations up to $250.

In response to Amazon's guarantee, eBay immediately offered its customers a similar guarantee, but not before Amazon gained free publicity from the media coverage of its guarantee. In 2003, eBay increased its guarantee to $500 in the hopes that it would induce new customers to buy at eBay auctions. The experiment worked well; in fact, eBay increased its guarantee again in 2004 to $1000. In 2005, eBay reduced its guarantee to $200 with a $25 deductible, but continued to offer a $1000 guarantee through its payment processing subsidiary PayPal. This change encourages bidders to use PayPal, yet still provides some protection for bidders who do not. Some eBay users have complained that the company does not act quickly on claims under the guarantee and does its best to avoid paying claims; however, the guarantee remains a powerful marketing tool. Buyers of more expensive items can protect themselves by using a third-party **escrow service**, which holds the buyer's payment until he or she receives and is satisfied with the purchased item. Escrow services are available through most auction sites. You will learn more about escrow services later in this chapter.

Amazon also used other strategies to compete with eBay. For example, Amazon established an online joint venture with Sotheby's, the famous British auction house, to hold online auctions of fine art, antiques, jewelry, and other high-value collectibles. Despite its years of effort, Amazon was unable to draw sellers and buyers in sufficient numbers and closed its general consumer auction site in 2006.

The success of eBay has inspired competition from a number of powerful and well-financed companies. Most of these competitors have met the same fate as Yahoo! and Amazon, failing after spending large amounts of money in their efforts. Future challengers to eBay will find that the economic structure of markets is biased against new entrants. Because markets become more efficient (yielding fairer prices to both buyers and sellers) as the number of buyers and sellers increases, new auction participants are inclined to patronize established marketplaces. Thus, existing auction sites, such as eBay, are inherently more valuable to customers than new auction sites. This basic economic fact, which economists call a **lock-in effect**, will make the task of creating other successful general consumer Web auction sites even more difficult in the future.

A somewhat ironic example of the lock-in effect exists in the Japanese general consumer auction market. In this market, unlike in the United States, Yahoo! was the first major company to offer online auctions. At the time (early 1999), Yahoo! did not charge fees to sellers. When eBay entered the Japanese market five months later, it charged fees and found few people interested in its services. Even when Yahoo! began charging fees in 2001 for its auctions, the lock-in effect preserved its strong lead in Japan. Today, Yahoo! Auctions holds more than 90 percent of the Japanese online auction market, while eBay's market share is less than 3 percent.

LEARNING FROM FAILURES

Auction Universe

One of the most promising new entrants into the general consumer auction business was Auction Universe. Times Mirror, the parent company of the *Los Angeles Times* newspaper, started Auction Universe in 1997 and then sold it in 1998 to a partnership of eight major newspaper companies (including Times Mirror itself) called **Classified Ventures**. These companies were concerned that classified advertising on the Web posed a threat to their newspapers' classified advertising, which is one of the most profitable elements in the newspaper business. Through their Classified Ventures partnership, these newspaper companies started their own Web sites for classified ads such as Apartments.com, Cars.com, and NewHomeNetwork.com. These sites earn revenue by charging for running ads, selling advertising on their pages, or both. Classified Ventures believed that the Auction Universe site could become an important and profitable part of its Web presence.

Auction Universe closed in August 2000, but Classified Ventures' classified ad sites continue to operate. The Auction Universe site was modeled on eBay and offered similar types of auctions and services for buyers and sellers. Some critics believed that the Auction Universe interface was more intuitive than eBay's and included a better search engine; however, the site failed to mount a sustained challenge to eBay's dominance. Even with major corporate sponsorship and a $10 million advertising campaign behind it, Auction Universe was unable to displace the advantage eBay obtained from the lock-in effect it has created for a large number of auction bidders and sellers.

Specialty Consumer Auctions

Rather than struggle to compete with a well-established rival such as eBay in the general consumer auction market, a number of firms have decided to identify special-interest market targets and create specialized Web auction sites that meet the needs of those market segments.

JustBeads.com is one example of an auction site that caters to buyers and sellers who are geographically dispersed but share highly focused interests. Other specialty consumer auction sites include **Cigarbid.com** and **Winebid**. These sites gain an advantage by identifying a strong market segment with readily identifiable products that are desired by people with relatively high levels of disposable income. Cigars and wine meet those requirements. These specialized consumer auctions occupy profitable niches, which allows them to coexist successfully with large general consumer sites, such as eBay.

Consumer Reverse Auctions

In the past, a number of companies have created sites that allow site visitors to describe items or services they wish to buy. The site then routes the visitor's request to a group of participating merchants who reply to the visitor by e-mail with offers to supply the item at a particular price. This type of offer is often called a **reverse bid**. The buyer can then accept the lowest offer or the offer that best matches the buyer's criteria. None of these sites were successful in developing a large enough following to interest merchants, so they have all closed.

Many people think of **Priceline.com** as a seller-bid auction site. Priceline.com allows site visitors to state a price they are willing to pay for airline tickets, car rentals, hotel rooms, and a few other services. If the price is sufficiently high, the transaction is completed. However, Priceline.com completes many of its transactions from an inventory that it has purchased from airlines, car rental agencies, and hotels.

Group Shopping Sites

Another type of business that the Internet made possible is the **group purchasing site**, or **group shopping site**. On these sites, the seller posts an item with a tentative price. As individual buyers enter bids on the item (these bids are agreements to buy one unit of that item, but no price is specified), the site operators negotiate with the seller to obtain a lower price. The posted price will decrease as the number of bids increases, but only if the number of bids increases. Thus, a group shopping site builds up the number of buyers sufficiently to encourage the seller to offer a quantity discount. The effect is similar to the outcome achieved by a reverse auction.

The types of products that work well for group shopping sites are branded products with well-established reputations. This allows buyers to feel confident that they are getting a good bargain and are not just getting a lower price for a low-quality product. Ideal products also have a high value-to-size ratio and are not perishable.

Two companies, Mercata and LetsBuyIt.com, operated major group shopping sites for several years; however, both closed their doors after failing to find consistent sources of products that sold well on their sites. They found that few sellers of products that are well suited to group shopping efforts—such as computers, consumer electronics, and small appliances—were willing to work with them. These sellers did not see any compelling advantage in offering reduced prices on their merchandise to Web sites that were probably cannibalizing sales in their existing marketing channels. They also worried about offending the regular distributors of their products by selling through group shopping sites.

In 2008, Andrew Mason and Eric Lefkofsky decided to give the group shopping business another try. Starting in Chicago, they launched a site called Groupon (a shortening of "group coupon"). The site offered one coupon offer (called a "groupon") per day in the city. A groupon requires a certain number of people to sign up for it or it does not become available to anyone. For example, a $50 dinner coupon redeemable at a specific restaurant might be sold for $30. The consumer gets a $50 dinner for $30. Groupon would keep approximately half the money paid by the consumer for the groupon ($15) and the remainder would go to the restaurant. Thus, the restaurant gets $15 for its $50 dinner, but it has a chance to impress a new customer and gain that customer's return business. Further, the restaurant makes no upfront cash outlay, as it would if it were purchasing advertising.

Groupon promotes its business using social networking sites such as Facebook and Twitter to make contacts with consumers and to spread the word about the groupon deal for the day. Groupon's current customer base is primarily female, so the bulk of its business is in the health, beauty, and fitness markets. Similar services are offered by LivingSocial and Gilt. Industry analysts expect that the continued success of these group buying sites will bring competition from larger companies such as eBay and Google.

Business-to-Business Auctions

Unlike consumer online auctions, business-to-business online auctions evolved to meet a specific existing need. Many manufacturing companies periodically need to dispose of unusable or excess inventory. Despite the best efforts of procurement and production management, businesses occasionally buy more raw materials than they need. Many times, unforeseen changes in customer demand for a product can saddle manufacturers with excess finished goods or spare parts.

Depending on its size, a firm typically uses one of two methods to distribute excess inventory. Large companies sometimes have liquidation specialists who find buyers for these unusable inventory items. Smaller businesses often sell their unusable and excess inventory to **liquidation brokers**, which are firms that find buyers for these items. Online auctions are the logical extension of these inventory liquidation activities to a new and more efficient channel, the Internet.

Two of the three emerging business-to-business Web auction models are direct descendants of these two traditional methods for handling excess inventory. In the large-company model, the business creates its own auction site that sells excess inventory. In the small-company model, a third-party Web auction site takes the place of the liquidation broker and auctions excess inventory listed on the site by a number of smaller sellers. The third business-to-business Web auction model resembles consumer online auctions. In this model, a new business entity enters a market that lacked efficiency and creates a site at which buyers and sellers who have not historically done business with each other can participate in auctions. An alternative implementation of this model occurs when a Web auction replaces an existing sales channel.

In the second business-to-business auction model, smaller firms sell their obsolete inventory through an independent third-party auction site. In some cases, these online auctions are conducted by the same liquidation brokers that have always handled the disposition of obsolete inventory. These brokers adapted to the changed environment and implemented electronic commerce to stay in business. One example is the **GoIndustry Dove Bid** site, established by the Ross-Dove Company, a traditional liquidation broker for many years. **Gordon Brothers Group**, another liquidation broker, has been selling the inventory of failed retailers since 1903. The company has used its expertise to launch or help others launch Web sites that liquidate retailer inventories.

A number of hospitals and other organizations are using online auctions to fill temporary employment openings. Health care workers, such as nurses, perform similar duties in specific health care settings in most hospitals. For example, the duties performed by an intensive care unit nurse are almost identical across hospitals. State regulations on nurse licensing require that nurses have similar levels of knowledge, skills, and abilities. Having similar job functions in workplaces and having similarly qualified persons working in those jobs allows both nurses and employers to treat the nursing function as a commodity. Therefore, nurses can easily work for a variety of employers and do not require long periods of training or learning procedures specific to a particular hospital. In the past, nurse agencies would coordinate placement, matching nurses who wanted to work particular days or shifts with hospitals and other health care organizations who had shifts to fill. The agency would earn a commission on each placement. Today, employers operate their own shift auctions. Nurses bid on the shifts they would prefer to work and the software manages the auctions. In an efficient matching of supply and demand, employers meet their staffing needs efficiently, nurses get to work when they want, and the agency fee is avoided.

Business-to-Business Reverse Auctions

In Chapter 5, you learned how businesses are creating various types of electronic marketplaces to conduct business-to-business (B2B) transactions. Many of these marketplaces include auctions and reverse auctions. Glass and building materials producer Owens Corning uses reverse auctions for items ranging from chemicals (direct materials) to conveyors (fixed assets) to pipe fittings (MRO). Owens Corning even held a reverse auction to buy bottled water. Asking its suppliers to bid has reduced the cost of those items by an average of 10 percent. Because Owens Corning buys billions of dollars worth of materials, fixed assets, and MRO items each year, the potential for cost savings is significant. Both the U.S. Navy and the federal government's General Services Administration use reverse auctions to acquire some of the billions of dollars worth of materials and supplies they purchase each year. Other companies that use reverse auctions include Agilent, Bechtel, Boeing, Raytheon, and Sony.

Not all companies are enthusiastic about reverse auctions, however. Some purchasing executives argue that reverse auctions cause suppliers to compete on price alone, which can lead suppliers to cut corners on quality or miss scheduled delivery dates. Others argue that reverse auctions can be useful for nonstrategic commodity items with established quality standards. Companies that have considered reverse auctions and decided not to use them include Cisco, Cubic, IBM, and Solar Turbines.

With compelling arguments on both sides, the advisability of using reverse auctions can depend on specific conditions that exist in a given company. A company can also determine whether to use reverse auctions based on guidelines that have emerged. For example, in some industry supply chains, the need for trust and long-term strategic relationships with suppliers makes reverse auctions less attractive. In fact, the trend in purchasing management over the last 30 years has been to build trust-based relationships that can endure for many years. Using reverse auctions replaces trusting relationships with a bidding activity that pits suppliers against each other and is seen by many purchasing managers as a step backward.

In some industries, suppliers are larger and more powerful than the buyers. In those industries, suppliers simply will not agree to participate in reverse auctions. If enough important suppliers refuse to participate, it is impossible to conduct reverse auctions. In industries where a high degree of competition exists among suppliers, however, reverse auctions can be an efficient way to conduct and manage the price bidding that would naturally occur in that market. Figure 6-7 lists the supply chain characteristics that support or discourage reverse auctions identified in research conducted by Dima Ghawi and the author of this book.

Supply Chain Characteristics That Support Reverse Auctions:

- Suppliers are highly competitive.
- Product features can be clearly specified.
- Suppliers are willing to reduce the margin they earn on this product.
- Suppliers are willing to participate in reverse auctions.

Supply Chain Characteristics That Discourage Reverse Auctions:

- Product is highly complex or requires regular changes in design.
- Product has customized features.
- Long-term strategic relationships are important to buyers and suppliers.
- Switching costs are high.

FIGURE 6-7   Supply chain characteristics and reverse auctions

Auction-Related Services

The growth of eBay and other auction sites has encouraged entrepreneurs to create businesses that provide auction-related services of various kinds. These include escrow services, auction directory and information services, auction software (for both sellers and buyers), and auction consignment services. This section describes each of these new industries that have arisen to meet the needs of auction participants.

Auction Escrow Services

A common concern among people bidding in online auctions is the reliability of the sellers. Surveys indicate that as many as 18 percent of all Web auction buyers either do not receive the items they purchased, or find the items to be different from the seller's representation in some significant way. About half of those buyers are unable to resolve their disputes to their satisfaction. When purchasing high-value items, buyers can use an escrow service to protect their interests.

You learned earlier in this chapter that an escrow service is an independent party that holds a buyer's payment until the buyer receives the purchased item and is satisfied that the item is what the seller represented it to be. Some escrow services take delivery of the item from the seller and perform the inspection for the buyer. In such situations, buyers give the escrow service authority to examine. Usually, escrow agents that perform this service are art appraisers, antique appraisers, and the like who are qualified to judge quality, usually with better judgment than the buyer. Escrow services do, however, charge fees ranging from 1 to 10 percent of the item's cost, subject to a minimum fee, typically between $5 and $50. The minimum fee provision can make escrow services too expensive for small purchases. Escrow services that handle Web auction transactions include Escrow.com and eDeposit. Some escrow firms also sell auction buyer's insurance, which can protect buyers from nondelivery and some quality risks. There have been cases of escrow fraud, especially in auctions of high-value items. The Better Business Bureau recommends that consumers determine whether an escrow service is licensed and bonded before using it. Consumers can do this by contacting the appropriate licensing agency in the state in which the escrow service is located. The Better Business Bureau recommends avoiding offshore escrow companies entirely.

Wary bidders in low-price auctions (for which the minimum escrow charges would be excessive) do have some other ways to protect themselves. One way is to check the seller's record on the auction site to see how the seller is rated. Also, some Web sites offer lists of auction sellers who have failed to deliver merchandise or who have otherwise cheated bidders in the past. These sites are operated as free services (often by bidders who have been cheated), so they sometimes contain unreliable information and they open and close periodically, but you can use your favorite search engine to locate sites that currently carry such lists.

Auction Directory and Information Services

Another service offered by some firms on the Web is a directory of auctions. AuctionBytes is an auction information site that publishes an e-mail newsletter with articles about developments in the online auction industry. It provides guidance for new auction participants and helpful hints and tips for more experienced buyers and sellers along with directories of online auction sites.

Price Watch is an advertiser-supported site on which those advertisers post their current selling prices for computer hardware, software, and consumer electronics items. Although this monitoring is a retail pricing service designed to help shoppers find the best price on new items, Web auction participants find it can help them with their bidding strategies.

Auction Software

Both auction buyers and sellers can purchase software to help them manage their online auctions. Sellers often run many auctions at the same time. Companies such as **AuctionHawk** and **Vendio** sell auction management software and services for both buyers and sellers. For sellers, these companies offer software and services that can help with or automate tasks such as image hosting, advertising, page design, bulk repeatable listings, feedback tracking and management, report tracking, and e-mail management. Using these tools, sellers can create attractive layouts for their pages and manage hundreds of auctions.

For buyers, a number of companies sell auction sniping software. **Sniping software** observes auction progress until the last second or two of the auction clock. Just as the auction is about to expire, the sniping software places a bid high enough to win the auction (unless that bid exceeds a limit set by the sniping software's owner). The act of placing a winning bid at the last second is called a **snipe**. Because sniping software synchronizes its internal clock to the auction site clock and executes its bid with a computer's precision, the software almost always wins out over a human bidder. The first sniping software, named Cricket Jr., was written by David Eccles in 1997. He sells the software on his **Cricket Sniping Software** site. A number of other sniping software sellers have entered the market—each claiming that its software will outbid other sniping software. Some sites offer sniping services; that is, the sniping software runs on their Web site and customers enter their sniping instructions on that site. Some of these companies offer subscriptions; others use a mixed-revenue model in which they offer some free snipes supported by advertising, but require payment for additional snipes.

Auction Consignment Services

Several entrepreneurs have identified yet another auction-related business that meets the needs of people and small businesses who want to use an online auction, but do not have the skills or the time to become a seller. These companies, called **auction consignment services**, take an item and create an online auction for that item, handle the transaction, and remit the balance of the proceeds after deducting a fee that ranges from 25 to 50 percent of the selling price obtained. Items that do not sell are returned or donated to charity. Auction consignment businesses include **ePowerSellers** and **iSold It**.

Summary

In this chapter, you learned how companies are now using the Web to do things that they have never done before, such as creating social networks, using mobile technologies to make sales and increase operational efficiency, operating auction sites, and conducting related businesses.

The Web's ability to bring together geographically dispersed people and organizations that share narrow interests has encouraged the development of virtual communities and social networks. Businesses are creating online communities using social networking features that connect them to their customers and suppliers. A growing number of businesses are exploiting the mobile commerce opportunities presented by smart phones and tablet devices that have high-bandwidth access to the Internet. As we enter the third wave of electronic commerce, individuals are using social networking sites, blogs, and microblogging tools for personal and business-related interactions. Companies are using internal social networking sites to communicate with employees and coordinate work across various organizational units.

You learned about the key characteristics of the seven major auction types, and learned how firms are using online auctions to sell goods to their customers and buy from their suppliers. Although some specialty sites do conduct significant auction activities, the consumer online auction business is dominated by eBay, at least in the United States. B2B auctions give companies a new and efficient way to dispose of excess inventory, and B2B reverse auctions provide an effective procurement tool under some conditions. A number of businesses offer ancillary services to Web users who participate in online auctions. These businesses include escrow services, auction directories and information sites, auction management software for both sellers and bidders, and auction consignment sites.

Key Terms

Apps	Microblogs
Ascending-price auction	Microlending
Auction consignment services	Minimum bid
Auctioneer	Minimum bid increment
Bidders	Mobile commerce (m-commerce)
Bids	Mobile wallet
Bulletin board systems (BBSs)	Monetizing
Consumer-to-business	Online community
Descending-price auctions	Open auction (open-outcry auction)
Double auction	Open-outcry double auction
Dutch auction	Open source software
English auction	Participatory journalism
Escrow service	Private valuations
First-price sealed-bid auction	Proxy bid
Group purchasing site (group shopping site)	Reserve price (reserve)
Idea-based networking	Reverse auction (seller-bid auction)
Idea-based virtual communities	Reverse bid
Jailbreaking	Rooting
Liquidation brokers	Sealed-bid auction
Lock-in effect	Second-price sealed-bid auction
Meetup	Shill bidders

Short messaging service (SMS)	Usenet newsgroups
Snipe	Vickrey auctions
Sniping software	Virtual community
Social networking site	Virtual learning network
Social shopping	Web community
Tablet devices	Winner's curse
Tweets	Wireless Application Protocol (WAP)
Usenet	Yankee auction

Review Questions

1. Write a paragraph in which you distinguish between a virtual community and a social networking Web site.

2. Identify a product that could be promoted using a social networking site such as Facebook. In about 100 words, explain why your chosen product would be a good candidate for a social networking-based promotion strategy.

3. In two or three paragraphs, describe the differences between a writing a blog and engaging in microblogging.

4. Briefly define participatory journalism, and then write one or two paragraphs about how your school newspaper might benefit from engaging in it.

5. In about 100 words, define stickiness and explain under what circumstances a social networking site would want to develop that characteristic.

6. In two or three paragraphs, outline at least three different ways in which a social networking site might monetize its visitors.

7. In one or two paragraphs, define the term "microlending."

8. In about 100 words, describe two or three specific apps that could use a smart phone's GPS capability. Be sure to make clear the benefit of using the GPS in the app in each case.

9. In two or three paragraphs, explain why the use of mobile wallets is less common in the United States than in other countries.

10. In two or three paragraphs, define the term "reserve price" and explain how the use of a reserve price can affect the progress and outcome of an auction.

11. In two or three paragraphs, describe how a Dutch auction works and explain why it might be a good auction method to use for purchase or sale of a company's stock.

12. In two or three paragraphs, explain how sniping software works.

Exercises

1. Google purchased YouTube in 2006 for $1.65 billion. In about 200 words, outline reasons that Google would have wanted to acquire a site such as YouTube and describe the benefits that the company obtained from its purchase.

2. Review both the Etsy and the We Love Etsy Web sites. In about 300 words, outline the elements of Etsy's Web site and business philosophy that make it a social networking site in addition to being an online business that sells goods.

3. Compare the apps offered in the **Apple Apps for iPhone** store to those offered in the **Android Market**. In about 100 words, present a comparison of the software applications offered in each store. You may also comment on the usability of the site.

4. Midland University, like most metropolitan universities, faces a chronic shortage of parking spaces on campus. Each stakeholder group in the typical university community (these groups include students, faculty, administrators, staff, and visitors) believes its members should have the top priority for parking spaces. You have been assigned to a university task force to study the problem. You decide that an annual online auction of parking spaces conducted on the university's intranet could provide a solution. In about 300 words, describe the elements of an annual online auction for parking spaces at Midland University. Be sure to include provisions for disabled persons and for those university employees who do not have regular access to computers in their typical work environment (such as janitors, physical plant maintenance workers, or gardeners).

5. Assume you work in the procurement department of a small aerospace parts manufacturer. Your company builds switches used to control heating and ventilation systems in large buildings. The parts your company buys must meet precise specifications and the parts are not generally interchangeable; that is, your company's engineers must work with your suppliers to design specific parts for particular systems. Your director of purchasing is interested in using online reverse auctions to buy these parts. In approximately 200 words, outline arguments for and against using online reverse auctions in this situation and conclude with a specific recommendation.

6. Some eBay users believe that the use of sniping software is unfair and that eBay should prohibit its use. In an essay of about 200 words, present facts and logical arguments that would convince eBay to prohibit the use of sniping software.

Cases

C1. Alibaba.com

In 1995, Jack Ma taught English in Hangzhou, China, a city near the economic center of Shanghai. Ma wanted to get into the business world, so he raised $2000 from relatives and friends to start Chinapage.com, one of the first Chinese online businesses. He followed that experience with a job at the Ministry of Foreign Trade and Economic Cooperation. He grew frustrated with the slow pace of the government bureaucracy and left after a year to start his own company again. He placed an ad on the Internet advertising a language translation service for companies that wanted to do business in China. Within two hours, he had received six e-mailed inquiries. About 60 percent of the Chinese economy is manufacturing, and 90 percent of manufacturing companies are small or midsized businesses. Ma began collecting information from Chinese manufacturing companies that wanted to do business internationally. He translated and organized the information, and then posted it on a B2B Web portal site he named Alibaba.com.

Alibaba.com has always concentrated on small and midsized businesses (SMBs). Ma believed that global companies spend most of their efforts on doing business with large companies. He sees China (and the rest of Asia) as having a different economic structure than the United States or Europe, where the economies are dominated by large companies. Ma believes that Alibaba.com's true opportunities lie in connecting SMBs around the world with SMBs in China. He argues that SMBs seldom have any sales channels outside their own country. To compensate, SMBs must travel extensively to meet suppliers and customers at exhibitions or trade fairs. Ma believes that Alibaba.com offers SMBs a reasonably priced alternative.

Foreign companies interested in buying from Chinese suppliers must register on Alibaba.com (buyer registration is free) before they can access the site's supplier database. Alibaba.com charges Chinese companies a membership fee of several thousand dollars for translating and listing their information. The site also lists foreign suppliers. These suppliers can list a small number of items at no charge; however, most choose to pay a small fee that pays for a credit check and allows them to be listed as TrustPass members on the site. The TrustPass designation provides assurance to Chinese companies that want to buy from these suppliers. By 2001, more than 1 million companies had registered with Alibaba.com. In 2003, the company reported its first profitable year, with a net income of $12 million. Since then, the company has grown steadily and continues to be profitable. Many of Alibaba.com's registered members are happy with the results they obtain, as indicated by the annual membership renewal rate, which exceeds 70 percent.

Alibaba.com, like all portal sites, suffered a setback during the 2001–2002 time period, but its fee-based revenue model allowed it to recover more quickly than portals that were dependent on advertising revenue. The company sees future growth in the continued expansion of trade between Chinese manufacturers and the rest of the world. Ma is also optimistic about the portal's potential for helping Chinese businesses connect with other Chinese businesses.

Required:

1. Alibaba.com was an early entrant into the B2B portal market in China. In about 100 words, explain how this might have created a lock-in effect, especially given the types of businesses the site attracts.

2. Alibaba.com currently charges foreign sellers an annual fee of about $400 for a TrustPass membership, but Chinese companies pay $8000 or more for their annual listings as China Gold Suppliers. In about 200 words, explain why the site has different listing charges for the two types of members and critically evaluate this practice.

3. You learned in Chapter 5 that large companies, such as General Electric and Sears, often require suppliers to follow specific rules if they want to do business (such as using EDI or even a specific EDI VAN). Alibaba.com currently focuses on connecting SMBs with each other. In about 200 words, discuss opportunities that might exist for Alibaba.com to become an intermediary in relationships between Chinese SMBs and large global companies such as General Electric and Sears.

4. In 2003, Alibaba.com launched Taobao.com to compete in the general consumer online auction market against eBay in China. After four years of an intensive and expensive battle, eBay withdrew from China completely. In about 200 words, describe the advantages Alibaba.com might have had over eBay in this new market, and then describe the advantages eBay might have had over Alibaba.com. Be sure to discuss lock-in effects where appropriate.

5. In 2005, Yahoo! paid $1 billion for a 40 percent interest in Alibaba.com. Yahoo! was interested in the company's Taobao.com auction site because Yahoo! had not been as successful as it would have liked in developing its own Chinese auction site. However, Yahoo! was also interested in using Alibaba.com's strong reputation in China to help it compete with Baidu.com, the top Chinese search engine site. In about 200 words, describe the ways in which Alibaba.com's reputation could help Yahoo! compete more effectively as a search engine and Web portal in China.

Note: Your instructor might assign you to a group to complete this case and might ask you to prepare a formal presentation of your results to your class.

C2. Old Metamora

Betty Shriver is the owner of Betty's Crystal, a small shop that sells collectible crystal and glass figurines. Betty's shop carries many items that she purchased from estate sales and regional auctions, but the shop also sells new crystal figurines from manufacturers such as Baccarat, Lalique, Orrefors, and Swarovski. The shop is located in Metamora, Indiana, which is a popular tourist destination for weekend travelers in the Midwest. The town of Old Metamora is a small historic area in a rural setting that is less than a day's drive from seven major metropolitan areas: Chicago, Cincinnati, Columbus, Detroit, Indianapolis, Louisville, and St. Louis.

The shop is very busy on weekends and during the spring and summer months when tourists flock to Old Metamora. In the early fall, the tourist traffic slows considerably, and in the winter months, the town becomes almost deserted. Two years ago, Betty began to pick up extra business during the off season by auctioning items on eBay. Not only did the auctions help keep inventory moving during the slow months, but Betty found that she was able to carry a wider selection of items in the store. In the past, she would see unusual items at estate sales and auctions that she feared would not sell quickly in the shop. Now Betty knows that any item that does not sell in the shop can be auctioned online quite easily. Another unexpected benefit of participating in online auctions is that Betty developed relationships with regular buyers of crystal figurines and with people who run collectibles stores in other parts of the country. Every auction involves at least two e-mails (one to confirm the final bid and another to confirm the payment). Many successful bidders also send e-mail messages to Betty when they receive the item with questions about the item, or just to thank Betty for sending the item so quickly. Some of these e-mail exchanges continue with discussions related to crystal figurines and other collectible items.

Betty's online auction experiences prompted her to consider expanding the online portion of her business. She has heard (from other shop owners) that eBay allows people to create online stores within the eBay site and that Amazon.com offers a similar service that lists seller's items on Amazon.com's regular product pages. She is also interested in creating a Web site that contains photos and descriptions of popular crystal figurines with additional information about how they are made. Betty also wants to include a list of figurines that are no longer manufactured (which makes them more valuable) and a guide to buying collectible crystal figurines that could help her customers and bidders on her auctions make more informed decisions as they add to their collections. She believes that such a site could attract a large number of people interested in crystal figurines. She wants to find ways to direct these site visitors to her auctions and her proposed Web store. Betty has hired you as a consultant to build on her ideas and to help her develop an expansion strategy for her online business activities.

Required:

1. Search for information about Amazon Marketplace and eBay Stores on the Web and in your library that will help you make a recommendation to Betty regarding which alternative would provide the best avenue for her online business expansion. Support your recommendation with relevant facts, including specific costs of operating each type of store and specific benefits that Betty could gain by using one or the other. Summarize your recommendation and supporting facts in a report to Betty of 400 words.

2. Outline a strategy that Betty could implement using a social networking site such as Facebook that might direct traffic to her Web site, to her auctions on eBay, and to her products for sale on Amazon.com. For each element in the strategy, provide an explanation of how it would help achieve Betty's goals. Summarize the social networking promotion strategy in a report to Betty of about 500 words.

Note: Your instructor might assign you to a group to complete this case and might ask you to prepare a formal presentation of your results to your class.

For Further Study and Research

Ankeny, J. 2009. "NTT DoCoMo Rolling Out Mobile Payments Program," *Fierce Mobile Content*, July 2. (http://www.fiercemobilecontent.com/story/ntt-docomo-rolling-out-mobile-payments-program/2009-07-02)

Baran, R. 2011. "Social Networking in China and The United States: Opportunities for New Marketing Strategy and Customer Relationship Management," *AFBE Journal*, 4(3), December, 464–481.

Barker, V. and H. Ota. 2011. "Mixi Diary Versus Facebook Photos: Social Networking Site Use Among Japanese and Caucasian American Females," *Journal of Intercultural Communication Research*, 40(1), 39–63.

Belson, K., R. Hof, and B. Elgin. 2001. "How Yahoo! Japan Beat eBay at Its Own Game," *Business Week*, June 4, 58.

Boyd, D. and N. Ellison. (2007). "Social Network Sites: Definition, History, and Scholarship," *Journal of Computer-Mediated Communication*, 13(1). (http://jcmc.indiana.edu/vol13/issue1/boyd.ellison.html)

Brandel, M. 2009. "Start Connecting With Mobile Customers," *Computerworld*, October 5, 19–22.

Breckenridge, M. 2008. "Old Meets New at Etsy," *Akron Beacon Journal*, March 6, D1.

Brohan, M. 2011. "Retailers Diving Into Mobile Commerce Are Coming Up With Significant Sales," *Internet Retailer*, September 30. (http://www.internetretailer.com/2011/09/30/internet-retailer-survey-mobile-commerce)

Burnham, K. 2009. "Scottrade: The Social Enterprise," *CIO*, November 1, 18.

Business Wire. 2008. "LookSmart Announces Final Results of Tender Offer for its Common Stock," February 21.

Carr, D. 2010. "Why Twitter Will Endure," *The New York Times*, January 1. (http://www.nytimes.com/2010/01/03/weekinreview/03carr.html)

Cassady, R. 1967. *Auctions and Auctioneering*. Berkeley, CA: University of California Press.

Chafkin, M. 2007. "How to Kill a Great Idea!" *Inc. Magazine*, June 1. (http://www.inc.com/magazine/20070601/features-how-to-kill-a-great-idea.html)

Chang, A. 2003. "Hospitals Auction Nursing Shifts Online," *The Boston Globe*, December 28, A28.

Chen, B. 2009. "Verizon Drafts Developers into Mobile Software War on Apple," *Wired News*, July 14. (http://www.wired.com/gadgetlab/2009/07/smartphone-war/)

Chen, K. and K. Qiu Haixu. 2004. "Chinese E-Commerce Sites Allow Small Firms to Reach Wider Base," *The Wall Street Journal*, February 25, A12.

Cheng, A. and J. Thaw. 2005. "Yahoo! Raises Stakes Higher in China With Alibaba Deal," *The Seattle Times*, August 22, C4.

Cohen, A. 2001. "The Sniper King," *On Magazine*, May.

Credit Union Management. 2007. "Focus on Microlending: Kiva Is People Helping People," May, 12.

Doebele, J. 2005. "Alibaba.com: Standing Up to eBay," *Forbes.com*, April 18. (http://www.forbes.com/business/forbes/2005/0418/050.html)

Dvorak, J. 2011. "Note to Google: Microsoft Had the Right Idea," *PC Magazine*, June 30. (http://www.pcmag.com/article2/0,2817,2387942,00.asp)

The Economist. 1997. "Going, Going…" May 31, 61.

The Economist. 2001. "We Have Lift-Off." February 3, 69–71.

Eisner, A. 2011. "Could Groupon's Deal Addicts Hurt Retailers This Year?" *Retrevo*, October 27. (http://www.retrevo.com/content/trackback/1911)

Epstein, Z. 2011. "Apple and Google Dominate Smartphone Space While Others Scramble," *BGR*, December 13. (http://www.bgr.com/2011/12/13/apple-and-google-dominate-smartphone-space-while-other-vendors-scramble/)

Ferraro, N. 2008. "Lending & Philanthropy 2.0," *InformationWeek*, February 4, 40.

Flandez, R. 2008. "Building an Online Community of Loyal and Vocal Users," *The Wall Street Journal*, March 6, B5.

Ghawi, D. and G. Schneider. 2004. "New Approaches to Online Procurement," *Proceedings of the Academy of Information and Management Sciences*, 8(2), October, 25–28.

Gilbert, J. and A. Kerwin. 1999. "Newspapers Carve Slice of Auction Pie," *Advertising Age*, 70(26), June 21, 32–34.

Hanlon, P. and J. Hawkins. 2008. "Expand Your Brand Community Online," *Advertising Age*, January 7, 14–15.

Heffernan, V. 2011. "The Old Internet Neighborhoods," *The New York Times*, July 10. (http://opinionator.blogs.nytimes.com/2011/07/10/remembrance-of-message-boards-past/)

Internet Retailer. 2010. "Online Liquidation Services." (http://www.internetretailer.com/vendors/online-liquidation-services/)

Intrator, Y. 2005. "The Trouble With Portals," *CIO Magazine Online*, May 9.

Kawakami, S. 2003. "China's Visionary B2B," *J@pan Inc*., May, 14–16.

Keegan, V. 2008. "Entrepreneurs Come Out of the Webwork," *The Guardian*, February 28, 4.

Kennedy, J. 1998. "Radio Daze," *Technology Review*, 101(6), November–December, 68–71.

Kolakowski, N. 2011. "Nokia Windows Phones Need U.S. Market, Symbian Customers," *eWeek*, October 27. (http://www.eweek.com/c/a/Mobile-and-Wireless/Nokia-Windows-Phones-Need-US-Market-Symbian-Customers-564760/)

MacMillan, D., P. Burrows, and S. Ante. 2009. "The App Economy," *Business Week*, November 2, 44–49.

Meece, M. 2011. "Making Short Work of Shopping for Tablet Users," *The New York Times*, December 7. (http://www.nytimes.com/2011/12/08/technology/personaltech/quick-and-easy-shopping-for-tablet-users.html)

Miller, C. 2011. "Another Try By Google to Take on Facebook," *The New York Times*, June 28. (http://www.nytimes.com/2011/06/29/technology/29google.html)

Miller, K. 2007. "An eBay for the Arts and Crafts Set," *Business Week*, July 23, 70.

Norris, F. 2004. "Google's Offering Proves Stock Auctions Can Really Work," *The New York Times*, August 23, C6.

Okazaki, S. and M. Yague. 2012. "Responses to an Adver-gaming Campaign on a Mobile Social Networking Site: An Initial Research Report," *Computers in Human Behavior*, 28(1), January, 78–86.

Petrecca, L. and B. Snyder. 1998. "Auction Universe Puts in $10 Mil Bid for Customers," *Advertising Age*, 43(8), October 26, 8.

Prochnow, D. 2010. "Creating Mobile Apps With a Point and a Click," *Popular Science*, January 22. (http://www.popsci.com/diy/article/2010-01/point-and-click-apps)

Purchasing. 2001. "What Top Supply Execs Say About Auctions," 130(12), June 21, S2–S3.

Quan, J. 1999. "Risky Business," *Rolling Stone*, March 4, 91–92.

Reuters. 2011. "Exclusive: Facebook Doubles First-half Revenue," September 7. (http://www.reuters.com/article/2011/09/07/us-facebook-idUSTRE7863YW20110907)

Rheingold, H. 1993. *The Virtual Community: Homesteading on the Electronic Frontier*. New York: HarperCollins.

Rheingold, H. 2002. *Smart Mobs*. Cambridge, MA: Basic.

Robins, W. 2000. "Auctions.com Now a Dot-Goner," *Editor & Publisher*, August 28, 6.

Sacco, A. 2009. "Paging Dr. BlackBerry: Smartphones Deliver EKGs for Faster Diagnoses," *CIO*, November 1, 15–16.

Schonfeld, E. 2011. "Google's YouTube Revenues Will Pass $1 Billion in 2012," *TechCrunch*, March 21. (http://techcrunch.com/2011/03/21/citi-google-local-youtube-1-billion/)

Seelye, K. 2005. "Why Newspapers Are Betting on Audience Participation," *The New York Times*, July 4, C2.

Spanbauer, S. 2008. "The Right Social Network for You," *PC World*, April, 105–110.

Stefano, T. 2007. "Social Networking: A Web 2.0 Revolution," *E-Commerce Times*, March 30. (http://www.ecommercetimes.com/story/56576.html)

Swift, M. 2011. "YouTube No Longer Google's Ugly Stepchild, With Revenue on the Rise," *The Columbus Dispatch*, March 28. (http://www.dispatch.com/content/stories/business/2011/03/28/youtube-no-longer-googles-ugly-stepchild-with-revenue-on-rise.html)

Tabuchi, H. 2011. "Facebook Wins Relatively Few Friends in Japan," *The New York Times*, January 10, B1.

Takahashi, T. 2010. "MySpace or Mixi? Japanese Engagement with Social Networking Sites in the Global Age," *New Media & Society*, 12(3), 453–475.

Thaler, W. 1994. *The Winner's Curse: Paradoxes and Anomalies of Economic Life*. Princeton, NJ: Princeton University Press.

Todras-Whitehall, E. 2005. "'Folksonomy' Carries Classifieds Beyond SWF and 'For Sale,'" *The New York Times*, October 5. (http://www.nytimes.com/2005/10/05/technology/techspecial/05ethan.html)

Tomchin, E. 2009. "EBay Alternatives Review: EBid, OnlineAuction and Overstock Auctions," *AuctionBytes*, March 1. (http://www.auctionbytes.com/cab/abu/y209/m03/abu0234/s03)

Tugend, A. 2009. "Losing Out After Winning an Online Auction," *The New York Times*, October 24. (http://www.nytimes.com/2009/10/24/technology/24shortcuts.html)

Ulanoff, L. 2011. "Android, Android, Everywhere," *PC Magazine*, 30(6), June 1, 38.

Vara, V. 2007. "Facebook Gets Help From Its Friends," *The Wall Street Journal*, June 22, B1–2.

Vickrey, W. 1961. "Counterspeculation, Auctions, and Competitive Sealed Tenders," *Journal of Finance*, 16(1), March, 8–37.

Wagner, M. 2009. *Smartphone App: What the Doctor Ordered*. Manhasset, NY: InformationWeek.

Wingfield, N. 2004. "Taking on eBay," *The Wall Street Journal*, September 13, R10.

Wireless Federation. 2009. "NTT DoCoMo's Credit Payment Subscriptions Reach 10 Million Mark," August 26. (http://wirelessfederation.com/news/17894-ntt-docomos-credit-payment-subscriptions-reach-10mn-mark/)

Zimmerman, E. 2007. "Investing in the Women of Ghana," *FSB: Fortune Small Business*, 17(4), May, 101–102.

CHAPTER 7

THE ENVIRONMENT OF ELECTRONIC COMMERCE: LEGAL, ETHICAL, AND TAX ISSUES

LEARNING OBJECTIVES

In this chapter, you will learn:

- How the legal environment affects electronic commerce activities
- What elements combine to form an online business contract
- How copyright, patent, and trademark laws govern the use of intellectual property online
- That the Internet has opened doors for online crime, terrorism, and warfare
- How ethics issues arise for companies conducting electronic commerce
- Ways to resolve conflicts between companies' desire to collect and use their customers' data and the privacy rights of those customers
- What taxes are levied on electronic commerce activities

INTRODUCTION

In 1999, **Dell Computer** and Micron Electronics (now doing business as **Micron Technology**), two companies that sold personal computers through their Web sites, agreed to settle U.S. Federal Trade Commission (FTC) charges that they had disseminated misleading advertising to their existing and potential customers. The advertising in question was for computer leasing plans that both companies had offered on their Web sites. The ads stated the price of the computer along with a monthly payment. Unfortunately for Dell and Micron, stating the monthly payment without disclosing full details of the lease plan is a violation of the Consumer Leasing Act of 1976. This law is implemented through a federal regulation that was written and is updated periodically by the Federal Reserve Board. This regulation, called Regulation M, was designed to require banks and other lenders to fully disclose the terms of leases so that consumers would have enough information to make informed financing choices when leasing cars, boats, furniture, and other goods.

Both Dell and Micron had included the required information on their Web pages, but FTC investigators noted that important details of the leasing plans, such as the number of payments and the fees due at the signing of the lease, were placed in a small typeface at the bottom of a long Web page. A consumer who wanted to determine the full cost of leasing a computer would need to scroll through a number of densely filled screens to obtain enough information to make the necessary calculations.

In the settlement, both companies agreed to provide consumers with clear, readable, and understandable information in their lease advertising. The companies also agreed to record-keeping and federal monitoring activities designed to ensure their compliance with the terms of the settlement.

Dell and Micron are computer manufacturers. It apparently did not occur to them that they needed to become experts in Regulation M, generally considered to be a banking regulation. Companies that do business on the Web expose themselves, often unwittingly, to liabilities that arise from today's business environment. That environment includes laws and ethical considerations that may be different from those with which the business is familiar. In the case of Dell and Micron, they were unfamiliar with the laws and ethics of the banking industry. The banking industry has a different culture than that of the computer industry—it is unlikely that a bank advertising manager would have made such a mistake.

As you will learn in this chapter, Dell and Micron are by no means the only Web businesses that have run afoul of laws and regulations. As companies move more of their operations online, they can find themselves subject to unfamiliar laws and different ethical frameworks much more rapidly than when they operated in familiar physical domains.

THE LEGAL ENVIRONMENT OF ELECTRONIC COMMERCE

Businesses that operate on the Web must comply with the same laws and regulations that govern the operations of all businesses. If they do not, they face the same penalties—including fines, reparation payments, court-imposed dissolution, and even jail time for officers and owners—that any business faces.

Businesses operating on the Web face two additional complicating factors as they try to comply with the law. First, the Web extends a company's reach beyond traditional boundaries. As you learned in Chapter 1, a business that uses the Web becomes an international business instantly. Thus, a company can become subject to many more laws more quickly than a traditional brick-and-mortar business based in one specific physical location. Second, the Web increases the speed and efficiency of business communications. As you learned in Chapters 3 and 4, customers often have much more interactive and complex relationships with online merchants than they do with traditional merchants. Further, the Web creates a network of customers who often have significant levels of interaction with each other. In Chapter 5, you learned how companies use online communications to facilitate complex strategic alliances and supply web relationships. These communication- and information-sharing supply chain channels also expose an organization's operations to other entities. Web businesses that violate the law or breach ethical standards can face rapid and intense reactions from large numbers of customers, vendors, and other stakeholders who become aware of the businesses' activities.

In this section, you will learn about the issues of borders, jurisdiction, and Web site content and how these factors affect a company's ability to conduct electronic commerce. You will also learn about legal and ethical issues that arise when the Web is used in the commission of crimes, terrorist acts, and even the conduct of war.

Borders and Jurisdiction

Territorial borders in the physical world serve a useful purpose in traditional commerce: They mark the range of culture and reach of applicable laws very clearly. When people travel across international borders, they are made aware of the transition in many ways. For example, exiting one country and entering another usually requires a formal examination of documents, such as passports and visas. In addition, both the language and the currency usually change upon entry into a new country. Each of these experiences, and countless others, are manifestations of the differences in legal rules and cultural customs in the two countries. In the physical world, geographic boundaries almost always coincide with legal and cultural boundaries. The limits of acceptable ethical behavior and the laws that are adopted in a geographic area are the result of the influences of the area's dominant culture. The relationships among a society's culture, laws, and ethical standards appear in Figure 7-1, which shows that culture affects laws directly and indirectly through its effect on ethical standards. The figure also shows that laws and ethical standards affect each other.

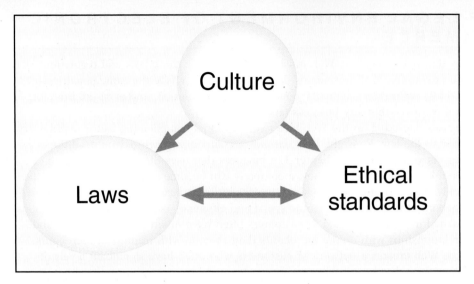

© Cengage Learning

FIGURE 7-1 Culture helps determine laws and ethical standards

The geographic boundaries on culture are logical; for most of our history, slow methods of transportation and conflicts among various nations have prevented people from travelling great distances to learn about other cultures. Both restrictions have changed in recent years, however, and now people can travel easily from one country to another within many geographic regions. One example is the European Union (EU), which allows free movement within the EU for citizens of member countries. Most of the EU countries (Great Britain being a notable exception) now use a common currency (the euro) instead of their former individual currencies (for example, French francs, German marks, and Italian lire). Legal scholars define the relationship between geographic boundaries and legal boundaries in terms of four elements: power, effects, legitimacy, and notice.

Power

Power is a form of control over physical space and the people and objects that reside in that space, and is a defining characteristic of statehood. For laws to be effective, a government must be able to enforce them. Effective enforcement requires the power both to exercise physical control over residents, if necessary, and to impose sanctions on those who violate the law. The ability of a government to exert control over a person or corporation is called **jurisdiction**.

Laws in the physical world do not apply to people who are not located in or do not own assets in the geographic area that created those particular laws. For example, the United States cannot enforce its copyright laws on a citizen of Japan who is doing business in Japan and owns no assets in the United States. Any assertion of power by the United States over such a Japanese citizen would conflict with the Japanese government's recognized authority over its citizens. Japanese citizens who bring goods into the United States to sell, however, are subject to applicable U.S. laws. A Japanese Web site that offers delivery of goods into the United States is, similarly, subject to applicable U.S. laws.

The level of power asserted by a government is limited to that which is accepted by the culture that exists within its geographic boundaries. Ideally, geographic boundaries, cultural groupings, and legal structures all coincide. When they do not, internal strife and civil wars can erupt.

Effects

Laws in the physical world are grounded in the relationship between physical proximity and the **effects**, or impact, of a person's behavior. Personal or corporate actions have stronger effects on people and things that are nearby than on those that are far away. Government-provided trademark protection is a good example of this. For instance, the Italian government can provide and enforce trademark protection for a business named Casa di Baffi located in Rome. The effects of another restaurant using the same name are strongest in Rome, somewhat less in geographic areas close to Rome, and even less in other parts of Italy. That is, the effects diminish as geographic distance increases. If someone were to open a restaurant in Kansas City and call it Casa di Baffi, the restaurant in Rome would experience few, if any, negative effects from the use of its trademarked name in Kansas City because it is so far away and because so few people would be potential customers of both restaurants. Thus, the effects of the trademark infringement would be controlled by Italian law because of the limited range within which such an infringement has an effect.

The characteristics of laws are determined by the local culture's acceptance or rejection of various kinds of effects. For example, certain communities in the United States require that houses be built on lots that are at least 5 acres. Other communities prohibit outdoor advertising of various kinds. The local cultures in these communities make the effects of such restrictions acceptable.

Once businesses began operating online, they found that traditional effects-based measures did not apply as well and that the laws based on these measures did not work well either. For example, France has a law that prohibits the sale of Nazi memorabilia. The effects of this law were limited to people in France and they considered it reasonable. U.S. laws do not include a similar prohibition because U.S. culture makes a different trade-off between the value of memorabilia (in general) and the negative cultural memory of Nazism. When U.S.-based online auction sites began hosting auctions of Nazi memorabilia, those sites were in compliance with U.S. laws. However, because of the international nature of the Web, these auctions were available to people around the world, including residents of France. In other words, the effects of U.S. culture and law were being felt in France. The French government ordered Yahoo! Auctions to stop these auctions. Yahoo! argued that it was in compliance with U.S. law, but the French government insisted that the effects of those Yahoo! auctions extended to France and thus violated French law. To avoid protracted legal actions over the jurisdiction issue, Yahoo! decided that it would no longer carry such auctions.

Legitimacy

Most people agree that the legitimate right to create and enforce laws derives from the mandate of those who are subject to those laws. In 1970, the United Nations passed a resolution that affirmed this idea of governmental legitimacy. The resolution made clear that the people residing within a set of recognized geographic boundaries are the ultimate source of legitimate legal authority for people and actions within those boundaries. Thus, **legitimacy** is the idea that those subject to laws should have some role in formulating them.

Some cultures allow their governments to operate with a high degree of autonomy and unquestioned authority. China and Singapore are countries in which national culture

285

The Environment of Electronic Commerce: Legal, Ethical, and Tax Issues permits the government to exert high levels of unchecked authority. Other cultures, such as those of the Scandinavian countries, place strict limits on governmental authority.

The levels of authority and autonomy with which governments of various countries operate vary significantly from one country to another. Online businesses must be ready to deal with a wide variety of regulations and levels of enforcement of those regulations as they expand their businesses to other countries. This can be difficult for smaller businesses that operate on the Web.

Notice

Physical boundaries are a convenient and effective way to announce the ending of one legal or cultural system and the beginning of another. The physical boundary, when crossed, provides **notice** that one set of rules has been replaced by a different set of rules. Notice is the expression of such a change in rules. People can obey and perceive a law or cultural norm as fair only if they are notified of its existence. Borders provide this notice in the physical world. The legal systems of most countries include a concept called constructive notice. People receive **constructive notice** that they have become subject to new laws and cultural norms when they cross an international border, even if they are not specifically warned of the changed laws and norms by a sign or a border guard's statement. Thus, ignorance of the law is not a sustainable defense, even in a new and unfamiliar jurisdiction.

This concept presents particular problems for online businesses because they may not know that customers from another country are accessing their Web sites. Thus, the concept of notice—even constructive notice—does not translate very well to online business. The relationship between physical geographic boundaries and legal boundaries in terms of these four elements is summarized in Figure 7-2.

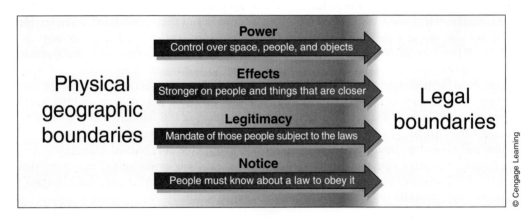

FIGURE 7-2   Physical geographic boundaries lead to legal boundaries

Jurisdiction on the Internet

The tasks of defining, establishing, and asserting jurisdiction are much more difficult on the Internet than they are in the physical world, mainly because traditional geographic boundaries do not exist. For example, a Swedish company that engages in electronic commerce could have a Web site that is entirely in English and a URL that ends in ".com," thus not indicating to customers that it is a Swedish firm. The server that hosts this company's Web page could be in Canada, and the people who maintain the Web site might work from their homes in Australia. If a Mexican citizen buys a product from the

Swedish firm and is unhappy with the goods received, that person might want to file a lawsuit against the seller firm. However, the world's physical border-based systems of law and jurisdiction do not help this Mexican citizen determine where to file the lawsuit. The Internet does not provide anything like the obvious international boundary lines in the physical world. Thus, the four considerations that work so well in the physical world—power, effects, legitimacy, and notice—do not translate very well to the virtual world of electronic commerce.

Governments that want to enforce laws regarding business conduct on the Internet must establish jurisdiction over that conduct. A **contract** is a promise or set of promises between two or more legal entities—people or corporations—that provides for an exchange of value (goods, services, or money) between or among them. If either party to a contract does not comply with the terms of the contract, the other party can sue for failure to comply, which is called **breach of contract**. Persons and corporations that engage in business are also expected to exercise due care and not violate laws that prohibit specific actions (such as trespassing, libel, or professional malpractice). A **tort** is an intentional or negligent action (other than breach of contract) taken by a legal entity that causes harm to another legal entity. People or corporations that wish to enforce their rights based on either contract or tort law must file their claims in courts with jurisdiction to hear their cases. A court has **sufficient jurisdiction** to hear a matter if it has both subject-matter jurisdiction and personal jurisdiction.

Subject-Matter Jurisdiction

Subject-matter jurisdiction is a court's authority to decide a particular type of dispute. For example, in the United States, federal courts have subject-matter jurisdiction over issues governed by federal law (such as bankruptcy, copyright, patent, and federal tax matters), and state courts have subject-matter jurisdiction over issues governed by state laws (such as professional licensing and state tax matters). If the parties to a contract are both located in the same state, a state court has subject-matter jurisdiction over disputes that arise from the terms of that contract. The rules for determining whether a court has subject-matter jurisdiction are clear and easy to apply. Few disputes arise over subject-matter jurisdiction.

Personal Jurisdiction

Personal jurisdiction is, in general, determined by the residence of the parties. A court has personal jurisdiction over a case if the defendant is a resident of the state in which the court is located. In such cases, the determination of personal jurisdiction is straightforward. However, an out-of-state person or corporation can also voluntarily submit to the jurisdiction of a particular state court by agreeing to do so in writing or by taking certain actions in the state.

One of the most common ways that people voluntarily submit to a jurisdiction is by signing a contract that includes a statement, known as a **forum selection clause**, that the contract will be enforced according to the laws of a particular state. That state then has personal jurisdiction over the parties who signed the contract regarding any enforcement issue that arises from the terms of that contract. Figure 7-3 shows a typical forum selection clause that might be used on a Web site.

> These terms of use shall be governed by and construed in accordance with the laws of the State of Washington, without regard to its conflict of laws rules. Any legal action arising out of this Agreement shall be litigated and enforced under the laws of the State of Washington. In addition, you agree to submit to the jurisdiction of the courts of the State of Washington, and that any legal action pursued by you shall be within the exclusive jurisdiction of the courts of King County in the State of Washington.

FIGURE 7-3   A typical forum selection clause

In the United States, individual states have laws that can create personal jurisdiction for their courts. The details of these laws, called **long-arm statutes**, vary from state to state, but generally create personal jurisdiction over nonresidents who transact business or commit tortious acts in the state. For example, suppose that an Arizona resident drives recklessly while in California and, as a result, causes a collision with another vehicle that is driven by a California resident. Due to the driver's tortious behavior in the state of California, the Arizona resident can expect to be called into a California court. In other words, California's long-arm statute gives its courts personal jurisdiction over the matter.

Businesses should be aware of jurisdictional considerations when conducting electronic commerce over state and international lines. In most states, the extent to which these laws apply to companies doing business over the Internet is unclear. Because these procedural laws were written before electronic commerce existed, their application to Internet transactions continues to evolve as more and more disputes arise from online commercial transactions. The trend in this evolving law is that the more business activities a company conducts in a state, the more likely it is that a court will assert personal jurisdiction over that company through the application of a long-arm statute.

One exception to the general rule for determining personal jurisdiction occurs in the case of tortious acts. A business can commit a tortious act by selling a product that causes harm to a buyer. The tortious act can be a **negligent tort**, in which the seller unintentionally provides a harmful product, or it can be an **intentional tort**, in which the seller knowingly or recklessly causes injury to the buyer. The most common business-related intentional torts involve defamation, misrepresentation, fraud, and theft of trade secrets. Although case law is rapidly developing in this area also, courts tend to invoke their respective states' long-arm statutes much more readily in the case of tortious acts than in breach of contract cases. If the matter involves an intentional tort or a criminal act, courts will assert jurisdiction more liberally.

Jurisdiction in International Commerce

Jurisdiction issues that arise in international business are even more complex than the rules governing personal jurisdiction across state lines within the United States. The exercise of jurisdiction across international borders is governed by treaties between the countries engaged in the dispute. Some of the treaties that the United States has signed with other countries provide specific determinations of jurisdiction for disputes that might arise. However, in most matters, U.S. courts determine personal jurisdiction for foreign companies and people in much the same way that these courts interpret the long-arm statutes in domestic matters. Non-U.S. corporations and individuals can be sued in U.S. courts if they conduct business or commit tortious acts in the United States. Similarly, foreign courts can enforce decisions against U.S. corporations or individuals through the U.S. court system if those courts can establish jurisdiction over the matter.

Courts asked to enforce the laws of other nations sometimes follow a principle called **judicial comity**, which means that they voluntarily enforce other countries' laws or judgments out of a sense of comity, or friendly civility. However, most courts are reluctant to serve as forums for international disputes. Also, courts are designed to deal with weighing evidence and making findings of right and wrong. International disputes often require diplomacy and the weighing of costs and benefits. Courts are not designed to do cost–benefit evaluations and cannot engage in negotiation and diplomacy. Thus, courts (especially U.S. courts) prefer to have the executive branch of the government (primarily the State Department) negotiate international agreements and resolve international disputes.

The difficulties of operating in multiple countries are faced by many large companies that do business online. For example, eBay, which had struggled to compete in China for many years, finally closed its operations in the country in 2006. eBay entered China in 2003 with a $30 million investment. In subsequent years, it poured another $250 million into acquisitions and advertising in China. But the effort to compete effectively against Alibaba.com's TaoBao consumer auction unit failed. Some observers believe that a Chinese cultural tendency to favor home-grown online services was primarily to blame for eBay's failure. But many others noted that the Chinese government made it difficult for eBay to operate in China by passing laws that favored companies that were majority-owned by Chinese entities and that blocked eBay's PayPal unit from operating in China. Some have even accused the Chinese government of intentionally blocking access to eBay's site for a few minutes each day so that Chinese competitors (some of which are owned, in part or completely, by the Chinese government) would appear to be more reliable. Because eBay was a foreign company, it was at a considerable disadvantage regarding government regulation and many have argued that this disadvantage was a larger factor in its failure than cultural issues.

The culture and government of China was also problematic for Google. In 2005, after going through the lengthy process of obtaining a government license to open a search engine site based in China (Google.cn; the company had operated Chinese language versions of Google.com for years), that license was revoked. The Chinese authorities questioned whether Google was operating a search engine (as permitted under the license) or a news service (under Chinese law, foreign owners are not permitted to operate online news services). Google worked hard to satisfy China's bureaucrats and was granted another operating license in 2007. After two years of operation under the new license, which included a number of conflicts between Google and the Chinese government over censorship, Google found that its computer systems in China had been hacked. Internal investigations concluded that the sophistication of the attack and its targets suggested that the Chinese government was involved in the attack. Specifically, the hackers had accessed the e-mail accounts of Chinese dissidents and human rights activists. In 2010, as a result of the attack and a general weariness with fighting with government censors, Google decided to close its operations in China.

Jurisdictional issues are complex and change rapidly. Any business that intends to conduct business online with customers or vendors in other countries should consult an attorney who is well versed in issues of international jurisdiction. However, there are a number of resources online that can be useful to non-lawyers who want to do preliminary investigation of a legal topic such as jurisdiction. The Harvard Law School's **Berkman Center for Internet & Society** Web site includes links to many current Internet-related legal issues and the *Berkeley Technology Law Journal* includes articles that analyze these topics. The **UCLA Online Institute for Cyberspace Law and Policy** contains an archive of legal reference materials published between 1995 and 2002, important years in the development of online law.

Conflict of Laws

In the United States, business is governed by federal laws, state laws, and local laws. Sometimes, these laws address the same issues in different ways. Lawyers call this situation a **conflict of laws**. Because online businesses usually serve broad markets that span many localities and many states, they generally look to federal laws for guidance. On occasion, this can lead to problems with state and local laws.

One online business that faced a serious conflict of laws problem was the online wine sales industry. Since the repeal of national Prohibition in 1933, all U.S. states and most local governments have enacted a myriad of laws that heavily regulate all types of alcoholic beverage sales. These laws govern when and where alcoholic beverages of various kinds can be sold, who can purchase them, and where they can be consumed.

The U.S. Constitution's Commerce Clause prohibits the states from passing laws that interfere with interstate commerce. However, the states do have the right to regulate matters pertaining to the health and welfare of their citizens. Under this right, most states have laws that require alcoholic beverages be sold through a regulated system of producers, wholesalers, and retailers. Some states allowed producers (such as wineries) to sell directly to the public, but only within that state. When online wine stores wanted to sell their products across state lines, they encountered these laws. Some states allowed the sales, others allowed the sales if the online store delivered to a licensed retailer in the destination state, and some states prohibited all sales by online stores not located within the state. This situation resulted in a classic conflict of laws.

State and local laws regulate the sale of alcoholic beverages in the interest of the health and welfare of the state's citizens, yet those same laws give in-state producers an advantage over out-of-state producers (in some states, in-state producers could sell directly without adding the markup of a retailer; in other states, out-of-state producers could not compete at all). When a state law gives an in-state business an advantage over an out-of-state business, the free flow of interstate commerce is impeded and courts often rule in such cases that the U.S. Constitution's Commerce Clause is violated.

For years, the online wine industry worked to find a way to resolve these issues with the states, but did not have much success. Finally, wineries filed suit on the Commerce Clause violation issue. In 2005, the U.S. Supreme Court voted 5–4 to strike down Michigan and New York laws that barred out-of-state wineries from selling directly to consumers. The online wine industry was happy with the outcome, as were wine lovers throughout the country who could buy wine directly from the more than 3400 wineries and online wine shops. Since then, the enthusiasm has dampened somewhat. The Supreme Court decision prohibits states from establishing laws that discriminate against out-of-state sellers; however, each state still can enforce laws limiting direct sales by all sellers and can specify that shipments originate within the state. After several years of trying to develop an online wine store, Amazon.com put its plans on indefinite hold in 2009, citing the difficulty of complying with the maze of state laws that govern online sales of alcoholic beverages. Since then, a number of smaller online wine stores have succeeded despite the legal barriers that have prevented large online retailers such as Amazon.com from entering the market. These smaller operations include vineyards that sell their own products as well as retailers that sell a variety of wines. You can learn more about the current state of legal challenges in this business at **Free the Grapes**, the Web site of a wine industry trade association that tracks developments in this area of online law.

Contracting and Contract Enforcement in Electronic Commerce

Any contract includes three essential elements: an offer, an acceptance, and consideration. The contract is formed when one party accepts the offer of another party. An **offer** is a commitment with certain terms made to another party, such as a declaration of willingness to buy or sell a product or service. An offer can be revoked as long as no payment, delivery of service, or other consideration has been accepted. An **acceptance** is the expression of willingness to take an offer, including all of its stated terms. **Consideration** is the agreed-upon exchange of something valuable, such as money, property, or future services. When a party accepts an offer based on the exchange of valuable goods or services, a contract has been created. An **implied contract** can also be formed by two or more parties that act as if a contract exists, even if no contract has been written and signed.

Creating Contracts: Offers and Acceptances

People enter into contracts on a daily, and often hourly, basis. Every kind of agreement or exchange between parties, no matter how simple, is a type of contract. Every time a consumer buys an item at the supermarket, the elements of a valid contract are met, for example, through the following sequence of actions:

1. The store invites offers for an item at a stated price by placing it on a store shelf.
2. The consumer makes an offer by indicating a willingness to buy the product for the stated price. For example, the consumer might take the item to a checkout station and present it to a clerk with an offer to pay.
3. The store accepts the customer's offer and exchanges its product for the consumer's payment at the checkout station. Both the store and the customer receive consideration at this point.

Contracts are a key element of traditional business practice, and they are equally important on the Internet. Offers and acceptances can occur when parties exchange e-mail messages, engage in electronic data interchange (EDI), or fill out forms on Web pages. These Internet communications can be combined with traditional methods of forming contracts, such as the exchange of paper documents, faxes, and verbal agreements made over the telephone or in person. The requirements for forming a valid contract in an electronic commerce transaction are met, for example, through the following sequence of actions:

1. The Web site invites offers for an item at a stated price by serving a Web page that includes information about the item.
2. The consumer makes an offer by indicating a willingness to buy the product for the stated price by, for example, clicking an "Add to Shopping Cart" button on the Web page that displays the item.
3. The Web site accepts the customer's offer and exchanges its product for the consumer's credit card payment on its shopping cart checkout page. The Web site obtains consideration at this point and the customer obtains consideration when the product is received (or downloaded).

As you can see, the basic elements of a consumer's contract to buy goods are the same whether the transaction is completed in person or online. Only the form of the offer and acceptance are different in the two environments. The substance of the offer, acceptance, and the completed contract are the same.

When a seller advertises goods for sale on a Web site, that seller is not making an offer, but is inviting offers from potential buyers. If a Web ad were considered to be a legal offer to form a contract, the seller could easily become liable for the delivery of more goods than it has available to ship. A summary of the contracting process that occurs in an online sale appears in Figure 7-4.

Step	Contract element	Participant	Action	
1.	Invites offers	Seller	Promotes product through Web page and states conditions under which offers will be accepted (for example, price and shipping terms)	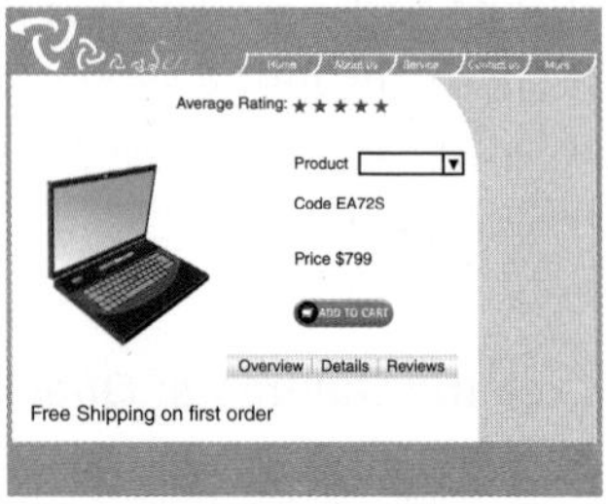
2.	Offer	Buyer	Clicks button to make offer to purchase product	
3.	Acceptance	Seller	Accepts buyer's offer, processes payment, and ships product	

© Cengage Learning

FIGURE 7-4 Contracting process in an online sale

When a buyer submits an order, which is an offer, the seller can accept that offer and create a contract. If the seller does not have the ordered items in stock, the seller has the option of refusing the buyer's order outright or counter offering with a decreased amount. The buyer then has the option to accept the seller's counteroffer.

Making a legal acceptance of an offer is quite easy to do in most cases. When enforcing contracts, courts tend to view offers and acceptances as actions that occur within a particular context. If the actions are reasonable under the circumstances, courts tend to interpret those actions as offers and acceptances. For example, courts have held that a number of different actions—including mailing a check, shipping goods, shaking hands, nodding one's head, taking an item off a shelf, or opening a wrapped package—are each, in some circumstances, legally binding acceptances of offers. An excellent resource for many of the laws concerning contracts, especially as they pertain to U.S. businesses, is the Cornell Law School Web site, which includes the full text of the **Uniform Commercial Code (UCC)**.

Click-Wrap and Web-Wrap Contract Acceptances

Most software sold today (either on CD or downloaded from the Internet) includes a contract that the user must accept before installing the software. These contracts, called **end-user license agreements (EULAs)**, often appear in a dialog box as part of the software installation process. When the user clicks the "Agree" button, the contract is deemed to be signed.

Years ago, when most software was sold in boxes that were encased in plastic shrink-wrap, EULAs were included on the box with a statement indicating that the buyer accepted the conditions of the EULA by removing the shrink-wrap from the box. This action was called a **shrink-wrap acceptance**. Today, a Web site user can agree to that site's EULA or its terms and conditions by clicking a button on the Web site (called a **click-wrap acceptance**) or by simply using the Web site (called a **Web-wrap acceptance** or **browser-wrap acceptance**).

Although many researchers and legal analysts have been critical of their use, U.S. courts have generally enforced the terms of EULAs to which users agreed using click-wrap or Web-wrap acceptances. Fewer cases have been adjudicated in the rest of the world. Although one case in Scotland (*Beta Computers v. Adobe Systems*) upheld a shrink-wrap acceptance, most European courts have been more likely to invalidate contract terms considered to be abusive or suspect under the Unfair Contract Terms European Union Directive and the consumer protection laws of many European countries, even if the user had reasonable notice.

Creating Written Contracts on the Web

In general, contracts are valid even if they are not in writing or signed. However, certain categories of contracts are not enforceable unless the terms are put into writing and signed by both parties. In 1677, the British Parliament enacted a law that specified the types of contracts that had to be in writing and signed. Following this British precedent, every state in the United States today has a similar law, called a **Statute of Frauds**. Although these state laws vary slightly, each Statute of Frauds specifies that contracts for the sale of goods worth more than $500 and contracts that require actions that cannot be completed within one year must be created by a signed writing. Fortunately for businesses and people who want to form contracts using electronic commerce, a writing does not require either pen or paper.

Most courts will hold that a **writing** exists when the terms of a contract have been reduced to some tangible form. An early court decision in the 1800s held that a telegraph transmission was a writing. Later courts have held that tape recordings of spoken words, computer files on disks, and faxes are writings. Thus, the parties to an electronic commerce contract should find it relatively easy to satisfy the writing requirement. Courts have been similarly generous in determining what constitutes a signature. A **signature** is any symbol executed or adopted for the purpose of authenticating a writing. Courts have held names on telegrams, telexes, faxes, and Western Union Mailgrams to be signatures. Even typed names or names printed as part of a letterhead have served as signatures. It is reasonable to assume that a symbol or code included in an electronic file would constitute a signature. As you will learn in Chapter 10, the United States now has a law that explicitly makes digital signatures legally valid for contract purposes.

Firms conducting international electronic commerce do not need to worry about the signed writing requirement in most cases. The main treaty that governs international sales of goods, Article 11 of the United Nations Convention on Contracts for the International Sale of Goods (CISG), requires neither a writing nor a signature to create a legally binding acceptance. You can learn more about the CISG and related topics in international commercial law at the **Pace University Law School CISG Database** Web site.

Implied Warranties and Warranty Disclaimers on the Web

Most firms conducting electronic commerce have little trouble fulfilling the requirements needed to create enforceable, legally binding contracts on the Web. One area that deserves attention, however, is the issue of warranties. Any contract for the sale of goods includes implied warranties. An **implied warranty** is a promise to which the seller can be held even though the seller did not make an explicit statement of that promise. The law establishes these basic elements of a transaction in any contract to sell goods or services. For example, a seller is deemed to implicitly warrant that the goods it offers for sale are fit for the purposes for which they are normally used. If the seller knows specific information about the buyer's requirements, acceptance of an offer from that buyer may result in an additional implied warranty of fitness, which suggests that the goods are suitable for the specific uses of that buyer. Sellers can also create explicit warranties by providing a detailed description of the additional warranty terms. It is also possible for a seller to create explicit warranties, often unintentionally, by making general statements in brochures or other advertising materials about product performance or suitability for particular tasks.

Sellers can avoid some implied warranty liability by making a warranty disclaimer. A **warranty disclaimer** is a statement declaring that the seller will not honor some or all implied warranties. Any warranty disclaimer must be conspicuously made in writing, which means it must be easily noticed in the body of the written agreement. On a Web page, sellers can meet this requirement by putting the warranty disclaimer in larger type, a bold font, or a contrasting color. To be legally effective, the warranty disclaimer must be stated obviously and must be easy for a buyer to find on the Web site. Figure 7-5 shows a portion of a sample warranty disclaimer for a Web site. The warranty disclaimer is printed in uppercase letters to distinguish it from other text on the page. This helps satisfy the requirement that the warranty disclaimer be easily noticed.

Disclaimers

WE DO NOT PROMISE THAT THIS WEB SITE OR ANY CONTENT, ELEMENT, OR FEATURE OF THIS SITE WILL BE ERROR-FREE OR UNINTERRUPTED, OR THAT ANY DEFECTS WILL BE CORRECTED, OR THAT YOUR USE OF THE SITE WILL PROVIDE SPECIFIC RESULTS. THE SITE AND ITS CONTENT ARE DELIVERED ON AN "AS-IS" BASIS. INFORMATION PROVIDED ON THE SITE IS SUBJECT TO CHANGE WITHOUT NOTICE. WE CANNOT ENSURE THAT ANY PROGRAMS, FILES OR OTHER DATA YOU DOWNLOAD FROM THE SITE WILL BE FREE OF VIRUSES OR DESTRUCTIVE FEATURES.

warranty disclaimer text is capitalized for emphasis

WE DISCLAIM ALL WARRANTIES, EXPRESS OR IMPLIED, INCLUDING ANY WARRANTIES OF ACCURACY, NON-INFRINGEMENT, MERCHANTABILITY AND FITNESS FOR A PARTICULAR PURPOSE. WE DISCLAIM ANY AND ALL LIABILITY FOR THE ACTS, OMISSIONS AND CONDUCT OF ANY THIRD PARTIES IN CONNECTION WITH OR RELATED TO YOUR USE OF THE SITE AND/OR ANY OF OUR SERVICES. YOU ASSUME TOTAL RESPONSIBILITY FOR YOUR USE OF THE SITE AND ANY LINKED SITES. YOUR SOLE REMEDY AGAINST US FOR DISSATISFACTION WITH THIS SITE OR ANY CONTENT CONTAINED ON THE SITE IS TO STOP USING THE SITE OR THE CONTENT. THIS LIMITATION OF RELIEF IS A PART OF THE BARGAIN BETWEEN THE PARTIES.

The above disclaimers apply to any damages, liability or injuries caused by any failure of performance, error, omission, interruption, defect of any kind, delay of operation or function, computer virus, communication failure, theft or destruction of or unauthorized access to, alteration of, or use, whether for breach of contract, tort, negligence or any other cause of action.

FIGURE 7-5   A Web site warranty disclaimer

Authority to Form Contracts

As explained previously in this section, a contract is formed when an offer is accepted for consideration. Problems can arise when the acceptance is issued by an imposter or someone who does not have the authority to bind the company to a contract. In electronic commerce, the online nature of acceptances can make it relatively easy for identity forgers to pose as others.

Fortunately, the Internet technology that makes forged identities so easy to create also provides the means to avoid being deceived by a forged identity. In Chapter 10, you will learn how companies and individuals can use digital signatures to establish identity in online transactions. If the contract is for any significant amount, the parties should require each other to use digital signatures to avoid identity problems. In general, courts will not hold a person or corporation whose identity has been forged to the terms of the contract; however, if negligence on the part of the person or corporation contributed to the forgery, a court may hold the negligent party to the terms of the contract. For example, if a company was careless about protecting passwords and allowed an imposter to enter the company's system and accept an offer, a court might hold that company responsible for fulfilling the terms of that contract.

Determining whether an individual has the authority to commit a company to an online contract is a greater problem than forged identities in electronic commerce. This issue, called **authority to bind**, can arise when an employee of a company accepts a contract and the company later asserts that the employee did not have authority to do so. For large transactions in the physical world, businesses check public information on file with the state of incorporation, or ask for copies of corporate certificates or resolutions, to establish the authority of persons to make contracts for their employers. These methods are available to parties engaged in online transactions; however, they can be time consuming and awkward. You will learn about some good electronic solutions, such as digital signatures and certificates from a certification authority, in Chapter 10.

Terms of Service Agreements

Many Web sites have stated rules that site visitors must follow, although most visitors are not aware of these rules. If you examine the home page of a Web site, you will often find a link to a page titled "Terms of Service," "Conditions of Use," "User Agreement," or something similar. If you follow that link, you find a page full of detailed rules and regulations, most of which are intended to limit the Web site owner's liability for what you might do with information you obtain from the site. These contracts are often called **terms of service (ToS)** agreements even when they appear under a different title. In most cases, a site visitor is held to the terms of service even if that visitor has not read the text or clicked a button to indicate agreement with the terms. The visitor is bound to the agreement by simply using the site, which is an example of the Web-wrap (or browser-wrap) acceptance you learned about earlier in this chapter. Figure 7-6 shows a typical Terms of Service agreement.

© Cengage Learning

FIGURE 7-6 Terms of Service agreement

USE AND PROTECTION OF INTELLECTUAL PROPERTY IN ONLINE BUSINESS

Online businesses must be careful with their use of intellectual property. **Intellectual property** is a general term that includes all products of the human mind. These products can be tangible or intangible. Intellectual property rights include the protections afforded to individuals and companies by governments through governments' granting of copyrights and patents, and through registration of trademarks and service marks. Depending on where they live, individuals may have a **right of publicity**, which is a limited right to control others' commercial use of an individual's name, image, likeness, or identifying aspect of identity. This right exists in most U.S. states but is limited by the provisions of the U.S. Constitution, specifically its First Amendment. Online businesses must take care to avoid deceptive trade practices, false advertising claims, defamation or product disparagement, and infringements of intellectual property rights by using unauthorized content on their Web sites or in their domain names. A number of legal issues can arise regarding the Web page content of electronic commerce sites. The most common concerns involve the use of intellectual property that is protected by other parties' copyrights, patents, trademarks, and service marks.

Copyright Issues

A **copyright** is a right granted by a government to the author or creator of a literary or artistic work. The right is for the specific length of time provided in the copyright law and gives the author or creator the sole and exclusive right to print, publish, or sell the work. Creations that can be copyrighted include virtually all forms of artistic or intellectual expression—books, music, artworks, recordings (audio and video), architectural drawings, choreographic works, product packaging, and computer software. In the United States, works created after 1977 are protected for the life of the author plus 70 years. Works copyrighted by corporations or not-for-profit organizations are protected for 95 years from the date of publication or 120 years from the date of creation, whichever is earlier.

The idea contained in an expression cannot be copyrighted. It is the particular form in which an idea is expressed that creates a work that can be copyrighted. If an idea cannot be separated from its expression in a work, that work cannot be copyrighted. For example, mathematical calculations cannot be copyrighted. A collection of facts can be copyrighted, but only if the collection is arranged, coordinated, or selected in a way that causes the resulting work to rise to the level of an original work. For example, the Yahoo! Web Directory is a collection of links to URLs. These facts existed before Yahoo! selected and arranged them into the form of its directory. However, most copyright lawyers would argue that the selection and arrangement of the links into categories probably makes the directory copyrightable.

In the past, many countries (including the United States) required the creator of a work to register that work to obtain copyright protection. U.S. law still allows registration, but registration is no longer required. A work that does not include the words "copyright" or "copyrighted," or the copyright symbol ©, but was created after 1989, is copyrighted automatically by virtue of the copyright law unless the creator specifically released the work into the public domain.

Most U.S. Web pages are protected by the automatic copyright provision of the law because they arrange the elements of words, graphics, and HTML tags in a way that creates an original work (in addition, many Web pages have been registered with the U.S. Copyright Office). This creates a potential problem because of the way the Web works. As you learned in Chapter 2, when a Web client requests a page, the Web server sends an HTML file to the client. Thus, a copy of the HTML file (along with any graphics or other files needed to render the page) resides on the Web client computer. Most legal experts agree that this copying is an allowable use of the copyrighted Web page.

U.S. copyright law includes an exemption from infringement actions for certain allowable uses of copyrighted works; the term for such uses is "fair use." The **fair use** of a copyrighted work includes copying it to use in specific restricted ways in criticism, comment, news reporting, teaching, scholarship, or research. The law's definition of fair use is intentionally broad and can be difficult to interpret. Figure 7-7 shows the text of the U.S. law that creates the fair-use exception.

Title 17, Chapter 1, § 107 of the United States Code

Limitations on exclusive rights: Fair use

Notwithstanding the provisions of sections 106 and 106A, the fair use of a copyrighted work, including such use by reproduction in copies or phonorecords or by any other means specified by that section, for purposes such as criticism, comment, news reporting, teaching (including multiple copies for classroom use), scholarship, or research, is not an infringement of copyright. In determining whether the use made of a work in any particular case is a fair use the factors to be considered shall include

(1) the purpose and character of the use, including whether such use is of a commercial nature or is for nonprofit educational purposes;
(2) the nature of the copyrighted work;
(3) the amount and substantiality of the portion used in relation to the copyrighted work as a whole; and
(4) the effect of the use upon the potential market for or value of the copyrighted work.

The fact that a work is unpublished shall not itself bar a finding of fair use if such finding is made upon consideration of all the above factors.

FIGURE 7-7   U.S. law governing the fair-use exception

As you can see in the figure, the law includes four specific factors that a court will consider in determining whether a specific use qualifies as a fair use. The first factor gives nonprofit educational uses a better chance at qualifying than commercial uses. The second factor allows the court to consider a painting using different standards than a sound recording. The third factor is often used to allow small sections of a work to qualify as fair use when the use of the entire work (or a substantial part of the work) might not qualify. The fourth factor, which is a deciding factor in most fair-use cases, allows the court to consider the amount of damage the use might cause to the value of the copyrighted work. The **University of Texas Copyright Crash Course** and the **Stanford Copyright & Fair Use** Web site are particularly helpful sources of information for making fair-use determinations. If you make fair-use of a copyrighted work for a school assignment, you should provide a citation to the original work to avoid charges of plagiarism.

Copyright law has always included elements, such as the fair-use exemption, that make it difficult to apply. The Internet has made this situation worse because it allows the immediate transmission of exact digital copies of many materials. In the case of digital music, the original Napster site provided a network that millions of people used to trade music files that they had copied from their CDs and compressed into MPEG version 3 format files, commonly referred to as MP3s. This constituted copyright infringement on a grand scale, and a group of music recording companies sued Napster for facilitating the individual acts of infringement.

Napster argued that it had only provided the "machinery" used in the copyright infringements—much as electronics companies manufacture and sell VCRs that might be used to make illegal copies of videotapes—and had not itself infringed on any copyrights. Both the U.S. District Court and the Federal Appellate Court held that Napster was liable for vicarious copyright infringement, even though it did not directly infringe any music recording companies' copyrights. An entity becomes liable for **vicarious copyright infringement** if it is capable of supervising the infringing activity and obtains a financial benefit from the infringing activity. Because Napster failed to monitor its network and indirectly profited (by selling advertising on its Web site) from the infringement, the company was held liable even though it did not itself transfer any copies. The courts shut down Napster and the company agreed to pay $26 million in copyright infringement damages before filing for bankruptcy. The **Napster** site that is owned and operated today by Best Buy offers legal music downloads to subscribers.

With the growth in popularity of portable music devices such as Apple's iPod, the demand for music in the MP3 (and similar) formats has continued to increase. The companies that sell music online today each have different rules and restrictions that come with the downloaded files. Some sites allow one copy to be installed on a portable music device. Others allow a limited number of copies to be installed. Still others allow unlimited copies, but only if the devices on which the copies are installed are owned by the person who downloaded the file.

The common practice of copying files from music CDs and placing those files on a portable music device, a smart phone, or a computer raises some interesting legal issues. This type of copying is governed in the United States by the fair-use provisions of the copyright laws, which you learned about earlier in this chapter. The fair-use provisions as they relate to copying music tracks are, at best, unclear and difficult to interpret. Some lawyers would argue that a person has the right under the fair-use provisions to make a backup copy of a music CD track, but other lawyers would disagree. A person who makes one copy for a portable music device, a second copy for a computer, and a third copy on a CD for backup purposes would be less likely to be protected under the fair-use provisions, but some lawyers would argue that all three uses should be protected.

Music that is purchased in digital form (as MP3 files, or through the Apple iTunes Store, for example) is often sold with specific restrictions on copying and sharing. Be sure to read and understand the terms under which you have purchased any digital music product before making copies, even for your own use.

Patent Issues

A **patent** is an exclusive right granted by the government to an individual to make, use, and sell an invention. In the United States, patents on inventions protect the inventor's rights for 20 years. An inventor may decide to patent the design of an invention instead of the invention itself, in which case the patent protects the design for 14 years. To be patentable, an invention must be genuine, novel, useful, and not obvious given the current state of technology. In the early 1980s, companies began obtaining patents on software programs that met the terms of the U.S. patent law. However, most firms that develop software to use in Web sites and for related transaction processing have not found the patent law to be very useful. The process of obtaining a patent is expensive and can take several years. Most developers of Web-related software believe that the technology in the software could become obsolete before the patent protection is secured, so they rely on copyright protection.

One type of patent has been of special interest to companies that do business online. A U.S. Court of Appeals ruled in 1998 that patents could be granted on "methods of doing business." The **business process patent**, which protects a specific set of procedures for conducting a particular business activity, is quite controversial. In addition to the Amazon.com patent on its 1-Click purchasing method (which you read about in Chapter 4), other Web businesses have obtained business process patents. The Priceline.com "name your own price" price-tendering system, About.com's approach to aggregating information from many different Web sites, and Cybergold's method of paying people to view its Web site have each received business process patents.

The ability of companies to enforce their rights under these patents is not yet clear. Many legal experts and business researchers believe that the issuance of business process patents grants the recipients unfair monopoly power and is an inappropriate extension of patent law. In 1999, Amazon.com sued Barnes & Noble for using a process on its Web site that was similar to the 1-Click method. The case was settled out of court in 2002, but the terms of the settlement were not disclosed.

The stakes in business process patent cases can be high. For example, a federal judge in 2007 entered a final judgment of $30 million against eBay in a business process patent case. MercExchange, a company that makes a business of buying patents and attempting to enforce them, had sued eBay for its use of a fixed price sales option that eBay calls "Buy It Now." MercExchange argued that several of its patents covered the business process of offering a fixed price option in an online auction. After winning the monetary damages, MercExchange continued to litigate the case, hoping to win an injunction that would prevent eBay from using the feature at all. In 2008, eBay agreed to buy three patents from MercExchange for an undisclosed sum to end the litigation.

Business process patents are common only in the United States. The intellectual property laws of most other countries do not permit patents to be issued for business processes. The appropriateness of business process patents is an issue that sparks intense debate among legal scholars and online business managers. To read an interesting discussion of both sides of the business process patent issue that includes exchanges between Jeff Bezos, founder of Amazon.com, and book publisher Tim O'Reilly, see the article posted at **My Conversation with Jeff Bezos**. The article concludes that business patents might be appropriate if their term is limited. There is some precedent for this position because current U.S. law includes a provision for a shorter time period in the case of design patents. Some would argue that a limited-term business process patent would be a logical extension of this policy.

Trademark Issues

A **trademark** is a distinctive mark, device, motto, or implement that a company affixes to the goods it produces for identification purposes. A **service mark** is similar to a trademark, but it is used to identify services provided. In the United States, trademarks and service marks can be registered with state governments, the federal government, or both. The name (or a part of that name) that a business uses to identify itself is called a **trade name**. Trade names are not protected by trademark laws unless the business name is the same as the product (or service) name. They are protected, however, under common law. **Common law** is the part of British and U.S. law established by the history of court decisions that has accumulated over many years. The other main part of British and U.S. law, called **statutory law**, arises when elected legislative bodies pass laws, which are also called statutes.

The owners of registered trademarks have often invested a considerable amount of money in the development and promotion of their trademarks. Web site designers must be very careful not to use any trademarked name, logo, or other identifying mark without the express permission of the trademark owner. For example, a company Web site that includes a photograph of its president who happens to be holding a can of Pepsi could be held liable for infringing on Pepsi's trademark rights. Pepsi can argue that the appearance of its trademarked product on the Web site implies an endorsement of the president or the company by Pepsi.

Domain Names and Intellectual Property Issues

Considerable controversy has arisen about intellectual property rights and Internet domain names. **Cybersquatting** is the practice of registering a domain name that is the trademark of another person or company in the hopes that the owner will pay huge amounts of money to acquire the URL. In addition, successful cybersquatters can attract many site visitors and, consequently, charge high advertising rates.

A related problem, called **name changing** (also called **typosquatting**), occurs when someone registers purposely misspelled variations of well-known domain names. These variants sometimes lure consumers who make typographical errors when entering a URL. For example, a person might easily type LLBaen.com instead of LLBean.com.

Since 1999, the U.S. Anticybersquatting Consumer Protection Act has prevented businesses' trademarked names from being registered as domain names by other parties. The law provides for damages of up to $100,000 per trademark. If the unauthorized registration of the domain name is found to be "willful," damages can be as much as $300,000.

Registering a generic name such as Wine.com is not cybersquatting. Registering a generic name is speculation that the name might one day become valuable and is completely legal. Disputes that arise when one person has registered a domain name that is an existing trademark or company name are settled by the World Intellectual Property Organization (WIPO). WIPO began settling domain name disputes in 1999 under its Uniform Domain Name Dispute Resolution Policy (UDRP). The problems of international jurisdiction made enforcement by the courts of individual countries cumbersome and ineffective. As an international organization, WIPO can transcend borders and provide rulings that will be effective in a global online business environment. Figure 7-8 shows the WIPO Domain Name Dispute Resolution information page.

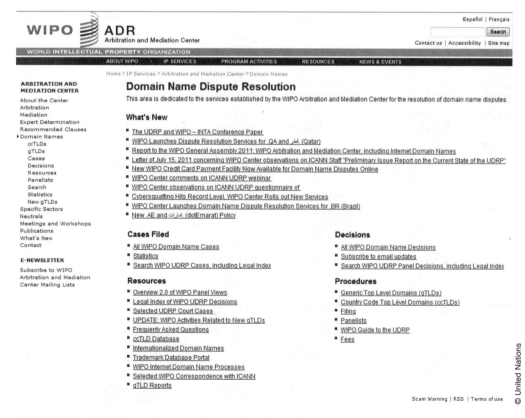

FIGURE 7-8 WIPO Domain Name Dispute Resolution information page

Disputes can arise when a business has a trademark that is a common term. If a person obtains the domain name containing that common term, the owner of the trademark must seek resolution at WIPO. In more than 90 percent of its cases, WIPO rules in favor of the trademark owner, but a win is never guaranteed.

In one example, three cybersquatters made headlines when they tried to sell the URL barrydiller.com for $10 million. Barry Diller, then the CEO of USA Networks, won a WIPO decision (*Barry Diller v. INTERNTCO Corp.*) that ordered the domain name transferred to him. The ruling established that a famous person's own name is a common law service mark. The WIPO panel in the Barry Diller case found that the cybersquatters had no legitimate rights or interests in the domain name and that they had registered the name and were using it in bad faith.

In another example, Gordon Sumner, who has performed music for many years as Sting, filed a complaint with WIPO because a Georgia man obtained the domain name www.sting.com and offered to sell it to Sting for $25,000; however, in this case, WIPO noted that the word "sting" was in common and general use and had multiple meanings other than as an identifier for the musician. WIPO refused to award the domain to Sumner. After the WIPO decision, Sumner purchased the domain name for an undisclosed sum. The musician's official Web site is now at www.sting.com.

Many critics have argued that the WIPO UDRP has been enforced unevenly and that many of the decisions under the policy have been inconsistent. One problem faced by those who have used the WIPO resolution service is that the WIPO decisions are not appealed to a single authority. Instead, the party losing in the WIPO hearing must find a court with jurisdiction over the dispute and file suit there to overturn the WIPO decision. No central authority maintains records of all WIPO decisions and appeals. This makes it very difficult for a trademark owner, a domain holder, or a lawyer for either party to anticipate how the UDRP will be interpreted in their specific cases.

You can learn more about WIPO UDRP decisions by reading the Harvard Law School's **Berkman Center UDRP Opinion Guide**. A list of UDRP decisions with links to the text of each decision appears on the **ICANN UDRP Proceedings** Web pages.

Another example of domain name abuse is name stealing. **Name stealing** occurs when someone other than a domain name's owner changes the ownership of the domain name. A **domain name ownership change** occurs when owner information maintained by a public domain registrar is changed in the registrar's database to reflect a new owner's name and business address. Once the domain name ownership is changed, the name stealer can manipulate the site, post graffiti on it, or redirect online customers to other sites—perhaps to sites selling competing products. The main purpose of name stealing is to harass the site owner because the ownership change can be reversed quickly when the theft is discovered; however, name stealing can cut off a business from its Web site for several days.

Protecting Intellectual Property Online

Several methods can be used to protect copyrighted digital works online, but they only provide partial protection. One technique uses a **digital watermark**, which is a digital code or stream embedded undetectably in a digital image or audio file. The digital watermark can be encrypted (you will learn more about encryption in Chapter 10) to protect its contents, or simply hidden among the digital information that makes up the image or recording. **Verance** is a company that provides, among other products, digital audio watermarking systems to protect audio files on the Internet. Its systems identify, authenticate, and protect intellectual property. They also enable companies to monitor, identify, and control the use of their digital audio or video recordings. The company also makes products that can alert users when telephonic conversations, audiovisual transcripts, or depositions have been altered.

Blue Spike produces a watermarking system that authenticates copyright and provides copy control. **Copy control** is an electronic mechanism for limiting the number of copies that one can make of a digital work. **Digimarc** is another company that provides watermark intellectual property protection software. Its products embed a watermark that allows any works protected by its system to be tracked across the Web. In addition, the watermark can link viewers to commerce sites and databases and can control software and playback devices. Digimarc's watermark also stores copyright information and links to the image's creator, which enables nonrepudiation of a work's authorship and facilitates selling and licensing the work online.

Defamation

A **defamatory** statement is a statement that is false and that injures the reputation of another person or company. If the statement injures the reputation of a product or service instead of a person, it is called **product disparagement**. In some countries, even a true and honest comparison of products may give rise to product disparagement. Because the difference between justifiable criticism and defamation can be hard to determine, commercial Web sites should consider the specific laws in their jurisdiction (and consider consulting a lawyer) before making negative, evaluative statements about other persons or products.

Web site designers should be especially careful to avoid potential defamation liability by altering a photo or image of a person in a way that depicts the person unfavorably. In most cases, a person must establish that the defamatory statement caused injury. However, most states recognize a legal cause of action, called **per se defamation**, in which a court deems some types of statements to be so negative that injury is assumed. For example, the court will hold inaccurate statements alleging conduct potentially injurious to a person's business, trade, profession, or office as defamatory per se—the complaining party need not prove injury to recover damages. Thus, online statements about competitors should always be carefully reviewed before posting to determine whether they contain any elements of defamation.

An important exception in U.S. law exists for statements that are defamatory but that are about a public figure (such as a politician or a famous actor). The law allows considerable leeway for statements that are satirical or that are valid expressions of personal opinion. Other countries do not offer the same protections, so operators of Web sites with international audiences do need to be careful.

Also, recall that defaming or disparaging statements must be false. This protects Web sites that include unfavorable reviews of products or services if the statements made are not false. For example, if a person reads a book and believes it to be terrible, that person can safely post a review on Amazon.com that includes assessments of the book's lack of literary value. Such statements of personal opinion are true statements and thus neither defamatory nor disparaging. Finally, in many U.S. states, use of an individual's name, photo, or other elements of personal identity can violate that individual's right of publicity. A company that does business in a jurisdiction that recognizes this right must be careful to obtain permission for any use of an individual's name, photo, likeness, or identifying characteristics on their Web sites.

Deceptive Trade Practices

The ease with which Web site designers can edit graphics, audio, and video files allows them to do many creative and interesting things. Manipulations of existing pictures, sounds, and video clips can be very entertaining. If the objects being manipulated are trademarked, however, these manipulations can constitute infringement of the trademark holder's rights. Fictional characters can be trademarked or otherwise protected. Many personal Web pages include unauthorized use of cartoon characters and scanned photographs of celebrities; often, these images are altered in some way. A Web site that uses an altered image of Mickey Mouse speaking in a modified voice is likely to hear from the Disney legal team.

Web sites that include links to other sites must be careful not to imply a relationship with the companies sponsoring the other sites unless such a relationship actually exists. For example, a Web design studio's Web page may include links to company Web sites that show good design principles. If those company Web sites were not created by the design studio, the studio must be very careful to state that fact. Otherwise, it would be easy for a visitor to assume that the linked sites were the work of the design studio.

In general, trademark protection prevents another firm from using the same or a similar name, logo, or other identifying characteristic in a way that would cause confusion in the minds of potential buyers of the trademark holder's products or services. For example, the trademarked name "Visa" is used by one company for its credit card and another company for its synthetic fiber. This use is acceptable because the two products are significantly different and few consumers of credit cards or synthetic fibers would likely be confused by the identical names. However, the use of very well-known trademarks can be protected for all products if there is a danger that the trademark might be diluted. Various state laws define **trademark dilution** as the reduction of the distinctive quality of a trademark by alternative uses. Trademarked names such as "Hyatt," "Trivial Pursuit," and "Tiffany," and the shape of the Coca-Cola bottle have all been protected from dilution by court rulings. Thus, a Web site that sells gift-packaged seafood and claims to be the "Tiffany of the Sea" risks a lawsuit from the famous jeweler asserting damages caused by trademark dilution.

Advertising Regulation

In the United States, advertising is regulated primarily by the **Federal Trade Commission (FTC)**. The FTC publishes regulations and investigates claims of false advertising. Its Web site includes a number of information releases that are useful to businesses and consumers. The Web page for the FTC's Bureau of Consumer Protection Business Center is shown in Figure 7-9. This page includes links to information about the FTC's advertising regulations.

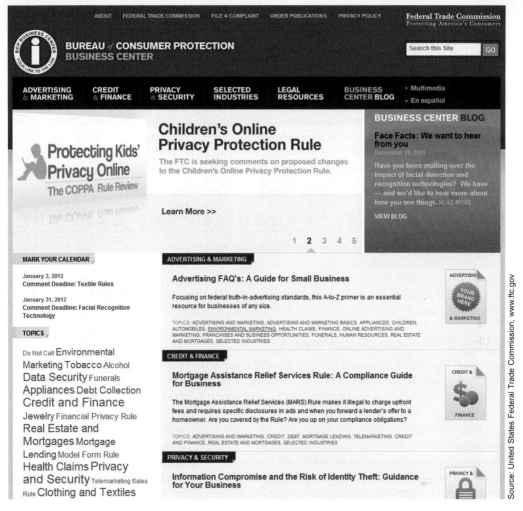

FIGURE 7-9   U.S. Federal Trade Commission Bureau of Consumer Protection Business Center page

Any advertising claim that can mislead a substantial number of consumers in a material way is illegal under U.S. law. In addition to conducting its own investigations, the FTC accepts referred investigations from organizations such as the Better Business Bureau. The FTC provides policy statements that can be helpful guides for designers creating electronic commerce Web sites. These policies include information on what is permitted in advertisements and cover specific areas such as these:

- Bait advertising
- Consumer lending and leasing
- Endorsements and testimonials
- Energy consumption statements for home appliances
- Guarantees and warranties
- Prices

Other federal agencies have the power to regulate online advertising in the United States. These agencies include the Food and Drug Administration (FDA), the Bureau of Alcohol, Tobacco, and Firearms (BATF), and the Department of Transportation (DOT).

The FDA regulates information disclosures for food and drug products. In particular, any Web site that is planning to advertise pharmaceutical products will be subject to the FDA's drug labeling and advertising regulations. The BATF works with the FDA to monitor and enforce federal laws regarding advertising for alcoholic beverages and tobacco products. These laws require that every ad for such products includes statements that use very specific language. Many states also have laws that regulate advertising for alcoholic beverages and tobacco products. The state and federal laws governing advertising and the sale of firearms are even more restrictive. Any Web site that plans to deal in these products should consult with an attorney who is familiar with the relevant laws before posting any online advertising for such products. The DOT works with the FTC to monitor the advertising of companies over which it has jurisdiction, such as bus lines, freight companies, and airlines.

ONLINE CRIME, TERRORISM, AND WARFARE

The Internet has opened up many possibilities for people to communicate and get to know each other better—no matter where in the world they live. The Internet has also opened doors for businesses to reach new markets and create opportunities for economic growth. It is sad that some people in our world have found the Internet to be a useful tool for perpetrating crimes, conducting terrorism, and even waging war.

Online Crime: Jurisdiction Issues

Crime on the Web includes online versions of crimes that have been undertaken for years in the physical world, including theft, stalking, distribution of pornography, and gambling. Other crimes, such as commandeering one computer to launch attacks on other computers, are new.

Law enforcement agencies have difficulty combating many types of online crime. The first obstacle they face is the issue of jurisdiction. As you learned earlier in this chapter, determining jurisdiction can be tricky on the Internet. Consider the case of a person living in Canada who uses the Internet to commit a crime against a person in Texas. It is unclear which elements of the crime could establish sufficient contact with Texas to allow police there to proceed against a citizen of a foreign country. It is possible that the actions that are considered criminal under Texas and U.S. law might not be considered so in Canada. If the crime is theft of intellectual property (such as computer software or computer files), the questions of jurisdiction become even more complex. You can learn more about online crime issues at the U.S. Department of Justice **Computer Crime & Intellectual Property Section** Web site.

The difficulty of prosecuting fraud perpetrators across international boundaries has always been an issue for law enforcement officials. The Internet has given new life to old fraud scams that count on jurisdictional issues to slow investigations of crimes. The advance fee fraud has existed in various forms for many years, and e-mail has made it inexpensive for perpetrators to launch large numbers of attempts to ensnare victims. In an **advance fee fraud**, the perpetrator offers to share the proceeds of some large payoff with the victim if the victim will make a "good faith" deposit or provide some partial funding first. The perpetrator then disappears with the deposit. In some online versions of this fraud, the perpetrator asks for identity information (bank account number, Social Security number, credit card number, and so on) and uses that information to steal the advance fee. Online advance fee frauds often victimize people who are less-sophisticated technology users and people who tend to trust unknown persons.

The most common online version of these schemes is the **Nigerian scam** (also called the **419 scam**, after the number of the section of the Nigerian penal code that specifies penalties for fraud in that country), in which the victim receives an e-mail from a Nigerian government official requesting assistance in moving money to a foreign bank account. The Financial Crimes Division of the U.S. Secret Service receives more than 100 reports each day about this type of fraud attempt.

Enforcing laws against the distribution of pornographic material has also been difficult because of jurisdiction issues. The distinction between legal adult material and illegal pornographic material is, in many cases, subjective and often difficult to make. The U.S. Supreme Court has ruled that state and local courts can draw the line based on local community standards. This creates problems for Internet sales. For example, consider a case in which questionable adult content is sold on a Web site located in Oregon to a customer who downloads the material in Georgia. A difficult question arises regarding which community standards might apply to the sale.

A similar jurisdiction issue arises in the case of online gambling. Many gambling sites are located outside the United States. If people in California use their computers to connect to an offshore gambling site, it is unclear where the gambling activity occurs. Several states have passed laws that specifically outlaw Internet gambling, but the jurisdiction of those states to enforce laws that limit Internet activities is not clear.

In 2008, the United States Department of the Treasury and the Federal Reserve Bank jointly issued regulations that implement the Unlawful Internet Gambling Enforcement Act (UIGEA) of 2006. As a federal law, the UIGEA gives clearer jurisdiction to law enforcement officers than any state law could. The law prohibits gambling businesses from knowingly accepting payments in connection with unlawful Internet gambling, including payments made through credit cards, electronic funds transfers, and checks. Under the UIGEA regulations, "unlawful Internet gambling" includes making bets using the Internet that are unlawful under any federal or state law in the jurisdiction where the bet or wager is initiated, received, or otherwise made.

The first major enforcement action under the regulations occurred in 2009, when federal authorities seized the bank accounts of some 27,000 online poker players, which contained more than $34 million. In 2011, the FBI arrested the founders of three major poker sites with large U.S. audiences on criminal gambling, bank fraud, and money laundering charges. The defendants were alleged to have circumvented the UIGEA by tricking some small U.S. banks into processing payments for them and bribing others to do the same. Prosecutors hope to recover more than $3 billion in gambling proceeds from the defendants.

Similar laws that restrict online gambling have been passed in other countries. However, some of these laws have been challenged as being discriminatory by the countries in which the online gambling companies operate. If a country's laws permit gambling within the country, but exclude foreign companies from providing gambling services (over the Internet), a basis exists for a discrimination complaint under the World Trade Organization's General Agreement on Trade and Services. The governments of Antigua and Barbados have each filed such complaints against the United States, arguing that the United States engaged in discriminatory trade practices by enforcing the UIGEA.

In 2011, the States of Illinois and New York proposed that they be permitted to use the Internet and out-of-state agents to sell lottery tickets to in-state adults. In response, the U.S. Department of Justice issued a memorandum opinion in which it reversed its long-held stand that virtually all forms of online gambling were illegal. The memorandum argued that state lotteries are not prohibited by federal law (specifically, the 1961 Wire Act, 18 U.S.C. 1084) because they do not involve wagering on sporting events. Because the underlying wagering is not illegal, the UIGEA (which requires the bets to be unlawful under federal or state law) does not apply. Gambling businesses and social networking sites were excited by the prospect of having locally sanctioned gambling on the Internet become legal and a number of state legislatures began drafting laws that would allow state governments and existing legal casinos to conduct non-sports gambling online.

New Types of Crime Online

As you learned in Chapter 6, the Internet made new types of business possible. The dark side of technological progress is that the Internet also made new types of crime possible. With these new types of crime, law enforcement officers often face difficulties when trying to apply laws that were written before the Internet became prevalent to criminal actions carried out on the Internet.

For example, most states have stalking laws that provide criminal penalties to people who harass, annoy, or alarm another person in a way that presents a credible threat. Many of these laws are triggered by physical actions, such as physically following the person targeted. The Internet gives a stalker the opportunity to use e-mail or chat room discussions to create the threatening situation. Laws that require physical action on the part of the stalker are not effective against online stalkers. Only a few states have passed laws that specifically address the problem of online stalking.

The Internet can amplify the effects of acts that, in the physical world, can be dealt with locally. For example, school playgrounds have long been the realm of bullying. Students who engaged in bullying were dealt with by school officials; only in extreme cases were such cases referred to law enforcement officials. Today, young people can use technology to harass, humiliate, threaten, and embarrass each other. These acts are called **cyberbullying**. Cyberbullying can include threats, sexual remarks, or pejorative comments transmitted on the Internet or posted on Web sites (social networking sites are often used for such postings). The perpetrator might also pose as the victim and post statements or media, such as photos or videos (often edited to cast the victim in an unfavorable light), that are intended to damage the victim's reputation. Because the Internet increases both the intensity and reach of these attacks, they are much more likely to draw the attention of law enforcement officials than bullying activities in the physical world.

Unfortunately, laws have not kept up with technology and many forms of stalking and cyberbullying are difficult to prosecute under criminal statutes. The victims of online harassment can, in many cases, file civil suits against the perpetrators for defamation, negligent misrepresentation, invasion of privacy, and inflicting emotional distress. Lawsuits against social media sites that host damaging content have been unsuccessful because such sites are generally not responsible for the content posted by individual members.

The practice of sending sexually explicit messages or photos using a mobile phone is called **sexting**. Sexting is a crime in many jurisdictions, even if the message is sent to a friend or acquaintance. A number of politicians, athletes, and other celebrities have been embarrassed by sexting activity. When young persons under the age of 18 transmit an explicit photo of themselves, they can create serious criminal liability for themselves and their recipients. In the United States and many other countries, the mere possession (regardless of intent) of explicit photos of a minor is a felony punishable by prison sentences and requires offenders to register as a sex offender.

An increasing number of companies have reported attempts by competitors and others to infiltrate their computer systems with the intent of stealing data or creating disruptions in their operations. Smaller companies are easier targets because they generally do not have strong security in place (you will learn more about security in electronic commerce in Chapter 10), but larger organizations are not immune to these attacks. In 2004, lawyer and computer expert Myron Tereshchuk was convicted for criminal extortion. Over a period of two years, he threatened MicroPatent, a patent and trademark services company, with disclosure of confidential client information unless the company paid him $17 million. MicroPatent spent more than $500,000 on legal and technical consultants during the investigation and devoted significant internal resources to the effort. MicroPatent's sales managers also had to spend a tremendous amount of time with clients, reassuring them that their confidential information (details of their pending patent and trademark applications, for example) had not been compromised. MicroPatent's experience was not unusual. According to a recent Computer Security Institute survey of 634 companies, the average loss due to unauthorized data access was more than $300,000 and the average loss due to information theft was more than $350,000. Another survey by *InformationWeek*/Accenture found that 78 percent of surveyed companies believed that they were more vulnerable because attackers were getting more sophisticated.

In 2010, the National Retail Federation joined with eBay and the FBI to combat retail crime organizations that specialize in stealing in bulk from physical stores and then selling the stolen goods online. In recent years, shoplifters who try to return stolen goods for refunds have been thwarted by store policies that require a receipt or ask for identification (to track persons who have many returns). The Internet has opened up a new way for these criminals to profit by selling the stolen goods online. By working with retailers, eBay can use its data tracking technology to identify auctions that offer stolen items and alert law enforcement officials who can investigate suspicious activity.

Although the Internet has made the work of law enforcement more difficult in many cases, there are exceptions. As police agencies become more experienced in using the Web, they have found that it can help track down the perpetrators of crime in some cases. A number of cases have been solved because criminals have bragged about elements of their crimes on social networking sites. From the Pennsylvania graffiti artists who posted photos of their work on their social network profiles to the California teens who firebombed an airplane hangar and uploaded a video of themselves in action, criminals who use the Internet are making it easy for police to track them down. In other cases, criminals leave clues in their online profiles that police can use to corroborate other evidence, as in the case of the suspected murderer who described his favorite murder weapon in his online profile. Although privacy watchdog groups have expressed concern about law enforcement officers randomly surfing the Web looking for leads, anything posted online is public information and is subject to their scrutiny.

Online Warfare and Terrorism

Many Internet security experts believe that we are at the dawn of a new age of terrorism and warfare that could be carried out or coordinated through the Internet. A considerable number of Web sites currently exist that openly support or are operated by hate groups and terrorist organizations. Web sites that contain detailed instructions for creating biological weapons and other poisons, discussion boards that help terrorist groups recruit new members online, and sites that offer downloadable terrorist training films now number in the thousands.

The U.S. Department of Homeland Security and international police agencies such as Interpol are devoting considerable resources to monitoring terrorist activities online. Historically, these agencies have not done a very good job of coordinating their activities around the world. The threat posed by global terrorist organizations that use the Internet to recruit members and to plan and organize terrorist attacks has motivated Interpol to update and expand its computer network monitoring skills and coordinate global antiterrorism efforts.

The Internet provides an effective communications network on which many people and businesses have become dependent. Although the Internet was designed from its inception to continue operating while under attack, a sustained effort by a well-financed terrorist group or rogue state could slow down the operation of major transaction-processing centers. As more business communications traffic moves to the Internet, the potential damage that could result from this type of attack increases. You will learn more about security threats and countermeasures for those threats in Chapter 10.

ETHICAL ISSUES

Companies using Web sites to conduct electronic commerce should adhere to the same ethical standards that other businesses follow. If they do not, they will suffer the same consequences that all companies suffer: the damaged reputation and long-term loss of trust that can result in loss of business. In general, advertising or promotion on the Web should include only true statements and should omit any information that could mislead potential customers or wrongly influence their impressions of a product or service. Even true statements have been held to be misleading when the ad omits important related facts. Any comparisons to other products should be supported by verifiable information. The next section explains the role of ethics in formulating Web business policies, such as those affecting visitors' privacy rights and companies' Internet communications with children.

Ethics and Online Business Practices

Online businesses are finding that ethical issues are important to consider when they are making policy decisions. Recall from Chapter 3 that buyers on the Web often communicate with each other. A report of an ethical lapse that is rapidly passed among customers can seriously affect a company's reputation. In 1999, *The New York Times* ran a story that disclosed Amazon.com's arrangements with publishers for book promotions. Amazon.com was accepting payments of up to $10,000 from publishers to give their books editorial reviews and placement on lists of recommended books as part of a cooperative advertising program. When this news broke, Amazon.com issued a statement that it had done nothing wrong and that such advertising programs were a standard part of publisher–bookstore relationships. The outcry on Internet newsgroups and mailing lists was overwhelming. Two days later—before most traditional media outlets had even reported the story—Amazon.com announced that it would end the practice and offer unconditional refunds to any customers who had purchased a promoted book. Amazon.com had done nothing illegal, but the practice appeared to be unethical to many of its existing and potential customers.

In early 1999, eBay faced a similar ethical dilemma. Several newspapers had begun running stories about sales of illegal items, such as assault weapons and drugs, on the eBay auction site. At this point in time, eBay was listing about 250,000 items each day. Although eBay would investigate claims that illegal items were up for auction on its site, eBay did not actively screen or filter listings before the auctions were placed on the site.

Even though eBay was not legally obligated to screen the items auctioned, and even though screening would be fairly expensive, eBay decided that screening for illegal and copyright-infringing items would be in the best long-run interest of eBay. The team decided that such a decision would send a signal about the character of the company to its customers and the public in general. eBay also decided to remove an entire category—firearms—from the site. Not all of eBay's users were happy about this decision—the sale of firearms on eBay, when done properly, is completely legal. However, eBay again decided that its overall image as an open and honest marketplace was so important to its future success that the company chose to ban all firearms sales.

In 2009, a number of software developers complained that the Apple Apps Store (which you learned about earlier in this book) was slow to approve software to be sold on its Web site. Apple responded that it had a responsibility to protect its customers (the owners of its iPhone product) from unscrupulous software vendors who might try to sell applications for the iPhone that do not function properly, crash the phone, or install malware. Apple argued that its testing and approval program was necessary to maintain customer confidence in its products, even though it had no legal obligation to perform such testing on software provided by third-party developers and sold on the Apps Store Web site.

An important ethical issue that organizations face when they collect e-mail addresses from site visitors is how the organization limits the use of the e-mail addresses and related information. In the early days of the Web, few organizations made any promises to visitors who provided such information. Today, most Web sites state the organization's policy on the protection of visitor information, but many do not. In the United States, organizations are not legally bound to limit their use of information collected through their Web sites. They may use the information for any purpose, including the sale of that information to other organizations. This lack of government regulation that might protect site visitor information is a source of concern for many individuals and privacy rights advocates. These concerns are discussed in the next section.

Privacy Rights and Obligations

The issue of online privacy is continuing to evolve as the Internet and the Web grow in importance as tools of communication and commerce. Many legal and privacy issues remain unsettled and are hotly debated in various forums. The **Electronic Communications Privacy Act of 1986** is the main law governing privacy on the Internet today. Of course, this law was enacted before the general public began its wide use of the Internet. The law was written to update an existing law that prevented the interception of audio signal transmissions so that any type of electronic transmissions (including, for example, fax or data transmissions) would be given the same protections. In 1986, people were not using the Internet to transmit commercially valuable data in any significant amount, so the law was written to deal primarily with interceptions that might occur on leased telephone lines.

In the United States, a number of laws have been enacted that address online privacy issues, but none have survived constitutional challenges. In 1999, the FTC issued a report that examined how well Web sites were respecting visitors' privacy rights. Although the FTC found a significant number of sites without posted privacy policies, the report concluded that companies operating Web sites were developing privacy practices with sufficient speed and that no federal laws regarding privacy were required at that time. Privacy advocacy groups responded to the FTC report with outrage and calls for legislation. The Direct Marketing Association (DMA), a trade association of businesses that advertise their products and services directly to consumers using mail, telephone, Internet, and mass media outlets, has established a set of privacy standards for its members. Critics note that past efforts by the DMA to regulate its members' activities have been less than successful and continue to push for privacy laws. The DMA lobbies legislators on behalf of its members, who generally do not want any privacy laws that would interfere with their business activities.

Ethics issues are significant in the area of online privacy because laws have not kept pace with the growth of the Internet and the Web. The nature and degree of personal information that Web sites can record when collecting information about visitors' page-viewing habits, product selections, and demographic information can threaten the privacy rights of those visitors. This is especially true when companies lose control of the data they collect on their customers (and other people). In recent years, many companies have made news headlines because they allowed confidential information about individuals to be released without the permission of those individuals. Examples include incidents such as:

- ChoicePoint (a company that compiles information about consumers) sold the names, addresses, Social Security numbers, and credit reports of more than 145,000 people to thieves who posed as legitimate businesses. More than 1000 fraud cases have been documented as a result of that privacy violation.
- Hackers broke into customer databases at DSW Shoe Warehouse and stole the credit card numbers, checking account numbers, and driver's license numbers of more than 1.4 million customers.
- A computer at Boston College was penetrated and the addresses and Social Security numbers of 120,000 alumni were exposed.

Not all privacy compromises are the work of external agents. Sometimes, companies just lose things. Examples include incidents such as:

- In 2005, Ameritrade, Bank of America, and Time Warner each reported that they had lost track of shipments containing computer backup tapes that held confidential information for hundreds of thousands of customers or employees.
- In 2008, Horizon Blue Cross Blue Shield of New Jersey reported that an employee's laptop computer had been stolen. The laptop contained the personal information (including Social Security numbers) of more than 300,000 individuals.

The number of security breaches leading to the loss of personal information continues to increase. In 2008, the Identity Theft Resource Center reported 446 incidents that exposed private information contained in more than 127 million records and projected that the upward trend in incidents will continue.

The Internet has also changed traditional assumptions about privacy because it allows people anywhere in the world to gather data online in quantities that would have been impossible a few years ago. For example, real estate transactions are a matter of public record in the United States. These transactions have been registered in county records for many years and have been available to anyone who wanted to go to the county recorder's office and spend hours leafing through large books full of handwritten records. Many counties have made these records available on the Internet, so now a researcher can examine thousands of real estate transaction records in hours without traveling to a single county office. Many privacy experts see this change in the ease of data access to be an important shift that affects the privacy rights of those who participate in real estate transactions. Because the Internet makes such data more readily available to a wider range of people, the privacy previously afforded to the participants in those transactions has been reduced.

Differences in cultures throughout the world have resulted in different expectations about privacy in electronic commerce. In Europe, for example, most people expect that information they provide to a commercial Web site will be used only for the purpose for which it was collected. Many European countries have laws that prohibit companies from exchanging consumer data without the express consent of the consumer. In 1998, the European Union adopted a Directive on the Protection of Personal Data. This directive codifies the constitutional rights to privacy that exist in most European countries and applies them to all Internet activities. In addition, the directive prevents businesses from exporting personal data outside the European Union unless the data will continue to be protected in accordance with provisions of the directive. The European Union and its member countries have consistently exhibited a strong preference for using government regulations to protect privacy. The United States has exhibited an opposite preference. U.S. companies, especially those in the direct mail marketing industry, have consistently and successfully lobbied to avoid government regulation and allow the companies to police themselves. Companies that do business internationally must be aware of these differences. For example, a U.S. company that does business in the European Union is subject to its privacy laws.

One of the major privacy controversies in the United States today is the opt-in vs. opt-out issue. Most companies that gather personal information in the course of doing business on the Web would like to be able to use that information for any purpose of their own. Some companies would also like to be able to sell or rent that information to other companies. No U.S. law currently places limits on companies' use of such information. Companies are, in general, also free to sell or rent customer information. An increasing number of U.S. companies do provide a way for customers who would like to restrict use of their personal information to do so. The most common policy used in U.S. companies today is an opt-out approach. In an **opt-out** approach, the company collecting the information assumes that the customer does not object to the company's use of the information unless the customer specifically chooses to deny permission (that is, to opt out of having their information used). In the less common **opt-in** approach, the company collecting the information does not use the information for any other purpose (or sell or rent the information) unless the customer specifically chooses to allow that use (that is, to opt in and grant permission for the use). Figure 7-10 shows an example Web page that presents a series of opt-in choices to site visitors. The Web site will not send any of these three items to a site visitor unless that visitor opts in by checking one or more boxes.

Many of our site visitors and customers enjoy receiving our newsletter, periodic notices of sales and special product offerings, and offers from other companies that we have chosen to ensure that they will be of interest to our site visitors. Please check the boxes below to add your e-mail address to our distribution list for any or all of these electronic mailings.

☐ Weekly e-mail newsletter

☐ Periodic notices of sales and special product offerings

☐ Offers from other companies

FIGURE 7-10 Example Web page showing opt-in choices

Figure 7-11 shows the opt-out approach. A Web site that uses the opt-out approach will send all three items to the site visitor unless the site visitor checks the boxes to indicate that the items are not wanted.

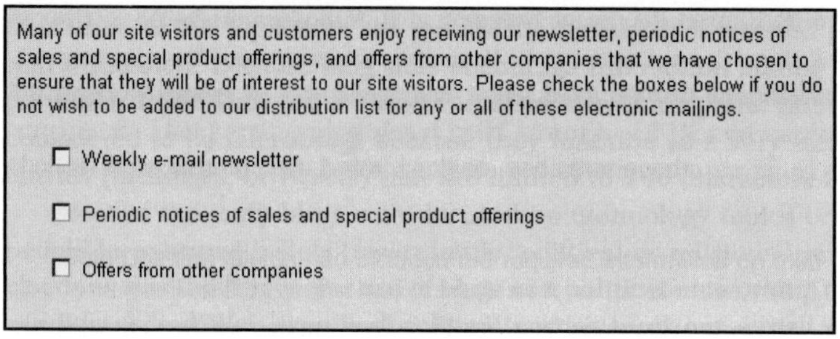

Many of our site visitors and customers enjoy receiving our newsletter, periodic notices of sales and special product offerings, and offers from other companies that we have chosen to ensure that they will be of interest to our site visitors. Please check the boxes below if you do not wish to be added to our distribution list for any or all of these electronic mailings.

☐ Weekly e-mail newsletter

☐ Periodic notices of sales and special product offerings

☐ Offers from other companies

FIGURE 7-11 Example Web page showing opt-out choices

As you can see, it is easy for site visitors to misread the text and make the wrong choice when deciding whether or not to check the boxes. Sites that use the opt-out approach are often criticized for requiring their visitors to take an affirmative action (checking the empty boxes) to prevent the site from sending items. Another approach to presenting opt-out choices is to use a page that includes checked boxes and instructs the visitor to "uncheck the boxes of the items you do not wish to receive." Most privacy advocates believe that the opt-in approach is preferable because it gives the customer privacy protection unless that customer specifically elects to give up those rights. Most U.S. businesses have traditionally taken the position that they have a right to use the information they collect unless the provider of the information explicitly objects. Some of these companies are changing to the opt-in approach, often at the prodding of privacy advocacy groups.

Until the legal requirements of privacy regulation become clearer, privacy advocates urge electronic commerce Web sites to be conservative in their collection and use of customer data. Mark Van Name and Bill Catchings, writing in *PC Week* in 1998, outlined four principles for handling customer data that provide a good outline for Web site administrators even today. These principles are as follows:

1. Use the data collected to provide improved customer service.
2. Do not share customer data with others outside your company without the customer's permission.
3. Tell customers what data you are collecting and what you are doing with it.
4. Give customers the right to have you delete any of the data you have collected about them.

Today, this list should also include a recommendation that customer data, once collected, be kept as secure as possible. A number of organizations are active in promoting privacy rights. You can learn more about current developments in privacy legislation and practices throughout the world by following the links to these organizations' Web sites that appear under the heading **Privacy Rights Advocacy Groups** in the Web Links.

DoubleClick

As you learned in Chapter 4, **DoubleClick** is one of the largest banner advertising networks in the world. DoubleClick arranges the placement of banner ads on Web sites. Like many other Web sites, DoubleClick uses **cookies**, which are small text files placed on Web client computers, to identify returning visitors.

Most visitors find the privacy risk posed by cookies to be acceptable. The Web servers at Amazon.com, for example, place Amazon.com cookies on the computers of visitors to the site so the visitors can be recognized when they return. This can be useful, for example, when a visitor who has placed several items in a shopping cart before being interrupted can return to Amazon.com later in the day and find the shopping cart intact. The Amazon.com Web server can read the client's Amazon.com cookie and find the shopping cart from the client's previous session. The Amazon.com server can read only its own cookies; it cannot read the cookies placed on the client computer by any other Web server.

There are two important differences between the Amazon.com scenario and what happens when DoubleClick serves a banner ad. First, the visitor usually does not know that the banner ad is coming from DoubleClick (and thus, does not know that the DoubleClick server could be writing a cookie to the client computer). Second, DoubleClick serves ads through Web sites owned by thousands of companies. As a visitor moves from one Web site to another, that visitor's computer can collect many DoubleClick cookies. The DoubleClick server can read all of its own cookies, gathering information from each one about which ads were served and the sites through which they were served. Thus, DoubleClick can compile a tremendous amount of information about a user's actions on the Web.

Even this amount of information collection would not trouble most people. DoubleClick can use the cookies to track a particular computer's connections to Web sites, but it does not record any identity information about the owner of that computer. Therefore, DoubleClick accumulates a considerable record of Web activity, but cannot connect that activity with a person.

In 1999, DoubleClick arranged a $1.7 billion merger with Abacus Direct Corporation. Abacus had developed a way to link information about people's Web behavior (collected through cookies such as those placed by DoubleClick's banner ad servers) to the names, addresses, and other information about those people that had been collected in an offline consumer database.

The reaction from online privacy protection groups was immediate and substantial. The FTC launched an investigation, the Internet's privacy issues e-mail lists and chat rooms buzzed with furious conversation and, in the end, DoubleClick abandoned its plans to integrate its cookie-generated data with the identity information in the Abacus database. Although DoubleClick is still one of the largest banner advertising networks, it had been counting on generating additional revenue by using the information in the combined database that it was unable to create.

When the FTC probe concluded two years later, DoubleClick was not charged with any violations of laws or regulations. The lesson here is that a company violates the Internet community's ethical standards at its own peril, even if the transgression does not break any laws.

Communications with Children

An additional set of privacy considerations arises when Web sites attract children and engage in some form of communication with those children. Adults who interact with Web sites can read privacy statements and make informed decisions about whether to communicate personal information to the site. The communication of private information (such as credit card numbers, shipping addresses, and so on) is a key element in the conduct of electronic commerce.

The laws of most countries and most sets of ethics consider children to be less capable than adults in evaluating information sharing and transaction risks. Thus, we have laws in the physical world that prevent or limit children's ability to sign contracts, get married, drive motor vehicles, and enter certain physical spaces (such as bars, casinos, tattoo parlors, and racetracks). Children are considered to be less able (or unable) to make informed decisions about the risks of certain activities. Similarly, many people are concerned about children's ability to read and evaluate privacy statements and then consent to providing personal information to Web sites. In 2006, MySpace hired a former federal prosecutor to serve as the site's security officer. MySpace was responding to concerns that participants in the social networking site, many of whom are under 18 and post personal information and photos, might be easy prey for sexual predators. MySpace regularly uses software that compares each registered participant against a database of known sex offenders and deletes the accounts of any it finds. However, most experts agree that no technology will ever protect as well as parental involvement in their children's online activities.

Under the laws of most countries, people under the age of 18 or 21 are not considered adults. However, those countries that have proposed or passed laws that specify differential treatment for the privacy rights of children often define "child" as a person below the age of 12 or 13. This approach complicates the issue because it creates two classes of nonadults.

In the United States, Congress enacted the Children's Online Protection Act (COPA) in 1998 to protect children from "material harmful to minors." This law was held to be unconstitutional because it unnecessarily restricted access to a substantial amount of material that is lawful, thus violating the First Amendment. Congress was more successful with the **Children's Online Privacy Protection Act of 1998 (COPPA)**, which provides restrictions on data collection that must be followed by electronic commerce sites aimed at children. This law does not regulate content, as COPA attempted to do, so it has not been successfully challenged on First Amendment grounds. In 2001, Congress enacted the Children's Internet Protection Act (CIPA). CIPA requires schools that receive federal funds to install filtering software on computers in their classrooms and libraries. Filtering software is used to block access to adult content Web sites. In 2003, the Supreme Court held that CIPA was constitutional.

Companies with Web sites that appeal to young people must be careful to comply with the laws governing their interactions with these young visitors. **Disney Online** is a site that appeals primarily to young children. The Disney Online registration page offers three choices to visitors who want to register with the site and receive regular communications and updates. The first registration choice is for adults, a second choice is for "teens," and a third choice is for "kids." The "kids" choice leads to a screen that asks for a parent's e-mail address so that Disney can invite the parent to set up a family account. The Disney.com registration page for "teens" asks for the visitor's name, birthday, and the e-mail address of a parent. Disney uses the birthday to calculate the visitor's age and, if the age is less than 13, Disney uses the parent's e-mail address to notify parents of their child's registration and to invite them to set up a family account. Family accounts are controlled by parents who can elect to allow family members who are under the age of 13 to use the site. By refusing to enroll any child under age 13 as a site subscriber, Disney Online meets the requirements of the COPPA law. Other sites that appeal to a young audience use similar techniques to limit unsupervised access to their Web pages. For example, Sanrio (the company that produces Hello Kitty and related products) asks for a birth date before allowing access to its English-language site that is directed at U.S. customers, **Sanriotown**. As shown in Figure 7-12, the site encourages visitors to notify the company that operates the site if they know a child who has gained access to the site in violation of COPPA.

Sanriotown.com does not collect personal information from persons under the age of 13. In order to ensure adherence to this policy, the opening page of our website asks for the date, month and year of birth of each visitor and denies further access to visitors whose birth date shows that they are under 13 years of age. If you believe that a child under 13 has gained access to the sanriotown.com site, or if you have any questions concerning sanriotown.com's privacy policy and practices, please contact us at:

Sanrio Digital (HK) Ltd
Unit 1109, Level 11, Cyberport 2
100 Cyberport Road
Hong Kong
Email: info@sanriotown.com

FIGURE 7-12   Sanrio's approach to COPPA compliance

TAXATION AND ELECTRONIC COMMERCE

Companies that do business on the Web are subject to the same taxes as any other company. However, even the smallest Web business can become instantly subject to taxes in many states and countries because of the Internet's worldwide scope. Traditional businesses may operate in one location and be subject to only one set of tax laws for years. By the time those businesses are operating in multiple states or countries, they have developed the internal staff and record-keeping infrastructure needed to comply with multiple tax laws. Firms that engage in electronic commerce must comply with these multiple tax laws from their first day of existence.

An online business can become subject to several types of taxes, including income taxes, transaction taxes, and property taxes. **Income taxes** are levied by national, state, and local governments on the net income generated by business activities. **Transaction taxes**, which include sales taxes, use taxes, and excise taxes, are levied on the products or services that the company sells or uses. Transaction taxes are also called **transfer taxes** because they arise when the ownership of a property or service is transferred to from one person or entity to another. **Property taxes** are levied by states and local governments on the personal property and real estate used in the business. In general, the taxes that cause the greatest concern for Web businesses are income taxes and sales taxes.

Nexus

A government acquires the power to tax a business when that business establishes a connection with the area controlled by the government. For example, a business that is located in Kansas has a connection with the state of Kansas and is subject to Kansas taxes. If that company opens a branch office in Arizona, it forms a connection with Arizona and becomes subject to Arizona taxes on the portion of its business that occurs in Arizona. This connection between a tax-paying entity and a government is called **nexus**. The concept of nexus is similar in many ways to the concept of personal jurisdiction discussed earlier in this chapter. The activities that create nexus in the United States are determined by state law and thus vary from state to state. Nexus issues have been frequently litigated, and the resulting common law is fairly complex. Determining nexus can be difficult when a company conducts only a few activities in or has minimal contact with the state. In such cases, it is advisable for the company to obtain the services of a professional tax advisor.

Companies that do business in more than one country face national nexus issues. If a company undertakes sufficient activities in a particular country, it establishes nexus with that country and becomes liable for filing tax returns in that country. The laws and regulations that determine national nexus are different in each country. Companies that sell through their Web sites do not, in general, establish nexus everywhere their goods are delivered to customers. Usually, a company can accept orders and ship from one state to many other states and avoid nexus by using a contract carrier such as FedEx or UPS to deliver goods to customers. Again, companies will find the services of a professional tax lawyer or accountant who has experience in international taxation to be valuable.

U.S. Income Taxes

The **Internal Revenue Service (IRS)** is the U.S. government agency charged with administering the country's tax laws. A basic principle of the U.S. tax system is that any verifiable increase in a company's wealth is subject to federal taxation. Thus, any company whose U.S.-based Web site generates income is subject to U.S. federal income tax. Furthermore, a Web site maintained by a company in the United States must pay federal income tax on income generated outside the United States. To reduce the incidence of double taxation of foreign earnings, U.S. tax law provides a credit for taxes paid to foreign countries. The IRS Web site's home page appears in Figure 7-13.

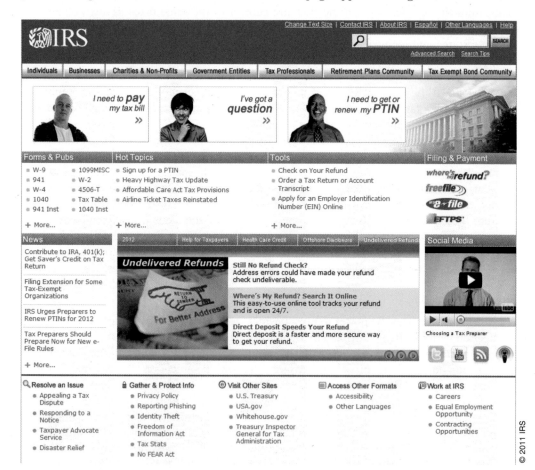

FIGURE 7-13   Internal Revenue Service home page

The IRS site includes links to downloadable tax forms, copies of IRS publications, current tax news, and other useful tax information. The home page offers links to sections of the Web site that are designed to help specific categories of site visitors.

Most states levy an income tax on business earnings. If a company conducts activities in several states, it must file tax returns in all of those states and apportion its earnings in accordance with each state's tax laws. In some states, the individual cities, counties, and other political subdivisions within the state also have the power to levy income taxes on business earnings. Companies that do business in multiple local jurisdictions must apportion their income and file tax returns in each locality that levies an income tax. The number of taxing authorities (which includes states, counties, cities, towns, school districts, water districts, and many other governmental units) in the United States exceeds 30,000.

U.S. State Sales Taxes

Most U.S. states levy a transaction tax on goods sold to consumers. This tax is usually called a sales tax. Businesses that establish nexus with a state must file sales tax returns and remit the sales tax they collect from their customers. If a business ships goods to customers in other states, it is not required to collect sales tax from those customers unless the business has established nexus with the customer's state. However, the customer in this situation is liable for payment of a use tax in the amount that the business would have collected as sales tax if it had been a local business.

A **use tax** is a tax levied by a state on property used in that state that was not purchased in that state. Most states' use tax rates are identical to their sales tax rates. In addition to property purchased in another state, use taxes are assessed on property that is not "purchased" at all. For example, lease payments on vehicles are subject to use taxes in most states. The leased vehicle is not purchased (in any state) but when it is used in the lessee's state, it incurs that state's use tax. In the past, few consumers filed use tax returns and few states enforced their use tax laws with regularity. However, an increasing number of states are providing a line on their individual income tax returns that asks people to report and pay their use tax for the year along with their state income taxes. Some states allow taxpayers to estimate their use tax liability; others require an exact statement of the use tax amount.

Larger businesses use complex software to manage their sales tax obligations. Not only are the sales tax rates different in the approximately 7500 U.S. sales tax jurisdictions (which include states, counties, cities, and other sales tax authorities), but the rules about which items are taxable also differ. For example, New York's sales tax law provides that large marshmallows are taxable (because they are "snacks"), but small marshmallows are not taxable (because they are "food").

Some purchasers are exempt from sales tax, such as certain charitable organizations and businesses buying items for resale. Thus, to determine whether a particular item is subject to sales tax, a seller must know where the customer is located, what the laws of that jurisdiction say about taxability and tax rate, and the taxable status of the customer.

The sales tax collection process in the United States is largely regarded as a serious problem. Even the Supreme Court, in one of its sales tax decisions more than 15 years ago, stated that the situation is needlessly confusing and encouraged Congress to act. Although a number of bills have been introduced over the years, none has become law.

A few states have enacted laws that require online retailers to collect and remit sales taxes on sales they make in their states, even though the online retailers do not have nexus with the state. Many more states have proposed or are considering such laws. These statutes are often called **Amazon laws** because they are directed at large online retailers, such as Amazon.com. The idea behind these laws is that online retailers have an unfair pricing advantage over local stores because they are not required to collect sales tax (although the purchasers are required to file and pay a use tax, such taxes are widely avoided and it is costly for states to pursue the violators). The laws are designed to remove the unfair advantage and collect sales tax revenue, which many states need to balance their budgets.

Many of the states have joined together through the National Governors Association and the National Conference of State Legislatures to create the **Streamlined Sales and Use Tax Agreement (SSUTA)**. The SSUTA would simplify state sales taxes by making the various state tax codes more congruent with each other while allowing each state to set its own rates. Each state must adopt the agreement, and once a state does adopt it, companies in the state can choose one of several simple procedures for collecting and remitting sales taxes nationwide. Most states have not yet adopted this agreement.

Import Tariffs

All countries in the world regulate the import and export of goods across their borders. In many cases, goods can only be imported into a country if a tariff is paid. A **tariff**, also called a **customs duty** or **duty**, is a tax levied on products as they enter the country. Countries have many reasons for imposing tariffs, and a complete discussion of tariffs and the role they play in international economics and foreign trade policy is beyond the scope of this book. Goods that are ordered online are subject to tariffs when they cross international borders. Even products that are delivered online (such as downloaded software) can be subject to tariffs. Many online shoppers have been surprised when an item they ordered from another country arrives with a bill from their government for the tariff.

European Union Value Added Taxes

The United States raises most of its revenue through income taxes. Other countries, especially those in the European Union (EU), use transaction taxes to generate most of their revenues. The Value Added Tax (VAT) is the most common transaction tax used in these countries. A VAT is assessed on the amount of value added at each stage of production. For example, if a computer keyboard manufacturer purchased keyboard components for $20 and then sold finished keyboards for $50, the value added would be $30. VAT is collected by the seller at each stage of the transaction. A product that goes through five different companies on its way to the ultimate consumer would have VAT assessed on each of the five sales. In most countries, VAT is calculated at the time of each intermediate sale and remitted to the country in which that sale occurs.

The EU enacted legislation concerning the application of VAT to sales of digital goods that became effective in 2003. Companies based in EU countries must collect VAT on digital goods no matter where in the EU the products are sold. This legislation has attracted the attention of companies based outside the EU that sell digital goods to consumers based in one or more EU countries. Under the law, non-EU companies that sell into the EU must now register with EU tax authorities and levy, collect, and remit VAT if their sales include digital goods delivered into the EU.

Summary

The legal concept of jurisdiction on the Internet is still unclear and ill defined. The relationship between geographic boundaries and legal boundaries is based on four elements: power, effects, legitimacy, and notice. These four elements have helped governments create the legal concept of jurisdiction in the physical world. Because the four elements exist in somewhat different forms on the Internet, the jurisdiction rules that work so well in the physical world do not always work well in the online world.

As in traditional commerce, contracts are a part of doing business on the Web and are established through various types of offers and acceptances. Any contract for the electronic sale of goods or services includes implied warranties. Many companies include contracts or rules on their Web sites in the form of terms of service agreements. Contracts can be invalidated when one of the parties to the transaction is an imposter; however, forged identities are becoming easier to detect through electronic security tools.

Seemingly innocent inclusion of photographs, whether manipulated or not, and other elements on a Web page can lead to infringement of trademarks, copyrights, or patents; defamation; and violation of publicity or privacy rights. An international administrative mechanism now exists for resolving domain name disputes that has reduced the need for lengthy and expensive litigation in many cases. Electronic commerce sites must be careful not to imply relationships that do not actually exist. Negative evaluative statements about entities, even when true, are best avoided given the subjective nature of defamation and product disparagement.

Unfortunately, some people use the Internet for perpetrating crimes, advocating terrorism, and even waging war. Law enforcement agencies have found it difficult to combat many types of online crime, and governments are working to create adequate defenses for online war and terrorism.

Web business practices such as collecting information and tracking consumer habits have led to questions of ethics regarding online privacy. Some countries are far more restrictive than others in terms of what type of information collection is acceptable and legal. Companies that collect personal information can use an opt-in policy, in which the customer must take an action to permit information collection, or an opt-out policy, in which the customer must take an action to prevent information collection. Opt-in policies are more protective of customers' privacy rights. Web businesses also must be careful when communicating with children. The laws of most countries require that parental consent be obtained before information is collected from children under the age of 13.

Companies that conduct electronic commerce are subject to the same laws and taxes as other companies, but the nature of doing business on the Web can expose companies to a large number of laws and taxes sooner than traditional companies usually face them. The international nature of all online business further complicates a firm's tax obligations. Although some legal issues are straightforward, others are difficult to interpret and follow because of the newness of electronic commerce and the unsettled nature of applicable law. The large number of government agencies that have jurisdiction and the power to tax makes it essential that companies doing business on the Web understand the potential liabilities of doing business with customers in those jurisdictions.

Key Terms

Acceptance

Advance fee fraud

Amazon law

Authority to bind

Breach of contract

Browser-wrap acceptance

Business process patent

Click-wrap acceptance

Common law

Conflict of laws

Consideration

Constructive notice

Contract

Cookies

Copy control

Copyright

Customs duty (duty)

Cyberbullying

Cybersquatting

Defamatory

Digital watermark

Domain name ownership change

Effects

End-user license agreement (EULA)

Fair use

Forum selection clause

Implied contract

Implied warranty

Income tax

Intellectual property

Intentional tort

Judicial comity

Jurisdiction

Legitimacy

Long-arm statute

Name changing

Name stealing

Negligent tort

Nexus

Nigerian scam (419 scam)

Notice

Offer

Opt-in

Opt-out

Patent

Per se defamation

Personal jurisdiction

Power

Product disparagement

Property tax

Right of publicity

Service mark

Sexting

Shrink-wrap acceptance

Signature

Statute of Frauds

Statutory law

Streamlined Sales and Use Tax Agreement (SSUTA)

Subject-matter jurisdiction

Sufficient jurisdiction

Tariff

Terms of service (ToS)

Tort

Trade name

Trademark

Trademark dilution

Transaction tax

Transfer tax

Typosquatting

Use tax

Vicarious copyright infringement

Warranty disclaimer

Web-wrap acceptance

Writing

Review Questions

1. Write two or three paragraphs in which you describe the role that culture plays in the development of a country's laws and ethical standards.

2. In the past, geographic borders have helped governments assert jurisdiction effectively. Write two paragraphs in which you describe one way the Internet has changed the role borders play in the determination of jurisdiction.

3. In one or two paragraphs, describe what a long-arm statute accomplishes and describe a situation in which an online retail business might become subject to such a statute.

4. Assume you have downloaded an app for your smart phone. In one or two paragraphs, describe how you and the app's seller have each obtained consideration in the completion of the transaction.

5. In two or three paragraphs, describe circumstances under which a company that is doing business online might want to use a warranty disclaimer.

6. In one or two paragraphs, describe the conditions under which a collection of facts can be copyrighted.

7. In two or three paragraphs, explain the differences among trademarks, service marks, and trade names. In your explanation, include at least one example of each.

8. In two or three paragraphs, explain how the WIPO's Domain Name Dispute Resolution process overcomes some of the jurisdictional issues that might arise if a United States court were to hear these types of disputes instead.

9. Explain what a digital watermark is in one or two paragraphs, and then write an additional paragraph in which you provide an example of an online business (other than the audio and video recording industries mentioned in the chapter) that might use digital watermarks.

10. Many advance fee frauds are targeted at older, often retired, persons. Write a paragraph or two in which you provide an example of an advance fee fraud, and then write another paragraph in which you explain why older persons might be more likely targets for this type of criminal activity.

Exercises

1. Use your favorite Web search engine to obtain a list of Web pages that include the words "privacy statement." Visit the Web pages on the search results list until you find a page that includes the text of a privacy statement. Print the page and turn it in with a report of about 200 words in which you answer the following questions:

    a. Does the site follow an opt-in or opt-out policy (or is the policy not stated clearly in the privacy statement)?

    b. Does the privacy statement include a specific provision or provisions regarding the collection of information from children?

    c. Does the privacy statement describe what happens to the collected personal information if the company goes out of business or is sold to another company (list these provisions, if any)?

    Close your report with one paragraph in which you evaluate the overall clarity of the privacy statement.

2. Companies that do business online can find themselves in legal trouble if they commit a crime, breach a contract, or engage in a tortious action. In about 200 words, provide an

online business example of each offense. As part of your answer, explain why you believe each action you describe is either a crime, a breach of contract, or a tort.

3. Your friend Alex has developed an energy drink that she would like to sell online. Prepare a report of about 200 words in which you outline for Alex the rules and guidelines she should follow in presenting claims about this product on her Web site. Use the link to the Federal Trade Commission: Advertising and Marketing on the Internet and your favorite search engine to gather information for your report.

4. Assume you are working for a company that sells jewelry online. The Marketing Department would like to send e-mails to customers who have purchased jewelry in the past. These e-mails would use information about the types of jewelry customers have purchased in the past and would offer them discounts on new designs of those types of jewelry. Write a memo of about 200 words in which you attempt to convince the marketing manager that the company should use an opt-in statement to request permission to send such e-mails. This opt-in request would appear when customers make their first purchase.

5. The merits of issuing business process patents have been vigorously debated by legal scholars and business researchers. One proposed solution to this debate would allow the issuance of business patents, but restrict the patent protection period to a short time, perhaps two or three years. In about 200 words, present logical and factual arguments that support the issuance of limited-term business process patents. Conclude your arguments with a policy recommendation.

6. Use your favorite search engine to find a Web site (other than Disney or Sanriotown) that is directed at young people. Examine the site to determine how it complies with COPPA. Test the site to ensure that it requires parental consent before it accepts information from children under the age of 13. Evaluate the site's compliance with COPPA in a report of about 200 words.

7. A number of U.S. states have proposed legislation called "Amazon laws." Use your favorite search engine to learn more about these laws. In about 200 words, critically evaluate the argument that such laws are necessary to protect local retailers from unfair price competition by large online retailers.

Cases

C1. Nissan.com

The Nissan Motor Company of Japan had sold its cars in the United States under the brand name Datsun for many years. In the late 1980s, the company changed its branding policy and began selling cars in the U.S. market with the name of Nissan. However, the company did not realize that the Web would become an important marketing tool and did not register the name nissan.com as soon as it became available.

Nissan was not the only auto company to miss an opportunity to register its brand's domain name early. General Motors had registered the domain gm.com in 1992, but it had not registered generalmotors.com. The company had to purchase that name from Gil Vanorder, who had registered it in 1997. Vanorder's site featured a cigar-smoking, uniform-wearing cartoon character named "General John C. Motors." Volkswagen (which had registered vw.com when it first became available) successfully sued Virtual Works (an ISP) to obtain the domain name vw.net. Other auto companies have purchased or sued (with mixed results) to obtain domain names that included their product brand names. DaimlerChrysler was able to purchase dodge.com in 2001 from the London financial software company that had registered it originally. Ford had to sue National A-1 Advertising to obtain the right to use lincoln.com. However, Ford was unsuccessful in its attempts to obtain mercury.com. That name was owned by the New York City information technology services company, Mercury Technologies, which is now owned by Hewlett-Packard.

In 1991, Uzi Nissan formed a company named Nissan Computer Corp. in North Carolina to sell computer hardware and provide related repair and consulting services. Nissan's company also offered networking hardware for sale, along with related services. In 1994, the company registered the name nissan.com. In 1996, the company registered the domain name nissan.net and began offering ISP services to individuals and companies at that Web site.

In 1995, he received a letter from a lawyer representing Nissan Motor Co. The letter requested information about how Nissan was planning to use the domain name nissan.com. Because he was operating a computer company and Nissan Motor Co. was an auto company, Nissan decided there would be no potential confusion in customers' minds about the relationship (or lack thereof) between Nissan Computer and Nissan Motor Co. Nissan did not respond to the letter. The lawyer did not follow up with any other contact, so Nissan considered the issue closed.

In 2000, Nissan Motor Co. sued Nissan Computer under the U.S. Anticybersquatting Consumer Protection Act for $10 million and the exclusive right to use the names nissan.com and nissan.net. Uzi Nissan argued in court that he was just using his family name (which is a common name in the Middle East) to which he had a basic right, that he had no intent to profit from the name (he was unwilling to sell it to Nissan Motor Co. at any price), and that there was little likelihood that his computer store would be confused in the minds of the consumers with the international auto company of the same name. Nissan Motor Co. argued that its brand name was so well known that any alternative use of the name would be confusing to consumers.

In 2002, opinions issued by the California Superior Court and the U.S. Ninth Circuit District Court held that Nissan Computer had not acted in bad faith when it acquired the disputed domain names. However, the court ruled that Nissan Computer could no longer use the domain names for commercial purposes because of the potential confusion it could create in the minds of consumers. Nissan Computer would have to find a different domain name for its business. The court also ordered that Nissan could not place any advertising on his Web sites at nissan.com or nissan.net and prohibited him from placing disparaging remarks or negative commentary about Nissan Motor Co. (or links to such remarks or commentary) on the two sites. The court did not, however, order the transfer of the two domain names to Nissan Motor Co. The Web Links for this chapter includes links to the Web sites operated today by Nissan Computer and Nissan Motor Co. In 2005, the U.S. Supreme Court refused to hear Nissan Motor Co.'s appeal of the lower court rulings, which allows them to stand.

Required:

1. U.S. courts sometimes appoint advisors (often called Special Masters) to help them decide cases that involve complex business or technical issues. Assume you are a business advisor to a court that is hearing an appeal of the *Nissan Motor Co. v. Nissan Computer Corp*. case. In about 200 words, explain why Nissan Motor Co. is so concerned about the use of these two domain names and how a monetary damages judgment of $10 million could be justified (if you do not believe that the monetary damages are justified, explain why).

2. In about 200 words, provide an outline of the ethics of the position taken by Uzi Nissan in this dispute.

3. In about 200 words, provide an outline of the ethics of the position taken by Nissan Motor Co. in this dispute.

4. If you believe that the courts' decisions in this case are fair to the parties and the general public, explain why in about 200 words. If you believe that the courts' decisions are not fair, outline a decision (in about 200 words) that you believe would be fair.

Note: Your instructor might assign you to a group to complete this case, and might ask you to prepare a formal presentation of your results to your class.

C2. Ellasaurus Products Enterprises

Ellen Carson is the author and illustrator of a successful series of children's books that chronicle the adventures of Ellasaurus, a 4-year-old orange dinosaur. Ellen has done well with the books, but her business advisors have told her that she could earn considerably more money by creating a merchandising business around the Ellasaurus character. Following this advice, she has created Ellasaurus Products Enterprises (EPE), a company that has begun developing and marketing Ellasaurus toys, stuffed animals, coloring books, pajamas, and Halloween costumes from its location in Flint, Michigan.

EPE has had some success in its attempts to get major retailers to stock the Ellasaurus product line, but Ellen is concerned that retailers might not be willing to take on a new and unproven product. She would like to create a Web site through which EPE could sell its merchandise directly to customers. She also sees the Web site as a way to build customer loyalty. Ellen envisions a site with a number of portal features in addition to the product sales. For example, she would like to offer online games, chat rooms, e-mail accounts, and other activities that would promote EPE products and her books.

The Ellasaurus book series appeals to children who are between 4 and 6 years old. Ellen expects the EPE product line to appeal to children in about the same age range. Ellen has visited sites such as **Hello Kitty** and **Nick Jr.**, which appeal to similar age groups, to get ideas for the site. She would like the site to be appealing to her main audience, but she would like to obtain registration information from site visitors so EPE can send e-mails with information about new products and Web site features to them.

Ellen plans to limit the Web site's merchandise sales to U.S. residents at first, but she hopes to begin selling internationally within a few years. The site will allow visitors from any country to register and participate in the online portal features.

Required:

1. Ellen will use some copyrighted illustrations from her books on the Web site. She will also include themes from the story lines of her books in some of the games that will be available (free) on the site to registered visitors. Prepare a report of about 300 words in which you discuss at least two intellectual property issues that might arise in the operation of the Web site.

2. In about 200 words, describe the ethical issues that Ellen faces because of the ages of her intended audience members.

3. In about 300 words, outline the laws with which the site must comply when it registers site visitors under the age of 13. Include recommendations regarding how Ellen can best comply with those laws.

4. In about 300 words, describe the sales tax liabilities to which the Web site will be exposed. Assume that Ellen will operate the site from the EPE headquarters location in Michigan and that EPE will manufacture the merchandise in Texas. The merchandise will be warehoused at EPE distribution centers in New Jersey, Ohio, and California.

Note: Your instructor might assign you to a group to complete this case, and might ask you to prepare a formal presentation of your results to your class.

For Further Study and Research

Alino, N. and G. Schneider, 2011. "European Union Value-added Taxes on International Sales of Digital Products," *Proceedings of the Academy of Legal, Ethical, and Regulatory Issues*, April, 1–6.

Angwin, J. and D. Bank. 2005. "Time Warner Alerts Staff to Lost Data: Files for 600,000 Workers Vanish During Truck Ride," *The Wall Street Journal*, May 3, A3.

Arrison, S. 2011. "California Shouldn't Follow NY's Internet Tax Plan," *TechNewsWorld*, January 26. (http://www.technewsworld.com/story/71725.html)

Bagby, J. and F. McCarty. 2003. *The Legal and Regulatory Environment of E-Business*. Cincinnati: Thomson South-Western.

Beta Computers (Europe), Ltd. v. Adobe Systems (Europe), Ltd. 1996 SLT 604; 1996 SCLR 587.

Barry Diller v. INTERNETCO Corp. 2001. WIPO Case No. D2000-1734, March 9. (http://www.wipo.int/amc/en/domains/decisions/html/2000/d2000-1734.html)

Barkacs, L., T. Dalton, G. Schneider, and C. Barkacs. 2004. "U.S. Sales Taxes on Internet Transactions: Historic Change is at the Door," *Journal of Accounting and Finance Research*, 12(5), 135–144.

Barnes, B. 2007. "Web Playgrounds of the Very Young," *The New York Times*, December 31. (http://www.nytimes.com/2007/12/31/business/31virtual.html)

Better Business Bureau. 2006. *Security & Privacy Made Simpler*. Arlington, VA: The Council of Better Business Bureaus.

Brilmayer, L. 1989. "Consent, Contract, and Territory," *Minnesota Law Review*, 74(1), 11–12.

Carver, B. 2010. "Why License Agreements Do Not Control Copy Ownership: First Sales and Essential Copies," *Berkeley Technology Law Journal*, 25, 1886–1954.

Cass, S. 2002. "Nissan v. Nissan," *IEEE Spectrum*, 39(10), October, 53–54.

Cathcart, R. 2008. "MySpace Is Said to Draw Subpoena in Hoax Case," *The New York Times*, January 10. (http://www.nytimes.com/2008/01/10/us/10myspace.html)

Claburn, T., M. Garvey, and V. Koen. 2005. "The Threats Get Nastier," *InformationWeek*, August 29, 34–41.

Clark, P. 2001. "Doubts Cloud DoubleClick's Repositioning," *B to B*, 86(15), August 28, 1–2.

Coll, S. and S. Glasser. 2005. "Terrorists Turn to the Web as Base of Operations," *The Washington Post*, August 7, A1.

Costello, A. 2010. "Facebook Lawsuit Dismissed," *Long Island Herald*, August 11. (http://www.liherald.com/stories/Facebook-lawsuit-dismissed,26966)

Crane, E. 2000. "Double Trouble," *Ziff Davis Smart Business*, 13(10), October, 62.

Damton, R. 2011. "A Digital Library Better Than Google's," *The New York Times*, March 23. (http://www.nytimes.com/2011/03/24/opinion/24darnton.html)

Digital Millennium Copyright Act. 1998. Public Law No. 105-304, 112 Statutes 2860.

Direct Marketing. 2001. "FTC Closes DoubleClick Investigation," 63(12), April, 18.

Federal Trade Commission (FTC). 1999. *Self-Regulation and Privacy Online: A Report to Congress*. Washington: FTC.

Fidler, S. 2007. "Terrorism Fight 'in Wrong Century,'" *Financial Times*, July 10, 4.

Foege, A. 2005. "Extortion.com," *Fortune Small Business*, September 1.

Foster, A. 2002. "Computer-Crime Incidents at 2 California Colleges Tied to Investigation Into Russian Mafia," *Chronicle of Higher Education*, June 24.

Granholm v. Heald 544 US 460(2005).

Greene, S. 2001. "Reconciling Napster with the Sony Decision and Recent Amendments to Copyright Law," *American Business Law Journal*, 39(1), Fall, 57–98.

Gregory, D., S. Roll, and W. Carlile. 2010. "Tough Economy, Waning Prospects for Federal Legislation May Increase Interest in Alternatives to Streamlined System," *BNA Daily Tax Report*, 225, November 24, J-1.

Hale, K. and R. McNeal. 2011. "Technology, Politics, and E-commerce: Internet Sales Tax and Interstate Cooperation," *Government Information Quarterly*, 28(2), 262–270.

Hamblen, M. 2003. "Regulatory Requirements Place New Burdens on IT: U.S. Firms Scramble to Comply with EU Tax," *Computerworld*, June 30, 1.

Hardesty, D. 2004. *Sales Tax and Electronic Commerce*. Larkspur, CA: ClickBank.

Hemphill, T. 2000. "DoubleClick and Consumer Online Privacy: An E-Commerce Lesson Learned," *Business & Society Review*, 105(3), Fall, 361–372.

Hulme, G. 2005. "Extortion Online," *InformationWeek*, September 13, 24–25.

Hwang, W. and J. Klosek. 2003. "Taxing the Sale of Digital Goods in Europe," *E-Commerce Law & Strategy*, 20(3), July 11, 1.

Ian, J. 2002. "The Internet Debacle: An Alternative View," *Performing Songwriter Magazine*, May. (http://www.janisian.com/)

Identity Theft Resource Center (ITRC). 2009. *2009 Breach List*. San Diego: ITRC.

Jones, K. 2007. "Sexual Predators: MySpace in the Middle," *Information Week*, May 21, 20.

Jordan, M. 2007. "Interpol Chief Calls U.K. Lax In Terror Fight; Failure to Share Data Also Cited," *Washington Post*, July 10, A11.

Journal of Internet Law. 2002. "Computer Firm's Use of Nissan.com Not Bad Faith Under Anticybersquatting Act," 6(1), July, 23.

Kaplan, C. 2002. "A Libel Suit May Decide E-Jurisdiction," *The New York Times*, May 27. (http://www.nytimes.com/2002/05/27/technology/27ELAW.html)

Kisiel, R. 2002. "Two Nissans Collide on Information Highway," *Automotive News*, December 16, 1IT–2IT.

Krim, J. 2004. "Justice Department to Announce Cyber-Crime Crackdown: Actions to Include Arrests, Subpoenas," *The Washington Post*, August 25, E5.

Lehman, P. and T. Lowry. 2007. "The Marshal of MySpace: How Hemanshu Nigam Is Trying to Keep the Site's 'Friends' Safe From Predators and Bullies," *Business Week*, April 23, 86.

Leonard, A. 2002. "Nissan vs. Nissan," *Salon.com*, June 3. (http://www.salon.com/tech/col/leon/2002/06/03/nissan/index.html)

Lessig, L. 2000. *Code and Other Laws of Cyberspace*. New York: Basic Books.

Levine, G. 2011. "Chances of Winning and Losing Domain Name Disputes," *UDRPCommentaries.com*, December 20. (http://www.udrpcommentaries.com/chances-of-winning-and-losing-domain-name-disputes/)

Levy, S. 2011. *In the Plex: How Google Thinks, Works, and Shapes Our Lives*. New York: Simon & Schuster.

Liptak, A. 2003. "U.S. Courts' Role in Foreign Feuds Comes Under Fire," *The New York Times*, August 3, 1.

Macdonald, E. 2011. "When is a Contract Formed by the Browse-wrap process?" *International Journal of Law and Information Technology*. (http://ijlit.oxfordjournals.org/content/early/2011/07/27/ijlit.ear009.short)

Mangalindan, M. 2007. "EBay Is Ordered to Pay $30 Million in Patent Rift," *The Wall Street Journal*, December 13, B4.

Manjoo, F. 2001. "Fine Print Not Necessarily in Ink," *Wired News*, April 6. (http://www.wired.com/news/business/0,1367,42858,00.html)

Maurer, H. and C. Lindblad. 2008. "Safer Networking," *Business Week*, January 28, 9.

Mitchell, K., D. Finkelhor, L Jones, and J. Wolak. 2012. "Prevalence and Characteristics of Youth Sexting: A National Study," *Pediatrics*, 129(1), January 1, 13–20.

Moringiello, J. and W. Reynolds. 2008. "Survey of the Law of Cyberspace: Electronic Contracting Cases 2007-2008," *The Business Lawyer*, 64(1), November, 199–218.

Murray, J. 2000. "E-Contracts Present Courts with Special Legal Challenges," *Purchasing*, 129(3), August 24, 119–120.

Nee, E. 2005. "Days of Wine and Roses," *CIO Insight*, July, 25–26.

Network Briefing Daily. 2002. "Amazon Settles 1-Click Patent Dispute," March 8, 3–4.

Newman, M. 2006. "MySpace.com Hires Official to Oversee Young Users' Safety," *International Herald Tribune*, April 13, 18.

Nigro, D. 2005. "Supreme Court Lifts Shipping Bans," *Wine Spectator*, 30(6), July 31, 12.

Nissan Motor Co. v. Nissan Computer Corp. 2002. 246 F.3d 675 (9th Cir.).

Null, C. 2009. "Amazon Likely to Scrap Wine Sales Program," *Today in Tech*, October 24. (http://tech.yahoo.com/blogs/null/153950)

O'Brien, T. 2005. "The Rise of the Digital Thugs," *The New York Times*, August 7, C1.

Oder, N. 2002. "COPA Ruling Offers Mixed Message," *Library Journal*, 127(11), June 15, 15.

Patchin, J. and S. Hinduja. 2008. "Offline Consequences of Online Victimization: School Violence and Delinquency," *Journal of School Violence*, 6(3), 89–112.

Phillips, D. 2003. "JetBlue Apologizes for Use of Passenger Records," *The Washington Post*, September 20, E1.

Popper, N. and T. Hue. 2011. "FBI Shuts Down Internet Poker Sites," *Los Angeles Times*, April 15. (http://articles.latimes.com/2011/apr/15/business/la-fi-poker-busts-20110416)

Porter, K. and S. Bradley. 1999. *eBay, Inc*. Case #9-700-007. Cambridge, MA: Harvard Business School.

Puzzanghera, J. 2011. "Justice Department Opinion Allows States to Offer Online Gambling," *Los Angeles Times*, December 27. (http://latimesblogs.latimes.com/money_co/2011/12/online-gambling-states-justice.html)

Reagle, J. 1999. "The Platform for Privacy Preferences," *Communications of the ACM*, 42(2), February, 48–51.

Richtel, M. 2004. "U.S. Steps Up Push Against Online Casinos by Seizing Cash," *The New York Times*, May 31, C1.

Romano, A. 2006. "Walking a New Beat: Surfing MySpace.com Helps Cops Crack the Case," *Newsweek*, April 24, 48.

Rustad, M. and M. V. Onufrio, 2010. "The Exportability of the Principles of Software: Lost in Translation?" *Hastings Science and Technology Law Journal*, 2(25), 25–80.

Sage, A. 2010. "Ebay, NRF to Take on Organized Retail Crime," *Reuters*, March 22. (http://www.reuters.com/article/idUSTRE62L0OR20100322)

Samborn, H. 2000. "Nibbling Away at Privacy," *ABA Journal*, 86(2), June, 26–27.

Samuelson, P. 2009. "Legally Speaking: When is a License Really a Sale," *Communications of the ACM*, 52(3), March, 27–29.

Schneider, G., L. Barkacs, and C. Barkacs. 2006. "Software Errors: Recovery Rights Against Vendors," *Journal of Legal, Ethical and Regulatory Issues*, 9(2), 61–67.

Schwanhausser, M. 2008. "EBay Patent Case Settled: It Owns 'Buy It Now' After Six-Year Battle," *San Jose Mercury News*. February 29.

Seitz, V. 2011. "Memorandum Opinion for the Assistant Attorney General, Criminal Division: Whether Proposals by Illinois and New York to Use The Internet and Out-of-State Transaction Processors to Sell Lottery Tickets to in-State Adults Violate the Wire Act," September 20. Washington, DC: U.S. Department of Justice. (http://www.justice.gov/olc/2011/state-lotteries-opinion.pdf)

Schultz, E. 2011. "Success in a Bottle: Wine Sites Finally Start to Win Over Web," *Advertising Age*, July 25. (http://adage.com/article/news/wine-websites-find-success-states-open-borders/228867)

Sherman, M. 2011. "Sixteen, Sexting, and a Sex Offender: How Advances in Cell Phone Technology Have Led to Teenage Sex Offenders," *Boston University Journal of Science and Technology Law*, 17(1), 138–161.

Smith, J. 2008. "New Rules for Banks Target Online Gambling," *The Washington Post*, November 13. (http://www.washingtonpost.com/wp-dyn/content/article/2008/11/12/AR2008111202668.html)

Stinson, J. 2007. "Interpol Chief Urges More Data Sharing, He Says Terrorism Information Should Flow Worldwide," *USA Today*, July 9, 9A.

Stone, M. 2001. "Court Dismisses Class Action Against eBay," *BizReport*, January 19. (http://www.bizreport.com/daily/2001/01/20010119-4.htm)

Tanford, J. 2005. "*Granholm v. Heald*: The Supreme Court Strikes Down Trade Barriers Against the Direct Sale of Wine," *Duke Law School: Supreme Court Online*, May. (http://www.law.duke.edu/publiclaw/supremecourtonline/commentary/gravhea.html)

Thomas, K. and C. McGee. 2012. "The Only Thing We Have to Fear Is... 120 Characters," *TechTrends*, 56(1), January-February, 19–33.

European Union. 1993. "Unfair Contract Terms Directive 93/13/EEC," April 5. (http://eur-lex.europa.eu/LexUriServ/LexUriServ.do?uri=CELEX:31993L0013:EN:NOT)

United Nations. 1970. "Declaration on Principles of International Law Concerning Friendly Relations and Cooperation Among States in Accordance with the Charter of the United Nations," *General Assembly Resolution*, #2625, 35th Session.

Van Alstine, P. 2004. "Federal Common Law in an Age of Treaties," *Cornell Law Review*, 89(892), 917–927.

Van Name, M. and B. Catchings. 1998. "Practical Advice About Privacy and Customer Data," *PC Week*, 15(27), July 6, 38.

Vara, V. and L. Chao. 2006. "EBay Steps Back From Asia, Will Shutter China Site," *The Wall Street Journal*, December 19. (http://online.wsj.com/article/SB116647579560853680.html)

Venezia, P. 2009. "Are Apple's App Store Policies Ruining Everything?" *InfoWorld*, November 16. (http://www.infoworld.com/t/mobile-applications/are-apples-app-store-policies-ruining-everything-353)

Vernor, Timothy S. v. Autodesk, Inc. 2011. No. 09-3596, Order (9th Cir. Jan 18).

Ward, B. and J. Sipior. 2011. "The Battle Over E-commerce Sales Taxes Heats Up," *Information Systems Management*, v28(4), 321–326.

Winkler, R. 2012. "Online Profits From Gambling in the Cards," *The Wall Street Journal*, January 3. (http://online.wsj.com/article/SB10001424052970203899504577130961317275678.html)

Whitlock, C. 2005. "Briton Used Internet As His Bully Pulpit," *The Washington Post*, August 8, A1.

Wild, C., S. Weinstein, N. MacEwan, and N. Geach. 2011. *Electronic and Mobile Commerce Law: An Analysis of Trade, Finance, Media and Cybercrime in the Digital Age*. Hertfordshire, UK: University of Hertfordshire Press.

Woo, S. and V. Vauhini. 2011. "Amazon Pursues Internet Tax Deal," *The Wall Street Journal*, September 9, B3.

World Intellectual Property Organization. 2011. "The Uniform Domain Name Dispute Resolution Policy and WIPO," August. (http://www.wipo.int/export/sites/www/amc/en/docs/wipointaudrp.pdf)

Ybarra, M., K. Mitchell, J. Wolak, and D. Finkelhor. 2006. "Examining Characteristics and Associated Distress Related to Internet Harassment: Findings From the Second Youth Internet Safety Survey," *Pediatrics*, 118(4), 1169–1177.

Zelinsky, E. 2011. "Lobbying Congress: Amazon Laws in the Lands of Lincoln and Mt. Rushmore," *State Tax Notes*, 60, 557–581.

PART 3

TECHNOLOGIES FOR ELECTRONIC COMMERCE

CHAPTER 8
Web Server Hardware and Software, 333

CHAPTER 9
Electronic Commerce Software, 371

CHAPTER 10
Electronic Commerce Security, 407

CHAPTER 11
Payment Systems for Electronic Commerce, 461

WEB SERVER HARDWARE AND SOFTWARE

INTRODUCTION

As you learned in earlier chapters, **Lands' End** was one of the most successful clothing retailers on the Web before it was acquired by Sears in 2003. Now, as a division of Sears, Lands' End continues to be a leader in adding features that attract customers to its Web site and keep those customers coming back. Behind the scenes at Lands' End, a team of experienced technology professionals implements new Web page features and performs the many regular maintenance tasks that are necessary to keep the Lands' End Web site running smoothly.

Lands' End closely monitors the performance of its Web site to make sure that customers have a consistent experience each time they visit the site. The Web site's technical team works hard to make sure that site visitors do not notice the Web site's operating characteristics. This goal has not always been easy to attain because the site's traffic volume has increased each year since the site opened. Also, regular major improvements to the Lands' End site keep the Web team busy.

Lands' End's specific goals for performance change as Web technologies improve. For example, the site management team has a target for the time it takes one of the site's Web pages to load on a visitor's computer. In the early days of the site, that target was 15 seconds. Today, the target is under one second. The Web site's technical team has always taken a conservative approach to operating the site so that the site can meet its performance goals more easily. For example, the technical team specifies the maximum and average sizes of Web pages and graphics files that the content team can use. In addition, the technical team must complete all major changes to the site (including thorough testing) before November 1 each year, prior to the holiday selling season. Retailers such as Lands' End make more than half of their sales in November and December, so they rarely take the chance of making major Web site changes during that time period.

The server hardware at Lands' End is a mix of Sun and IBM computers that are managed by another computer that allocates incoming Web traffic. Some of the Web site's advanced features, such as the graphics-intensive My Virtual Model, are created on a separate set of computers. These computers are all located at the Lands' End division headquarters in a small town near Madison, Wisconsin. The computers run a UNIX-based operating system called Solaris and a version of the Apache Web server software, about which you will learn more in this chapter. Although the Lands' End technical team writes some of the software that it uses to monitor the Web site's performance, the company also uses the services of Keynote Systems. Keynote's software can measure how fast particular pages load or how rapidly transactions are completed at various times of the day and in various locations around the world. The Lands' End technical team uses this information to fine-tune its hardware configuration.

By paying close attention to the details, the technical team at Lands' End keeps the Web site operating at or above expected levels. The technical team's goal is to prevent customers from being distracted from their shopping experiences by the operation of the site.

WEB SERVER BASICS

As you learned in Chapter 2, Web servers are computers that are designed to provide public access to files that are rendered as Web pages on visitors' computers. In this chapter, you will learn about these computers and the software they use to deliver Web sites. Web sites that have many visitors must use a large number of these Web server computers to deliver Web page files efficiently. Operating large numbers of computers requires synchronization of their activities, dividing the workload that each computer must carry. You will learn about these elements of Web site operation in this chapter as well.

When people use Web browser software to become part of the Web, their computers become Web client computers on a worldwide client/server network. Client/server architectures are used in LANs, WANs, and the Web. In a client/server architecture, the client computers typically request services, such as printing, information retrieval, and database access, from the server, which processes the clients' requests. The computers that perform the server function usually have more memory and larger, faster disk drives than the client computers they serve. Recall from Chapter 2 that Web browser software (for example, Microsoft Internet Explorer or Mozilla Firefox) is the software that makes computers work as Web clients. Thus, a Web browser is also called Web client software.

The Internet connects many different types of computers and other devices, each running different types of operating system software. The ability of a network to connect devices that use different operating systems is called **platform neutrality**. Because Web software is platform neutral, it lets these computers communicate with each other easily and effectively. This platform neutrality was a critical factor in the rapid spread and widespread acceptance of the Web. Before the Internet's platform neutrality, the computers that were connected to each other using leased phone lines either had to run the same operating system software or they required translation software that allowed each computer to communicate with the other one. Figure 8-1 shows how the Web's platform neutrality provides multiple interconnections among a wide variety of client and server computers.

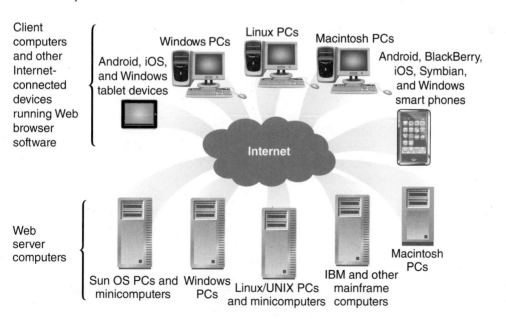

FIGURE 8-1   Platform neutrality of the Web

The main job of a Web server computer is to respond to requests from Web client computers. The three main elements of a Web server are the hardware (computers and related components), operating system software, and Web server software. All three of these elements must work together to provide sufficient capacity in a given situation.

After most companies have decided on the goals they want to accomplish with their Web sites, they begin developing their sites by estimating the number of visitors they expect to have, how many pages those visitors will view during an average visit, how large those pages will be (including graphics and other page elements), and the likely maximum number of simultaneous visitors.

In the early days of electronic commerce, Web sites were collections of individual pages about the site's product or service offerings. Today, Web sites often deliver customized pages in response to customers' specific needs. You will learn how sites do that in the next section.

Dynamic Content Generation

A **dynamic page** is a Web page whose content is shaped by a program in response to user requests, whereas a **static page** is an unchanging page retrieved from a file (or, more typically, a set of files) on a Web server. On a Web site that is a collection of HTML pages, the content on the site can be changed only by editing the HTML in the pages. This is cumbersome and does not allow customized pages to be produced in response to specific queries from site visitors.

Dynamic content is information constructed in response to a Web client's request. Dynamic content can give the user an interactive experience with the Web site. The text, graphics, form fields, and other Web page elements can change in response to user input or other variables. For example, if a Web client inquires about the status of an existing order by entering a customer number or order number into a form, the Web server generates a dynamic Web page based on the customer information stored in the company's database, thus fulfilling the client's request. A dynamic page is a specific response to the requester's query that is assembled from information stored in a company's back-end databases and internal data on the Web site.

Dynamic content can be created using two basic approaches. In the first approach, called **client-side scripting**, software operates on the Web client (the browser) to change what is displayed on the Web page in response to a user's actions (such as mouse clicks or keyboard text input). In client-side scripting, changes are generated within the browser using software such as JavaScript or Adobe Flash. The Web client retrieves a file from the Web server that includes code (JavaScript, for example). The code instructs the Web client to request specific page elements from the Web server and dictates how they will be displayed in the Web browser window. This approach is often used to manage the activity displayed on a Web page by various media elements (audio, video, changing graphics or text). Client-side scripting emerged on the Web in 1996, when the JavaScript language became widely available.

In the second approach, called **server-side scripting**, a program running on a Web server creates a Web page in response to a request for specific information from a Web client. The content of the request can be determined by several things, including text that a user has entered into a Web form in the browser, extra text added to the end of a URL, the type of Web browser making the request, or simply the passage of time. For example, if you are logged into an online banking site and do not enter any text or click anywhere on the page for a few minutes, you might find that the Web server ends your connection and sends a page to your browser indicating that "your session has expired."

A number of Web programming languages and frameworks have evolved that allow site designers to generate dynamic Web pages and make them interactive. In dynamic page-generation technologies, server-side scripts are mixed with HTML-tagged text to create the dynamic Web page. Microsoft developed the first widely used server-side dynamic page-generation technology, called **Active Server Pages (ASP)**. The current version of that technology is called **ASP.NET**. ASP allows Web programmers to use their choice of programming languages, such as VBScript, Jscript, or Perl. Sun Microsystems developed a similar technology called **JavaServerPages (JSP)**. Java, a programming language created by Sun, can also be used to produce dynamic pages. Such server-side programs are called **Java servlets**. The open-source Apache Software Foundation sponsored a third alternative called the **Hypertext Preprocessor (PHP)**. Yet another alternative is available from Adobe in its **ColdFusion** product. These server-side languages generally use the **Common Gateway Interface (CGI)**, which was introduced in 1993 as a standard way of interfacing external applications with Web servers. In its first applications, CGI was used to connect existing databases to Web servers, which allowed users all over the world to access those databases from their Web browsers. That is, CGI provided a gateway allowing users to enter remote databases.

Two dynamic page-generation tools that have become popular in recent years include AJAX and Ruby on Rails. **AJAX** (asynchronous JavaScript and XML) is a development framework that can be used to create interactive Web sites that look like applications running in a Web browser. Most dynamic Web pages must reload in their entirety if any page content changes. AJAX lets programmers create Web pages that will update asynchronously by exchanging small amounts of data with the server while the remainder of the Web page continues to be displayed in the browser. Because the entire Web page does not reload with every change, the user experience is improved. Google Maps is an example of a dynamic page that is generated using Ajax. **Ruby on Rails** is another Web development framework that lets programmers create dynamic Web pages that present users with an interface that looks like application software running in a Web browser. **Python** is a scripting language that can also be used in dynamic Web page generation.

Multiple Meanings of "Server"

As you learned in Chapter 2, computers that are connected to the Internet and make some of their contents publicly available using the HTTP protocol are called Web servers. Unfortunately, the term "server" is used in many different ways by information systems professionals. These multiple uses of the term can be confusing to people who do not have a strong background in computer technology. You are likely to encounter a number of different uses of the word "server."

A **server** is any computer used to provide (or "serve") files or make programs available to other computers connected to it through a network (such as a LAN or a WAN). The software that the server computer uses to make these files and programs available to the other computers is often called **server software**. Sometimes this server software is included as part of the operating system that is running on the server computer. Thus, some information systems professionals informally refer to the operating system software on a server computer as server software, a practice that adds considerable confusion to the use of the term "server."

Some servers are connected through a router to the Internet. As you learned in Chapter 2, these servers can run software, called Web server software, that makes files on those servers available to other computers on the Internet. When a server computer is connected to the Internet and is running Web server software (usually in addition to the server software it runs to serve files to client computers on its own network), it is called a Web server.

Similar terminology issues arise for server computers that perform e-mail processing and database management functions. Recall that the server computer that handles incoming and outgoing e-mail is usually called an **e-mail server**, and the software that manages e-mail activity on that server is frequently called e-mail server software. The server computer on which database management software runs is often called a **database server**. The computer on which a company runs its accounting and inventory management software is sometimes called a **transaction server**.

Thus, the word "server" is used to describe several types of computer hardware and software, all of which might be found in a typical electronic commerce operation. The only way to determine which server people are talking about when they use the term is from the context or by asking a clarifying question. If you hear a computer technician say, "The server is down today," the problem might be in the hardware, the software, or a combination of the two.

Web Client/Server Architectures

In Chapter 2, you learned how the Web is software that runs on the Internet. In this section, you will learn more about how Web client and Web server software work. When a person uses a Web browser to visit a Web site, the Web browser (also known as a Web client) requests files from the Web server at the company or organization that operates the Web site. Using the Internet as the transportation medium, the request is formatted by the browser using HTTP and sent to the server computer. When the server receives the request, it retrieves the file containing the Web page or other information that the client requested, formats it using HTTP, and sends it back to the client over the Internet.

When the requested information—a file containing the text and markup tags of a Web page, in this instance—arrives at the client computer, the Web browser software determines that the information is an HTML page. It displays the page on the client machine according to the directions defined in the page's HTML code. This process repeats as the client requests, the server responds, and the client displays the result. Sometimes, a single client request results in dozens or even hundreds of separate server responses to locate and deliver information. A Web page containing many graphics and other objects can be slow to appear in the client's Web browser window because each page element (each graphic or multimedia file) requires a separate request and response.

The basic Web client/server model is a two-tier model because it has only one client and one server. All communication takes place on the Internet between the client and the server. Of course, other computers are involved in forwarding packets of information across the Internet, but the messages are created and read only by the client and the server computers in a **two-tier client/server architecture**. Figure 8-2 shows how a Web client and a Web server communicate with each other in a two-tier client/server architecture.

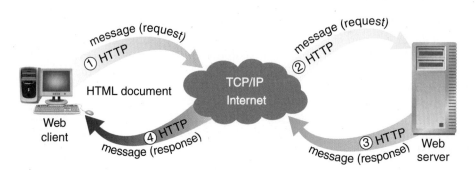

FIGURE 8-2   Message flows in a two-tier client/server network

The message that a Web client sends to request a file or files from a Web server is called a **request message**. A typical request message from a client to a server consists of three major parts:

- Request line
- Optional request headers
- Optional entity body

The **request line** contains a command, the name of the target resource (a file name and a description of the path to that file on the server), and the protocol name and version number. Optional **request headers** can contain information about the types of files that the client will accept in response to this request. Finally, an optional **entity body** is sometimes used to pass bulk information to the server.

When the server receives the request message, it executes the command included in the message (in this case, it sends a particular Web page file back to the client). The server does this by retrieving the Web page file from its disk (or another disk on a network to which it is connected) and then creating a properly formatted **response message** to send back to the client. A server's response consists of three parts that are identical in structure to a request message: a response header line, one or more response header fields, and an optional entity body. In the response, however, each part has a slightly different function than it does in the request. The **response header line** indicates the HTTP version used by the server, the status of the response (whether the server found the file that the client wanted), and an explanation of the status information. Response header fields follow the response header line. A **response header field** returns information describing the server's attributes. The entity body returns the HTML page requested by the client machine.

Although the two-tier client/server architecture works well for the delivery of Web pages, a Web site that delivers dynamic content and processes transactions must do more than respond to requests for Web pages. A **three-tier architecture** extends the two-tier architecture to allow additional processing (for example, collecting the information from a database needed to generate a dynamic Web page) to occur before the Web server responds to the Web client's request. The third tier often includes databases and related software applications that supply information to the Web server. The Web server can then use the output of these software applications when responding to client requests, instead of just delivering a Web page.

A good example of services supported by a database in a three-tier architecture is a catalog-style Web site with search, update, and display functions. Assume that a user requests a display of an online specialty food store's exotic fruit selections. The client request is formulated into an HTTP message by the Web browser (tier 1), sent over the Internet to the Web server, and examined by the Web server. The Web server (tier 2) analyzes the request and determines that responding to the request requires the help of the server's database. The server sends a request to the database management software (tier 3) to search for, retrieve, and return all information about exotic fruit in the company's catalog database. The database information flows back through the database management software system to the server, which formats the response into an HTML document and sends that document inside an HTTP response message back to the client over the Internet. Figure 8-3 shows an overview of information flows in a three-tier architecture. Numbers on the flow arrows indicate the order in which the messages flow over the indicated paths.

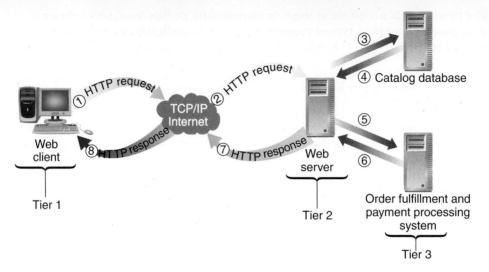

FIGURE 8-3   Message flows in a three-tier client/server network

Architectures that have four, five, or even more tiers divide into separate tiers the software applications and the databases and database management programs that work with those software applications. Also, some sites have software applications that generate information (a fourth tier) that feeds into other software applications or databases (in the third tier) that in turn generate information for the Web server to turn into Web pages (in the second tier), which then go to the requesting client (in the first tier). Architectures that have more than three tiers are often called **n-tier architectures**. N-tier systems can track customer purchases stored in shopping carts, look up sales tax rates, keep track of customer preferences, update in-stock inventory databases, and keep the company catalog current.

SOFTWARE FOR WEB SERVERS

Some Web server software can run on only one computer operating system, while some can run on several operating systems. In this section, you will learn about the operating system software used on most Web servers and the Web server software itself. You also will learn about other programs, such as Internet utilities and e-mail software, that companies often run on Web servers or other computers as part of electronic commerce operations.

Operating Systems for Web Servers

Operating system tasks include running programs and allocating computer resources such as memory and disk space to programs. Operating system software also provides input and output services to devices connected to the computer, including the keyboard, monitor, and printers. A computer must have an operating system to run programs. For large systems, the operating system has even more responsibilities, including keeping track of multiple users logged on to the system and ensuring that they do not interfere with one another.

Most Web server software runs on Microsoft Windows Server products, Linux, or other UNIX-based operating systems such as FreeBSD. Some companies believe that Microsoft server products are simpler for their information systems staff to learn and use than UNIX-based systems. Other companies worry about the security weaknesses caused by the tight integration between application software and the operating system in Microsoft products. UNIX-based Web servers are more widely used, and many industry experts believe that UNIX is a more secure operating system on which to run a Web server.

Linux is an open-source operating system that is fast, efficient, and easy to install. An increasing number of companies that sell computers intended to be used as Web servers include the Linux operating system in default configurations. Although Linux can be downloaded free from the Web, most companies buy it through a commercial distributor. These commercial distributions of Linux include additional software, such as installation utilities, and a support contract for the operating system. Commercial Linux distributors that sell versions of the operating system with utilities for Web servers include Mandriva, Red Hat, and SuSE Linux Enterprise. Canonical sells technical support and services for the Ubuntu Linux distribution. Oracle sells Web server hardware along with its UNIX-based operating system, Solaris. You can learn more about open-source software at the Open Source Initiative Web site.

Web Server Software

This section describes the two most commonly used Web server programs, Apache HTTP Server and Microsoft Internet Information Server (IIS). Other Web server software products are used by online businesses, including Oracle iPlanet, nginx (pronounced "engine-x"), and lighttpd (pronounced "lighty"). Some large online businesses have written their own Web server software; for example, Google runs Google Web Server with the Linux operating system on its millions of server computers.

These popularity rankings were compiled through surveys done by Netcraft, a networking consulting company in Bath, England, known throughout the world for its Web server surveys. Netcraft continually conducts surveys to tally the number of Web sites in existence and measure the relative popularity of Internet Web server software. Figure 8-4 shows the use of Web server software by active sites in January 2012.

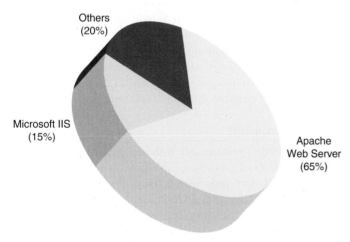

Source: Netcraft Web Surveys, http://www.netcraft.com

FIGURE 8-4 Percent of active Web sites that use major Web server software products

The Netcraft Web server surveys show that the market share of Web server software has stabilized in recent years. Apache generally holds over half of the market, and Microsoft IIS usually holds between 10 and 20 percent of the market. Because Google operates so many server computers, Google Web Server usually ranks high in the Netcraft surveys—even though Google is the only company that uses it.

Apache HTTP Server

Apache is an ongoing group software development effort. Rob McCool developed Apache while he was working at the University of Illinois at the National Center for Supercomputing Applications (NCSA) in 1994. Other Web site developers around the world created their own extensions to the server and formed an e-mail group so that they could coordinate their changes (known as "patches") to the system. The system consisted of the original core system with a lot of patches—thus, it became known as "a patchy" server, or simply, "Apache." The Apache Web server is currently available on the Web at no cost as open-source software.

Apache HTTP Server has dominated the Web since 1996 because it is free, performs efficiently, and has a large number of knowledgeable users who contribute technical advice to online discussion forums, wikis, and blogs. A number of companies sell support services for Apache for organizations that want the additional security; however, most Apache installations are supported by the organization's own technical staff using the free online help that is available. Apache runs on many operating systems (including FreeBSD-UNIX, HP-UX, Linux, Microsoft Windows, SCO-UNIX, and Solaris) and the hardware that supports them.

Microsoft Internet Information Server

Microsoft Internet Information Server (IIS) comes bundled with current versions of Microsoft Windows Server operating systems. IIS is used on many corporate intranets because many companies have adopted Microsoft products as their standard products. Some small sites running personal Web pages also use IIS, as do some of the largest online business sites on the Web. IIS itself is free; however, the Microsoft Windows Server operating system software with which it is packaged can range in cost from under $1000 for a small business running one or two servers to many thousands of dollars for large organizations running many servers (details are complicated; the Microsoft Windows Server pricing guide is a document of more than 50 pages).

IIS, as a Microsoft product, is designed to run only on Windows server operating systems. IIS supports the use of ASP, ActiveX Data Objects, and SQL database queries. IIS's inclusion of ASP provides an application environment in which HTML pages, ActiveX components, and scripts can be combined to produce dynamic Web pages.

Finding Web Server Software Information

People who want to know the type of operating system and Web server software that a Web site is running can visit the **Netcraft** Web site. On Netcraft's home page is a link named "What's that site running?" that leads to a page with a search function. Visitors can use that search function to find out what operating system and what Web server software a specific site is now running and what the site ran in the past.

ELECTRONIC MAIL (E-MAIL)

Although the Web, with its interactions between Web servers and clients, is the most important technology used in electronic commerce today, many buyers and sellers also use e-mail to gather information, execute transactions, and perform other online business tasks. E-mail originated in the 1970s on the ARPANET. Although the goals of the ARPANET were to control weapons systems and transfer research files, general communications uses emerged on the network. As you learned in Chapter 2, in 1972, ARPANET researcher Ray Tomlinson wrote a program that could send and receive messages over the network. Today, e-mail is the most popular form of business communication—far surpassing the telephone, conventional mail, and fax in volume.

E-Mail Benefits

Not only was e-mail one of the first Internet applications, it was also one reason that many people were originally attracted to the Internet. E-mail conveys messages from one destination to another in a few seconds. Messages can contain plain text, or they can contain character formatting similar to word-processing programs.

One useful feature of e-mail is that documents, pictures, movies, worksheets, or other information can be sent along with the message itself. These **attachments** are frequently the most important part of the message. A business e-mail message attachment might contain an invoice, a 200-page wholesale catalog, or a set of Web pages that describe the company's products. Many electronic commerce sites use e-mail to confirm the receipt of customer orders and then the shipment of items ordered. Software vendors can also use e-mail to send information about a purchase to the buyer. As you learned in Chapter 3, many online stores use e-mail to announce specials, sales, or to keep in touch with customers.

E-Mail Drawbacks

Despite its many benefits, e-mail does have some drawbacks. One annoyance associated with e-mail is the amount of time that businesspeople spend answering their e-mail today. Researchers have found that most managers can deal with e-mail messages at an average rate of about five minutes per message. Some messages can be deleted within a few seconds, but those are balanced by the e-mails that require the manager to spend much more time finding facts, checking files, making phone calls, and doing other tasks as part of answering e-mail. Researchers have found that most people (not including those people who answer e-mails as a full-time job) begin to resent the time that e-mail consumes when they start getting more than 20 or 30 messages a day. At that point, the average person is spending about two hours a day answering e-mail.

A second major irritation brought by e-mail is the **computer virus**, more simply known as a **virus**, which is a program that attaches itself to another program and can cause damage when the host program is activated. Recall that e-mail messages can carry attachments. Although attached files usually carry useful information, they can contain viruses. Using virus protection software and dealing with e-mailed security threats is a cost that all must bear for the convenience of using e-mail. You will learn more about computer viruses and other threats that can be transmitted through e-mail (and how to control them) in Chapter 10.

As you learned in Chapter 2, the most frustrating and expensive problem associated with e-mail today is the issue of unsolicited commercial e-mail, also known as UCE or **spam**. This nagging problem is discussed in the next section.

Spam

Figure 8-5 shows the rapid increase in the proportion of all e-mail entering business e-mail servers that is spam. The sheer magnitude of the spam problem is hard to believe. During one 24-hour period in 2009 (the peak year for spam), researchers estimated that 220 billion spam e-mail messages were sent. Researchers who track spam believe that spam growth has leveled off and that currently available technical solutions will continue to reduce the amount of spam as a percentage of total e-mail traffic in the future.

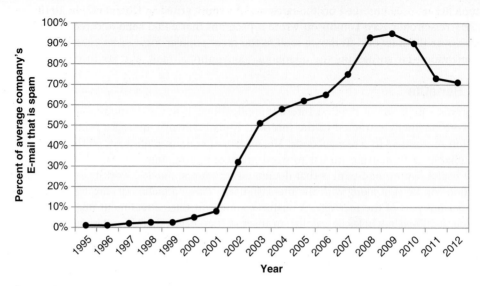

Sources: Symantec *Intelligence Reports, Spam and Phishing Reports*, and *Spam Reports*; www.symantec.com

FIGURE 8-5 Spam as a proportion of all business e-mail

A number of companies now offer software that organizations can run on their e-mail server computers to limit the amount of spam that gets through to their employees. Although individual users can install client-based spam-filtering programs on their computers or set filters that might be available within their e-mail client software, most companies find it more effective and less costly to eliminate spam before it is downloaded to user computers. These antispam efforts and software products can help limit the annoyance and cost of spam.

Solutions to the Spam Problem

As spam has become a serious problem for all users of e-mail, the methods used to limit spam and its effects have taken various forms. Some of these approaches require the passing of laws, and some require technical changes in the mail-handling systems of the Internet. Other approaches can be implemented under existing laws and with current technologies, but only if large numbers of organizations and businesses cooperate. A few tactics that reduce spam can be undertaken by individual e-mail users. In the sections that follow, you will learn more about each of these approaches to controlling the spam problem.

Individual User Antispam Tactics

One way individuals can limit spam is to reduce the likelihood that a spammer can automatically generate their e-mail addresses. Many organizations create e-mail addresses for their employees by combining elements of each employee's first and last names. For example, small companies often combine the first letter of an employee's first name with the entire last name to generate e-mail addresses for all employees. Larger companies often use employees' entire first and last names because they are likely to have both a Jane Smith and a Judy Smith working for them. A spam sender able to obtain an employee list can generate potential e-mail addresses using the names on the list. If no employee list is available, the spam sender might simply generate logical combinations of first initials (or names) and common last names. The cost of sending e-mail is so low that a spammer can afford to send thousands of e-mails to randomly generated addresses in the hope that a few of them are valid. By using an e-mail address that is more complex, such as xq7yy23@mycompany.com, individuals can reduce the chances that a spammer can randomly generate his or her address. Of course, such an address is hard to remember, which somewhat defeats the purpose of e-mail as a convenient way to communicate.

A second way to reduce spam is to control the exposure of an e-mail address. Spammers use software robots to search the Internet for character strings that include the @ character, which appears in every e-mail address. These robots search Web pages, discussion boards, chat rooms, and any other online source that might contain e-mail addresses. Again, the spammer can afford to send thousands of messages to e-mail addresses gathered in this way. Even if only one or two people respond, the spammer can earn a profit because the cost of sending e-mail messages is so low.

Some individuals use multiple e-mail addresses to thwart spam. They use one address for display on a Web site, another to register for access to Web sites, another for shopping accounts, and so on. If a spammer starts using one of these addresses, the individual can stop using it and switch to another. Many Web hosting services include a large number (often 100 to 200) of e-mail addresses as part of their service, so this can be a useful tactic for people or small businesses with their own Web sites.

These three strategies focus on limiting spammers' access to or use of an e-mail address. Other approaches use one or more techniques that filter e-mail messages based on their contents.

Basic Content Filtering

All content-filtering solutions require software that identifies content elements in an incoming e-mail message that indicate the message is (or is not) spam. The content-filtering techniques differ in which content elements they examine, whether they look for indications that the message is (or is not) spam, and how strictly they apply the rules for classifying messages. Most basic content filters examine the e-mail headers (From, To, Subject) and look for indications that the message might be spam. The software that performs the filtering task can be placed on individual users' computers (called **client-level filtering**) or on mail server computers (called **server-level filtering**). Server-level filtering can be implemented on an ISP's mail server, an individual company's mail server, or both. Also, many individuals that have ISP and/or company mail servers that filter their e-mail also install client-level filters on their computers. Spam that gets through one filter can be trapped by another filter.

The most common basic content-filtering techniques are black lists and white lists. A **black list spam filter** looks for From addresses in incoming messages that are known to be spammers. The software can delete the message or put it into a separate mailbox for review. A black list spam filter can be implemented at the individual, organization, or ISP level. Several organizations, such as the Spam and Open Relay Blocking System collect black lists and make them available to ISPs and company e-mail administrators. Other groups, such as the Spamhaus Project, track known spammers and publish lists of the mail servers they use. Some of these are free services; others charge a fee. The biggest drawback to the black list approach is that spammers frequently change their e-mail servers, which means that a black list must be continually updated to be effective. This updating requires that many organizations cooperate and communicate information about known spammers. In addition to its black list, the Spamhaus Project maintains a list of known spammers on its site. These are individuals and companies who have had their services terminated by an ISP for spam-related violations of an acceptable use policy more than three times. The Spamhaus Project provides detailed information about those on this list to law enforcement agencies.

A **white list spam filter** examines From addresses and compares them to a list of known good sender addresses (for example, the addresses in an individual's address book). A white list filter is usually applied at the individual user level, although it is possible to do the filtering at the organization level if the e-mail administrator has access to all individuals' address books (some companies mandate such access for security purposes). The main drawback to this approach is that it filters out any incoming messages sent by unknown parties, not just spam. Because the number of **false positives** (messages that are rejected but should not have been) can be very high for white list filters, the rejected e-mails are always placed into a review mailbox instead of being deleted.

White list and black list approaches can be used in client-level or server-level filters, but both have serious drawbacks. To overcome these drawbacks, the two approaches are often used together or with other content-filtering approaches to achieve an acceptable level of filtering without an excessive false positive rate.

Challenge-Response Content Filtering

One content-filtering technique uses a white list as the basis for a confirmation procedure. This technique, called **challenge-response**, compares all incoming messages to a white list. If the message is from a sender who is not on the white list, an automated e-mail response is sent to the sender. This message (the challenge) asks the sender to reply to the e-mail (the response). The reply must contain a response to a challenge presented in the e-mail.

These challenges are designed so that a human can respond easily, but a computer would have difficulty formulating the response. For example, a challenge might include a picture of a fruit bowl and would ask the sender to respond with the number of apples in the bowl. This prevents a spammer from setting up a computer that receives challenges and answers them (the program would have difficulty identifying and counting the number of apples). It would be inefficient for a spammer to hire a human to respond to thousands of challenges. Most implementations also include an audio alternative for visually impaired users. To learn more about this technique, you can visit the **CAPTCHA Project** site at Carnegie Mellon University. An example of a challenge that uses distorted letters and numbers (in this case, 5BM6HW3F) is shown in Figure 8-6.

Type the code shown

FIGURE 8-6 Example of a challenge that uses distorted letters and numbers

The major drawback to challenge-response systems is that they can be abused. For example, a perpetrator could send out thousands of e-mails to recipients that use challenge-response systems. If the perpetrator includes the victim's e-mail as the From address in those e-mails, the victim will be bombarded by the automated challenges sent out by the challenge-response systems of the recipients. What is worse, the potential damage of this tactic becomes greater as more e-mail servers install challenge-response systems.

Another issue with challenge-response systems could arise if they were to become widespread. Most mail that any individual receives from unknown senders is spam. A challenge-response system sends a challenge message to every unknown sender. That is, for every spam message received, a second e-mail is sent. A challenge-response system thus doubles the amount of useless e-mail messages that must be handled by the Internet's infrastructure. If everyone were to use a challenge-response system, the Internet capacity wasted by spam would approximately double. Because challenge-response systems require users to change their behavior, and because they do not provide an immediate and significant benefit (the benefit is spam reduction over time), these systems have not become very widely used.

Advanced Content Filtering

Advanced content filters that examine the entire e-mail message can be more effective than basic content filters that only examine the message headers or the IP address of the e-mail's sender. Creating effective content filters can be challenging. For example, a company might want to delete any e-mail message that includes the word "sex." If the company deletes all e-mails containing that character string, they will unintentionally delete all e-mailed orders from customers in the town of Essex.

Many advanced content filters operate by looking for spam indicators throughout the e-mail message. When the filter identifies an indicator in a message, it increases that message's spam "score." Some indicators increase the score more than others. Indicators can be words, word pairs, certain HTML codes (such as the code for the color white, which makes part of the message invisible in most e-mail clients), and information about where a word occurs in the message. Unfortunately, as soon as spam filter vendors identify a good set of indicators, spammers stop including those indicators in their messages.

One type of advanced content filter that is based on a branch of applied mathematics called Bayesian statistics shows some promise of staying one step ahead of the spammers. **Bayesian revision** is a statistical technique in which additional knowledge is used to revise earlier estimates of probabilities. In software that contains a **naïve Bayesian filter** (the most common type in use today), the software begins by not classifying any messages. The user reviews messages and indicates to the software which messages are spam and which are not. The software gradually learns (by revising its estimates of the probability that a message element appears in a spam message) to identify spam messages.

After seeing a few dozen messages classified, the naïve Bayesian filter can successfully classify spam messages about 80 percent of the time. As the filter continues to work, the user reviews its classifications and tells the software when it makes a mistake. After classifying a few hundred messages (and being corrected by the user when it errs), a naïve Bayesian filter typically reaches correct spam classification rates above 95 percent. Although these filters are highly effective and have low false positive rates, they must be trained, which takes time. The training is best done by each individual user because one person's spam can be another person's important message. Having users train their own filters provides the most rapid training and the best results. Most organizations do not currently use naïve Bayesian filters because they require attention by individual e-mail users. However, naïve Bayesian filters can be installed on some client computers (such as those used by people who receive large amounts of e-mail) in organizations that also use other techniques (such as white list or black list filters) at the server level.

Research on naïve Bayesian filters began appearing in published research papers in 2002. An open-source software development project led by John Graham-Cumming released one of the first functional Bayesian filter products for individual users that year, named **POPFile**. POPFile is a program that runs on individual client computers and works with many different e-mail clients (including Microsoft Outlook and Mozilla Thunderbird) to provide content filtering. Because it is open-source software, POPFile is free (although the project team welcomes donations). POPFile does require that e-mail be retrieved using a Post Office Protocol (POP) connection, so it cannot be used with most Web-based e-mail accounts such as Yahoo! or Hotmail. Newer releases of some e-mail client software, such as Mozilla Thunderbird, now include naïve Bayesian filtering tools that work the same way POPFile does.

Naïve Bayesian filters are very effective client-level filters, but they do not work well as server-level filters. The content that is common in one person's spam might be common in another person's valid e-mail; therefore, one user's reclassifications tend to cancel out those of other users. This prevents the filter from building its accuracy to high levels. One good alternative for organizations is to use black list filters at the server level combined with white list and naïve Bayesian filters at the client level. The major drawback of any client-level filtering approach is that it requires individual users to update their own filters regularly. Although it takes less time to update a filter than to delete hundreds of spam messages, it still does take time.

Legal Solutions

A number of U.S. jurisdictions have passed laws that provide penalties for the sending of spam. In January 2004, the U.S. CAN-SPAM law (the law's name is an acronym for "Controlling the Assault of Non-Solicited Pornography and Marketing") went into effect. Researchers who track the amount of spam noted a drop in the percentage of all e-mail that was spam in February and March of that year. However, by April, the rate was back up. It appears that spammers slowed down their activities immediately after the effective date of CAN-SPAM to see if a broad federal prosecution effort would occur. When the threat did not materialize, the spammers went right back to work.

The CAN-SPAM law is the first U.S. federal government effort to legislate controls on spam. It regulates all e-mail messages sent for the primary purpose of advertising or promoting a commercial product or service, including messages that promote the content displayed on a Web site. The law's main provisions include:

- *Misleading address header information*: E-mail headers and routing information, including the originating domain name and e-mail address, must be accurate and must identify the person who sent the e-mail.
- *Deceptive subject headers*: The e-mail's subject line cannot mislead the recipient about the contents or subject matter of the message.
- *Clear and conspicuous notice of message nature*: The e-mail must contain a clear and conspicuous notice that the message is an advertisement or solicitation and that the recipient can opt out of receiving further commercial e-mail from the sender.
- *Physical postal address*: The e-mail must include the sender's valid physical postal address.
- *Mandatory provision of an opt-out mechanism*: The e-mail must include a return e-mail address or another Internet-based response mechanism that allows a recipient to ask not to be sent future e-mail messages. These requests must be honored. The message may include a menu of choices that allows a recipient to opt out of certain types of messages, but one option on

the menu must be an option to stop sending all commercial messages of any type.

- *Effectiveness of opt-out mechanism*: Opt-out requests must be honored within 10 business days. Any opt-out mechanism offered must be able to process opt-out requests for at least 30 days after the e-mail is sent. Once an opt-out request has been received, the sender is prohibited from helping any other entity send e-mail to the opt-out address or from having another entity send e-mail on the sender's behalf to that address.
- *Transfer of e-mail addresses*: Once a recipient has submitted an opt-out request, the sender is prohibited from selling or transferring that e-mail address to any other entity.

The law also prohibits misleading address header information in transaction-related e-mail messages. For example, an e-mail that facilitates a transaction or that updates a customer regarding a business transaction would fall under this provision. Each violation of a provision of the law is subject to a fine of up to $11,000. Additional fines are assessed for those who violate one of the preceding provisions and do one or more of the following:

- Harvest e-mail addresses from Web sites or Web services that have published a notice prohibiting the transfer of e-mail addresses for the purpose of sending e-mail.
- Send e-mail messages to addresses that have been generated by combining names, letters, or numbers into multiple combinations and permutations.
- Use scripts or other automated tools to register for multiple e-mail or user accounts that are then used to send commercial e-mail.
- Relay e-mails through a computer or network without the permission of the computer's or network's owner.

Thus, a successful prosecution could cost the convicted spammer a considerable amount of money. The law further provides for criminal penalties, including imprisonment, for commercial senders of e-mail who do or conspire to do any of the following:

- Use another person's or entity's computer to send commercial e-mail from or through it without the computer owner's permission.
- Use a computer to relay or retransmit multiple commercial e-mail messages with the intent to deceive or mislead recipients or an Internet access service about the origin of the messages.
- Send multiple e-mail messages that contain false header information.
- Present false identification when registering for multiple e-mail accounts or domain names.
- Falsely represent themselves as owners of multiple IP addresses that are used to send commercial e-mail messages.

You can learn more about the law on the **U.S. Federal Trade Commission CAN-SPAM Law** information pages. The FTC issues new rules from time to time under the law. To obtain current updates on those rules, visit the **U.S. Federal Trade Commission OnGuardOnline.gov Spam** information site. The home page of that site is shown in Figure 8-7.

OnGuardOnline.gov

Search OnGuardOnline.gov | Español

STOP | THINK | CONNECT

Avoid Scams

Secure Your Computer

Protect Kids Online

Be Smart Online

Video and Media

OnGuard Online Blog

Vea esta página en español

Spam

Unwanted commercial email – also known as "spam" – can be annoying. Worse, it can include bogus offers that could cost you time and money. Take steps to limit the amount of spam you get, and treat spam offers the same way you would treat an unimited telemarketing sales call. Don't believe promises from strangers. Learn to recognize the most common online scams.

Related Items

- Avoiding Online Scams
- Malware

How Can I Reduce the Amount of Spam I Get?

Use an email filter.

Check your email account to see if it provides a tool to filter out potential spam or to channel spam into a bulk email folder. You might want to consider these options when you're choosing which Internet Service Provider (ISP) or email service to use.

Limit your exposure.

You might decide to use two email addresses — one for personal messages and one for shopping, newsletters, chat rooms, coupons and other services. You also might consider using a disposable email address service that forwards messages to your permanent account. If one of the disposable addresses begins to receive spam, you can shut it off without affecting your permanent address.

Also, try not to display your email address in public. That includes on blog posts, in chat rooms, on social networking sites, or in online membership directories. Spammers use the web to harvest email addresses.

Check privacy policies and uncheck boxes.

Check the privacy policy before you submit your email address to a website. See if it allows the company to sell your email to others. You might decide not to submit your email address to websites that won't protect it.

When submitting your email address to a website, look for pre-checked boxes that sign you up for email updates from the company and its partners. Some websites allow you to opt out of receiving these mass emails.

Choose a unique email address.

Your choice of email addresses may affect the amount of spam you receive. Spammers send out millions of messages to probable name combinations at large ISPs and email services, hoping to find a valid address. Thus, a common name such as jdoe may get more spam than a more unique name like j26d0e34. Of course, there is a downside - it's harder to remember an unusual email address.

How Can I Help Reduce Spam for Everyone?

Hackers and spammers troll the internet looking for computers that aren't protected by up-to-date security software. When they find unprotected computers, they try to install hidden software – called malware – that allows them to control the computers remotely.

Many thousands of these computers linked together make up a "botnet ," a network used by spammers to send millions of emails at once. Millions of home computers are part of botnets. In fact, most spam is sent this way.

Don't let spammers use your computer.

You can help reduce the chances that your computer will become part of a botnet:

- **Use good** computer security practices **and disconnect from the internet when you're away from your computer**. Hackers can't get to your computer when it's not connected to the internet.
- **Be cautious about opening any attachments or downloading files from emails you receive**. Don't open an email attachment — even if it looks like it's from a friend or coworker — unless you are expecting it or you know what it is. If you send an email with an attached file, include a message explaining what it is.
- **Download free software only from sites you know and trust**. It can be appealing to download free software – like games, file-sharing programs, and customized toolbars. But remember that free software programs may contain malware.

Detect and get rid of malware.

It can be difficult to tell if a spammer has installed malware on your computer, but there are some warning signs:

- Your friends may tell you about weird email messages they've received from you.
- Your computer may operate more slowly or sluggishly.
- You may find email messages in your sent folder that you didn't send.

If your computer has been hacked or infected by a virus, disconnect from the internet right away. Then take steps to remove malware.

OnGuard Online.gov

Privacy Policy
Contact Us
File a Complaint
About Us

  Homeland Security

 Department of Commerce

 Information Assurance Support Environment

© U.S. Federal Trade Commission

FIGURE 8-7 U.S. Federal Trade Commission Spam information site home page

The CAN-SPAM law has allowed U.S. prosecutors to bring a number of successful cases against spammers, including cases in which damages were assessed in the hundreds of millions of dollars. Some of the more notorious spammers have been sent to prison. Spammers' appeals of these decisions, usually based on the argument that spam is protected speech under the First Amendment, have been consistently rejected by the courts.

These successes have helped stem the tide of spam over the past few years. However, many spammers use mail servers located in countries that do not have (and that are unlikely to adopt) antispam laws. As you learned in Chapter 7, the issues of jurisdiction can be unclear for businesses that operate online. Even if a plaintiff is successful in court, enforcement of court-ordered fines or collection of damages can be difficult. Spammers can also evade cease-and-desist orders because they can move their operations from one server to another in minutes. Many spammers forward mail through servers that they have hijacked (you will learn more about threats to servers in Chapter 10).

In a decision that disappointed many in the information technology community, the FTC refused to create a do-not-spam list that would have been modeled after its do-not-call list, which has been reasonably successful in limiting marketers' phone calls.

Legal solutions to the spam problem have achieved only limited success in reducing spam because it is expensive for governments to prosecute spammers. To become cost effective, prosecutors must be able to identify spammers easily (to reduce the cost of bringing an action against them) and must have a greater likelihood of winning the cases they file (or must see a greater social benefit to winning). The best way to make spammers easier to find has been to make technical changes in the e-mail transport mechanism in the Internet's infrastructure.

Technical Solutions

The Internet was not designed to do many of the things it does today. It was not designed to be secure, to process transactions, or to handle billions of e-mail messages. As you learned in Chapter 2, Internet e-mail was an incidental afterthought in a system designed to transfer large files from one researcher to another. As it was originally designed, and as it operates today, the Internet did not include any mechanisms for ensuring that the identity of an e-mail sender would always be known to the e-mail's recipient.

At least one technical strategy for fighting spam exploits a weakness in the original design of the Internet. The Internet protocol that governs communication among servers on the Internet (including e-mail servers) was designed to be a polite set of rules. When one computer on the Internet sends a message to another computer, it will wait to receive an acknowledgment that the message has been received before sending more messages. In the ordinary course of Internet communications, the acknowledgment messages come back in far less than a second. If a computer is set to send the acknowledgment back more slowly, the originating computer will slow down because it must continue to scan for the acknowledgment (which consumes some of its processing power) and it will not send any more messages to that address until it does receive the acknowledgment.

To use this characteristic of the Internet messaging rules to counter spam, the defending company must develop a way to identify computers that are sending spam. Some vendors, such as IBM, sell software and access to a large database that tracks such computers continually. Other vendors sell software that identifies multiple e-mail messages coming from a single source in rapid succession (as would happen if a spammer were sending spam to everyone at a particular company). Once the spamming computer is identified, the software delays sending the message acknowledgments. It can also launch a return attack, sending e-mail messages back to the computer that originated the suspected spam. This practice is called **teergrubing**, which is from the German word for "tar pit." The objective is to ensnare the spam-sending computer in a trap that drags down its ability to send spam. Although many organizations use teergrubing as part of their spam defense strategy, some are concerned that launching a counterattack might violate laws that were enacted, ironically enough, to punish spammers.

Most industry observers agree that the ultimate solution to the spam problem will come when new e-mail protocols are adopted that provide absolute verification of the source of each e-mail message. This will require all mail servers on the Internet to be upgraded. The new protocols have not yet been written, so this solution is several years away.

Proposals for identification standards have been made by Time Warner's AOL division, Microsoft, Yahoo!, and other companies and organizations. The Internet Engineering Task Force (IETF) working group that is responsible for e-mail standards has rejected some of these proposals, but has stated its commitment to working out a set of standards that will accomplish sender authentication.

The most effective technical solutions to the spam problem have been the coordinated efforts of large Internet users to identify the sources of spam and to block them. As more and more spamming activity moves to countries that have lax regulations regarding spam, it has become easier to identify and block these users. Most industry experts agree that the recent reduction in the levels of spam are a result of these efforts. You can learn more about current developments in spam control and find the most recent statistics on the percentage of e-mail that is spam at the **Symantec Intelligence Reports** Web site.

WEB SITE UTILITY PROGRAMS

In addition to Web server software, people who develop Web sites work with a number of utility programs, or tools. TCP/IP supports a wide variety of these utility programs. Some of these programs run on the Web server itself, while others run on the client computers that Web developers use when they are creating Web sites. E-mail was one of the earliest Internet utility programs and it has become one of the most important. In earlier chapters, you learned how companies are using e-mail as a key element in their electronic commerce strategies. You will learn about several of these programs and see examples of how they work.

Finger and Ping Utilities

Finger is a program that runs on UNIX operating systems and allows a user to obtain some information about other network users. A Finger command yields a list of users who are logged on to a network, or reports the last time a user logged on to the network. Many organizations have disabled the Finger command on their systems for privacy and security reasons. For example, if you send a Finger command to a server at www.microsoft.com, you receive no response. Some e-mail programs have the Finger program built into them, so you can send the command while reading your e-mail.

A program called **Ping**, short for **Packet Internet Groper**, tests the connectivity between two computers connected to the Internet. Ping provides performance data about the connection between Internet computers, such as the number of computers (hops) between them. It sends two packets to the specified address and waits for a reply. Network technicians sometimes use Ping to troubleshoot Internet connections. Many freeware and shareware Ping programs are available on the Internet.

Tracert and Other Route-Tracing Programs

Tracert (TRACE RouTe) sends data packets to every computer on the path (Internet) between one computer and another computer and clocks the packets' round-trip times. This provides an indication of the time it takes a message to travel from one computer to another and back, ensures that the remote computer is online, and pinpoints any data traffic congestion. Route-tracing programs also calculate and display the number of hops between computers and the time it takes to traverse the entire one-way path between machines.

Figure 8-8 shows a route traced from a Cox Cable network in Connecticut to one of the BBC's Web servers in London using the Tracert program on a Windows PC.

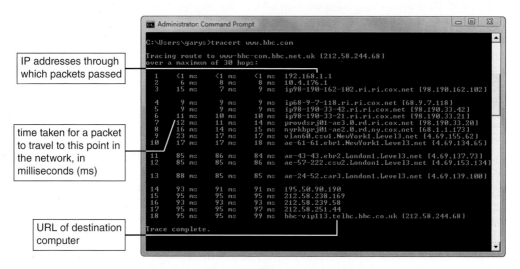

FIGURE 8-8 Tracing a path between two computers on the Internet

By looking at the first column in the figure, you can see that the route included 18 hops and took just under one-tenth of a second (which is 100 milliseconds) to travel the entire length of the transmission path. The Windows Tracert program sends three test packets; the speeds for each packet are shown in milliseconds in the second, third, and fourth columns in the figure. The last column shows either the URL or the IP address of each computer through which the packets passed.

Telnet and FTP Utilities

Telnet is a program that allows a person using one computer to access files and run programs on a second computer that is connected to the Internet. This remote login capability can be useful for running older software that does not have a Web interface. Several Telnet client programs are available as free downloads on the Internet, and Microsoft Windows systems include a Telnet client called Telnet.exe. Telnet lets a client computer give commands to programs running on a remote host, allowing for remote troubleshooting or system administration. Telnet programs use a set of rules called the **Telnet protocol**. Some Web browsers function as a Telnet client. A user can enter "telnet://" followed by the domain name of the remote host. As more companies place information on Web pages, which are accessible through any Web browser, the use of Telnet will continue to decrease.

The **File Transfer Protocol (FTP)** is the part of the TCP/IP rules that defines the formats used to transfer files between TCP/IP-connected computers. FTP can transfer files one at a time, or it can transfer many files at once. FTP also provides other useful services, such as displaying remote and local computers' directories, changing the current client's or server's active directory, and creating and removing local and remote directories. FTP uses TCP and its built-in error controls to copy files accurately from one computer to another.

Accessing a remote computer with FTP requires that the user log on to the remote computer. A number of FTP client programs exist; however, many people just use their Web browser software. Typing the protocol name, ftp://, before the domain name of the remote computer establishes an FTP connection. Users who have accounts on remote computers can log on to their accounts using the FTP client. FTP establishes contact with the remote computer and logs on to the account on that computer.

An FTP connection to a computer on which the user has an account is called **full-privilege FTP**. Another way to access a remote computer is called anonymous FTP. **Anonymous FTP** allows the user to log on as a guest. By entering the username "anonymous" and an e-mail address as a password, users can read and copy files that are stored on the remote computer.

Indexing and Searching Utility Programs

Search engines and indexing programs are important elements of many Web servers. Search engines or search tools search either a specific site or the entire Web for requested documents. An indexing program can provide full-text indexing that generates an index for all documents stored on the server. When a browser requests a Web site search, the search engine compares the index terms to the requester's search term to see which documents contain matches for the requested term or terms. More advanced search engine software (such as that used by the popular search engine site Google) uses complex relevance ranking rules that consider things such as how many other Web sites link to the target site. Many Web server software products also contain indexing software. Indexing software can often index documents stored in many different file formats.

Data Analysis Software

Web servers can capture visitor information, including data about who is visiting a Web site (the visitor's URL), how long the visitor's Web browser viewed the site, the date and time of each visit, and which pages the visitor viewed. This data is placed into a Web **log file**. As you can imagine, the file grows very quickly—especially for popular sites with thousands of visitors each day. Careful analysis of the log file can be fruitful and reveal many interesting facts about site visitors and their preferences. To make sense of a log file, you must run third-party Web log file analysis programs. These programs summarize log file information by querying the log file and either returning gross summary information, or accumulating details that reveal how many visitors came to the site per day, hour, or minute, or which hours of the day were peak loading times. Popular Web log file analysis programs include products by **Adobe SiteCatalyst**, **Urchin from Google**, and **WebTrends**.

Link-Checking Utilities

One function that is important to Web site managers is the ability to check the links on their sites. Over time, the Web sites to which a given page links can change their URLs or even disappear. A **dead link**, when clicked, displays an error message rather than a Web page. Maintaining a site that is free of dead links is vital because visitors who encounter too many dead links on a site might jump to another site. Web-browsing customers are just a click away from going to a competitor's site if they become annoyed with an errant Web link. The undesirable situation of a site that contains a number of links that no longer work is sometimes derisively called **link rot**.

A **link checker** utility program examines each page on the site and reports any URLs that are broken, seem broken, or are in some way incorrect. It can also identify orphan files. An **orphan file** is a file on the Web site that is not linked to any page. Other important site management features include script checking and HTML validation. Some management tools can locate error-prone pages and code, list broken links, and e-mail maintenance results to site managers.

Some Web site development and maintenance tools, such as Adobe's Dreamweaver, include link-checking features. Most link-checking programs, however, are separate utility programs. One of these link-checking programs, Elsop LinkScan, is available in a demo version as a free download. The results of the link checker either appear in a Web browser or are e-mailed to a recipient. Besides checking links, Web site validation programs sometimes check spelling and other structural components of Web pages.

LinxCop is one of several reverse link checkers available. A **reverse link checker** checks on sites with which a company has entered a link exchange program (which you learned about in Chapter 4) and ensures that link exchange partners are fulfilling their obligation to include a link back to the company's Web site.

Remote Server Administration

With **remote server administration** software, a Web site administrator can control a Web site from any Internet-connected computer. It is convenient for an administrator to be able to monitor server activity and manipulate the server from wherever he or she happens to be. LabTech Software and NetMechanic are two companies that sell software that includes remote administration functions along with link-checking, HTML troubleshooting, site-monitoring, and other utility programs that can be useful in managing the operation of a Web site.

WEB SERVER HARDWARE

Organizations use a wide variety of computer brands, types, and sizes to host their online operations. Very small companies can run Web sites on desktop PCs. Most electronic commerce Web sites are operated on computers designed specifically for the task of Web site hosting, however.

Server Computers

Web server computers generally have more memory, larger (and faster) hard disk drives, and faster processors than the typical desktop or notebook PCs with which you are probably familiar. Many Web server computers use multiple processors; very few desktop PCs have more than one processor. Because Web server computers use faster and higher-capacity hardware elements (such as memory and hard disk drives) and use more of these elements, they are usually much more expensive than workstation PCs. Today, a high-end desktop PC with a fast processor, sufficient memory, a large hard disk, and monitor costs between $1000 and $1500. A company might be able to buy a low-end Web server computer for about the same amount of money, but most companies spend between $2000 and $100,000 on an individual Web server and expect it to have a useful lifespan of three to five years. Large organizations that use thousands of servers can spend millions of dollars on their server hardware. Companies that sell Web server hardware, such as Dell, Gateway, Hewlett Packard, and Oracle, all have configuration tools on their Web sites that allow visitors to design their own Web servers.

Although some Web server computers are housed in freestanding cases, most are installed in equipment racks. These racks are usually about 6 feet tall and 19 inches wide. They can each hold from five to ten midrange server computers. An increasingly popular server configuration involves putting small server computers on a single computer board and then installing many of those boards into a rack-mounted frame. These servers-on-a-card are called **blade servers**, and some manufacturers now make them so small that more than 300 of them can be installed in a single 6-foot rack. Each blade server costs between $500 and $6000, depending on its components. Figure 8-9 shows a set of rack-mounted blade servers.

FIGURE 8-9   Rack-mounted blade servers

Recall that the fundamental job of a Web server is to process and respond to Web client requests that are sent using HTTP. For a client request for a Web page, the server program finds and retrieves the page, creates an HTTP header, and appends the HTML document to it. For dynamic pages, the server uses an architecture with three or more tiers that uses other programs, receives the results from the back-end process, formats the response, and sends the pages and other objects to the requesting client program. IP-sharing, or a virtual server, is a feature that allows different groups to share a single Web server's IP address. A **virtual server** or **virtual host** is a feature that maintains more than one server on one machine. This means that different groups can have separate domain names, but all domain names refer to the same physical Web server.

Web Servers and Green Computing

The use of large collections of computers, especially powerful computers such as Web servers, requires significant amounts of electrical power to operate. Although much of this electrical power is used to operate the servers themselves, a substantial portion of it is used to cool the rooms in which the servers reside. Large computers generate tremendous amounts of heat. Efforts to reduce the environmental impact of large computing installations are called **green computing**. Companies that operate large numbers of Web server computers are finding some very interesting ways to minimize the impact of using so much electricity and the heat that it generates.

In 2009, Google opened a server facility in Finland in a building that was previously used as a paper mill. This installation is located near the coastline and is built over granite tunnels that draw in seawater that Google uses instead of electric-powered air conditioning to dissipate the heat generated by the servers. The low average temperatures in Finland reduce the overall need for cooling as well.

In 2011, Facebook began work on a Web server facility in Lulea, Sweden. Lulea is just 60 miles south of the Arctic Circle and Facebook plans to use the outside air to cool its servers. A nearby river has a dam with hydroelectric power generation that can provide inexpensive electricity to operate the servers themselves. Facebook will use these servers to handle the increased traffic resulting from its expansion into European markets.

Hewlett-Packard uses cool air available in the high altitudes of the Rocky Mountains in its Fort Collins, Colorado, server facility. FedEx and Harris Corporation have also used natural cooling in their U.S. Web server installations.

All of these efforts reduce the impact that online businesses have on the planet's limited energy resources. They can also provide substantial energy cost savings for the companies that use these strategies.

Web Server Performance Evaluation

Benchmarking Web server hardware and software combinations can help in making informed decisions for a system. **Benchmarking**, in this context, is testing that is used to compare the performance of hardware and software.

Elements affecting overall server performance include hardware, operating system software, server software, connection speed, user capacity, and type of Web pages being delivered. When evaluating Web server performance, a company should know exactly what factors are being measured and ensure that these are important factors relative to the expected use of the Web server. Another factor that can affect a Web server's performance is the speed of its connection. A server on a T3 connection can deliver Web pages to clients much faster than on a T1 connection.

The number of users the server can handle is also important. This can be difficult to measure because results are affected by the bandwidth of the Internet connection between the server and the client, and by the sizes of the Web pages delivered. Two factors to evaluate when measuring a server's Web page delivery capability are throughput and response time. **Throughput** is the number of HTTP requests that a particular hardware and software combination can process in a unit of time. **Response time** is the amount of time a server requires to process one request. These values should be well within the anticipated loads a server can experience, even during peak load times.

One way to choose Web server hardware configurations is to run tests on various combinations, remembering to consider the system's scalability. Of course, you need to have the hardware and software set up to do this, so it is difficult to evaluate potential configurations that you have not yet purchased. Independent testing labs such as **Mindcraft** test software, hardware systems, and network products for users. Its site contains reports and statistics comparing combinations of application server platforms, operating systems, and Web server software products. A not-for-profit company that develops benchmarks for servers is the **Standard Performance Evaluation Corporation**.

Anyone contemplating purchasing a server that will handle heavy traffic should compare standard benchmarks for a variety of hardware and software configurations. Customized benchmarks can give Web site managers guidelines for modifying file sizes, cache sizes, and other parameters.

Companies that operate more than one Web server must decide how to configure servers to provide site visitors with the best service possible. The various ways that servers can be connected to each other and to related hardware, such as routers and switches, are called **server architectures**.

Web Server Hardware Architectures

Earlier in this chapter, you learned that electronic commerce Web sites can use two-tier, three-tier, or n-tier architectures to divide the work of serving Web pages, administering databases, and processing transactions. Some electronic commerce sites are so large that more than one computer is required within each tier. For example, large electronic commerce Web sites must deliver millions of individual Web pages and process thousands of customer and vendor transactions each day.

Administrators of these large Web sites must plan carefully to configure their Web server computers, which can number in the hundreds or even thousands, to handle the daily Web traffic efficiently. These large collections of servers are called **server farms** because the servers are often lined up in large rooms, row after row, like crops in a field. One approach, sometimes called a **centralized architecture**, is to use a few very large and fast computers. A second approach is to use a large number of less-powerful computers and divide the workload among them. This is sometimes called a **distributed architecture** or, more commonly, a **decentralized architecture**. These two different approaches to Web site architecture are shown in Figure 8-10.

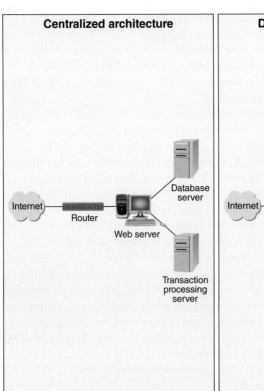
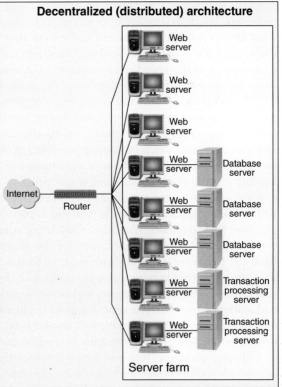

FIGURE 8-10   Centralized and decentralized Web site architectures

Each approach has benefits and drawbacks. The centralized approach requires expensive computers and is more sensitive to the effects of technical problems. If one of the few servers becomes inoperable, a large portion of the site's capability is lost. Thus, Web sites with centralized architectures must have adequate backup plans. Any server problem, no matter how small, can threaten the operation of the site. The decentralized architecture spreads that risk over a large number of servers. If one server becomes inoperable, the site can continue to operate without much degradation in capability. The smaller servers used in the decentralized architecture are less expensive than the large servers used in the centralized approach. That is, the total cost of 100 small servers is usually less than the cost of one large server with the same capacity as the 100 small servers. However, the decentralized architecture does require additional hubs or switches to connect the servers to each other and to the Internet. Most large decentralized sites use load-balancing systems, which cost additional money, to assign the workload efficiently. Load-balancing systems are described in the following section of this chapter.

LEARNING FROM FAILURES

Web Servers at eBay

The online auction site eBay is very popular, as you have learned in earlier chapters. Indeed, it is so popular that its Web servers deliver hundreds of millions of pages per day. These pages are a combination of static HTML pages and dynamically generated Web pages. The dynamic pages are created from queries run against eBay's Oracle database, in which it keeps all of the information about all auctions that are under way or have closed within the most recent 30 days. With millions of auctions under way at any moment, this database is extremely large. The combination of a large database and high-transaction volume makes eBay's Web server operation an important part of the company's success and a potential contributor to its failure. The servers at eBay failed more than 15 times during the first five years (1995–2000) of the company's life. The worst series of failures occurred during May and June of 2000, when the site went down four times. One of these failures kept the site offline for more than a day—a failure that cost eBay an estimated $5 million. The company's stock fell 20 percent in the days following that failure.

At that point, eBay decided it needed to make major changes in its approach to Web server configuration. Many of eBay's original technology staff had backgrounds at Oracle, a company that has a tradition of selling large databases that run on equally large servers. Further, the nature of eBay's business—any visitor might want to view information about any auction at any time—led eBay management initially to implement a centralized architecture with one large database residing on a few large database server computers. It also made sense to use similar hardware to serve the Web pages generated from that database.

In mid-2000, following the worst site failure in its history, eBay decided to move to a decentralized architecture. This was a tremendous challenge because it meant that the single large auction information database had to be replicated across groups, or clusters, of Web and database servers. However, eBay realized that using just a few large servers had made it too vulnerable to the failure of those machines. Once eBay completed the move to decentralization, it found that adding more capacity was easier. Instead of installing and configuring a large server that might have represented 15 percent or more

Continued

of the site's total capacity, clusters of six or seven smaller machines could be added that represented less than one percent of the site's capacity. Routine periodic maintenance on the servers also became easier to schedule.

The lesson from eBay's Web server troubles is that the architecture should be carefully chosen to meet the needs of the site. Web server architecture choices can have a significant effect on the stability, reliability, and, ultimately, the profitability of an electronic commerce Web site.

Load-Balancing Systems

A **load-balancing switch** is a piece of network hardware that monitors the workloads of servers attached to it and assigns incoming Web traffic to the server that has the most available capacity at that instant in time. In a simple load-balancing system, the traffic that enters the site from the Internet through the site's router encounters the load-balancing switch, which then directs the traffic to the Web server best able to handle the traffic. Figure 8-11 shows a basic load-balancing system.

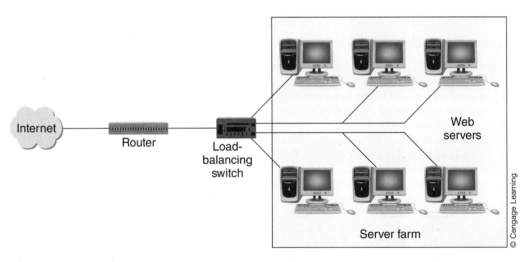

FIGURE 8-11   Basic load-balancing system

In more complex load-balancing systems, the incoming Web traffic, which might enter from two or more routers on a larger Web site, is directed to groups of Web servers dedicated to specific tasks. In the complex load-balancing system that appears in Figure 8-12, the Web servers have been gathered into groups of servers that handle delivery of static HTML pages, servers that coordinate queries of an information database, servers that generate dynamic Web pages, and servers that handle transactions.

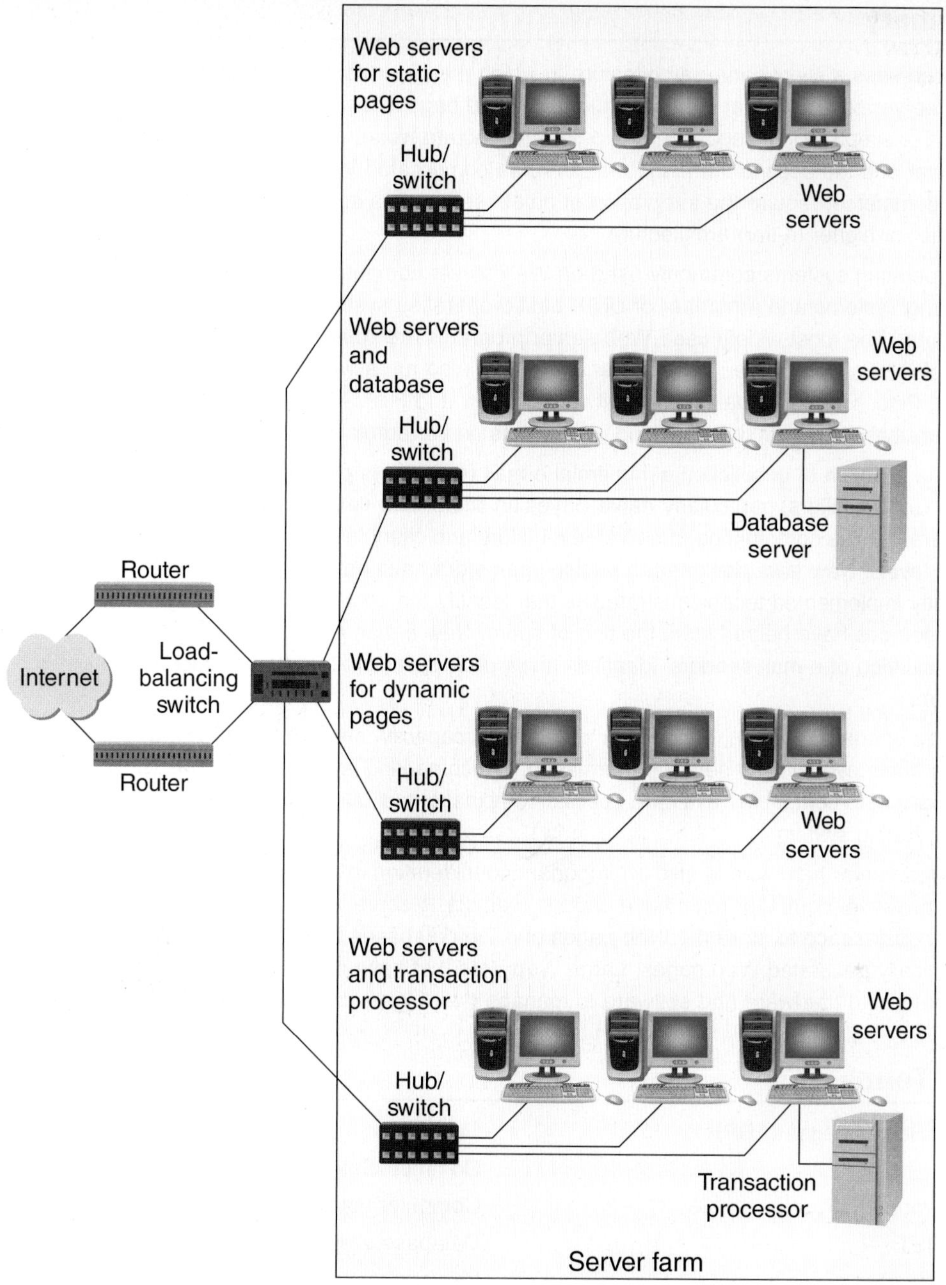

FIGURE 8-12   Complex load-balancing system

Load-balancing switches and the software that helps them do their work usually cost about $2000 for a simple system. Larger and more complex systems usually cost $20,000 to $40,000.

Summary

The Web uses a client/server architecture in which the client computer requests a Web page and a server computer that is hosting the requested page locates and sends a page back to the client. For simple HTTP requests, a two-tier architecture works well. The first tier is the client computer and the second tier is the server. More complicated Web interactions, such as electronic commerce, require the integration of databases and payment-processing software in a three-tier or higher (n-tier) architecture.

Operating systems commonly used on Web server computers include Microsoft server operating systems and a number of UNIX-based operating systems such as SunOS, FreeBSD, and Linux. The most widely used Web server programs are Apache HTTP Server and Microsoft Internet Information Server. Web server computers also run a variety of utility programs such as Finger, Ping, Tracert, e-mail server software, Telnet, and FTP. Most Web servers also have software that helps with link checking and remote server administration tasks.

The problem of unsolicited commercial e-mail (spam) has grown dramatically in recent years. Content filters, particularly naïve Bayesian filters, can deal with the problem. Organizations are using a combination of server-level filters and client-level filters to reduce spam to tolerable levels. New laws designed to punish spammers have not stemmed the tide of spam. Recently implemented technical strategies that identify the source of spam e-mails and block those sources have helped stem the tide of spam. New e-mail protocols that provide absolute authentication of e-mail senders' identities show promise as tools to help reduce future levels of spam.

The operating system, connection speed, user capacity, and the type of pages that the site serves affect overall Web server performance. Benchmarking software and consulting firms that use it can help companies evaluate specific combinations of Web server hardware, software, and operating systems.

Web server hardware is also an important consideration in the design of an online business site. The server computer must have enough memory to serve Web pages to all site visitors and enough disk space to store the Web pages and the databases that store the elements of dynamically generated Web pages. Large Web sites that have many Web server computers use load-balancing hardware and software to manage their high-activity volumes.

Key Terms

Active Server Pages (ASP)	ColdFusion
AJAX	Common Gateway Interface (CGI)
Anonymous FTP	Computer virus
ASP.NET	Database server
Attachment	Dead link
Bayesian revision	Decentralized architecture
Benchmarking	Distributed architecture
Black list spam filter	Dynamic content
Blade server	Dynamic page
Centralized architecture	E-mail server
Challenge-response	Entity body
Client-level filtering	False positive
Client-side scripting	File Transfer Protocol (FTP)

Finger

Full-privilege FTP

Green computing

Hypertext Preprocessor (PHP)

JavaServer Pages (JSP)

Java servlet

Link checker

Link rot

Load-balancing switch

Log file

Naïve Bayesian filter

N-tier architecture

Orphan file

Ping (Packet Internet Groper)

Platform neutrality

Python

Remote server administration

Request header

Request line

Request message

Response header field

Response header line

Response message

Response time

Reverse link checker

Ruby on Rails

Server

Server architecture

Server farm

Server-level filtering

Server software

Server-side scripting

Spam

Static page

Teergrubing

Telnet

Telnet protocol

Three-tier architecture

Throughput

Tracert

Transaction server

Two-tier client/server architecture

Virtual host

Virtual server

Virus

White list spam filter

Review Questions

1. Write a paragraph in which you explain the advantages of having platform neutrality in a network.

2. Briefly describe the two approaches used to create dynamic Web pages and, in one or two paragraphs, explain how each works.

3. In a paragraph or two, outline the different meanings of the term "Web server."

4. In about 100 words, list and briefly explain the elements contained in a Web client request message.

5. Many large organizations use Apache Web server software even though it is not sold by a well-known company that provides ongoing support. Write a paragraph in which you explain why organizations are willing to do this.

6. Write a paragraph in which you explain the difference between client-level and server-level e-mail filtering.

7. In a paragraph, describe how a challenge-response content-filtering technique works to limit spam e-mails from being delivered.

8. In about 100 words, explain the purpose and use of a route-tracing program.

9. In a paragraph, explain how a blade server differs from any other server computer.

10. Identify the benefits and costs of using a decentralized instead of a centralized server architecture in an online business operation. Summarize your findings in about 100 words.

11. Most electronic commerce Web sites use a three-tier client/server architecture. In about 100 words, explain why they do and briefly describe what happens in the third tier of most electronic commerce Web sites.

Exercises

1. Using your favorite search engine, find at least two companies that provide technical support for users of the Apache Web server software. Learn what services they provide and, if possible, what they charge. Review their Web sites to learn more about the companies and summarize your findings in a report of about 200 words.

2. You created a Web site for International Paper Products and Pulp, complete with links to other pages on your site and to pages on the Internet. Bob Pardee, your supervisor, wants you to check periodically that the links on the corporate site are still valid. Instead of purchasing and installing a link-checking program, you decide to investigate online link checkers (Web sites that allow you to enter a Web site's root or home address and then check all the links that emanate from that site). Use W3C Link Checker or Elsop LinkScan Quick Check to check the links on any site of your choice. Print a few pages of the report and be prepared to turn them in to your instructor. Be patient. The program can take some time to complete its work—especially on a Web page that has a large number of links.

3. Assume you are planning Web server computer capacity for a business that has 5000 business customers and sells about 1200 different products. Each customer buys between 10 and 50 items two or three times each month. The business has 300 employees, 200 of whom regularly interact with the company's online sales system. As you learn more about disk storage options for the Web server, you learn that many companies selling Web servers offer a configuration option for controlling those computers' disk drives called "RAID." Using the Web and your library, investigate the purpose of RAID controllers. Learn what these controllers do and how they do it. Form an opinion regarding the suitability of these controllers in the Web server you are planning. Summarize your findings in a 300-word briefing report suitable for presentation to a nontechnical manager.

4. Using the Web and your library, identify at least three different load-balancing products. For each of those products, write a paragraph in which you describe the product, its features, any limitations you can identify, and its price. Learn what these controllers do and how they do it. Summarize your findings in a 400-word briefing report suitable for presentation to a nontechnical manager.

Cases

C1. Microsoft and the People's Republic of China

Software piracy has been a major challenge for software makers such as Microsoft that want to sell software in the global marketplace. Laws that protect intellectual property vary from country to country, and the laws in many countries provide little or no protection. Governments in developing countries are reluctant to increase the protections afforded by their intellectual property laws because they see no point in passing laws that protect the profits of foreign corporations by imposing higher costs on their struggling local businesses and citizens. In the late 1990s, after years of holding firm on its global pricing, Microsoft began to offer significant discounts on its software to governments, small businesses, and individuals in developing countries. It also provided discounts on Windows operating systems software that was installed in new personal computers manufactured in developing countries. Microsoft donated software licenses to schools in developing countries. Just as these efforts were beginning to show some results, however, Microsoft faced a new threat to its global market position—open-source software.

Open-source operating system software, such as Linux, gives governments and businesses in developing countries a way to avoid paying any server software licensing fees to Microsoft. In 2000, the Brazilian state of Penambuco became the first governmental entity to pass a law that requires the use of open-source software on all computers used for state business. Shortly thereafter, the Brazilian state of Rio Grande do Sul passed a similar law that requires the use of open-source software in all of the state's offices and in all privately operated utilities. In 2003, IBM realized the potential for open-source consulting business in the country and opened several centers for the development of Linux-based application software in Brazil. Concerned about a Latin American open-source domino effect, Microsoft embarked on a public relations campaign in the region that included increased advertising spending and donations to public schools. In 2002, Peru was considering passing a law that would require public schools to use open-source software. Microsoft founder Bill Gates flew to Peru and, with great public fanfare, donated $550,000 to the schools that would have been affected by the legislation. The law was quietly dropped from consideration by the legislature shortly thereafter. In 2004, Microsoft announced that it would donate $1 billion in cash and software over five years through the United Nations Development Program to not-for-profit organizations in 45 countries.

Most industry observers believe that Microsoft's largest non-U.S. market today is the People's Republic of China (PRC). Although the PRC generates about $300 million in licensing revenue for Microsoft, more than 90 percent of all Microsoft products used in China today are pirated. Bootleg copies of the company's latest products can be purchased on the street for a few dollars. Thus, Microsoft believes that converting users to paid licenses could increase its PRC licensing revenues tremendously. As the PRC moves from being a less-developed country toward becoming a major economic power in the world, Microsoft sees an opportunity to increase its licensing revenue in the country. In the past, Microsoft has used a global antipiracy strategy that relied on identifying users of pirated software and threatening those users with legal action, but the company is changing its approach in developing markets such as Latin America and Asia. In the PRC, Microsoft's near-term goal is to develop a market for full-price software licenses that includes large business and government customers. Its new approach focuses more on recruiting major PRC business organizations as customers and less on sending threatening letters to users of pirated Microsoft software.

In developing its business in the PRC, Microsoft faces a number of challenges. Juliet Wu, former general manager of Microsoft China, published a book in 2000 that was highly critical of the company. The book was widely read in the PRC and received many good reviews. PRC officials have often criticized Microsoft for many things ranging from high prices to the company's use of Taiwanese programmers (the PRC does not officially recognize Taiwan as an independent nation separate from the PRC). Government officials in the PRC are also concerned about security. Microsoft has always maintained that the code to its software products is a trade secret and has refused to allow its publication or distribution. Companies that develop software that runs on Microsoft Windows, for example, must sign a nondisclosure agreement with Microsoft to obtain information they need about how Windows operates so they can make their software compatible with it. Many PRC officials believe that Microsoft, as a U.S.-based company, might include secret code in its software that would allow the U.S. government to enter PRC government computers undetected in a time of international conflict or war. At a very basic level, the ideology of the PRC's socialist government is a polar opposite to the highly competitive capitalist principles that have driven Microsoft to success. But the greatest challenge that Microsoft must overcome in the PRC is the attraction of open-source software.

Open-source programs' code is public; thus, it cannot include secret code. The PRC has established a record of preference for open-source software; for example, its growing personal computer manufacturing industry ships most of its domestic production with the Linux operating system. Also, the PRC's national lottery, post office, and social security systems all run on Linux operating systems. Since 2003, the Procurement Center of the State Council has required that any computer purchased by the government must be delivered with PRC-produced software only.

In the face of these challenges, Microsoft has worked hard to deliver its message that open-source software can result in higher total costs because even though it is free, it requires more effort to install, maintain, and update than Microsoft products. In large organizations, this effort results in extra hours worked and thus, extra costs. Microsoft also argues that open-source software's publicly available program code makes it a greater security risk. According to Microsoft, attackers can easily learn how any open-source program works and develop strategies for attacking the software when it is running on publicly accessible computers, such as Web servers.

Required:

1. Assume that you are on the staff of a PRC legislator. Outline the arguments that you would use to support a law that required all government agencies to use only open-source software on their Web servers.

2. Assume that you are working for the marketing department of Microsoft China. Develop a detailed list of briefing points that would help your salespeople convince top executives of large PRC companies to use Windows operating system software on their Web servers.

3. Assume that you are working for the business system's analysis department in IBM's PRC division, which offers both Microsoft Windows and Linux consulting services to PRC businesses and government offices. Develop a checklist that IBM analysts could use in consulting projects that could help advise clients as they make a choice between Windows or Linux operating system software for their Web servers.

4. Companies such as RedHat, Novell (with its SuSE distribution), and others offer Linux operating system software for sale. Although Linux is available at no cost from various sources, these companies charge a fee for installation and configuration help. They also offer service contracts to help users maintain and upgrade the software on a continuing basis. Briefly outline the strategies that these companies might use to expand their market share in the PRC.

Note: Your instructor might assign you to a group to complete this case, and might ask you to prepare a formal presentation of your results to your class.

C2. Random Walk Shoes

Amy Lawrence, the owner of Random Walk Shoes, has asked you to help her as she launches her company's first Web site. In college, Amy was a business major with an artistic bent. She helped to pay her way through college by decorating sneakers with her hand-painted designs. Her business grew through word of mouth and through her participation in crafts fairs. By the time she earned her degree, Amy was running a successful business from her dorm room.

Amy expanded her sales efforts to include crafts fairs in nearby towns. She hired two college students to work for her, and she convinced several area gift shops to stock samples of her merchandise. The gift shops were not an ideal retail outlet for her products, however. Most people who want to buy decorated sneakers want to choose specific designs or have special designs created just for them. Customers also want to choose the specific shoes on which the design is placed. One of Amy's student workers suggested that she consider selling her products on the Web.

Realizing that the Web would give Random Walk Shoes a chance to reach a much wider audience and would allow customers to choose design-shoe combinations, Amy began gathering information and developing estimates about her planned Web activity. She bought a digital camera and took several hundred pictures of shoes, designs, and shoe-design combinations. She then hired a local Web designer to create sample pages for the Web site, including catalog pages that contained the digital images.

When the Web designer had completed a prototype of the site, Amy worked with the designer to calculate page sizes (including the images). The average page size was 100 KB. Amy and her employees then navigated the prototype site several hundred times to develop an estimate of how many pages an average visitor would download. They concluded that an average site visitor would visit 23 pages during each visit. Amy worked with the Web designer to develop estimates of the activity they expect to occur on the Web site during its first two years of operation. These estimates include:

- The database of Web page information (including the images) will require about 400 MB of disk space.
- The database management software itself will require about 200 MB of disk space.
- The shopping cart software will require about 200 MB of disk space.
- About 8000 customers will visit the site during the first month, and site traffic will grow about 20 percent each month during the first two years.
- The site should accommodate a peak traffic load of 3000 visitors at one time.

Amy wants to include features on the site that are similar to those found on competing sites (a list of links to businesses that sell customized shoes on the Web is included in the Online Companion for your reference). Amy wants the site to provide a good experience for visitors. If the site is successful, it will generate sufficient revenue to allow an upgrade after two years. However, she does not want to spend more money than is necessary to get the site up and keep it running for the next two years.

Required:

1. Determine the features and capacities (RAM, disk storage, processor speed) that Amy should include in the Web server computer she will need for her site. Summarize your purchase recommendation in a one-page memorandum to Amy. You may include information from vendors' sites (such as **Dell**, **Hewlett Packard**, or **Oracle**) as an appendix to your memorandum.

2. Consider the advantages and disadvantages of each major operating system that Amy might use on the new Web server computer. In a one-page memorandum to Amy, make a specific recommendation and support it with facts and a logical argument. If you do not believe that one operating system is clearly superior for this application, explain why.

3. Consider the advantages and disadvantages of each major Web server software package for accomplishing the goals that Amy has for this site. In a one-page memorandum to Amy, make a specific recommendation regarding which Web server software package she should use. Provide an explanation that supports your recommendation.

Note: Your instructor might assign you to a group to complete this case and might ask you to prepare a formal presentation of your results to your class.

For Further Study and Research

Andrews, P. 2003. "Courting China," *U.S. News & World Report*, November 24, 44–45.

Ante, S. 2001. "Big Blue's Big Bet on Free Software," *Business Week*, December 10, 78–79.

Asay, M. 2007. "Study: 95 Percent of All E-mail Sent in 2007 Was Spam," December 12. (http://www.cnet.com/8301-13505_1-9831556-16.html)

Babcock, C. 2007. "Linux on Half of All New Servers? Red Hat's Got Plans," *Information Week*, November 12, 30.

Bradner, S. 2008. "Irrelevant Victories in the War on Spam," *Network World*, March 24, 30.

Business Week Online. 2004. "China and Linux: Microsoft, Beware!" November 15.

Chabrow, E. 2005. "In The Fight Against Spam, A Few Knockouts: Microsoft Wins $7 Million Spam Settlement; Complaints From AOL Members Drop 85%," *InformationWeek*, August 15, 34.

Chen, L. 2008. "Four Tips on Load Balancing," *Communications News*, 45(5), May, 14.

The Computer & Internet Lawyer. 2009. "Court Orders Spammers to Give Up $3.7 Million in US SAFE Web Case," 26(9), September, 26–27.

Epstein, J. 2004. "Standing Up to Redmond," *Latin Trade*, 12(6), June, 19.

Galli, P. 2004. "New IBM Unit to Target Emerging Markets," *eWeek*, 21(30), July 26, 9–10.

Graham, P. 2003. "Better Bayesian Filtering," Paul Graham, January. (http://www.paulgraham.com/better.html)

Gross, G. 2004. "Judge Awards ISP $1 Billion in Spam Damages," *Computerworld*, December 20. (http://www.computerworld.com/governmenttopics/government/legalissues/story/0,10801,98421,00.html)

Henderson, N. 2011. "Noise Filter: Google's Seawater-cooled Finland Data Center," *Web Host Industry Review*, June 6. (http://www.thewhir.com/web-hosting-news/noise-filter-googles-seawater-cooled-finland-data-center)

Hess, K. 2012. "The Seven Best Servers for Linux," *ServerWatch*, January 6. (http://www.serverwatch.com/server-trends/the-7-best-servers-for-linux.html)

Hitchcock, J. 2009. "Is Spam Here to Stay?" *Information Today*, 26(3), March, 1, 44.

Ibrahim, A. and I. Osman. 2012. "A Behavioral Spam Detection System," *Future Computer, Communication, Control and Automation*, 119, 77–81.

Information Week. 2004. "AOL Reports Big Drop in Spam," December 27. (http://www.informationweek.com/story/showArticle.jhtml?articleID=56200528)

Lakka, S., C. Michalakelis, D. Varoutas, and D. Martakos. 2012. "Exploring the Determinants of the OSS Market Potential: The Case of the Apache Web Server," *Telecommunications Policy*, 36(1), 51–68.

Marsono, M., M. El-Kharashi, and F. Gebali. 2009. "Targeting Spam Control on Middleboxes: Spam Detection Based on Layer-3 E-mail Content Classification," *Computer Networks*, 53(6), April, 835–848.

Moore, J. 2012. "Nginx Edges Microsoft in Server Battle," *Info Boom*, January 9. (http://www.theinfoboom.com/articles/nginx-edges-microsoft-in-server-battle/)

Pavlov, O., N. Melville, and R. Plice. 2005. "Mitigating the Tragedy of the Digital Commons: The Problem of Unsolicited Commercial E-mail," *Communications of the AIS*, 2005(16), 73–90.

PC World, 2005. "Spam Law Test," 23(1), January, 20–22.

Potter, N. 2011. "Facebook Plans Server Farm in Sweden; Cold Is Great for Servers," *ABC News*, October 27. (http://abcnews.go.com/Technology/facebook-plans-server-farm-arctic-circle-sweden/story?id=14826663#.TxJovtSm8vY)

Schafer, S. 2004. "Microsoft's Cultural Revolution," *Newsweek*, June 28, E10–12.

Shen, X. 2005. "Intellectual Property and Open Source: A Case Study of Microsoft and Linux in China," *International Journal of IT Standards & Standardization Research*, 3(1), January–June, 21–43.

Smalley, E. 2011. "2011: The Year Data Centers Turned Green," *Wired*, December 30. (http://www.wired.com/wiredenterprise/2011/12/green-data-centers-of-2011/)

Stone, B. 2009. "Spam Back to 94% of All E-mail," *The New York Times*, March 31. (http://bits.blogs.nytimes.com/2009/03/31/spam-back-to-94-of-all-e-mail/)

Symantec. 2009. *State of Spam*. Mountain View, CA: Symantec. (http://eval.symantec.com/mktginfo/enterprise/other_resources/b-state_of_spam_report_06-2009.en-us.pdf)

Symantec, 2011. *Symantec Intelligence Report*. Mountain View, CA: Symantec. (http://www.symantec.com/connect/sites/default/files/SYMCINT_2011_11_November_FINAL-en.pdf)

Wagner, M. and T. Kemp. 2001. "What's Wrong with eBay?" *InternetWeek*, January 15, 1–2.

White, B. 2008. "New Routers Catch the Eyes of IT Departments," *The Wall Street Journal*, March 25, B7.

Xiaobai, S. 2005. "Developing Country Perspectives on Software: Intellectual Property and Open Source, a Case Study of Microsoft and Linux in China," *International Journal of IT Standards & Standardization Research*, 3(1), January–June, 21–43.

Xinhua, 2004. "Microsoft Teams Up with China's Leading Server and Solutions Supplier," November 9.

CHAPTER 9

ELECTRONIC COMMERCE SOFTWARE

LEARNING OBJECTIVES

In this chapter, you will learn:

- How to find and evaluate Web-hosting services
- What the basic and advanced functions of electronic commerce software are and how they work
- How the size of a business affects its choice of electronic commerce software
- Which electronic commerce software works well for midsize to large businesses
- Which electronic commerce software works well for larger businesses that have an existing information technology infrastructure
- How electronic commerce software works with other software to perform business functions

INTRODUCTION

Many luxury clothing and jewelry items are sold online today. Some are sold directly through the manufacturers' Web sites, but most are sold through well-known retail merchandisers that have online stores. Of course, high-fashion brand goods sell for luxury prices in most cases. Shoppers who would like to buy these items, but cannot always afford them, might look for bargains at outlet stores. **Gilt Groupe** offers an interesting angle on the outlet store idea. At noon each day, its Web site lists a selection of designer clothes and other luxury brand items for sale at deep discounts. The store, which acquires the items through its network of high-end suppliers, sells only the listed items and only for 36 hours (or until an item sells out, which happens frequently). By midnight the next day, the sale is over and a new selection is listed the following noon.

This "limited time" element is designed to create a buying frenzy in which shoppers experience the excitement of a sale combined with the satisfaction of getting a true bargain. Shoppers must apply for a membership to be eligible to view and purchase the sale items, which adds to the feeling of exclusivity.

The operation of this Web site requires software that can display the items in an attractive way, process the sale transactions efficiently, and track information about customers and what they are buying. The tracking element is very important for Gilt because it helps the site negotiate purchases of highly desirable items from name-brand manufacturers. To get a sampling of the best new designs and innovative products, Gilt offers to share with its vendors information it gathers about how many of each item it sells and how rapidly the items sell.

Because Gilt compresses the lifetime of the sale event into 36 hours, it collects data about customer demand that a traditional retailer might not get for weeks, or even months. High-fashion product suppliers find this information to be valuable—so valuable, in fact, that they are willing to sell Gilt a sampling of their inventories at very low prices.

In this chapter, you will learn about the kinds of software that sites like Gilt use to make their revenue models work, including software that enables catalog display of goods, shopping cart functions, and transaction processing activities. In addition, you will learn about the type of software that Gilt uses to analyze sales and transmit that analysis to its buyers and to the suppliers of its luxury goods. That information helps suppliers fine-tune their production so they are making more of the items that are likely to sell better in their regular sales outlets.

WEB HOSTING ALTERNATIVES

When companies need to incorporate electronic commerce components, they may opt to run servers in-house; this approach is called **self-hosting**. This is the option used most often by large companies. Other companies, especially midsize and smaller companies, often decide that a third-party Web-hosting service provider is a better choice than self-hosting. Many small Web stores use a third-party host provider for both Web services and electronic commerce functions, particularly when the Web site is small or the company sells a limited number of products.

As you learned in Chapter 2, a number of companies, called Internet service providers (ISPs), are in the business of providing Internet access to companies and individuals. Many of these companies offer Web-hosting services as well. To distinguish themselves from companies that provide only Internet access services, these hosting service firms sometimes call themselves something other than ISPs. Because the hosting services they offer are designed to help companies conduct electronic commerce, these hosting service firms sometimes call themselves **commerce service providers (CSPs)**. These firms often offer Web server management and rent application software (such as databases, shopping carts, and content management programs) to businesses; thus, these companies also sometimes call themselves **managed service providers (MSPs)** or **application service providers (ASPs)**. Despite the increasing variety of acronyms, many companies that provide some or all of these additional services still call themselves ISPs.

Service providers offer clients hosting arrangements that include shared hosting, dedicated hosting, and co-location. **Shared hosting** means that the client's Web site is on a server that hosts other Web sites simultaneously and is operated by the service provider at its location. With **dedicated hosting**, the service provider makes a Web server available to the client, but the client does not share the server with other clients of the service provider. In both shared hosting and dedicated hosting, the service provider owns the server hardware and leases it to the client. The service provider is responsible for maintaining the Web server hardware and software, and provides the connection to the Internet through its routers and other network hardware. In a **co-location** (also spelled **collocation** and **colocation**) service, the service provider rents a physical space to the client to install its own server hardware. The client installs its own software and maintains the server. The service provider is responsible only for providing a reliable power supply and a connection to the Internet through its routers and other networking hardware. You can find service providers by using your favorite search engine and search terms such as "Web hosting services."

When making Web server–hosting decisions, a company should ask whether the hardware platform and software combination can be upgraded when the traffic on its Web site increases. A company's Web server requirements are directly related to its electronic commerce transaction volume and Web site traffic. The best hosting services provide Web server hardware and software combinations that are **scalable**, which means they can be adapted to meet changing requirements when their clients grow.

BASIC FUNCTIONS OF ELECTRONIC COMMERCE SOFTWARE

Because electronic commerce sites vary so greatly in terms of size, purpose, audience, and other factors, a vast range of software and hardware products are available for building electronic commerce sites. Sites with minimal needs can use externally hosted stores that provide software tools to build an online store on a host's site. At the other end of the range are sophisticated electronic commerce software suites that can handle high-transaction volumes and include a broad assortment of features and tools.

The type of electronic commerce software an organization needs depends on several factors, with size and budget being the primary drivers. One of the most important factors is the expected size of the enterprise and its projected traffic and sales. A high-traffic electronic commerce site with thousands of catalog inquiries each minute requires different software than a small online shop selling a dozen items. Another determining factor is budget. Creating an online store can be much less expensive than building a chain of retail stores. The start-up cost of an electronic commerce operation can be much lower than the cost of creating a brick-and-mortar sales and distribution channel that includes warehouses and multiple retail outlets. A traditional store requires a physical location with leases, employees, utility payments, and maintenance. The cost of creating the infrastructure for an online business can be much lower.

Another early decision is whether the company should use an external host or host the electronic commerce site in-house. Companies that have an existing information technology (IT) staff of programmers, Web designers, and network engineers are more likely to choose an in-house hosting approach. If a company does not have or cannot easily hire people with the skills required to set up and maintain an electronic commerce site, it can outsource all or part of the job to a service provider. Companies that are located outside major metropolitan areas and want to host sites themselves must also determine whether their Internet connections have sufficient bandwidth to handle the volume of activity their business might generate. In many cases, these companies find that they are not close enough to a major Internet access point or that their connections do not have sufficient bandwidth to handle large volumes of traffic efficiently. Even if these companies have employees with the necessary skills, they might decide to use a service provider to host their electronic commerce sites. All electronic commerce software must provide the following elements:

- A catalog display
- Shopping cart capabilities
- Transaction processing

Larger and more complex electronic commerce sites also use software that adds other features and capabilities to the basic set of commerce tools. These additional software components can include:

- Middleware that integrates the electronic commerce system with existing company information systems that handle inventory control, order processing, and accounting
- Enterprise application integration
- Web services
- Integration with enterprise resource planning (ERP) software
- Supply chain management (SCM) software

- Customer relationship management (CRM) software
- Content management software
- Knowledge management software

Capabilities required by most online business sites are described in the following sections. The more advanced functions used by larger and more comprehensive sites are covered later in this chapter.

Catalog Display Software

A catalog organizes the goods and services being sold. To further organize its offerings, a retailer may break them down into departments. As in a physical store, merchandise in an online store can be grouped within logical departments to make locating an item, such as a camping stove, simpler. Web stores often use the same department names as their physical counterparts. In most physical stores, each product is kept in only one place. A Web store has the advantage of being able to include a single product in multiple categories. For example, running shoes can be listed as both footwear and athletic gear.

A small commerce site can have a very simple static catalog. A **catalog** is a listing of goods and services. A **static catalog** is a simple list written in HTML that appears on a Web page or a series of Web pages. To add an item, delete an item, or change an item's listing, the company must edit the HTML of one or more pages. Larger commerce sites are more likely to use a dynamic catalog. A **dynamic catalog** stores the information about items in a database, usually on a separate computer that is accessible to the server that is running the Web site itself. A dynamic catalog can feature multiple photos of each item, detailed descriptions, and a search tool that allows customers to search for an item and determine its availability. The software that implements a dynamic catalog is often included in larger electronic commerce software packages; however, some companies write their own software to link their existing databases of product information to their Web sites. Both types of catalog (static and dynamic) are located in the third tier of the Web site architecture that you learned about in Chapter 8.

In addition to the large online businesses that everyone knows about, there are many smaller online stores that operate successfully and provide specialized products or products that appeal to smaller audiences. Figure 9-1 shows the home page of Teak Wood Patio Furniture, which is such a Web site. As you can see, the company specializes in the sale of a specific type of patio furniture that appeals to a particular small market segment. The site includes information about how teak is harvested in a sustainable manner and shows detailed pictures of the furniture sold by the company. This site uses simple, inexpensive electronic commerce software that provides all of the essential features needed to sell online, including a static catalog.

FIGURE 9-1 Small electronic commerce site

Small online stores (those that sell fewer than 100 items) can often get by with a simple list of products or categories. The organization of the items on the Web site is not particularly important. Companies that offer only a small number of items can provide a photo of each item on the Web page that is a link to more information about the product. A static catalog is sufficient for their needs. Larger electronic commerce sites require the more sophisticated navigation aids and better product organization tools that are a part of dynamic catalogs.

Good sites give buyers alternative ways to find products. Besides offering a well-organized catalog, large sites with many products can provide a search engine that allows customers to enter descriptive search terms, such as "men's shirts," so they can quickly find the Web page containing what they want to purchase. Remember the most important rule of all commerce: Never stand in the way of a customer who wants to buy something.

Shopping Cart Software

In the early days of electronic commerce, shoppers selected items they wanted to purchase by filling out online forms, which required a shopper to manually enter product descriptions and item numbers, along with other information, into online ordering systems. This system was awkward and error-prone for customers ordering more than one or two items at a time. Figure 9-2 illustrates the problems that shoppers faced with forms-based ordering systems.

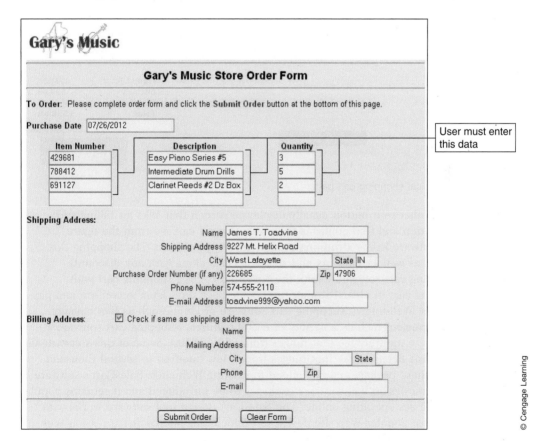

© Cengage Learning

FIGURE 9-2   Using a form to enter an order

Because forms-based ordering is cumbersome and error-prone, only a few of the smallest online stores still use it. Shopping carts are now the standard method for processing sales on most electronic commerce sites. As you learned in Chapter 4, a shopping cart, also sometimes called a shopping bag or shopping basket, keeps track of the items the customer has selected and allows customers to view the contents of their carts, add new items, or remove items. To order an item, the customer simply clicks a button or link near the item's description that indicates "add to cart" or similar language. All of the details about the item, including its price, product number, and other identifying information, are stored automatically in the cart. If a customer later changes his or her mind about an item, he or she can view the cart's contents and remove the unwanted items. When the customer is ready to conclude the shopping session, the click of a button executes the purchase transaction. Figure 9-3 shows a typical shopping cart page at a site that sells tools.

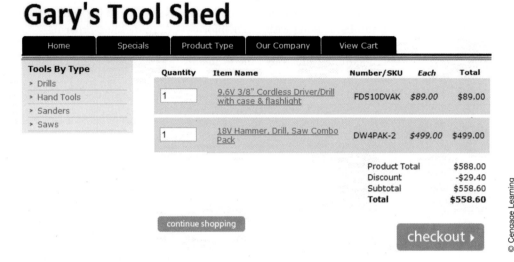

FIGURE 9-3   Typical shopping cart page

Clicking the checkout button usually displays a screen that asks for billing and shipping information and that confirms the order. As you can see from the figure, the shopping cart software keeps a running total of each type of item. The shopping cart calculates a total as well as sales tax and shipping costs along with any discounts.

Some shopping cart software allows the customer to fill a shopping cart with purchases, put the cart in virtual storage, and come back days later to confirm and pay for the purchases. Well-known shopping cart software products such as **BigCommerce**, **SalesCart**, and **Volusion**, include a variety of useful features. Shopping cart software generally requires a monthly fee that ranges from $50 to $300. Some of this software also requires a one-time licensing fee that ranges from a few hundred to several thousand dollars. Most of these packages can be added to existing Web sites. SalesCart's software was one of the first design tool software packages to be introduced and it remains popular with small businesses operating online. The SalesCart design tool software works with several of the most popular Web site design tools, including Microsoft Expression Web, FrontPage, and Adobe Dreamweaver, as described on the SalesCart home page, shown in Figure 9-4.

FIGURE 9-4 SalesCart shopping cart software page

Because the Web is a stateless system that does not retain information from one transmission or session to another, shopping cart software must store information for a specific shopper to retrieve later. Furthermore, it must distinguish one shopper from another so that purchases are not mixed up. One way that shopping cart software uniquely identifies users and stores information about their choices is to create cookies; which, as you learned in earlier chapters, are bits of information stored on a client computer. When a customer returns to the shopping site that issued a particular cookie, the shopping cart software reads the cookie from the customer's computer and uses the information stored there to retrieve the customer's shopping information from the seller's server computer.

If a shopper's Web browser software is set to refuse cookies, sites can use another way to preserve shopping cart information from one browser session to another. Some shopping cart software packages, such as ShopSite, do this by assigning each shopper a temporary number. This number is added to the end of the URL that appears in the browser's address bar and persists as the shopper navigates from one Web site to another. When the customer returns, the URL still contains shopping cart information that the Web server can interpret. When the shopper closes the browser, the temporary number is discarded and is no longer available, even if the customer later reopens the browser and returns to the same Web site. Thus, this approach is only partially successful in retaining shopper information from one site visit to another.

LEARNING FROM FAILURES

PDG Software

PDG Software is a company based in Tucker, Georgia, that sells electronic commerce software to companies that operate small and midsize electronic commerce Web sites. PDG sells shopping cart software, auction software, and a number of other packages. Although it sells some of its software directly to the companies that use it, most of its sales are through resellers—firms that use PDG software as part of Web sites that they design, build, and deliver to customers as complete units.

An attacker discovered a vulnerability in the PDG software that would allow intruders to gain access to PDG shopping cart software installed on a retailer's Web site and open the file that contained customer names, contact information, and credit card numbers. PDG developed a patch that its customers could use to repair the software the same day it found out about the intrusions. PDG posted the patch on its Web site so that companies using the software could download and install the patch. Both PDG and the FBI issued press releases immediately to warn users of the problem with the shopping cart software and encourage them to obtain the patch. Unfortunately, many users of PDG shopping cart software had purchased it as part of a complete electronic commerce Web site. These users were, in many cases, unaware that they were running the PDG shopping cart software.

Because it took so long—several months, in some cases—to find and contact the companies using the software, online offenders were able to exploit this vulnerability and collect thousands of credit card numbers. In most cases such as this, the difficulty of finding the sites that are running the vulnerable software helps slow down the attackers. Unfortunately, in this case, the intruder who discovered the opening also found that entering a specific word in a search engine's search expression would instantly return a list of the thousands of sites running the PDG software.

Most of the Web sites found out about the problem when their customers called them, suspicious because their credit card information had been compromised. The lesson from this failure is that companies that operate electronic commerce Web sites must know the source of the software used in creating and maintaining their sites and must monitor news about the security of that software.

Transaction Processing

Transaction processing occurs when the shopper proceeds to the virtual checkout counter by clicking a checkout button. Then the electronic commerce software performs any necessary calculations, such as volume discounts, sales tax, and shipping costs. At checkout, the customer's Web browser software and the seller's Web server software both switch into a secure state of communication. You will learn more about how Web clients and servers establish these secure communication states in Chapter 10. Figure 9-5 shows how the three key functions of a basic electronic commerce Web site (catalog display, shopping cart, and transaction processing) are combined in the site's architecture.

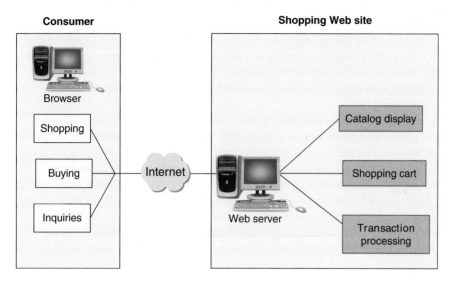

FIGURE 9-5   Basic electronic commerce Web site architecture

Although a basic online store's electronic commerce software can generate reports that summarize sales and inventory shipped, most midsize and larger companies use an accounting software package to record sales and inventory movements. To integrate effectively with accounting software, the electronic commerce software must communicate with that accounting software, which typically runs on other computers in the seller's network. When an item is sold online, the electronic commerce software must communicate that fact to both the sales and inventory management modules in the accounting software.

Computing sales taxes and shipping costs are also important parts of online sales. Sales tax rates and shipping rates can change often, so Web site managers must either monitor and update the rates continually or use software that updates the rates automatically. Shipping companies such as FedEx and UPS offer software to shippers that integrates with electronic commerce software to ensure that the rates they have are current. Other calculation complications include provisions for coupons, special promotions, and time-sensitive offers; for example, "purchase a round-trip ticket before the end of the month and receive a 50 percent discount."

In larger companies, the integration of the Web site's transaction processing into the accounting and operation-control systems of the company can be very complex. The next section discusses some of the advanced functions that larger companies look for in electronic commerce software.

HOW ELECTRONIC COMMERCE SOFTWARE WORKS WITH OTHER SOFTWARE

In this section, you will learn about the features that larger companies need in their electronic commerce software. Although there are exceptions, such as Amazon.com and Buy.com, most large companies that have electronic commerce operations also have substantial business activity that is not related to electronic commerce. Thus, integrating electronic commerce activities into the company's other operations is very important. A basic element of any large company's information system is its collection of databases.

Databases

A **database** is a collection of information that is stored on a computer in a highly structured way. The rules a business establishes about its database structure are carefully thought out and take into account how the company does business (its **business rules**) and how the company can reduce the likelihood that errors and inconsistencies will develop in the database.

A **database manager** (or **database management software**) is software that makes it easy for users to enter, edit, update, and retrieve information in the database. The most commonly used low-end database manager is Microsoft Access. More complex database managers that can handle larger databases and can perform more functions at higher speeds include IBM DB2, Microsoft SQL Server, and Oracle. Companies with very large databases that have operations in many locations must make most (or all) of their data available to users in those locations. Large information systems that store the same data in many different physical locations are called **distributed information systems**, and the databases within those systems are called **distributed database systems**. The complexity of these systems leads to their high cost.

Most companies use commercial database products; however, an increasing number of companies and other organizations are using MySQL, which was developed and is maintained by a community of programmers on the Web. Similar to the Linux operating system you learned about in earlier chapters, **MySQL** is open-source software, even though it was developed by a Swedish company (MySQL AB), which is now owned by Oracle. Oracle sells annual subscriptions for MySQL support and maintenance services.

Except for small sites offering only a few products, companies should determine the level of database support provided by any electronic commerce software they are considering. Most online stores that sell many products use a database that stores product information, including size, color, type, and price details. Usually, the database that serves an online store is the same one that is used by the company's existing sales operations. It is usually better to have one database serving the two sales functions (online and in-store retail, for example) because it eliminates the errors that can occur when running parallel but distinct databases. If a company has existing inventory and product databases, then it should consider only electronic commerce software that supports these systems. The details of database design and operation can become quite complex and are beyond the scope of this book. You can learn more about databases by taking courses in database design and database programming.

Middleware

Larger companies usually establish the connections between their electronic commerce software (that is, their catalog display, shopping cart, and transaction processing software) and their accounting and inventory management databases or applications by using middleware. **Middleware** is software that takes information about sales and inventory shipments from the electronic commerce software and transmits it to accounting and inventory management software in a form that these systems can read. For example, the sales module of an accounting system might be designed to accept the input of a telephone salesperson. The salesperson enters the product numbers, quantities, and shipping method into the sales module by using a keyboard while talking to the customer on the phone. Middleware would extract information about a sale from the Web site's shopping cart software and enter it directly into the accounting software's sales module without requiring that a person re-enter the information.

Some large companies that have sufficient IT staff write their own middleware; however, most companies purchase middleware that is customized for their businesses by the middleware vendor or a consulting firm. Thus, most of the cost of middleware is not the software itself, but the consulting fees needed to make the software work in a given company. Making a company's information systems work together is called **interoperability** and is an important goal of companies when they install middleware.

The total cost of a middleware implementation can range from $50,000 to several million dollars, depending on the complexity of the company's underlying operations and its existing information systems. Major middleware vendors include Broadvision, IBM Tivoli Software, and Informatica.

Enterprise Application Integration

A program that performs a specific function, such as creating invoices, calculating payroll, or processing payments received from customers, is called an **application program**, **application software**, or, more simply, an **application**. An **application server** is a computer that takes the request messages received by the Web server and runs application programs that perform some kind of action based on the contents of the request messages. The actions that the application server software performs are determined by the rules used in the business. These rules are called **business logic**. An example of a business rule is the following: When a customer logs in, check the password entered against the password file in the database.

In many organizations, the business logic is distributed among many different applications that are used in different parts of the organization. In recent years, many IT departments have devoted significant resources to the creation of links among these scattered applications so that the organization's business logic can be interconnected. The creation and management of these links is called **application integration** or **enterprise application integration**. The integration is accomplished by programs that transfer information from one application to another. For example, a program might transfer information from order entry systems in several different divisions to a single accounts receivable and sales system that integrates all enterprise-wide sales activity. In many cases, the data formats in the various programs are different and the transfer programs must edit and reformat the data before transferring it. Increasingly, programmers are using XML data feeds to move data from one application to another in enterprise integration implementations.

Application servers are usually grouped into two types: page-based and component-based systems. **Page-based application systems** return pages generated by scripts that include the rules for presenting data on the Web page with the business logic. Common page-based server systems include Adobe ColdFusion, JavaServer Pages (JSP), Microsoft Active Server Pages (ASP), and Hypertext Preprocessor (PHP).

Page-based systems work well for Web sites with low to moderate activity levels; however, these systems combine the page presentation logic with the business logic. The combination of presentation and business logic makes these systems hard to revise and update once they reach a higher level of complexity.

To avoid this problem, larger businesses often prefer to use a **component-based application system** that separates the presentation logic from the business logic. Each component of logic is created and maintained separately, which makes updating and changing elements of the system much easier on large systems that are built and maintained by teams of programmers. The most common component-based systems used on the Web are **Enterprise JavaBeans (EJBs)**, **Microsoft Component Object Model (COM)**, and the Object Management Group's **Common Object Request Broker Architecture (CORBA)**.

Integration with ERP Systems

Many B2B Web sites must be able to connect to existing information systems such as enterprise resource planning software. **Enterprise resource planning (ERP)** software packages are business systems that integrate all facets of a business, including accounting, logistics, manufacturing, marketing, planning, project management, and treasury functions.

The two major ERP vendors are **Oracle** and **SAP**. A typical installation of ERP software costs between $1 million and $50 million; thus, companies that are already running these systems have made a significant investment in them and require that their electronic commerce and EDI operations integrate with them.

Figure 9-6 shows a typical architecture for a B2B Web site that connects to several existing information systems, including the ERP system within the company and its trading partners' systems through EDI connections.

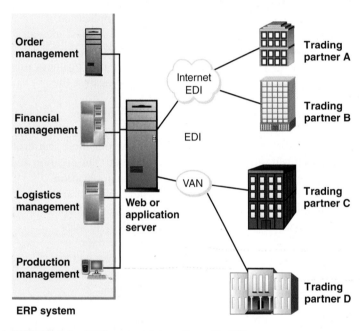

FIGURE 9-6 ERP system integration with EDI

© Cengage Learning

Web Services

Companies are using the Internet to connect specific software applications at one organization directly to software applications at other organizations. The W3C defines **Web services** as software systems that support interoperable machine-to-machine interaction over a network. In other words, a Web service is a set of software and technologies that allow computers to use the Web to interact with each other directly, without human operators directing the specific interactions.

For example, a handbag manufacturer's computers can contact its customers' computers to learn which of its products are selling well. Once it obtains this information from a number of the company's customers, the computer can adjust manufacturing schedules, increasing the production of some handbags and reducing the production of others.

A general name for the ways programs interconnect with each other is **application program interface (API)**. When the interaction is done over the Web, the techniques are called **Web APIs**. Web services use Web APIs of various types, as you will learn later in this section.

What Web Services Can Do

Companies are using Web services to offer improved customer service and reduce costs. In some companies, Web services are used to transmit the XML-tagged data from one application to another in enterprise application integration efforts. In other applications, Web services provide data feeds between two different companies. Many companies that have used Web services to accomplish application integration have found it to be less expensive to implement than older approaches that required programmers to write or adapt multiple middleware software programs. Here are some examples of specific Web services implementations:

- J.P. Morgan Chase & Co., a major investment bank, uses Web services in its investment information portal. The Web services pull information, such as general economic forecasts, financial analyses of specific companies, industry forecasts, and financial markets results, into continually updated online reports that customers can obtain on the J.P. Morgan Chase portal site. The bank's customers could obtain all of this information themselves, but the aggregation is a service that the bank provides. The information flow in this case is from the bank to its customers.
- Nationwide Building Society, a mortgage company in Swindon, England, uses a Web services tool to automate its communications with mortgage application service companies. These service companies obtain information from consumers who want mortgages and then forward the information in a prescribed XML format to Nationwide. The Nationwide Web services software reformats the submission and submits it to Nationwide's enterprise computer system. When a lending decision has been reached, the Web services tool conveys the decision back to the mortgage application service company. This Web services approach has reduced costs and decreased turnaround time for loan decisions at Nationwide.
- CUNA Mutual Group sells services to credit unions throughout the United States from its headquarters in Madison, Wisconsin. These services include everything from check clearing to construction management. CUNA provides many of its services by running programs on old computer systems that have been in operation for years. Instead of reprogramming everything so it could be accessible on the Web, CUNA created a Web services layer that takes information from the old computer systems and generates Web pages that its customers can use to obtain those services.
- The MSN Money site buys stock quotes from the Interactive Data Corporation, which delivers them, computer-to-computer, using Web services. If you view an MSN Money stock quote page, you can see the Interactive Data Real-Time Services acknowledgement for those stock quotes (along with those of other Web services providers that contributed to the page) near the bottom of the Web page under the heading "Data Providers."

How Web Services Work

A key element of the Web services approach is that programmers can write software that accesses these units of business application logic without knowing the details of how each unit is implemented. Web services can be mixed and matched with other Web services to execute a complex business transaction. Thus, Web services allow programs written in different languages on different platforms to communicate with each other and accomplish transaction processing and other business tasks.

The common format of this machine-to-machine communication was originally HTML; however, most Web services implementations now use XML. As you learned in Chapter 2, organizations can use XML to mark up content with agreed-upon sets of descriptive tags.

The first Web services were nothing more than sources of information. The Web services model allowed programmers to incorporate these information sources into software applications. For example, a company that wanted to collect all of its financial management information into one spreadsheet could use Web services to obtain information about bank account and loan balances, stock portfolio holdings, and current interest rates on financial instruments. Commonly available spreadsheet software can then be used to create a spreadsheet model that uses the information supplied by those Web services to update itself automatically. Some of the information might be available as a Web service at no cost; other information access might require a subscription. But Web services can make accessing the information much easier and more efficient.

A more advanced example would be a company that uses purchasing software to help manage that activity. The purchasing software can use Web services to obtain price information from a variety of vendors. After the purchasing agent reviews the price and delivery information and authorizes the purchase, the software can submit the order and track it until the shipment is received. On the other side of this transaction, the vendor's software can use Web services (in addition to providing price and delivery information) to check the buyer's credit and contract with a freight company to handle the shipment. As Web services become more sophisticated, they can make the decisions rather than simply providing information to people who then make decisions.

Web Services Specifications

The first widely used approach to Web services was **Simple Object Access Protocol (SOAP)**, which is a message-passing protocol that defines how to send marked-up data from one software application to another across a network. Implementing SOAP uses three rule sets (usually called protocols or specifications) that let programs work with formatted (using XML or HTML) data flows. The communication rules are included in the SOAP specification. The other two specifications are the **Web Services Description Language (WSDL)**, which is used to describe the characteristics of the logic units that make up specific Web services, and the **Universal Description, Discovery, and Integration Specification (UDDI)**, which identifies the locations of Web services and their associated WSDL descriptions.

Programmers use the information in a WSDL description to modify an application program so it can connect to a Web service. WSDL descriptions also allow programs to configure themselves so they can connect to multiple Web services. Programmers (and the programs themselves) use UDDI to find the location of Web services before they can interpret their characteristics (described in WSDL) or communicate with them (using SOAP).

Much of the data in SOAP applications is stored and transmitted in XML format. Because there are so many variations of XML in use today, data-providing and data-using partners must agree on which XML implementation to use. SOAP-based Web services often include quality of service and service-level specifications on which applications developers at each company can rely. In many cases, each Web services subscriber must work out a detailed agreement (specifying service levels, quality of service standards, and so on) with each Web services provider.

The SOAP set of protocols was the first approach to implementing Web services to be widely used (and continues to be widely used in large corporate information management applications). Another approach to Web services implementation that was developed later, but which uses a somewhat simpler structure, has become more common in Web services implementations, as you will learn in the next section.

REST and RESTful Design

Roy Fielding, one of the authors of the original HTTP specification, wrote his doctoral dissertation in 2000 on the subject of network-based software architectures. In the dissertation, he outlined a principle called **Representational State Transfer (REST)**, that describes the way the Web uses networking architecture to identify and locate Web pages and the elements (graphics, audio clips, and so on) that make up those Web pages. Designers of Web services who found SOAP to be unnecessarily complex for the applications they were building turned to Fielding's REST idea and began using it as a structure for their work.

Web services that are built on the REST model are said to use **RESTful design** and are sometimes called **RESTful applications**. A RESTful application transfers structured information from one Web location to another. This structured information can be any type of media, but it is most often an XML-tagged data set. RESTful applications can also transfer HTML- or XHTML-tagged data. The Web service is made available at a specific address (much as a Web page is made available at its URL) and can be accessed by any other computer that has a Web browser function.

More than half of all Web services applications today are RESTful applications. Probably the most widely used is the **Atom Publishing Protocol**, a blogging application that simplifies the blog publishing process and makes its functions available as a Web service so other computers can interact with blog content. You can see examples of Web services that use RESTful design at ProgrammableWeb.

ELECTRONIC COMMERCE SOFTWARE FOR SMALL AND MIDSIZE COMPANIES

In this section, you will learn about software that small and midsize businesses can use to implement online business Web sites. In most cases, these companies can create a Web site that stands alone in its business activities (primarily promotion and sales activities) and does not need to be coordinated completely with the business's other activities, which would include human resources, purchasing, and so on.

Basic Commerce Service Providers

Using a service provider's shared or dedicated hosting services instead of building an in-house server or using a co-location service means that the staffing burden shifts from the company to the Web host. CSPs have the same advantages as ISP hosting services, including spreading the cost of a large Web site over several "renters" hosted by the service. The biggest single advantage—low cost—occurs because the host provider has already purchased the server and configured it. The host provider has to worry about keeping it working through lightning storms and power outages.

CSPs offer free or low-cost electronic commerce software for building electronic commerce sites that are then kept on the CSP's server. Services in this category usually cost less than $20 per month, and the software is built into the CSP's site, allowing companies to immediately begin building and storing a storefront using the Web interface of the software. These services are designed for small online businesses selling only a few items (usually no more than 50) and having relatively low transaction volumes (fewer than 20 transactions per day). **Gate.com** is an example of a CSP. Gate.com offers businesses comprehensive hosting services, including shared hosting, dedicated hosting, and co-location services. **ProHosting.com** and **1&1 Internet** are other examples of Web-hosting companies serving the small and midsize company market. Because these companies offer a variety of services, they might be called ISPs, CSPs, MSPs, or ASPs by different users, depending on the service they are seeking. Yahoo! offers a wide range of Web-hosting and electronic commerce services for companies of all sizes. Its commerce services, which include site design, an online payment function, order processing and shipping, and marketing programs, are all described on its **Yahoo! Small Business Merchant Solutions** Web page, which appears in Figure 9-7.

FIGURE 9-7   Yahoo! Small Business Merchant Solutions page

Mall-Style Commerce Service Providers

Mall-style commerce service providers (CSPs) provide small businesses with an Internet connection, Web site creation tools, and little or no banner advertising clutter. These service providers charge a low monthly fee and may also charge one-time setup fees. Some of these providers also charge a percentage of or fixed amount for each customer transaction. These Web hosts also provide online store design tools, storefront templates, an easy-to-use interface, and Web page-generation capabilities and page maintenance.

Mall-style CSPs provide shopping cart software or the ability to use another vendor's shopping cart software. They also provide payment-processing services so the online store can accept credit cards.

In the early days of the Web, many mall-style CSPs were offering their services. Some even provided free Web site hosting in exchange for displaying ads on the sites. Today, the main mall-style CSP that remains in business is **eBay Stores**. You can open an eBay Basic Store for a monthly fee that is less than $20, although Premium and Anchor Store owners pay substantially higher fees.

Another mall-style option for selling online is Amazon.com, which allows an individual to sell certain used items (such as books) on the same page that Amazon.com lists the new product. Instead of the eBay Stores approach, in which each small merchant has its own store, Amazon.com lets merchants display their offerings product by product, mixed in with all of the other items Amazon.com offers for sale. Amazon.com charges a fee for each item sold and takes a percentage of the selling price. The percentage varies depending on the type of product being sold. For businesses that want to sell more than a few items, Amazon offers its Pro Merchant program, which waives the per-item fee, but charges a monthly subscription fee of about $40.

Both basic and mall-style CSPs usually provide data-mining capabilities that search through site data collected in log files. Data mining, which you learned about in Chapter 4, can help businesses find customers with common interests and discover previously unknown relationships among the data. Reports can indicate problematic pages in a store's design where, for example, a large number of customers get stuck and then leave the Web site. Other facts that data-mining reports can reveal include the number of pages an average customer must load and display before locating the merchandise he or she wants. If customers have to load too many pages, they might become impatient and leave without making a purchase.

Estimating Operating Expenses for a Small Web Business

A small business owner who wants to open a small online business activity would normally expect to spend between $500 and $5000 to become operational using either a basic CSP or a mall-style CSP. These estimates assume that the business will offer fewer than 100 items for sale and that the business already owns a computer and has Internet access for that computer. Figure 9-8 shows the estimated ranges of first-year expenses that a small business owner might incur to put this type of store on the Web.

	Cost estimates	
Operating costs	Low	High
Initial site setup fee	$ 0	$ 200
Annual CSP maintenance fee (12 x $20 to $150)	240	1800
Domain name registrations	0	300
Scanner for photo conversion or digital camera	100	900
Photo editing software	60	800
Occasional HTML and site design help	100	800
Merchant credit card setup fees	0	200
Total first-year costs	$500	$5000

FIGURE 9-8 Approximate costs to put a small store online

The estimates shown in the figure omit ongoing payment-processing charges, which might average 50 cents per transaction and 2 percent of each sale's total. Most new merchants estimate that payment processing will cost between 3 percent and 5 percent of dollar sales. You will learn more about payment-processing options for online businesses in Chapter 11. The costs shown are average low and high estimates for each item. Depending on which CSP and electronic commerce software options are chosen, the actual costs could be somewhat lower or considerably higher. For example, some CSPs include free registration for several domains when a store signs up for a one-year or longer contract for services.

Contrast the preceding costs with comparable estimated costs for self-hosting a Web site. Setup and Web site maintenance costs include equipment, communications, physical location, and staff. Equipment—a server and networking gear—has a one-time cost ranging from $2000 to $20,000. A high-bandwidth Internet connection (see Chapter 2) costs between $600 and $12,000 per year. A server must be housed in a room that is both secure and convenient to communications access. The cost to secure a small room, properly air-condition it, and install a chemical fire extinguishing system could easily reach $5000 a year. A self-hosted system requires an information technology staff that is familiar with Web programming and scripting languages, electronic commerce packages, and database management systems. Technicians will likely be required to monitor and maintain equipment. Staff costs could range from $50,000 to $100,000 annually. In total, annual operating costs for self-hosting will generally run between $60,000 and $100,000. Companies should carefully compare their estimates of self-hosting costs with the fees charged by hosting services that provide similar capabilities.

The costs previously discussed are for a small online business site. Costs for larger sites are much more difficult to estimate. The cost of integrating the Web site with the existing systems of the company is often the largest element of the total cost. Midsize businesses typically incur start-up costs ranging from $100,000 to $500,000 and recurring annual costs of about half that amount. Large businesses typically spend between $1 million and $50 million to launch an electronic commerce site and then spend another 50 percent of the launch cost every year to operate, maintain, and improve the site. You will learn more about managing the costs of Web site implementation and operation for large organizations in Chapter 12. Next, you will learn about midrange electronic commerce software packages that are suitable for running larger businesses.

ELECTRONIC COMMERCE SOFTWARE FOR MIDSIZE TO LARGE BUSINESSES

This section includes a discussion of software that midsize and large companies can use to implement electronic commerce features on their Web sites. It also includes an outline of Web site development tools that can be used for that purpose and an overview of three specific midrange electronic commerce software products that are representative of the types of products available.

These midrange packages allow the merchant to have explicit control over merchandising choices, site layout, internal architecture, and remote and local management options. In addition, the midrange and basic electronic commerce packages differ on price, capability, database connectivity, software portability, software customization tools, and computer expertise required of the merchant.

Web Site Development Tools

Although they are more often used for creating small business sites, it is possible to construct the elements of a midrange electronic commerce Web site using the Web page creation and site management tools you learned about in Chapter 2.

After creating the Web site with these development tools, the designer can add purchased software elements, such as shopping carts and content management software, to the site. The final step is to create the middleware that connects the site to the company's existing product and transaction-processing databases.

Buying and using midrange electronic commerce software is significantly more expensive than using one of the CSPs described in the previous section, with annual costs ranging from $2000 to $50,000. Midrange software traditionally offers connectivity to database systems that store catalog information. Having the catalog stored in a database simplifies updates and changes. Several of the midrange systems provide connections into existing inventory and ERP systems. This can yield savings because there is no need to run duplicate inventory systems, and the cost of the existing systems is spread across several software systems.

Three midrange electronic commerce systems are described in this section. They are representative of the whole group yet are different from one another in important ways. The systems are Intershop Enfinity, WebSphere Commerce Suite by IBM, and Commerce Server by Microsoft.

Intershop Enfinity

Intershop Enfinity provides search and catalog capabilities, electronic shopping carts, online credit card transaction processing, and the ability to connect to existing back-end business systems and databases. Intershop Enfinity has setup wizards and good catalog and data management tools. It provides many built-in storefront templates. Management and editing of a storefront are done through a Web browser—either locally at the server or remotely through any Internet connection. The products inventory management module tracks inventory levels and allows merchants to view the quantity of items available, create a list of inventory transactions, and enter new products into the inventory. Discount rules are also easy to enter. Merchants define the business rules for a discount and dates during which special discounts apply. Bundled with the software is a database management system. Alternatively, Enfinity can work with DB2 (IBM's relational database) or Oracle databases. The software includes an automated e-mail facility that can send order confirmations to customers. Enfinity includes support for secure transactions. A wide variety of site and customer reports are available to track Web page visits and customer activities.

IBM WebSphere Commerce Professional

IBM produces WebSphere Commerce Professional, which is a family of electronic commerce packages. IBM WebSphere is a set of software components that provides software suitable for midsize to large businesses to sell goods and services on the Internet. It includes catalog templates, setup wizards, and advanced catalog tools to help companies create attractive and efficient electronic commerce sites. WebSphere Commerce Professional can be used both for business-to-business and business-to-consumer applications and provides a smooth connection to existing corporate systems, such as inventory databases and procurement systems.

WebSphere Commerce products run on many different operating systems. Merchants can begin with a small store and then move up to a bigger, more capable store as necessary. A wizard leads the merchant through the process of creating a starter store.

Once that is up and working, more functionality can be added by executing commands and writing code. With the basic pages built, the merchant can populate the catalog with products, prices, and product pictures. The WebSphere Commerce Professional Edition also accommodates electronic download products, such as audio tracks or software.

WebSphere offers a large collection of functions, utility programs, and commands that allow a merchant to create a customized online store experience. However, JavaScript, Java, or C++ expertise is required. Typical of commerce programs in this class, WebSphere can connect to existing databases and other legacy systems through DB2 or Oracle databases. A single store or several different stores can be administered from the same browser-based interface. A large number of midrange electronic commerce sites use WebSphere software. The system has all the standard electronic commerce features, including tools for a shopping cart, e-mail notifications upon sale completion, secure transaction support, promotions and discounting, shipment tracking, links to legacy accounting systems, and browser-based local and remote administration. A typical installation of WebSphere Commerce Professional Edition costs between $50,000 and $300,000, depending on how many servers will be running the product and which options are purchased with the software.

Microsoft Commerce Server

Microsoft **Commerce Server** allows businesses to sell products or services on the Web using tools such as user profiling and management, transaction processing, product and service management, and target audience marketing. Commerce Server has wizards that can help users build a site in several steps, but program code must be written to make the software meet specific user needs. The Microsoft Visual Studio .NET tools are bundled with the software and allow companies to customize the sites they build.

Like other midrange electronic commerce software, Commerce Server has tools that help companies engage the customer (through marketing and advertising), complete an order, and analyze the sales information after the sale. Commerce Server also includes tools for advertising, promotions, cross-selling, and customer targeting and personalization.

Commerce Server provides many predefined reports for analyzing site activities and product sales data. The system provides several storefront templates, wizards for setting up and initializing a store, and database connections. It also provides a shopping cart, confirms completed sales transactions by e-mail, and supports secure transactions. It can connect to existing accounting systems, and the administrator can oversee the site through a Web browser. The product only runs on systems that are running the Windows Server operating system and the SQL Server database system. Commerce Server licenses cost between $7100 and $21,000 per processor, but the required operating system and database software licenses can add another $7000 per processor. Licensing a typical installation of Microsoft Commerce Server usually runs between $30,000 and $300,000.

ELECTRONIC COMMERCE SOFTWARE FOR LARGE BUSINESSES

Larger businesses require many of the same advanced capabilities as midsize firms, but the larger firms need to handle higher transaction loads. In addition, they need dedicated software applications to deal with specific elements of their online business. In this section, you will learn about electronic commerce software that has higher transaction-load capability, and you will learn about software that accomplishes specific tasks in large businesses, such as customer relationship management, supply chain management, content management, and knowledge management.

The distinction between midrange and large-scale electronic commerce software is much clearer than the one between basic systems and midrange systems. The telltale sign is price. Other elements, such as extensive support for business-to-business commerce, also indicate that the software is in this category. Commerce software in this class is sometimes called **enterprise-class software**. The term "enterprise" is used in information systems to describe a system that serves multiple locations or divisions of one company and encompasses all areas of the business or enterprise. Enterprise-class electronic commerce software provides tools for both B2B and B2C commerce. In addition, this software interacts with a wide variety of existing systems, including database, accounting, and ERP systems. As electronic commerce has become more sophisticated, large companies have demanded that their Web sites and supporting information infrastructure do more things. The cost of these enterprise systems for large companies ranges from $200,000 for basic systems to $10 million and more for comprehensive solutions.

Enterprise-Class Electronic Commerce Software

Enterprise-class electronic commerce software running large online organizations usually requires several dedicated computers—in addition to the Web server system and any necessary firewalls. Examples of enterprise-class products that can be used to run a large online business with high transaction rates include IBM WebSphere Commerce Enterprise, Oracle E-Business Suite, and several products from Broadvision.

Enterprise-class software typically provides tools for linking to and supporting supply and purchasing activities. A large part of B2B commerce is ordering supplies from trading or business partners and issuing the appropriate documents (or EDI transaction sets), such as purchase orders. For a selling business, e-business software provides standard electronic commerce activities, such as secure transaction processing and fulfillment, but it can also do more. For instance, it can interact with the firm's inventory system and make the proper adjustments to stock, issue purchase orders for needed supplies when they reach a critically low point, and generate other accounting entries in ERP, legacy accounting, or file systems. In contrast, both basic and midrange electronic commerce packages usually require an administrator to check inventory manually and place orders explicitly for items that need to be replenished.

In B2C situations, customers use their Web browsers to locate and browse a company's catalog. For electronic goods (software, research papers, music tracks, and so on), customers can download the items directly from the site, or they can complete order forms and have the hard-copy versions of the products shipped to them. The Web server is linked to back-end systems, including a database management system, a merchant server, and an application server. The database usually contains millions of rows of information about products, prices, inventory, user profiles, and user purchasing history. The history provides a way to recommend to a user on a return visit related items that he or she might wish to purchase. A merchant server houses the e-business system and key back-end software. It processes payments, computes shipping and taxes, and sends a message to the fulfillment department when it must ship goods to a purchaser. Figure 9-9 shows a typical enterprise-class electronic commerce architecture.

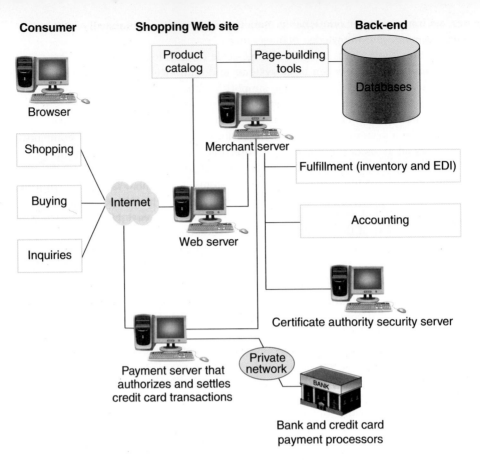

Consumer

Browser

Shopping

Buying — Internet

Inquiries

Shopping Web site

Product catalog — Page-building tools

Merchant server

Web server

Back-end

Databases

Fulfillment (inventory and EDI)

Accounting

Certificate authority security server

Payment server that authorizes and settles credit card transactions

Private network

Bank and credit card payment processors

© Cengage Learning

FIGURE 9-9 Typical enterprise-class electronic commerce architecture

Large companies also use additional specialized software to accomplish particular objectives that are not met by existing comprehensive electronic commerce software packages. For example, a company that wants to deliver entertainment (music or videos) directly to consumers' mobile devices might use **OpenMarket** software, a product designed to deliver and charge for that specific type of content in a mobile environment.

As you learned in Chapter 5, companies are using the Web to integrate their supply chains. As a result, enterprise-class commerce Web sites must include or work with supply chain management software.

In Chapter 6, you learned about companies that were building social networking elements into their sites to engage their customers and suppliers. A part of that strategy is providing useful, fresh content to attract site visitors. This need has given rise to software that automatically manages and rotates content on Web sites. Some companies have even developed software that helps them manage the knowledge that exists in their businesses.

In Chapters 3 and 4, you learned that companies are storing data about site visitors in large databases and analyzing it to improve their relationships with those customers. These clickstreams track the path a visitor takes through a Web site, including which pages were viewed, the amount of time spent on each page, and the sequence in which pages were viewed. Thus, large electronic commerce sites must include customer relationship management software.

An enterprise-class Web site often includes several of these types of software packages in its design. The next sections of this chapter discuss software that works with electronic commerce software in large companies to help those companies achieve all of their electronic commerce objectives.

Content Management Software

Large companies are finding new ways to use the Web to share information among their employees, customers, suppliers, and partners. **Content management software** helps companies control the large amounts of text, graphics, and media files that have become crucial to doing business. The increased use of social media and networking as part of online business operations (which you learned about in Chapter 6) has made content management even more important for all kinds of Web sites.

Content management software should be tested before committing to it. The testing should ensure that company employees find the software's procedures for performing regular maintenance (for example, adding new categories of products and new items to existing product pages) to be straightforward. The software should also facilitate typical content creation tasks, such as adding sale-item specials.

Companies that need many different ways to access corporate information—for example, product specifications, drawings, photographs, or lab test results—often choose to manage the information and access to that information using content management software. The leading providers of content management software include IBM and Oracle, which provide the software as components in other enterprise software packages, and several smaller companies that provide stand-alone content management software. Content management software generally costs between $50,000 and $500,000, but it can cost three or four times that much to customize, configure, and implement.

Knowledge Management Software

An increasing number of large companies have achieved cost savings by using content management software. Most content management software is designed to help companies manage information that, until recently, was stored in paper reports, schedules, analyses, and memos. Although the cost reductions that can be obtained by moving mountains of paper into an electronic format are significant, some companies have begun to understand that the true value of those documents is in the information contained in them. Thus, they began the search for systems that would help them manage the knowledge itself, rather than the documentary representations of that knowledge. The software that has been developed to meet that goal is called **knowledge management (KM) software**.

KM software helps companies do four main things: collect and organize information, share the information among users, enhance the ability of users to collaborate, and preserve the knowledge gained through the use of information so that future users can benefit from the learning of current users. KM software includes tools that read electronic documents (in formats such as Microsoft Word or Adobe PDF), scanned paper documents, e-mail messages, and Web pages. KM software often includes powerful search tools that use proprietary semantic and statistical algorithms to help users find the content, human experts, and other resources that can aid them in their research and decision-making tasks. Early KM systems often disrupted the flow of users' work. Today, KM systems collect knowledge elements by extracting them from the normal interactions users have with information.

The major software vendors have KM software offerings, including IBM and Microsoft SharePoint. Smaller companies, such as CustomerVision, also offer KM software and technologies. Total costs for a KM software implementation, including hardware, software licenses, and consultant fees, can range from $10,000 to $1 million or more.

Supply Chain Management Software

Supply chain management (SCM) software helps companies to coordinate planning and operations with their partners in the industry supply chains of which they are members. SCM software performs two general types of functions: planning and execution. Most companies that sell SCM software offer products that include both components, but the functions are quite different. SCM planning software helps companies develop coordinated demand forecasts using information from each participant in the supply chain. SCM execution software helps with tasks such as warehouse and transportation management. The two major firms offering SCM software are **JDA Software** and **Logility**.

Common supply chain management software components include those that manage demand planning, supply planning, and demand fulfillment. Demand planning components examine customers' buying patterns and generate continually updated forecasts. The supply planning component coordinates distribution logistics, inventory-level forecasting, collaborative procurement, and supply allocations. The demand fulfillment component handles execution activities, including order management, customer verification, backlog control, and order fulfillment.

Most supply chain management software was developed for manufacturing firms that wanted to manage inventory purchases and manufacturing processes. JDA Software had a successful line of software products for managing retail order entry and the sales side of inventory control. In 2006, it decided to expand its product line to include supply chain management tools. Rather than develop its own software from the ground up, it purchased Manugistics, which had a full line of supply management, demand management, and transportation and logistics management software. The company later purchased i2 Technologies to obtain its software for supply planning and demand fulfillment. JDA Software now offers software that companies can use to manage every operation in the supply chain, from raw materials purchase to the delivery of finished products to consumers.

The cost of SCM software implementations varies tremendously depending on how many locations (retail stores, wholesale warehouses, distribution centers, and manufacturing plants) are in the supply chain. For example, a retailer with 500 stores might pay between $2 million and $10 million for an SCM package that includes both planning and execution functions, but a wholesaler with only three or four distribution centers might be able to install an SCM product for under $500,000.

Customer Relationship Management Software

You learned about the philosophy and techniques of customer relationship management (CRM) in Chapter 4. The goal of CRM is to understand each customer's specific needs and then customize a product or service to meet those needs. The idea is that a customer whose needs are being met exactly is willing to pay more for the goods or services that they need. Although companies of all sizes can practice CRM techniques, large companies can afford to buy and implement software products that automate many CRM functions.

Customer relationship management (CRM) software must obtain data from operations software that conducts activities such as sales automation, customer service center operations, and marketing campaigns. The software must also gather data about customer activities on the company's Web site and any other points of contact the company has with its existing and potential customers. CRM software uses this data to help managers conduct analytical activities, such as gathering business intelligence, planning marketing strategies, customer behavior modeling, and customizing the products and services to meet the needs of specific customers or categories of customers.

In its most basic form, CRM uses information about customers to sell them more (or more profitable) goods or services. More advanced CRM is about delivering extremely attractive and positive experiences regularly to customers. CRM can be very important in maintaining customer loyalty in businesses where the purchase process is long and complex. Companies that design and install custom machinery, software products, or office workflow systems often find themselves involved in these types of long and complex processes. CRM software can help maintain positive and consistent contacts with multiple employees at the purchasing company.

In the early days of CRM software implementation (approximately 1996 to 2000), companies spent many millions of dollars to buy CRM systems that promised to monitor and improve relationships with existing customers. Most of these systems were focused on giving companies the information they needed to identify changing customer preferences and respond very quickly to those changes. By responding quickly, companies hoped that they would be able to gain sales that might otherwise be lost to competitors that could respond better to the new customer preferences. In addition to gaining sales, the use of CRM software was expected to help retain customers and reduce the need to spend money on marketing to find new customers. The goal was to instantly make available perfect information about all customer behaviors from all customer-interaction points throughout the company.

Many companies that were early implementers of CRM found that they did not achieve the benefits they had expected, which led to a slowing of CRM sales in the early 2000s. In addition, a number of industry analysts declared CRM to be just another fleeting business fad.

Most companies, however, realized that investing large amounts of money in complete restructurings of their customer-interaction strategies was ineffective. Despite CRM consultants' claims, a single large-scale CRM implementation was not going to solve all of their problems in one fell swoop. Companies stopped thinking of CRM software as a tool for changing their overall customer strategy and instead began using CRM software to solve smaller and more specific problems. For example, a cable company might use CRM to track service outages and repair team responses in real time, but would not expect the CRM system to calculate the profitability of on-demand video services on a continual basis.

One of the most popular targets for these focused CRM applications has been call center operations. By examining problems that arise in their call centers, many companies have identified specific applications where CRM software can improve response times, accuracy, and effectiveness.

In addition to general CRM software, many companies use small, precisely focused software to address specific problems. For example, online clothing retailer Bluefly uses customer experience monitoring software from Tealeaf to identify specific issues in its operations. It identified a technical problem in its shopping cart software (the software was not displaying an error message it was supposed to present under a certain set of conditions) that was causing almost half of its customers who encountered the issue to abandon their shopping carts. By monitoring such metrics as shopping cart abandonments, product returns, product page views, and user session characteristics, these tools can help companies examine specific elements of the customer experience and make changes to their Web sites that can increase their effectiveness and profitability.

Some companies create their own CRM software using outside consultants and their own IT staffs. In recent years, software vendors have increased the quality and variety of their offerings and today, most companies are likely to buy a CRM software package rather than create their own. Siebel Systems was the first company to specialize in CRM software and it had gained a large share of the market. In 2005, Oracle bought Siebel and merged its operations into its existing CRM business, called **Oracle CRM On Demand**. Other major software firms have created products in this market, including **SAP CRM**. Prices for these systems start at around $25,000 (on average, about $1500 per user); large implementations can cost millions of dollars.

Salesforce.com has grown rapidly as a vendor of CRM software. It was one of the first companies to offer any type of software as an application that customers could access through their Web browsers. That is, users did not need to install the software on their own servers. The buyer's employees simply log in to the CRM vendor's Web site and use the software.

Cloud Computing

The Salesforce.com practice of replacing a company's investment in computing equipment by selling Internet-based access to its own computing hardware and software is called **cloud computing**, and it has become an important new force in the software industry. Cloud computing allows companies to gain the benefits of software without having to install computing hardware and maintain it. As the software changes, users do not need to upgrade or reconfigure their servers. The vendor manages the entire software installation at its sites. The software user pays a subscription fee, which can be as low as a few hundred dollars per user per year.

Cloud computing has become a popular cost-reduction strategy for many companies of all sizes. Small companies are happy to avoid spending time and money investigating and evaluating complex technology choices. Midsize companies use cloud computing to avoid tying up substantial capital in computing infrastructure. Larger companies use cloud computing to gain flexibility in launching new operations and to provide a cushion that helps them handle unexpected large volumes of transactions.

Summary

In this chapter, you learned about electronic commerce software for small, midsize, and large businesses and the functions provided by each software type. The electronic commerce software a company chooses depends on its size, objectives, and budget and requires making major decisions. A company must first choose between paying a service provider to host the site and self-hosting. External hosting options include shared hosting, dedicated hosting, and co-location. Many hosting companies offer comprehensive services to merchants, such as databases, shopping carts, and content management, in addition to basic Web-hosting services.

Key elements of all electronic commerce software include catalogs, shopping carts, and transaction-processing capabilities. Companies can use Web services to get their information systems to work across organizational boundaries.

Small enterprises that are just starting an electronic commerce initiative might use a basic commerce service provider (CSP). Basic CSP and mall-style hosting services for small businesses provide a range of standard features, including tools for quickly creating storefronts, catalogs, and transaction processing. These packages are usually wizard and template driven.

If a company already has computing equipment and staff in place, purchasing a midrange electronic commerce software package provides more control over the site and allows for expansion. Midrange software can interact with database software to create dynamic catalogs and shopping carts and handle order processing.

Large enterprises that have high transaction rates, B2B partnerships, or a significant investment in ERP and other existing information systems need to invest in larger, more customizable systems that can provide needed features and flexibility. These packages can include customer relationship management, supply chain management, content management, and knowledge management capabilities, or they can work with dedicated software that performs these functions. A growing number of software vendors offer their products as a cloud subscription service rather than as software that must be installed on users' servers. This software is accessed through a Web browser and saves users the costs and trouble of maintaining server hardware and managing software upgrades.

Key Terms

Application integration

Application program (application)

Application program interface (API)

Application server

Application service providers (ASPs)

Application software (application)

Atom Publishing Protocol

Business logic

Business rules

Catalog

Cloud computing

Co-location (collocation, colocation)

Commerce service providers (CSPs)

Component-based application system

Content management software

Customer relationship management (CRM) software

Database

Database manager (database management software)

Dedicated hosting

Distributed database systems

Distributed information systems

Dynamic catalog

Enterprise application integration

Enterprise-class software

Enterprise resource planning (ERP)

Interoperability

Knowledge management (KM) software

Mall-style commerce service providers

Managed service providers (MSPs)

Middleware

Page-based application system

Representational State Transfer (REST)

RESTful applications

RESTful design

Scalable

Self-hosting

Shared hosting

Simple Object Access Protocol (SOAP)

Static catalog

Supply chain management (SCM) software

Transaction processing

Universal Description, Discovery, and Integration (UDDI) specification

Web APIs

Web services

Web Services Description Language (WSDL)

Review Questions

1. In one or two paragraphs, define "shared hosting," "dedicated hosting," and "co-location."

2. Write a paragraph in which you identify the three key elements that must exist in any electronic commerce software package.

3. In one or two paragraphs, explain the difference between a static and a dynamic catalog and describe situations in which you would use each.

4. In about 100 words, outline the elements included in the transaction processing function of an electronic commerce site.

5. In two or three paragraphs, explain why database software is an important element in an electronic commerce site.

6. Provide a brief definition of the term "middleware." In one or two paragraphs, explain why middleware can be difficult to write and test.

7. In a paragraph or two, describe the role of XML in enterprise application integration.

8. In about 100 words, summarize the advantages and disadvantages of using a mall-style commerce service provider such as eBay Stores or Amazon.com's Pro Merchant program instead of operating a stand-alone electronic commerce site.

9. In a paragraph, outline how the REST approach to Web services differs from that used in protocols such as SOAP.

10. In one or two paragraphs, explain the functions of content management and knowledge management software.

11. Write a paragraph in which you explain the appeal of cloud computing.

Exercises

1. Your friend Faye Borthick wants to set up a small Web site devoted to gardening. She believes her many years of experience in gardening give her an understanding of the kinds of gardening tools, fertilizers, soil-amendment products, herbicides, pesticides, and plants that appeal to the serious gardener. Right now Faye does not want to sell anything, although she might change her mind in the future. She merely wants to display pages of plant photography, write and store short how-to papers for novice gardeners, and provide links to other gardening tips on the Web. She wants your advice on whether to self-host the Web site or use an ISP (or CSP) to start her endeavor. Use your favorite search engine to locate information on the cost of using a service provider to host a Web site. Then, estimate

what a small Web site might cost in terms of the minimal configuration of hardware and software. Estimate the design and development costs and the annual maintenance costs. Then, select one of the Web server programs. Estimate the cost of a Web connection. Write a 200-word summary of everything you think Faye needs to know to use either of the two options (she builds it or she uses a service provider) for creating her site.

2. Annette Jackson owns a small crafts store in central Missouri. She wants to expand her store's reach outside the region to increase her profits and simultaneously reduce her inventory. Annette's teenage daughter suggested that she consider selling online to expand her total sales. She asked you to help her estimate how much it might cost in the first year to create a simple store with a catalog of about 100 items. Annette wants you to investigate two CSP offerings and report back to her what you find. Because her store is small, limit your research to basic commerce and mall-style services. Annette would like to consider the following information for the two CSP offerings you examine:

- Costs: initial setup fee, monthly fee, and transaction fees
- Amount of disk space the CSP would provide for Annette's 100-item store
- Promotion and marketing opportunities
- Customer communications capabilities, such as automated e-mail confirmation of orders
- Shopping cart or other order entry mechanism
- Storefront-building wizards for creating a new store
- Upload capabilities for product names, descriptions, images, and costs (Can they be uploaded from files or databases, or must the merchant enter each item individually?)
- Existence of an online user manual for the merchant

Use your favorite search engine to find CSPs that might meet Annette's needs. Produce a report of about 400 words summarizing your findings.

3. Using your library or the Web, find an article that describes a successful application of Web services. In about 200 words, describe how the company implemented the Web services application and explain why using Web services was better than using an alternative approach to solve the problem.

4. Visit the product Web sites to learn more about two of the knowledge management software products discussed in the chapter. In a report of about 200 words addressed to the president of a local university, explain how that university could benefit from an implementation of knowledge management software.

5. Two major providers of cloud computing services are Amazon.com and Google. Each of these providers offers a different set of options and is pursuing a somewhat different strategy. Amazon.com offers cloud computing services through its Amazon Web Services (AWS) division. AWS lets its customers use the power of AWS's millions of servers to buy the computing capacity they need on a short-term basis. Companies that use AWS must provide their own applications, databases, and content; but AWS offers an instantly available platform that can handle application hosting, Web hosting, backup and storage, and content delivery. With Google Apps for Business, the company focuses its cloud computing services on providing functional replacements for software that a company might need to buy and for which it must manage usage and licensing. For example, Google offers its Gmail product to replace a company's e-mail server system, and offers productivity applications (Google Documents) to replace other vendors' offerings of word-processing, spreadsheet, presentation, and database software. Instead of selling the computing power, Google sells the right to use its software. Use your favorite search engine and resources in

your library to learn more about Amazon.com's AWS services and Google's cloud computing services. Prepare a report of about 400 words in which you compare and contrast the two companies' offerings. Be sure to include a discussion of the differences in the sales strategies adopted by Amazon.com and Google for their cloud computing service offerings.

Cases

C1. Ingersoll-Rand Club Car Division

Ingersoll-Rand is a $9 billion diversified manufacturing company that sells its products worldwide. Its well-known brands include Ingersoll-Rand tools and portable power generators, Bobcat construction equipment, Thermo King refrigerated transport systems, Dexter and Schlage locks, and ARO industrial fluids equipment. The company's Club Car division manufactures and sells a variety of small electric cart vehicles to golf courses and industrial users. The division also sells a rough-terrain version designed for farmers, ranchers, construction workers, and recreational users.

In 2001, the Club Car division was experiencing a sales decline. The downturn in the general economy was affecting golf courses, which, in turn, were reducing the size and frequency of their golf cart orders. Club Car had a general sense that this major market segment was causing their revenues to decline, but their information systems were not providing enough data about exactly which sales were being most affected by the economic downturn.

Club Car sales managers relied on their sales representatives for information about likely future sales. Sales forecasting was a matter of judgment, guesswork, and a few spreadsheet software models scattered throughout the regional sales offices. The sales representatives had little influence on how the carts were customized for particular customer segments or for individual customers.

The company decided it needed better information about all of its sales and marketing activities, so it spent more than $2 million to install a comprehensive CRM system. This system was designed to automate the entire customer sales cycle: prospect evaluation, proposal writing, product configuration, and order entry. However, the users at Club Car division found the new system difficult to use and therefore were reluctant to spend much time learning how to use it. Thus, the promised benefits of improved productivity and more detailed reports were not forthcoming. Sales managers did not see the ultimate benefits that the system might provide. Salespeople found that the new system was requiring them to spend time entering data into the system rather than seeing customers. The order entry staff found the system to be cumbersome and unfamiliar.

When Club Car's president realized that the CRM system was not delivering on its promise, he had the management team go back and re-examine the key elements in the division's customer relationships and asked them to choose one or two issues that needed attention. The management team identified two major issues. First, the order entry process required the time of salespeople and order entry staff, but it did not include any interaction with customers. Second, the division was not producing accurate and timely sales forecasts.

In 2002, Club Car division relaunched its CRM efforts and focused on these two problem areas. The new effort included the sales representatives in redesigning the order entry process. The division was able to reduce the data entry time and effort required, especially the time of salespeople. Salespeople have remote access to the system, so they can work on site with customers to configure the carts to the customers' exact specifications. Salespeople can obtain pricing information and explore various alternatives with customers while they are at the customer's site. They can also examine manufacturing schedules and provide more accurate delivery date estimates. All of this remote, real-time information access helps salespeople close deals and increase sales volume and profitability.

Sales forecasts are more accurate now because the information about sales orders is automatically collected when the sales representatives close sales at the customers' sites. The CRM system combines this real-time sales order information with general industry information on cart demand, cart replacement cycles, and economic trends in their customers' industries. The increased accuracy of sales forecasts allows the company to create more stable production schedules, which means that more customers receive their carts on the delivery date they were promised.

Required:

1. List the types of information that Club Car division's CRM system makes available to sales representatives in the field. For each type of information, briefly explain how salespeople's remote access to that type of information can help them close sales on their customers' sites.

2. In the CRM relaunch, Club Car division focused on two CRM elements. In about 200 words, explain why this approach would work better, in general, than implementing a comprehensive CRM system that could track all of the division's sales activities and related information in real time.

3. In about 100 words, describe the benefits Club Car division might obtain by using cloud-based CRM software.

Note: Your instructor might assign you to a group to complete this case, and might ask you to prepare a formal presentation of your results to your class.

C2. Web Services for State Government

You are a member of the Web site management team of a state government. You have worked on all of the state's Web sites from time to time and have managed the launch of four major sites and the redesign and relaunch of two others. Some of the Web sites on which you have worked include electronic commerce features such as order acceptance, payment processing, and purchasing.

You report to Anne Nelson, the state's CIO. Anne asked you to lead a project to explore the potential uses of Web services in carrying out state government activities. She scheduled a formal briefing at which you will present an overview of Web services technology. You will also outline specific applications of Web services technologies to specific tasks that the state either currently performs or that it might perform in the future.

Anne knows that the state has many current and potential applications that could use Web services technologies, so she asked you to focus on four specific areas of state government in your briefing. At the briefing, you will address the directors of four state departments: the Attorney General's Department of Corporation Records, the Tax Administration and Collection Department, the Department of Motor Vehicles, and the Department of Fish and Wildlife Management.

The Attorney General's Department of Corporation Records maintains the official records of corporations chartered by the state or holding licenses to do business in the state. In addition to the original charter or license, companies must file annual reports that include the names and addresses of corporate directors and officers, the amount of company stock issued or redeemed during the year, and the current address of the company.

The Tax Administration and Collection Department is responsible for accepting income tax, personal property tax, and sales tax return filings of companies and individuals. The department also processes payments of these taxes and authorizes the State Treasurer to issue refunds that are due to taxpayers who have overpaid their taxes. This department currently provides tax forms and instructions in Adobe PDF format on its Web site. It also maintains an extensive frequently asked questions (FAQs) list on the site.

The Department of Motor Vehicles issues driver's license renewals and vehicle registration renewals (for cars, trucks, and boats) and accepts auto dealerships' monthly reports of vehicles purchased or sold on its Web site. The site also includes extensive collections of information about motor vehicle laws and administrative rulings that visitors can review to ensure they are in compliance.

The Department of Fish and Wildlife Management provides downloadable applications for hunting and fishing licenses on its site. Current hunting and fishing license holders can renew their licenses and pay their annual fees on the Web site. Companies that have state-issued permits to undertake logging or mining operations can file their monthly activity reports on the department's Web site, too.

Anne suggests that you review current IT trade publications (both in print and on the Web) to learn more about Web services applications that have been implemented in government agencies. She also recommends that you examine a number of other state Web sites to see how they are performing these tasks.

Required:

1. Prepare a briefing report of about 200 words in which you describe Web services technology in a way that will be understandable to the four department directors. These directors are experienced administrators, but they are not technology experts.

2. Prepare a briefing report that outlines opportunities for the use of Web services in each department. Include about 100 words for each department.

3. Prepare an analysis of costs and benefits for two major applications of Web services that you identify. In this setting, a benefit can arise from an increase in revenue, a reduction in expense, an improvement in the quality of service provided, or an increase in the speed with which a service is provided. This report should be directed to Anne and should include an implementation recommendation (whether the state should implement or should not implement) for each of the two Web service applications you identified. This report should be about 300 words in length.

Note: Your instructor might assign you to a group to complete this case, and might ask you to prepare a formal presentation of your results to your class.

For Further Study and Research

Abate, C. 2002. "Going Once, Going Twice . . . Sold!" *Smart Business*, 15(4), May, 72–76.

Al-Shammary, D. and I. Khalil. 2012. "Redundancy-aware SOAP Message Compression and Aggregation for Enhanced Performance," *Journal of Network and Computer Applications*, 35(1), 365–381.

Barrett, V. 2010. "Salesforce.com: The Web's Big Upstart," *Forbes*, December 6, 1–3.

Benslimane, D., S. Dustdar, and A. Sheth. 2008. "Services Mashups: The New Generation of Web Applications," *IEEE Internet Computing*, 12(5), 13–15.

Berfield, S. 2009. "Susan Lyne on Gilt.com's Pleasures and Pressures," *Business Week*, December 14, 17–18.

Birman, K. 2012. "CORBA: The Common Object Request Broker Architecture," 249–269. In *Guide To Reliable Distributed Systems*. London: Springer.

Blair, G. and P. Grace. 2012. "Emergent Middleware: Tackling the Interoperability Problem," *IEEE Internet Computing*, 16(1), 78–82.

Boucher-Ferguson, R. 2007. "Salesforce Under Pressure," *eWeek*, November 26, 28.

Bruno, E. 2007. "SOA, Web Services, and RESTful Systems," *Dr. Dobb's Journal*, 32(7), July, 32–37.

Corredor, I., J. Martinez, M. Familiar, and L. Lopez. 2012. "Knowledge-aware and Service-oriented Middleware for Deploying Pervasive Services," *Journal of Network and Computer Applications*, March, 35(2), 562–576.

Cowley, S. 2005. "Salesforce.com Battles Rivals," *Network World*, 22(23), June 13, 31–32.

CRM Magazine. 2007. "Software AG Is Set to Acquire webMethods," 11(6), June, 16.

Deloitte Development. 2011. *Tech Trends 2011: The Natural Convergence of Business and IT*. New York: Deloitte Development.

Ferguson, G. 2002. "Have Your Objects Call My Objects," *Harvard Business Review*, 80(6), June, 138–143.

Gartner, Inc. 2007. *Magic Quadrant for Enterprise Content Management*. Stamford, CT: Gartner, Inc.

Guernsey, L. 2003. "On the Web, Without Wasting Time," *The New York Times*, May 6, G10.

Henschen, D. 2011. "Salesforce.com Steps Past CRM Into Social Marketing," *InformationWeek*, November 30. (http://www.informationweek.com/news/software/enterprise_apps/232200417)

Hoover, J. 2008. "Microsoft Extends SQL Server to the Web with Data Services," *Intelligent Enterprise*, 11(3), March, 1.

Ismail, A., S. Patil, and S. Saigal. 2002. "When Computers Learn to Talk: A Web Services Primer," *McKinsey Quarterly*, Special Edition (Issue 2), June, 70–78.

Jayachandran, S., S. Sharma, P. Kaufman, and P. Raman. 2005. "The Role of Relational Information Processes and Technology Use in Customer Relationship Management," *Journal of Marketing*, 69(4), October, 177–192.

Karande, A., V. Chunekar, and B. Meshram. 2011. "Working of Web Services Using BPEL Workflow in SOA," *Advances in Computing, Communication, and Control*, 125, 143–149.

Karpinski, R. 2008. "Web Services in Action," *Telephony*, 248(4), March 17, 6.

Kay, R. 2007. "Representational State Transfer (REST)," *Computerworld*, 41(32), August 6, 40.

Morochove, R. 2008. "Choosing a Host for Your E-commerce Site," *PC World*, 26(4), April, 36.

Nash, K. 2008. "How to Do CRM Online: Three Big Ideas for 2008," *CIO Magazine*, January 2. (http://www.cio.com/article/168353/How_To_Do_CRM_Online_Three_Big_Ideas_for_2008)

Payne, A. and P. Frow. 2005. "A Strategic Framework for Customer Relationship Management," *Journal of Marketing*, 69(4), October, 167–176.

RESTwiki. 2009. *REST in Plain English*. November 19. (http://rest.blueoxen.net/cgi-bin/wiki.pl?RestInPlainEnglish)

Rigby, D. and D. Ledingham. 2004. "CRM Done Right," *Harvard Business Review*, 82(11), November, 118–127.

Scribner, K. and S. Seely. 2009. *Effective REST Services via .NET*. Boston: Addison-Wesley.

Sharma, R. and M. Sood. 2011. "A Model-driven Approach to cloud SaaS Interoperability," *International Journal of Computer Applications*, 30(8), September, 1–8.

Siebel Systems. 2004. *Ingersoll-Rand Maximizes Customer Focus*. San Mateo, CA: Siebel Systems. (http://www.siebel.com/downloads/case_studies/)

Tedeschi, B. 2005, "Small Internet Retailers Are Using Web Tools to Level the Selling Field," *The New York Times*, December 19. (http://www.nytimes.com/2005/12/19/technology/19ecom.html)

Wan, P., J. Zhi, L. Liu, and G. Cai. 2008. "Building Toward Capability Specifications of Web Services Based on an Environment Ontology," *IEEE Transactions on Knowledge and Data Engineering*, 20(4), April, 547–562.

Wang, W. and W. Liu. 2011. "Study on the Integration of ERP and APS Based on CORBA Static Invocation," *IEEE International Conference on Service Operations, Logistics, and Informatics*, Beijing, July, 172–176.

Waxer, C. 2009. "Bluefly's Bug Zapper," *CIO Magazine*, December 1, 22.

Zhu, Y., K. Chen, X. Guo, and Y. He. 2011. "Management Information Ontology Middleware and Its Needs Guidance Technology," *Recent Advances in Computer Science and Information Engineering*, 125, 415–421.

406

CHAPTER **10**

ELECTRONIC COMMERCE SECURITY

LEARNING OBJECTIVES

In this chapter, you will learn:

- What security risks arise in online business and how to manage them
- How to create a security policy
- How to implement security on Web client computers
- How to implement security in the communication channels between computers
- How to implement security on Web server computers
- What organizations promote computer, network, and Internet security

INTRODUCTION

Large business and government Web sites are constantly under attack by a variety of potential intruders, ranging from computer-savvy high school students to highly trained espionage workers employed by competing businesses or other governments. For example, the U.S. Pentagon reports that its computers are scanned by potential attackers thousands of times every hour. These attackers are continually looking for a way to break through computer security defenses in the hopes of finding any information that could help their employers embarrass, disable, or hurt competitors or enemies. The software that potential attackers use to scan computers is widely available; therefore, government agencies, companies, organizations, and even individuals can expect that their computers are scanned frequently as well.

In 2009, several incidents provided examples of these issues. During the U.S. July 4 holiday and continuing for more than a week after, a series of attacks on U.S. and South Korean Web sites was launched from networks that included more than 200,000 computers located all over the world. These attacks, which targeted both government and business Web sites in both countries, shut down the sites for several hours and included attempts (none reported to be successful) to gather sensitive data. These attacks occurred just a few weeks after U.S. President Barack Obama had announced the creation of a new government agency devoted to defending the country against cyber-terrorism, including attacks of exactly this nature. Although investigators believed that the attacks were the work of operatives of the North Korean government, they were not able to identify defini-tively those responsible for the attack.

Later in 2009, an attack was successful in obtaining an 11-page file that contained a briefing on defensive military operations that would be undertaken by the United States and South Korea if war were to break out with North Korea. A South Korean military officer had left a USB device containing the plans plugged into his computer when he switched the computer from a restricted-access military network to the Internet. Within minutes, an attacker accessed the document and stole a copy of the briefing. Investigators traced the attack to an IP address that is owned by the Chinese government, which had leased it to North Korea. Both governments denied any involvement in the theft.

In this chapter, you will learn how companies and governments protect themselves from attacks that are intended to shut down their Web sites or gain entry to data stored or transmitted in the course of their operational activities. Because the threats are constantly changing, and because the attackers are highly motivated and, in many cases, highly trained, the challenges are constant and dynamic.

ONLINE SECURITY ISSUES OVERVIEW

In the early days of the Internet, one of its most popular uses was electronic mail. Despite e-mail's popularity, business users of e-mail have been concerned about security issues. For example, a business rival might intercept e-mail messages for competitive gain.

Another fear was that employees' nonbusiness correspondence might be read by their supervisors, with negative repercussions. These were significant and realistic concerns.

Today, the stakes are much higher. In addition to e-mail, people all over the world use the Internet and the Web for shopping and conducting all types of financial transactions ranging from an individual buying an item on eBay using PayPal to a large company making a vendor payment through a VPN. These advances make security a concern for all users.

A common worry of Web shoppers is that their credit card numbers might be stolen as they travel across the Internet. Although online wiretapping does occur, it is far more likely that a credit card number will be stolen from a computer on which it is stored after being transmitted over the Internet. Recent surveys show that more than half of all Internet users have at least "some concern" about the security of their credit card numbers in electronic commerce transactions.

As you learned in Chapter 7, people are concerned about personal information they provide to companies over the Internet. Increasingly, people doubt that these companies have the willingness and the ability to keep customers' personal information confidential. This chapter examines security in the context of electronic commerce, presenting an introduction to important security problems and some solutions to those problems.

Origins of Security on Interconnected Computer Systems

Data security measures date back to the time of the Roman Empire, when Julius Caesar coded information to prevent enemies from reading secret war and defense plans carried by his Roman legions. Many modern electronic security techniques were developed for wartime use. The U.S. Department of Defense was the main driving force behind early security requirements and more recent advances. In the late 1970s, the Defense Department formed a committee to develop computer security guidelines for handling classified information on computers. The result of that committee's work was *Trusted Computer System Evaluation Criteria*, known in defense circles as the "Orange Book" because its cover was orange. It spelled out rules for mandatory access control—the separation of confidential, secret, and top secret information—and established criteria for certification levels for computers ranging from D (not trusted to handle multiple levels of classified documents at once) to A1 (the most trustworthy level).

When businesses began using computers, they adopted many of the military's security methods. They established security by using physical controls over access to computers. Alarmed doors and windows, guards, security badges to admit people to sensitive areas, and surveillance cameras were the tools used to secure computers. Back then, interactions between people and computers were limited to the terminals of large mainframe computers. There were very few networks of computers and those that existed did not extend outside the organization to which they belonged. Thus, computer security could be accomplished by managing the activities of the few people who had access to terminals or the computer room.

Since the early years of computing, both the population of computer users and the number of ways those users access computing resources have increased tremendously. It is no longer a simple matter to determine who is using a computing resource. A user in South Africa could be using a computer in California. Security tools and methods have evolved and are used today to protect computers and the electronic assets they store. The transmission of valuable information, such as electronic receipts, purchase orders, payment data, and order confirmations, has drastically increased the need for comprehensive plans and techniques for dealing with the risks posed by weaknesses in computer security.

Computer Security and Risk Management

Computer security is the protection of assets from unauthorized access, use, alteration, or destruction. There are two general types of security: physical and logical. **Physical security** includes tangible protection devices, such as alarms, guards, fireproof doors, security fences, safes or vaults, and bombproof buildings. Protection of assets using nonphysical means is called **logical security**. Any act or object that poses a danger to computer assets is known as a **threat**. A **countermeasure** is a procedure that recognizes, reduces, or eliminates a threat. The extent and expense of countermeasures can vary, depending on the importance of the asset at risk.

Threats that are deemed low risk and unlikely to occur can be ignored when the cost to protect against the threat exceeds the value of the protected asset. For example, it would make sense to protect from tornadoes a computer network in Oklahoma, where there is significant and regular tornado activity. However, a similar network located in Maine would not require the same protection because tornadoes are extremely rare in Maine. The risk management model shown in Figure 10-1 illustrates four general actions that an organization could take, depending on the impact (cost) and the probability of the physical threat. In this model, a tornado in Oklahoma would be in quadrant II, whereas a tornado in Maine would be in quadrant IV.

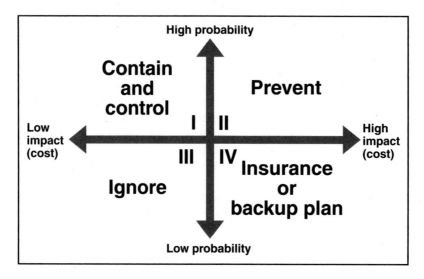

FIGURE 10-1   Risk management model

The same sort of risk management model applies to protecting Internet and electronic commerce assets from both physical and electronic threats. Examples of the latter include impostors, eavesdroppers, and thieves. An **eavesdropper**, in this context, is a person or device that can listen in on and copy Internet transmissions. People who write programs or manipulate technologies to obtain unauthorized access to computers and networks are called **crackers** or **hackers**.

A cracker is a technologically skilled person who uses their skills to obtain unauthorized entry into computers or network systems—usually with the intent of stealing information or damaging the information, the system's software, or even the system's hardware. Originally, the term hacker was used to describe a dedicated programmer who enjoyed writing complex code that tested the limits of technology. Although the term hacker is still used in a positive way by computer professionals who make a strong distinction between the terms hacker and cracker, the media and the general public usually use the term to describe those who use their skills for ill purposes. Some people also use the terms **white hat hacker** and **black hat hacker** to make a distinction between good hackers and bad hackers.

To implement a good security scheme, organizations must identify risks, determine how to protect threatened assets, and calculate how much to spend to protect those assets. In this chapter, the primary focus in risk management protection is on the central issues of identifying the threats and determining the ways to protect assets from those threats, rather than on the protection costs or value of assets.

Elements of Computer Security

Computer security is generally considered to include three main elements: secrecy, integrity, and necessity (also known as denial of service). **Secrecy** refers to protecting against unauthorized data disclosure and ensuring the authenticity of the data source. **Integrity** refers to preventing unauthorized data modification. **Necessity** refers to preventing data delays or denials (removal).

Secrecy is the best known of the computer security elements. Every month, news media report on break-ins to government computers or theft of stolen credit card numbers that are used to order goods and services.

Integrity threats are reported less frequently and are less well known to the general public. An **integrity violation** occurs, for example, when an e-mail message is intercepted and its contents are changed before it is forwarded to its original destination. That is, the integrity of the message has been violated. In this particular exploit, which is called a **man-in-the-middle exploit**, the contents of the e-mail are often altered in a way that changes the message's original meaning.

Necessity violations can take several forms, but always involve preventing or delaying access to data. Either delaying a message or destroying it can have serious consequences. For example, assume that a message sent at 10:00 a.m. to an online stockbroker includes an order to purchase 1000 shares of IBM at market price. If the stockbroker does not receive the message until 2:30 p.m. because an attacker delays it and IBM's stock price has increased by $3, the buyer loses $3000. Other necessity violations include activities such as overwhelming a business Web site with inquiries from automated fake customers so that genuine customers cannot access the Web site.

Establishing a Security Policy

Any organization concerned about protecting its electronic commerce assets should have a **security policy** in place. A security policy is a written statement describing which assets to protect and why they are being protected, who is responsible for that protection, and which behaviors are acceptable and which are not. The policy primarily addresses physical security, network security, access authorizations, virus protection, and disaster recovery. The policy develops over time and is a living document that the company management and security personnel must review and update at regular intervals.

Good security guidelines provide that organizations must protect assets from unauthorized disclosure, modification, or destruction. However, military security policy differs from commercial policy because military applications stress separation of multiple levels of security. Corporate information is usually classified as either "public" or "company confidential." A typical security policy concerning confidential company information could be as simple as "do not reveal confidential company information to anyone outside the company."

411

Most organizations follow a five-step process when creating a security policy. These steps include:

1. Determine which assets must be protected from which threats. For example, a company that stores customer credit card numbers might decide that those numbers are an asset that must be protected.
2. Determine who should have access to various parts of the system or specific information assets. In many cases, some of those users who need access to some parts of the system (such as suppliers, customers, and strategic partners) are located outside the organization.
3. Identify resources available or needed to protect the information assets while ensuring access by those who need it.
4. Using the information gathered in the first three steps, the organization develops a written security policy.
5. Following the written policy, the organization commits resources to building or buying software, hardware, and physical barriers that implement the security policy. For example, if a security policy disallows unauthorized access to customer information (such as credit card numbers or credit history), then the organization must either create or purchase software that guarantees end-to-end secrecy for electronic commerce customers.

A comprehensive plan for security should protect a system's privacy, integrity, and availability (necessity) and authenticate users. When these goals are used to create a security policy for an electronic commerce operation, they should be selected to satisfy the list of requirements shown in Figure 10-2. These requirements provide a minimum level of acceptable security for most electronic commerce operations.

Requirement	Meaning
Secrecy	Prevent unauthorized persons from reading messages and business plans, obtaining credit card numbers, or deriving other confidential information.
Integrity	Enclose information in a digital envelope so that the computer can automatically detect messages that have been altered in transit.
Availability	Provide delivery assurance for each message segment so that messages or message segments cannot be lost undetectably.
Key management	Provide secure distribution and management of keys needed to provide secure communications.
Nonrepudiation	Provide undeniable, end-to-end proof of each message's origin and recipient.
Authentication	Securely identify clients and servers with digital signatures and certificates.

© Cengage Learning

FIGURE 10-2   Requirements for secure electronic commerce

WindowSecurity.com is a good source of information about security policies. Its Network Security Library includes a number of white papers that provide guidance on how to craft a workable security policy. Information Security Policy World is another Web site that provides information about security policy matters.

Although absolute security is difficult to achieve, organizations can create enough barriers to deter most intentional violators. With good planning, organizations can also reduce the impact of natural disasters or terrorist acts. Integrated security means having all security measures working together to prevent unauthorized disclosure, destruction, or modification of assets. A good security policy should address the following:

- *Authentication*: Who is trying to access the site?
- *Access control*: Who is allowed to log on to and access the site?
- *Secrecy*: Who is permitted to view selected information?
- *Data integrity*: Who is allowed to change data?
- *Audit*: Who or what causes specific events to occur, and when?

In this chapter, you will explore these security policy issues with a focus on how they apply to electronic commerce in particular. The electronic commerce security topics in this chapter are organized to follow the transaction-processing flow, beginning with the consumer and ending with the Web server (or servers) at the electronic commerce site. Each logical link in the process includes assets that must be protected to ensure security: client computers, the communication channel on which the messages travel, and the Web servers, including any other computers connected to the Web servers.

SECURITY FOR CLIENT COMPUTERS

Client computers, usually PCs, must be protected from threats that originate in software and data that are downloaded to the client computer from the Internet. In this section, you will learn that active content delivered over the Internet in dynamic Web pages can be harmful. Another threat to client computers can arise when a malevolent server site masquerades as a legitimate Web site. Users and their client computers can be duped into revealing information to those Web sites. This section explains these threats, describes how they work, and outlines some protection mechanisms that can prevent or reduce the threats they pose to client computers.

Cookies and Web Bugs

The Internet provides a type of connection between Web clients and servers called a stateless connection. In a **stateless connection**, each transmission of information is independent; that is, no continuous connection (also called an **open session**) is maintained between any client and server on the Internet. Earlier in this book, you learned that cookies are small text files that Web servers place on Web client computers to identify returning visitors. Cookies also allow Web servers to maintain continuing open sessions with Web clients. An open session is necessary to do a number of things that are important in online business activity. For example, shopping cart and payment processing software both need an open session to work properly. Early in the history of the Web, cookies were devised as a way to maintain an open session despite the stateless nature of Internet connections. Thus, cookies were invented to solve the stateless connection problem by saving information about a Web user from one set of server–client message exchanges to another.

There are two ways of categorizing cookies: by time duration and by source. The two kinds of time-duration cookie categories include **session cookies**, which exist until the Web client ends the connection (or "session"), and **persistent cookies**, which remain on the client computer indefinitely. Electronic commerce sites use both kinds of cookies. For example, a session cookie might contain information about a particular shopping visit and a persistent cookie might contain login information that can help the Web site recognize visitors when they return to the site on subsequent visits. Each time a browser moves to a different part of a merchant's Web site, the merchant's Web server asks the visitor's computer to send back any cookies that the Web server stored previously on the visitor's computer.

Another way of categorizing cookies is by their source. Cookies can be placed on the client computer by the Web server site, in which case they are called **first-party cookies**, or they can be placed by a different Web site, in which case they are called **third-party cookies**. A third-party cookie originates on a Web site other than the site being visited. These third-party Web sites usually provide advertising or other content that appears on the Web site being viewed. The third-party Web site providing the advertising is often interested in tracking responses to their ads by visitors who have already seen the ads on other sites. If the advertising Web site places its ads on a large number of Web sites, it can use persistent third-party cookies to track visitors from one site to another. Earlier in this book, you learned about DoubleClick and similar online ad placement services that perform this function.

The most complete way for Web site visitors to protect themselves from revealing private information or being tracked by cookies is to disable cookies entirely. The problem with this approach is that useful cookies are blocked along with the others, requiring visitors to enter information each time they revisit a Web site. The full resources of some sites are not available to visitors unless their browsers are set to allow cookies. For example, most distance learning software used by schools to deliver online courses does not work properly in student Web browsers unless cookies are enabled.

Web users can accumulate large numbers of cookies as they browse the Internet. Most Web browsers have settings that allow the user to refuse only third-party cookies or to review each cookie before it is accepted. Figure 10-3 shows the dialog box that can be used to manage stored cookies in the Mozilla Firefox Web browser.

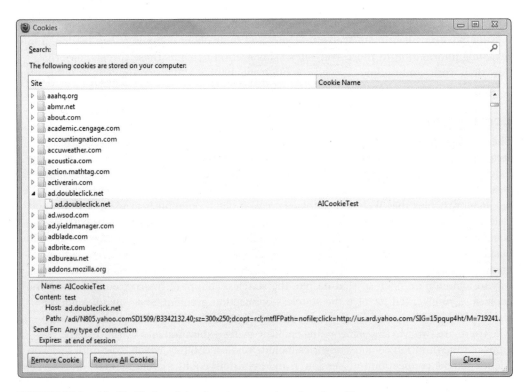

FIGURE 10-3   Mozilla Firefox dialog box for managing stored cookies

You can learn more about cookies at the Cookie Central Web site, which includes current news on cookie developments and answers common questions about cookies. Most of the electronic commerce software packages you learned about in Chapter 9 include features Web site managers can use to analyze Internet traffic at their sites. These services also provide information to Web sites about who visits their sites and what sites the visitors came from.

Some advertisers send images (from their third-party servers) that are included on Web pages but are too small to be visible. A **Web bug** is a tiny graphic that a third-party Web site places on another site's Web page. When a site visitor loads the Web page, the Web bug is delivered by the third-party site, which can then place a cookie on the visitor's computer. A Web bug's only purpose is to provide a way for a third-party Web site (the identity of which is unknown to the visitor) to place cookies from that third-party site on the visitor's computer. The Internet advertising community sometimes calls Web bugs "clear GIFs" or "1-by-1 GIFs" because the graphics can be created in the GIF format with a color value of "transparent" and can be as small as 1 pixel by 1 pixel.

Active Content

Until the debut of executable Web content, Web pages could do little more than display content and provide links to related pages with additional information. The widespread use of active content has changed the situation. **Active content** refers to programs that are embedded transparently in Web pages and that cause action to occur. For example, active content can display moving graphics, download and play audio, or implement Web-based spreadsheet programs. Active content is used in electronic commerce to place items into a shopping cart and compute a total invoice amount, including sales tax, handling, and shipping costs. Developers use active content because it extends the functionality of HTML and moves some data processing chores from the busy server machine to the user's client computer. Unfortunately, because active content elements are programs that run on the client computer, active content can damage the client computer. Thus, active content can pose a threat to the security of client computers.

Active content is provided in several forms. The best-known active content forms are cookies, Java applets, JavaScript, VBScript, and ActiveX controls. Other ways to provide Web active content include graphics, Web browser plug-ins, and e-mail attachments. Most Web browsers allow the user to disable both Java and JavaScript individually or together. Some users do so to avoid the threats posed by allowing them to operate. However, many Web sites use these active content tools to provide important functionality to users, so this is not always an effective threat-reduction technique.

JavaScript and VBScript are **scripting languages**; they provide scripts, or commands, that are executed on the client. An **applet** is a small application program. Applets typically run within the Web browser. Active content is launched in a Web browser automatically when that browser loads a Web page containing active content. The applet downloads automatically with the page and begins running. Some browsers include tools that can limit the actions taken by JavaScript applets. For example, the Options dialog box in Mozilla Firefox has an Advanced JavaScript Settings dialog box (shown in Figure 10-4) in which you can specify the types of JavaScript actions your browser may execute.

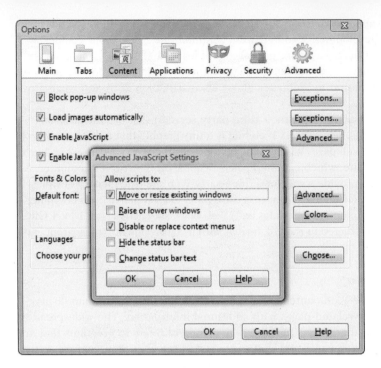

FIGURE 10-4　Advanced JavaScript settings in Mozilla Firefox

Because active content modules are embedded in Web pages, they can be completely invisible when you visit a page containing them. Crackers intent on doing mischief to client computers can embed malicious active content in these seemingly innocuous Web pages. This delivery technique is called a Trojan horse. A **Trojan horse** is a program hidden inside another program or Web page that masks its true purpose. The Trojan horse could snoop around a client computer and send back private information to a cooperating Web server—a secrecy violation. The program could alter or erase information on a client computer—an integrity violation. Zombies are equally threatening. A **zombie** is a Trojan horse that secretly takes over another computer for the purpose of launching attacks on other computers. The computers running the zombie are also sometimes called zombies. When a Trojan horse (or other type of virus) has taken over a large number of computers (and thus made them into zombies), the person who planted the virus can take control of all the computers and form a **botnet** (short for **robotic network**, also called a **zombie farm** when the computers in the network are zombies) that can act as an attacking unit, sending spam or launching denial-of-service attacks against specific Web sites.

Java Applets

Java is a programming language developed by Sun Microsystems that is used widely in Web pages to provide active content. The Web server sends the Java applets along with Web pages requested by the Web client. In most cases, the Java applet's operation will be visible to the site visitor; however, it is possible for a Java applet to perform functions that would not be noticed by the site visitor (such as reading, writing, or erasing files on the site visitor's computer). The client computer then runs the programs within its Web browser. Java can also run outside the confines of a Web browser. Java is platform independent; that is, it can run on many different computers. This "develop once, deploy everywhere" feature reduces development costs because only one program needs to be developed for all operating systems.

Java adds functionality to business applications and can handle transactions and a wide variety of actions on the client computer. That relieves an otherwise busy server-side program from handling thousands of transactions simultaneously. Once downloaded, embedded Java code can run on a client's computer and damage the computer, run a Trojan horse, or turn the computer into a zombie.

To counter this threat, the Java sandbox security model was developed. The **Java sandbox** confines Java applet actions to a set of rules defined by the security model. These rules apply to all untrusted Java applets. **Untrusted Java applets** are those that have not been established as secure. When Java applets are run within the constraints of the sandbox, they do not have full access to the client computer. For example, Java applets operating in the sandbox cannot perform file input, output, or delete operations. This prevents secrecy (disclosure) and integrity (deletion or modification) violations. You can follow the Web Links to the Cerias - Java Security Page maintained by the Center for Education and Research in Information Assurance and Security (CERIAS) to learn more about Java applet security.

JavaScript

JavaScript is a scripting language developed by Netscape to enable Web page designers to build active content. Despite the similar-sounding names, JavaScript is based only loosely on Sun's Java programming language. Supported by popular Web browsers, JavaScript shares many of the structures of the full Java language. When a user downloads a Web page with embedded JavaScript code, it executes on the user's (client) computer.

Like other active content vehicles, JavaScript can be used for attacks by executing code that destroys the client's hard disk, discloses the e-mail stored in client mailboxes, or sends sensitive information to the attacker's Web server. JavaScript code can also record the URLs of Web pages a user visits and capture information entered into Web forms. For example, if a user enters credit card numbers while reserving a rental car, a JavaScript program could copy the credit card number. JavaScript programs, unlike Java applets, do not operate under the restrictions of the Java sandbox security model.

Unlike Java applets, a JavaScript program cannot commence execution on its own. To run an ill-intentioned JavaScript program, a user must start the program. For example, a site with a retirement income calculator might require a visitor to click a button to see a retirement income projection. Once the user clicks the button, the JavaScript program starts and does its work.

ActiveX Controls

An **ActiveX** control is an object that contains programs and properties that Web designers place on Web pages to perform particular tasks. ActiveX components can be constructed using many different programming languages, but the most common are C++ and Visual Basic. Unlike Java or JavaScript code, ActiveX controls run only on computers with Windows operating systems.

When a Windows-based Web browser downloads a Web page containing an embedded ActiveX control, the control is executed on the client computer. Other ActiveX controls include Web-enabled calendar controls and Web games. The ActiveX page at Download.com contains a comprehensive list of ActiveX controls.

The security danger with ActiveX controls is that once they are downloaded, they execute like any other program on a client computer. They have full access to all system resources, including operating system code. An ill-intentioned ActiveX control could reformat a user's hard disk, rename or delete files, send e-mails to all the people listed in the user's address book, or simply shut down the computer. Because ActiveX controls have full access to client computers, they can cause secrecy, integrity, or necessity violations. The actions of ActiveX controls cannot be halted once they begin execution. Most Web browsers can be configured to provide a notice when a Web site attempts to download and install an ActiveX control (or other software). An example of such a warning is shown in Figure 10-5. This dialog box appears in the Internet Explorer Web browser when a user begins downloading the IconDeveloper ActiveX control (a tool for converting graphics files into icons).

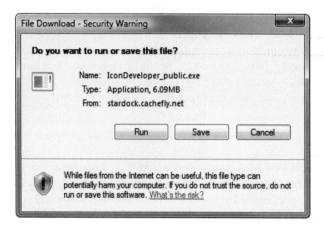

FIGURE 10-5 ActiveX control download warning dialog box in Internet Explorer

Graphics and Plug-Ins

Graphics, browser plug-ins, and e-mail attachments can harbor executable content. Some graphics file formats have been designed specifically to contain instructions on how to render a graphic. That means that any Web page containing such a graphic could be a threat because the code embedded in the graphic could harm to a client computer. Similarly, browser **plug-ins**, which are programs that enhance the capabilities of browsers, handle Web content that a browser cannot handle. Plug-ins are normally beneficial and perform tasks for a browser, such as playing audio clips, displaying movies, or animating graphics. Apple's QuickTime, for example, is a plug-in that downloads and plays movies stored in a special format.

Plug-ins can also pose security threats to a client computer. Users download these plug-in programs and install them so their browsers can display content that cannot be included in HTML tags. Popular plug-ins include Adobe Flash Player, Apple's QuickTime Player, Microsoft Silverlight, and RealNetworks' RealPlayer.

In 1999, *The New York Times* reported that RealNetworks had been using its RealPlayer plug-in to gather information surreptitiously from users. Downloaded and installed easily from the Internet, RealPlayer was recording user information such as the RealPlayer user's name, e-mail address, country, ZIP code, computer operating system, and other details. RealPlayer used the Internet connection to send the information it had gathered back to RealNetworks. Soon after the discovery, and after considerable public embarrassment, RealNetworks issued a statement that a software patch was available that users could install to prevent the RealNetworks software from collecting and transmitting their information.

Many plug-ins execute commands buried within the media being manipulated. This opens the door to the possibility that someone intent on doing harm could embed commands within a seemingly innocuous video or audio clip. The ill-intentioned commands hidden within the object that the plug-in is interpreting could damage a client computer by erasing some (or all) of its files.

Viruses, Worms, and Antivirus Software

The potential dangers lurking in e-mail attachments get a lot of news coverage and are the most familiar to the general population. E-mail attachments provide a convenient way to send nontext information over a text-only system—electronic mail. Attachments can contain word-processing files, spreadsheets, databases, images, or virtually any other information you can imagine. Most programs, including Web browser e-mail programs, display attachments by automatically executing an associated program; for example, the recipient's Excel program reads an attached Excel workbook file and opens it, or Word opens and displays a Word document. Although this activity itself does not cause damage, Word and Excel macro viruses inside the loaded files can damage a client computer and reveal confidential information when those files are opened.

A virus is software that attaches itself to another program and can cause damage when the host program is activated. A **worm** is a type of virus that replicates itself on the computers that it infects. Worms can spread quickly through the Internet. A **macro virus** is a type of virus that is coded as a small program, called a macro, and is embedded in a file. You have probably read about e-mail attachment–borne virus attacks and might have experienced one or more in your own e-mail use.

Although the history of e-mailed viruses dates back to the 1980s, the first virus to become major news in the mainstream media was the ILOVEYOU virus, also known as the "love bug," and its variants in 2000. The ILOVEYOU virus was eventually traced to a 23-year-old computer science student who lived in the Philippines. The virus spread through the Internet with amazing speed as an e-mail message. It infected the computer of anyone who opened the e-mail attachment and clogged e-mail systems with thousands of copies of the useless e-mail message. The virus spread quickly because it automatically sent itself to as many as 300 addresses stored in a computer's Microsoft Outlook address book. Besides replicating itself explosively through e-mail, the virus caused other harm, destroying digital music and photo files stored on the target computers. The ILOVEYOU virus also searched for other users' passwords and forwarded that information to the original perpetrator. Within days, the virus spread to 40 million computers in more than 20 countries and caused an estimated $9 billion in damages—most of it in lost worker productivity.

In 2001, the incidences of virus and worm attacks increased. With more than 40,000 reported security violations occurring that year, the parade of attacks included Code Red and Nimda virus–worm combinations, each affecting millions of computers and costing billions of dollars to clean up. Both Code Red and Nimda are examples of a **multivector virus**, so-called because they can enter a computer system in several different ways (vectors). Even though Microsoft issued security patches that should have stopped the Code Red virus–worm, it continued to propagate throughout the Internet in 2002. Both the original Code Red virus and a variant called Code Red 2 infected thousands of new computers during the year.

New virus–worm combinations also appeared in 2002 and 2003, including a version of the Code Red virus called Bugbear. Bugbear was spread through Microsoft Outlook e-mail clients. The person receiving the e-mail did not even have to click an attachment to run the malicious code—Bugbear started itself through a security loophole in the connection between Outlook and the Internet Explorer browser. Of course, Microsoft issued a security patch for the browser, but many users did not install the patch (or, in many cases, did not even know about it). When launched, Bugbear first checked to see if the computer was running antivirus software. **Antivirus software** detects viruses and worms and either deletes them or isolates them on the client computer so they cannot run. If antivirus software existed on the system, Bugbear attempted to destroy it. Then it installed a Trojan horse program on the computer that let attackers access the computer through the Internet and upload or download files at will. Bugbear would then send out e-mail messages with attachments that would infect the recipients. It did not create its own e-mail messages but took previously sent e-mail messages that were on the computer and re-sent them to different addresses. This often fooled recipients because the e-mail messages had subject headers that seemed normal and did not hint that the e-mail might contain a virus. Bugbear was difficult to eliminate from an infected computer because it gave its own files a randomly generated name; thus, the virus files had different names on every infected computer.

In 2005 and 2006, Zotob was unleashed on the world. One of a new breed of Trojan horse–worm combination threats, Zotob scans the ports of potential target computers and attacks those that have a specific security flaw. Zotob is designed to be helpful to identity thieves and perpetrators of corporate espionage. Once Zotob has infected a target computer, it logs keystrokes and captures screens with the goal of stealing logins, passwords, and even software keys that a user enters to register and activate newly installed programs. Zotob can make the target computer a zombie so it can be used to send spam or launch attacks on other computers.

In 2007, the Storm virus appeared. Storm appears as an e-mail message telling of an interesting news story with a related video clip included as an attachment. The attachment contains the virus, which allows a remote computer to take over the infected computers and form a botnet.

Beginning in 2008 and continuing into 2009, a similar virus named Conficker has been extremely successful. Conficker is believed to have infected nearly 10 million computers. Antivirus vendors and Microsoft have issued patches and updates that eradicate the virus, but it reinstalls itself and has proven to be quite resilient. The size of the ongoing infection has caused great concern, and a number of Internet service providers, computer security firms, and online businesses have formed the **Conficker Working Group** to monitor the virus. The large number of computers that remain infected provide a constant threat because they could be activated remotely at any time to launch a major attack on any Web site in the world.

In 2009 and 2010, new viruses that are designed specifically to hijack users' online banking sessions were introduced or emerged. URLzone waits on the infected computer until a user logs into an account at one of the financial institutions that the virus is programmed to recognize. The virus accesses the bank account at the same time the unknowing victim does and transfers money from the victim's accounts to co-conspirators who take a cut and then use the money to buy goods shipped to a foreign address, where the perpetrator sells the goods and disappears. Clampi is a similar online banking virus that has been infecting computers for years but only became active in 2009. Clampi is reported to recognize more than 4000 different bank, broker, and other financial institution logins.

In 2010, a new use of the Trojan horse–worm combination attack was introduced. For the first time, a Trojan horse that spread through a computer operating system (in this case, Microsoft Windows) was designed to target industrial equipment. The specific target in this case was control systems made by German industrial giant Siemens. These systems were used in many different industrial settings, but the target of the 2010 attack appeared to be systems that controlled Iranian uranium enrichment operations.

In 2011, two existing Trojan horse–worm exploits, Zeus and SpyEye, were combined to create a series of new variants that targeted bank account information stored on computers. These new variants are not visible in Microsoft Windows Task Manager, hide their files from regular Windows Explorer searches, and hide their registry keys. They can intercept credit card or online banking data entered into a Web browser and transmit it to the perpetrator. Figure 10-6 summarizes some of the major viruses, worms, and Trojan horses that have plagued Internet users over the years.

Year	Name	Type	Description
1986	Brain	Virus	Written in Pakistan, this virus infects floppy disks used in personal computers at that time. It consumes empty space on the disks, preventing them from being used to store data or programs.
1988	Internet Worm	Worm	Robert Morris, Jr., a graduate student at Cornell University, wrote this experimental, self-replicating, self-propagating program and released it onto the Internet. It replicated faster than he had anticipated, crashing computers at universities, military sites, and medical research facilities throughout the world.
1991	Tequila	Virus	Tequila writes itself to a computer's hard disk and runs any time the computer is started. It also infects programs when they are executed. Tequila originated in Switzerland and was mostly transmitted through Internet downloads.
1992	Michelangelo	Trojan horse	Set to activate on March 6 (Michelangelo's birthday), this Trojan horse overwrites large portions of the infected computer's hard disk.
1993	SatanBug	Virus	SatanBug infects programs when they run, causing them to fail or perform incorrectly. SatanBug was designed to interfere with antivirus programs so they cannot detect it.
1996	Concept	Virus, Worm	One of the first viruses to be written in the Microsoft Word macro language, Concept travels with infected Word document files. When an infected document is opened, Concept places macros in the Word default document template, which infects any new Word document created on that computer.
1999	Melissa	Virus, Worm	Melissa is a Microsoft Word macro virus that spreads by e-mailing itself automatically from one user to another. It inserts comments from "The Simpsons" television show and confidential information from the infected computer. Melissa spread throughout the world in a few hours. Many large companies were inundated by Melissa. For example, Microsoft closed down its e-mail servers to prevent the spread of this virus within the company.
2000	ILOVEYOU	Virus, Worm	Arrives attached to an e-mail message with the subject line "ILOVEYOU" and infects any computer on which the attachment is opened. It sends itself to addresses in any Microsoft Outlook address book it finds on the infected computer. The virus destroys music and photo files stored on the infected computers. When it was launched, it clogged e-mail servers in many large organizations and slowed down the operation of the entire Internet.
2001	Code Red	Virus, Worm, Trojan horse	Code Red can infect Web servers and personal computers. It defaces Web pages and can be transmitted from Web servers to personal computers. It can give hackers control over Web server computers. Code Red can reinstall itself from hidden files after it is removed.

FIGURE 10-6 Major viruses, worms, and Trojan horses

Year	Name	Type	Description
2001	Nimda	Virus, Worm	Nimda modifies Web documents and certain programs on the infected computer. It also creates multiple copies of itself using various file-names. It can be transmitted by e-mail, a LAN, or from a Web server to a Web client.
2002	BugBear	Virus, Worm, Trojan horse	BugBear is spread through e-mail and through local area networks. It identifies antivirus software and attempts to disable it. BugBear can log keystrokes and store them for later transmission through a Trojan horse program that it installs on the infected computer. This program gives hackers access to the computer and allows file uploads and downloads.
2002	Klez	Virus, Worm	Klez is transmitted as an e-mail attachment and overwrites files, creates hidden copies of the original files, and attempts to disable antivirus software.
2003	Slammer	Worm	Slammer's primary purpose was to demonstrate how rapidly a worm could be transmitted on the Internet. It infected 75,000 computers in its first 10 minutes of propagation.
2003	Sobig	Trojan horse	Sobig turns infected computers into spam relay points. Sobig transmits mass e-mails with copies of itself to potential victims.
2004	MyDoom	Worm, Trojan horse	MyDoom turns the infected computer into a zombie that will participate in a denial of service attack on a specific company's Web site.
2004	Sasser	Virus, Worm	Written by a German high school student, Sasser finds computers with a specific security flaw and then infects them. The infected computers are slowed by the virus, often to the point that they must be rebooted.
2005	Zotob	Worm, Trojan horse	Zotob peforms port scans and infects computers that appear to have a specific security flaw. Once installed on a target computer, Zotob can log keystrokes, capture screens, and steal authentication credentials and CD software keys. Infected computers can also be used as zombies for mass mailing or attacking other computers.
2006	Nyxem	Worm, Trojan horse	Nyxem disables security and file-sharing software. It destroys files created by Microsoft Office programs. Nyxem activates on the third of each month and spreads itself by mass mailing.
2006	Leap	Worm, Virus	Leap (also called Oompa-Loompa) infects programs that run on the Macintosh OX X operating system. Delivered over the iChat instant messaging system, it can only spread within a specific network.
2007	Storm	Worm, Trojan horse	Storm gathers infected computers into a botnet from which it launches spam. It is spread as an e-mail containing phony news clips with an attachment that it alleges is a news film.

FIGURE 10-6   Major viruses, worms, and Trojan horses (continued)

Year	Name	Type	Description
2008	Conficker	Worm, Trojan horse	Conficker has not been used in any significant way, but it is able to reinstall itself and remains on more than 7 million computers. If activated, it could launch a devastating barrage of spam e-mail or a crippling denial-of-service attack on any Web site in the world.
2009	Clampi	Worm, Trojan horse	Activated in 2009 after lying dormant for years, Clampi captures username and password information for more than 4000 bank, broker, and other financial institution Web sites. It forwards information to perpetrators who can use it to purchase goods or transfer funds from victims' accounts.
2009	URLzone	Worm, Trojan horse	URLzone monitors user activity and hijacks the session when the victim logs into a financial institution Web site that it is programmed to recognize. It then transfers money from the victim's accounts to confederates, who take their cut and then buy goods shipped to a foreign address used by the perpetrator. The perpetrator sells the goods and moves on.
2010	Stuxnet	Worm, Trojan horse	Stuxnet spreads through Microsoft Windows, but targets industrial software and equipment built by Siemens. The first worm designed to attack such systems, experts believe it was created for the purpose of damaging Iranian uranium enrichment systems.
2010	VBManie	Virus, Trojan horse	A virus transmitted by e-mail messages with the subject header "here you have." The message states that the attachment is "The Document I told you about."
2011	Anti-spyware 2011	Virus, Trojan horse	Posing as an antivirus program, Antispyware 2011 actually attacks and disables security features of antivirus programs already installed on the victim's computer. It also blocks Internet access so the disabled antivirus program cannot obtain updates that might restore it.
2011	ZeuS/ SpyEye variants	Worm, Trojan horse	These two Trojans were merged to create a series of new variants designed to attack mobile banking information stored on computers.

FIGURE 10-6   Major viruses, worms, and Trojan horses *(continued)*

Symantec and McAfee, among other companies, keep track of viruses and sell antivirus software. You can follow the Web Links (**Symantec Security Response** and **McAfee Virus Information**) to find descriptions of thousands of viruses. Antivirus software is only effective if the antivirus data files are kept current. The data files contain virus-identifying information that is used to detect viruses on a client computer. Because new viruses appear regularly, users must be vigilant and update their antivirus data files regularly so that the newest viruses are recognized and eliminated. Some Web e-mail systems, such as Yahoo! Mail, let users scan attachments using antivirus software before downloading e-mail. In these cases, the antivirus software is run by the Web site and the user does not need to take any action to keep the software updated.

Microsoft Internet Information Server

As you learned in Chapter 8, Internet Information Server (IIS) is Microsoft's Web server software. Microsoft supplies IIS with the versions of its Windows server operating systems that are suitable for use in operating electronic commerce Web sites.

In August 2001, Microsoft faced an uncomfortable situation that many U.S. manufacturing companies have experienced with recalled, defective products. Microsoft executives stood by at a news conference while a U.S. government official announced to reporters that there was a serious flaw in a Microsoft product. The director of the FBI's National Infrastructure Protection Center was warning reporters that the Code Red worm, which was spreading through the Internet for the third time in as many weeks, was a serious threat to the continued operation of the Internet.

The Code Red worm exploits a vulnerability in the Microsoft IIS Web server software. When the worm was first identified, Microsoft quickly made a patch available on its Web site. Microsoft also announced that Web server installations that had kept current with all of the updates and patches that Microsoft had issued would not be subject to attack by the worm.

Many Microsoft customers were outraged by these statements, noting that Microsoft had issued more than 40 software patches in the first half of 2001 and 100 or more patches in each of several prior years. IIS users complained that keeping the software current was virtually impossible and called for Microsoft to deliver software that was more secure when first installed.

Many IIS users began to consider switching to other Web server software. Gartner, Inc., a major IT consulting firm, recommended to its clients that they seriously consider alternatives to IIS for their critical Web server installations. Many industry observers and software engineers agree that Microsoft was a victim of its own success. It had created a very popular and complex piece of software. It is extremely difficult to ensure that no bugs exist in complex software products, and the popularity of the software made it an attractive target for crackers—one worm could bring down many of the servers operating on the Internet. These two factors, plus the likelihood that many IIS servers would not have all of the available security upgrades installed, combined to make it an irresistible target for a worm creator.

Microsoft has struggled to gain the confidence of large corporate IT departments. The company has worked hard to convince users that its operating system software is reliable and trustworthy. For example, when Microsoft introduced version 7 of IIS in 2008, it announced that its architecture had been changed so that users could install only the modules they needed to reduce the software's "attack surface."

The Code Red worm attack on its Web server software was a major setback in its reputation-building effort. Since that attack, a number of security weaknesses have been identified in IIS and patched by Microsoft. The news reports that inevitably accompany these patches have created a continuing public relations issue for the company. You can review the Microsoft Safety & Security Center through the Web Links to see how Microsoft deals with ongoing concerns that its software is secure in the face of attacks that are both regular and frequent.

Digital Certificates

One way to control threats from active content is to use digital certificates. A **digital certificate** or **digital ID** is an attachment to an e-mail message or a program embedded in a Web page that verifies that the sender or Web site is who or what it claims to be. In addition, the digital certificate contains a means to send an encrypted message—encoded so others cannot read it—to the entity that sent the original Web page or e-mail message. In the case of a downloaded program containing a digital certificate, the encrypted message identifies the software publisher (ensuring that the identity of the software publisher matches the certificate) and indicates whether the certificate has expired or is still valid. The digital certificate is a **signed** message or code. Signed code or messages serve the same function as a photo on a driver's license or passport. They provide proof that the holder is the person identified by the certificate. Just like a passport, a certificate does not imply anything about either the usefulness or quality of the downloaded program. The certificate only supplies a level of assurance that the software is genuine. The idea behind certificates is that if the user trusts the software developer, signed software can be trusted because, as proven by the certificate, it came from that trusted developer.

Digital certificates are used for many different types of online transactions, including electronic commerce, electronic mail, and electronic funds transfers. A digital ID verifies a Web site to a shopper and, optionally, identifies a shopper to a Web site. Web browsers or e-mail programs exchange digital certificates automatically and invisibly when requested to validate the identity of each party involved in a transaction.

When a browser has established secure communication with a Web site; that is, when the name of the site appears in the browser's address bar in a different color text or with a colored background (the exact notification methods vary from browser to browser), the user can double-click the site name to display the Web site's security information and its digital certificate. Figure 10-7 displays the digital certificate owned by Cengage Learning and used on its CengageBrain site.

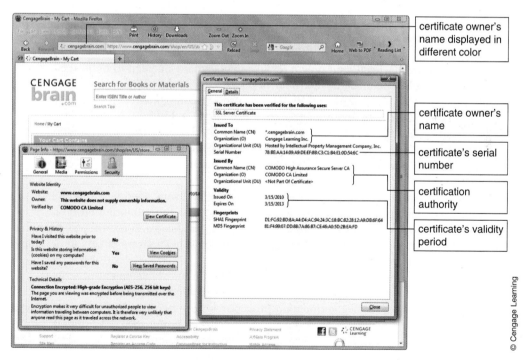

FIGURE 10-7 Cengage Learning's digital certificate information displayed in Firefox browser

A digital certificate for software is an assurance that the software was created by a specific company. The certificate does not attest to the quality of the software, just to the identity of the company that published it. Digital certificates are issued by a

certification authority (CA). A CA can issue digital certificates to organizations or individuals. A CA requires entities applying for digital certificates to supply appropriate proof of identity. Once the CA is satisfied, it issues a certificate. Then, the CA signs the certificate, and its stamp of approval is affixed in the form of a public encryption key (you will learn more about encryption later in this chapter—in encryption, a message is encoded into an unintelligible form that only the proper recipient can convert back into the original message). The public encryption key "unlocks" the certificate for anyone who receives the certificate attached to the publisher's code. Digital certificates cannot be forged easily. A digital certificate includes six main elements, including:

- Certificate owner's identifying information, such as name, organization, address, and so on
- Certificate owner's public encryption key
- Dates between which the certificate is valid
- Serial number of the certificate
- Name of the certificate issuer
- Digital signature of the certificate issuer

A **key** is a number—usually a long binary number—that is used with the encryption algorithm to "lock" the characters of the message being protected so that they are undecipherable without the key. Longer keys usually provide significantly better protection than shorter keys. In effect, the CA is guaranteeing that the individual or organization that presents the certificate is who or what it claims to be.

Identification requirements vary from one CA to another. One CA might require a driver's license for individuals' certificates; others might require a notarized form or fingerprints. CAs usually publish identification requirements so that any Web user or site accepting certificates from each CA understands the stringency of that CA's validation procedures. Only a small number of CAs exist because the certificates issued are only as trustworthy as the CA itself, and only a few companies have decided to build the reputation needed to be a successful seller of digital certificates. Two of the most commonly used CAs are **Thawte** and **VeriSign**, but other companies such as **Comodo, DigiCert, Entrust, GeoTrust,** and **RapidSSL.com** also offer CA services. In 2011, Symantec (the company that sells the well-known Norton antivirus and security programs) bought VeriSign; in 2012, the company began identifying holders of VeriSign digital certificates as "Norton Secured, powered by VeriSign." The digital certificate for Cengage Learning (shown in Figure 10-7) was issued by Comodo.

Although the use of digital certificates increased Web users' confidence in online shopping and banking sites, some CAs were performing just the minimum level of verification on certificate applicants before issuing the certificates. A growing concern that fraudulent Web sites might be obtaining digital certificates led a group of CAs to develop a more stringent set of verification steps.

In 2008, the higher standards for verification led to the establishment of stricter criteria and an assurance of consistent application of verification procedures. CAs that followed these more extensive verification procedures were permitted to issue a new type of certificate called a **Secure Sockets Layer-Extended Validation (SSL-EV) digital certificate.** To issue an SSL-EV certificate, a certification authority must confirm the legal existence of the organization by verifying the organization's registered legal name, registration number, registered address, and physical business address. The CA must also verify the organization's right to use the domain name and that the organization has authorized the request for an SSL-EV certificate.

You can tell if you are visiting a Web site that has an SSL-EV certificate by looking at the address window of your browser. In Firefox, the site's verified organization name appears in the address window to the left of the URL in green text. In Internet Explorer, the background of the address window turns green and the verified name of the organization appears to the right of the URL and alternates with the name of the certification authority, as shown in Figure 10-8.

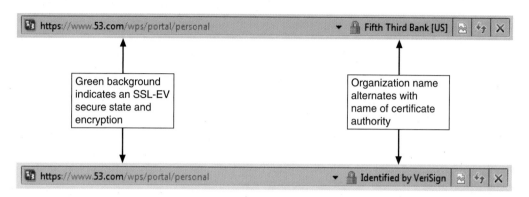

FIGURE 10-8   Internet Explorer address window display for an SSL-EV Web site

Annual fees for digital certificates range from about $200 to more than $1500, depending on the features they include (such as encryption strength, or the SSL-EV designation) and whether they are purchased alone or with certificates for other Web sites owned by the same company. Digital certificates expire after a period of time (often one year). This built-in limit provides protection for both users and businesses. Limited-duration certificates guarantee that businesses and individuals must submit their credentials for reevaluation periodically. The expiration date appears in the certificate itself and in the dialog boxes that browsers display when a Web page or applet that has a digital certificate is about to be opened. Certificates become invalid on their expiration dates or when they are revoked by the CA. If the CA determines that a Web site has delivered malicious code or has otherwise violated the terms to which it agreed, the CA will refuse to issue new certificates to that site and revoke existing certificates.

Steganography

The term **steganography** describes the process of hiding information (a command, for example) within another piece of information. This information can be used for malicious purposes. Frequently, computer files contain redundant or insignificant information that can be replaced with other information. This other information resides in the background and is undetectable by anyone without the correct decoding software. Steganography provides a way of hiding an encrypted file within another file so that a casual observer cannot detect that there is anything of importance in the container file. In this two-step process, encrypting the file protects it from being read, and steganography makes it invisible.

Many security analysts believe that the terrorist organization Al Qaeda used steganography to hide attack orders and other messages in images that its confederates posted on Web sites in preparation for the attacks of September 11, 2001. Messages hidden using steganography are extremely difficult to detect. This fact, combined with the fact that there are millions of images on the Web, makes the use of steganography by global terrorist organizations a deep concern for governments and security professionals.

The Web Links include a link to a site with more information about **Steganography and Digital Watermarking**.

Physical Security for Clients

In the past, physical security was a major concern for large computers that ran important business functions such as payroll or billing; however, as networks (including intranets and the Internet) have made it possible to control important business functions from client computers, concerns about physical security for client computers have become greater. Many of the physical security measures used today are the same as those used in the early days of computing; however, some interesting new technologies have been implemented as well.

Devices that read fingerprints are now available for personal computers. These devices, which cost less than $100, provide much stronger protection than traditional password approaches. In addition to fingerprint readers, companies can use other biometric security devices that are more accurate and, of course, cost more. A **biometric security device** is one that uses an element of a person's biological makeup to perform the identification. These devices include writing pads that detect the form and pressure of a person writing a signature, eye scanners that read the pattern of blood vessels in a person's retina or the color levels in a person's iris, and scanners that read the palm of a person's hand (rather than just one fingerprint) or that read the pattern of veins on the back of a person's hand.

Client Security for Mobile Devices

As more and more people use mobile devices, such as smart phones and tablets, to access the Internet, concern for the security of these devices increases proportionally. Security issues related to mobile client devices can be simple, such as the physical threat of losing a phone or tablet device. They can also be more complex, such as an attack by a Trojan horse, a virus, or an app that shares your personal information.

The first step to take in securing a mobile device is to set up a password for access to the phone. This can prevent or at least delay a thief who has stolen your device from obtaining private information you have stored on it.

Almost all mobile devices include software that allows the owner to initiate a remote wipe if the device is stolen. A **remote wipe** clears all of the personal data stored on the device, including e-mails, text messages, contact lists, photos, videos, and any type of document file. If a mobile device does not include remote wipe software, it can be added as an app. Most corporate e-mail servers include the ability to do a remote wipe of any employee's mobile devices through the e-mail synchronization software installed on the devices.

Web sites that contain malware can infect mobile devices just as easily as they can client computers. Text messages and e-mails with attached viruses and Trojan horses can infect smart phones and tablet devices also. Thus, an increasing number of users are installing antivirus software on their mobile devices.

Apps that contain malware or that collect information from the mobile device and forward it to perpetrators are called **rogue apps**. The Apple App Store tests apps before they are authorized for sale to weed out rogue apps. The Android Market does not screen for rogue apps as extensively as Apple; however, all Android apps must request permission from the user to access any specific information stored on the device. The app will request these permissions when the user installs the app. To avoid rogue Android apps, experts advise mobile device users to read reviews of any app they are thinking about installing and not to be in a rush to install newly available apps that have not been installed by very many users yet. They also recommend avoiding app stores other than the Android Market.

COMMUNICATION CHANNEL SECURITY

The Internet serves as the electronic connection between buyers (in most cases, clients) and sellers (in most cases, servers). The most important thing to remember as you learn about communication channel security is that the Internet was not designed to be secure. Although the Internet has its roots in a military network, that network was not designed to include any significant security features. It was designed to provide redundancy in case one or more communications lines were cut. In other words, the goal of the Internet's packet-switching design was to provide multiple alternative paths on which critical military information could travel. The military always sends sensitive information in an encrypted form so that the content of messages traveling over any network—even if intercepted—remains secret. The security of messages traversing the military predecessors to the Internet was provided by software that operated independently of the network to encrypt messages. As the Internet developed, it did so without any significant security features that became a part of the network itself.

Today, the Internet remains largely unchanged from its original, insecure state. Message packets on the Internet travel an unplanned path from a source node to a destination node. A packet passes through a number of intermediate computers on the network before reaching its final destination. The path can vary each time a packet is sent between the same source and destination points. Because users cannot control the path and do not know where their packets have been, it is possible that an intermediary can read the packets, alter them, or even delete them. That is, any message traveling on the Internet is subject to secrecy, integrity, and necessity threats. This section describes these problems in more detail and outlines several solutions for those problems.

Secrecy Threats

Secrecy is the security threat that is most frequently mentioned in articles and the popular media. Closely linked to secrecy is privacy, which also receives a great deal of attention. Secrecy and privacy, though similar, are different issues. Secrecy is the prevention of unauthorized information disclosure. **Privacy** is the protection of individual rights to nondisclosure. The **Privacy Council**, which helps businesses implement smart privacy and data practices, created an extensive Web site that addresses privacy—covering both business and legal issues. Secrecy is a technical issue requiring sophisticated physical and logical mechanisms, whereas privacy protection is a legal matter. A classic example of the difference between secrecy and privacy is e-mail.

A company might protect its e-mail messages against secrecy violations by using encryption. Secrecy countermeasures protect outgoing messages. E-mail privacy issues address whether company supervisors should be permitted to read employees' messages randomly. Disputes in this area center around who owns the e-mail messages: the company or the employees who sent them. The focus in this section is on secrecy, preventing unauthorized persons from reading information they should not be reading.

One significant threat to electronic commerce is theft of sensitive or personal information, including credit card numbers, names, addresses, and personal preferences. This kind of theft can occur any time anyone submits information over the Internet because it is easy for an ill-intentioned person to record information packets (a secrecy violation) from the Internet for later examination. The same problems can occur in e-mail transmissions. Software applications called **sniffer programs** provide the means to record information that passes through a computer or router that is handling Internet traffic. Using a sniffer program is analogous to tapping a telephone line and recording a conversation. Sniffer programs can read e-mail messages and unencrypted Web client–server message traffic, such as user logins, passwords, and credit card numbers.

Periodically, security experts find electronic holes, called backdoors, in electronic commerce software. A **backdoor** is an element of a program (or a separate program) that allows users to run the program without going through the normal authentication procedure for access to the program. Programmers often build backdoors into programs while they are building and testing them to save the time it would take to enter a login and password every time they open the program. Sometimes programmers forget to remove backdoors when they are finished writing the program; other times, programmers intentionally leave a backdoor.

A backdoor allows anyone with knowledge of its existence to cause damage by observing transactions, deleting data, or stealing data. For example, a security consulting firm found that Cart32, a widely used shopping cart program, had a backdoor through which credit card numbers could be obtained by anyone with knowledge of the backdoor. This backdoor resulted from a programming error and not an intentional effort (and Cart32 provided a software patch that closed the backdoor immediately), but customers of the merchants who used Cart32 had their credit card numbers exposed to hackers around the world until those merchants applied the patch.

Credit card number theft is an obvious problem, but proprietary corporate product information or prerelease product data sheets mailed to corporate branches can be intercepted and passed along easily, too. Confidential information can be considerably more valuable than information about credit cards, which usually have spending limits. Stolen corporate information, such as blueprints, product formulas, or marketing plans, can be worth millions of dollars.

Here is an example of how an online eavesdropper might obtain confidential information. Suppose a user logs on to a Web site that contains a form with text boxes for name, address, and e-mail address. When the user fills out those text boxes and clicks the submit button, the information is sent to the Web server for processing. Some Web servers obtain and track that data by collecting the text box responses and placing them at the end of the server's URL (which appears in the address box of the user's Web browser). This long URL (with the text box responses appended) is included in all HTTP request and response messages that travel between the user's browser and the server.

So far, no violations have occurred. Suppose, however, that the user decides not to wait for a response from the server. Instead, the user visits another Web site. The server at this second Web site might be set up to collect Web demographics. If it is, it logs the URL from which the user just came by capturing it from the HTTP request message that the browser sends. Web sites use this URL logging technique for the completely legitimate purpose of identifying sources of customer traffic. However, any employee at the second site who has access to the server log can read the part of the URL that includes the information entered into those text boxes on the first site, thus obtaining that user's confidential information.

Web users continually reveal information about themselves when they use the Web. This information includes IP addresses and the type of browser being used. Such data exposure is a secrecy breach. Several Web sites offer an anonymous browser service that hides personal information from sites visited. These sites provide a measure of secrecy to Web surfers who use them by replacing the user's IP address with the IP address of the anonymous Web service on the front end of any URLs that the user visits. When the Web site logs the site visitor's IP address, it logs the IP address of the anonymous Web service rather than that of the visitor, which preserves the visitor's privacy.

Using such a service can make anonymous Web surfing possible, but tedious, because each URL that the user wants to visit must be typed into the text box on the anonymous Web service's home page. To make the process easier, companies such as **Anonymizer** provide browser plug-in software that users can download and install for an annual subscription fee. **ShadowSurf.com** provides a free anonymous browser service online.

Integrity Threats

An integrity threat, also known as **active wiretapping**, exists when an unauthorized party can alter a message stream of information. Unprotected banking transactions, such as deposit amounts transmitted over the Internet, are subject to integrity violations. Of course, an integrity violation implies a secrecy violation because an intruder who alters information can read and interpret that information. Unlike secrecy threats, where a viewer simply sees information he or she should not, integrity threats can cause a change in the actions a person or corporation takes because a mission-critical transmission has been altered.

Cybervandalism is an example of an integrity violation. **Cybervandalism** is the electronic defacing of an existing Web site's page. The electronic equivalent of destroying property or placing graffiti on objects, cybervandalism occurs whenever someone replaces a Web site's regular content with his or her own content. Recently, several cases of Web page defacing involved vandals replacing business content with pornographic material and other offensive content.

Masquerading or **spoofing**—pretending to be someone you are not, or representing a Web site as an original when it is a fake—is one means of disrupting Web sites. **Domain name servers (DNSs)** are the computers on the Internet that maintain directories that link domain names to IP addresses. Perpetrators can use a security hole in the software that runs on some of these computers to substitute the addresses of their Web sites in place of the real ones to spoof Web site visitors.

For example, a hacker could create a fictitious Web site masquerading as www. widgets.com by exploiting a DNS security hole that substitutes the hacker's fake IP address for Widgets.com's real IP address. All subsequent visits to Widgets.com would be redirected to the fictitious site. There, the hacker could alter any orders to change the number of widgets ordered and redirect shipment of those products to another address. The integrity attack consists of intercepting and altering an order and then passing it to the real company's Web server. The Web server is unaware of the integrity attack and simply verifies the consumer's credit card number and passes on the order for fulfillment.

Major electronic commerce sites that have been the victims of masquerading attacks in recent years include Amazon.com, AOL, eBay, and PayPal. Some of these schemes combine spam with spoofing. The perpetrator sends millions of spam e-mails that appear to be from a legitimate company. The e-mails contain a link to a Web page that is designed to look exactly like the company's site. The victim is encouraged to enter username, password, and sometimes even credit card information. These exploits, which capture confidential customer information, are called **phishing expeditions**. The most common victims of phishing expeditions are users of online banking and payment system (such as PayPal) Web sites. You will learn more about the phishing problem and the measures banks and other companies are taking to combat it in Chapter 11.

Necessity Threats

The purpose of a **necessity threat**, which usually occurs as a **delay, denial, or denial-of-service (DoS) attack**, is to disrupt normal computer processing, or deny processing entirely. A computer that has experienced a necessity threat slows processing to an intolerably slow speed. For example, if the processing speed of a single ATM transaction slows from one or two seconds to 30 seconds, users will abandon ATMs entirely. Similarly, slowing any Internet service drives customers to competitors' Web or commerce sites—possibly discouraging them from ever returning to the original commerce site. In other words, slower processing can render a service unusable or unattractive. For example, an online newspaper that reports three-day-old news is worth very little. The Internet Worm attack of 1998, which disabled thousands of computer systems that were connected to the Internet, was the first recorded example of a DoS attack.

Attackers can use the botnets you learned about earlier in this chapter to launch a simultaneous attack on a Web site (or a number of Web sites) from all of the computers in the botnet. This form of attack is called a **distributed denial-of-service (DDoS) attack**. The attack on U.S. and South Korean government and business Web sites you learned about at the beginning of this chapter was a DDoS attack.

DoS attacks can remove information altogether, or delete information from a transmission or file. One denial attack targeted PCs that have Quicken (an accounting program) installed. The perpetrator's computer was able to take control of Quicken and use that program's electronic payment capability to divert money to the perpetrator's bank account. In another DoS attack against high-profile electronic commerce sites such as Amazon.com and Yahoo!, the attackers used a botnet to send a flood of data packets to the sites. This overwhelmed the sites' servers and choked off legitimate customers' access. Prior to the attack, perpetrators located vulnerable computers and loaded them with the software that attacked the commerce sites.

Threats to the Physical Security of Internet Communications Channels

The Internet was designed from its inception to withstand attacks on its physical communication links. Recall from Chapter 2 that the main purpose of the U.S. government research project that led to the development of the Internet was to provide an attack-resistant technology for coordinating military operations. Thus, the Internet's packet-based network design precludes it from being shut down by an attack on a single communications link on that network.

However, an individual user's Internet service can be interrupted by destruction of that user's link to the Internet. Few individual users have multiple connections to an ISP. However, larger companies and organizations (and ISPs themselves) often do have more than one link to the main backbone of the Internet. Typically, each link is purchased from a different network access provider. If one link becomes overloaded or unavailable, the service provider can switch traffic to another network access provider's link to keep the company, organization, or ISP (and its customers) connected to the Internet.

Threats to Wireless Networks

As you learned in Chapter 2, networks can use wireless access points (WAPs) to provide network connections to computers and other mobile devices within a range of several hundred feet. If not protected, a wireless network allows anyone within that range to log in and have access to any resources connected to that network. Such resources might include any data stored on any computer connected to the network, networked printers, messages sent on the network, and, if the network is connected to the Internet, free access to the Internet. The security of the connection depends on the **Wireless Encryption Protocol (WEP)**, which is a set of rules for encrypting transmissions from the wireless devices to the WAPs.

Companies that have large wireless networks are usually careful to turn on WEP in devices, but smaller companies and individuals who have installed wireless networks in their homes often do not turn on the WEP security feature. Many WAPs are shipped to buyers with a default login and password already set. Companies that install these WAPs sometimes fail to change that login and password. This has given rise to a new avenue of entry into networks.

In some cities that have large concentrations of wireless networks, attackers drive around in cars using their wireless-equipped laptop computers to search for accessible networks. These attackers are called **wardrivers**. When wardrivers find an open network (or a WAP that has a common default login and password), they sometimes place a chalk mark on the building so that other attackers will know that an easily entered wireless network is nearby. This practice is called **warchalking**. Some warchalkers have even created Web sites that include maps of wireless access locations in major cities around the world. Companies can avoid becoming targets by simply turning on WEP in their access points and changing the logins and passwords to something other than the manufacturers' default settings.

An early victim of a wireless attack, Best Buy was using wireless point-of-sale (POS) terminals in some of its stores. The wireless POS terminals could be moved easily from one area of the store to another, and they helped Best Buy handle large customer flows better than it could using only fixed POS terminals. Unfortunately, Best Buy had not enabled WEP on these terminals. A customer who had just purchased a wireless card for his laptop decided to launch a sniffer utility program on the laptop in his car in the parking lot. The customer was able to intercept data from the POS terminals, including transaction details and what he said looked like credit card numbers. Best Buy stopped using the wireless POS terminals when the story appeared on several Web sites and newswire services.

Encryption Solutions

Encryption is the coding of information by using a mathematically based program and a secret key to produce a string of characters that is unintelligible. The science that studies encryption is called **cryptography**, which comes from a combination of the two Greek words *krypto* and *grapho*, which mean "secret" and "writing," respectively. That is, cryptography is the science of creating messages that only the sender and receiver can read.

Cryptography is different from steganography, which makes text undetectable to the naked eye. Cryptography does not hide text; it converts it to other text that is visible but does not appear to have any meaning. What an unauthorized reader sees is a string of random text characters, numbers, and punctuation.

Encryption Algorithms

A program that transforms normal text, called **plain text**, into **cipher text** (the unintelligible string of characters) is called an **encryption program**. The logic behind an encryption program that includes the mathematics used to do the transformation from plain text to cipher text is called an **encryption algorithm**. There are a number of different encryption algorithms in use today. Some have been developed by the U.S. government and others have been developed by IBM and other commercial enterprises. You can learn more about the development of encryption algorithms, including an evaluation of currently available algorithms, by consulting a Web security textbook (see, for example, the Mackey reference in the For Further Study and Research section at the end of this chapter).

Messages are encrypted just before they are sent over a network or the Internet. Upon arrival, each message is decoded, or **decrypted**, using a **decryption program**—a type of encryption-reversing procedure. Encryption algorithms are considered so vitally important to preserving security within the United States that the National Security Agency has control over their dissemination. Some encryption algorithms are considered so important that the U.S. government has banned publication of details about them. Currently, it is illegal for U.S. companies to export some of these encryption algorithms. The Freedom Forum Online contains a number of articles on lawsuits and legislation surrounding encryption export laws. Critics consider publication restrictions a freedom of speech issue. If you are interested in reading more about the latest arguments in the ongoing debates over freedom of speech and export law, search the **Freedom Forum** using the keyword "encryption" as the search term.

One property of encryption algorithms is that someone can know the details of the algorithm and still not be able to decipher the encrypted message without knowing the key that the algorithm used to encrypt the message. The resistance of an encrypted message to attack attempts depends on the size (in bits) of the key used in the encryption procedure. A 40-bit key is currently considered to provide a minimal level of security. Longer keys, such as 128-bit keys, provide much more secure encryption. A sufficiently long key can help make the security unbreakable.

The type of key and associated encryption program used to lock a message, or otherwise manipulate it, subdivides encryption into three functions:

1. Hash coding
2. Asymmetric encryption
3. Symmetric encryption

Hash Coding

Hash coding is a process that uses a **hash algorithm** to calculate a number, called a **hash value**, from a message of any length. It is a fingerprint for the message because it is almost certain to be unique for each message. Good hash algorithms are designed so that the probability of two different messages resulting in the same hash value, which would create a **collision**, is extremely small. Hash coding is a particularly convenient way to tell whether a message has been altered in transit because its original hash value and the hash value computed by the receiver will not match after a message is altered.

Asymmetric Encryption

Asymmetric encryption, or **public-key encryption**, encodes messages by using two mathematically related numeric keys. In 1977, Ronald Rivest, Adi Shamir, and Leonard Adleman invented the RSA Public Key Cryptosystem while they were professors at MIT. Their invention revolutionized the way sensitive information is exchanged. In their system, one key of the pair, called a **public key**, is freely distributed to the public at large—to anyone interested in communicating securely with the holder of both keys. The public key is used to encrypt messages using one of several different encryption algorithms. The second key—called a **private key**—belongs to the key owner, who keeps the key secret. The owner uses the private key to decrypt all messages received.

Here is an example showing how asymmetric encryption works: If Herb wants to send a message to Allison, he obtains Allison's public key from any of several well-known public places. Then, he encrypts his message to Allison using her public key. Once the message is encrypted, only Allison can read the message by decrypting it with her private key. Because the keys are unique, only one secret key can open the message encrypted with a corresponding public key, and vice versa. Reversing the process, Allison can send a private message to Herb using Herb's public key to encrypt the message. When he receives Allison's message, Herb uses his private key to decrypt the message and then read it. If they are sending e-mail to one another, the message is secret only while in transit. Once a message is downloaded from the mail server and decoded, it is stored in plain text on the recipient's machine for all to view.

One of the most popular technologies used to implement public-key encryption today is called **Pretty Good Privacy (PGP)**. PGP was invented in 1991 by Phil Zimmerman, who charged businesses for use of PGP but allowed individuals to use PGP at no cost. PGP is a set of software tools that can use several different encryption algorithms to perform public-key encryption. The PGP business was purchased by Network Associates in 1997 and sold back to the product's developers, who formed the PGP Corporation in 2002. Today, individuals can download free versions of PGP for personal use from the PGP International site. Individuals can use PGP to encrypt their e-mail messages to protect them from being read if they are intercepted on the Internet. The Symantec (which bought the original PGP Corporation) site sells PGP licenses to businesses that want to use the technology to protect business communication activities.

Symmetric Encryption

Symmetric encryption, also known as **private-key encryption**, encodes a message with an algorithm that uses a single numeric key, such as 456839420783, to encode and decode data. Because the same key is used, both the message sender and the message receiver must know the key. Encoding and decoding messages using symmetric encryption is very fast and efficient. However, the key must be guarded. If the key is made public, then all messages sent previously using that key become vulnerable, and the keys must be changed.

It can be difficult to distribute new keys to authorized parties while maintaining security and control over the keys. The catch is that to transmit anything privately (including a new secret key), it must be encrypted. Another problem with private keys is that they do not work well in large environments such as the Internet. Each pair of users on the Internet who wants to share information privately must have their own private key. That results in a huge number of key–pair combinations.

In secure environments such as the defense sector, using private-key encryption is simpler, and it is the prevalent method to encode sensitive data. Distribution of classified information and encryption keys is often used in military applications. It requires guards (two-person control) and secret transportation plans. The **Data Encryption Standard (DES)** is a set of encryption algorithms adopted by the U.S. government for encrypting sensitive or commercial information and is the most widely used private-key encryption system. The size of DES private keys must be increased regularly because researchers use increasingly fast computers to break them. For example, the Electronic Frontier Foundation's Deep Crack key breaker used a network of 100,000 PCs to break a DES-encrypted test message in under 23 hours.

As a result of these key-breaking experiments, the U.S. government began using a stronger version of the Data Encryption Standard, called **Triple Data Encryption Standard (Triple DES or 3DES)**. In 2001, the U.S. government developed a more secure encryption standard called the **Advanced Encryption Standard (AES)**. Today, most U.S. government agencies and many private businesses use AES with various key bit lengths. AES uses longer bit lengths to increase the difficulty of cracking its keys, just as the DES methods do.

Comparing Asymmetric and Symmetric Encryption Systems

Public-key (asymmetric) systems provide several advantages over private-key (symmetric) encryption methods. First, the combination of keys required to provide private messages between enormous numbers of people is small. If n people want to share secret information with one another, then only n unique public-key pairs are required—far fewer than an equivalent private-key system. Second, key distribution is not a problem. Each person's public key can be posted anywhere and does not require any special handling to distribute. Third, public-key systems make implementation of digital signatures possible. This means that an electronic document can be signed and sent to any recipient with nonrepudiation. That is, with public-key techniques, it is not possible for anyone other than the signer to produce the signature electronically; in addition, the signer cannot later deny signing the electronic document.

Public-key systems have disadvantages. One disadvantage is that public-key encryption and decryption are significantly slower than private-key systems. This extra time can add up quickly as individuals and organizations conduct commerce on the Internet. Public-key systems do not replace private-key systems but serve as a complement to them. Public-key systems are used to transmit private keys to Internet participants so that additional, more efficient communication can occur in a secure Internet session. Figure 10-9 shows a graphical comparison of the hash coding, private-key, and public-key encryption methods: Figure 10-9a shows hash coding; Figure 10-9b depicts private-key encryption; and Figure 10-9c illustrates public-key encryption.

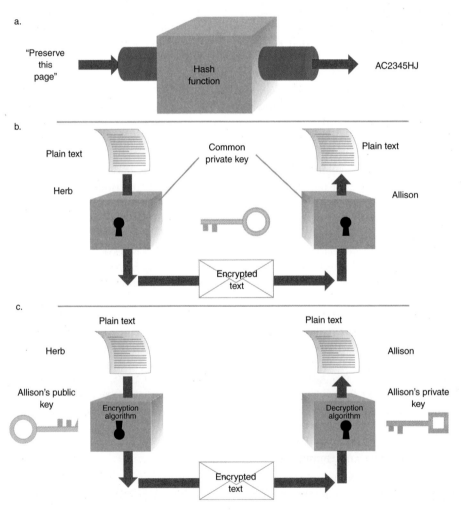

© Cengage Learning

FIGURE 10-9 Comparison of (a) hash coding, (b) private-key, and (c) public-key encryption

Several encryption algorithms exist that can be used with secure Web servers. The U.S. government approves the use of several of these inside the United States. Electronic commerce Web servers can accommodate most of these algorithms because they must be able to communicate with a wide variety of Web browsers.

The **Secure Sockets Layer (SSL)** system developed by Netscape Communications and the Secure Hypertext Transfer Protocol (S-HTTP) developed by CommerceNet are two protocols that provide secure information transfer through the Internet. SSL and S-HTTP allow both the client and server computers to manage encryption and decryption activities between each other during a secure Web session.

SSL and S-HTTP have different goals. SSL secures connections between two computers, and S-HTTP sends *individual* messages securely. Encryption of outgoing messages and decryption of incoming messages happens automatically and transparently with both SSL and S-HTTP.

Secure Sockets Layer (SSL) Protocol

SSL provides a security "handshake" in which the client and server computers exchange a brief burst of messages. In those messages, the client and server agree on the level of security to be used for exchange of digital certificates and other tasks. Each computer identifies the other. After identification, SSL encrypts and decrypts information flowing between the two computers. This means that information in both the HTTP request and any HTTP response is encrypted. Encrypted information includes the URL the client is requesting, any forms containing information the user has completed (which might include sensitive information such as a login, a password, or a credit card number), and HTTP access authorization data, such as usernames and passwords. In short, *all* communication between SSL-enabled clients and servers is encoded. When SSL encodes everything flowing between the client and server, an eavesdropper receives only unintelligible information.

SSL can secure many different types of communication between computers in addition to HTTP. For example, SSL can secure FTP sessions, enabling private downloading and uploading of sensitive documents, spreadsheets, and other electronic data. SSL can secure Telnet sessions in which remote computer users can log on to corporate host machines and send their passwords and usernames. The protocol that implements SSL is HTTPS. By preceding the URL with the protocol name HTTPS, the client is signifying that it would like to establish a secure connection with the remote server.

Secure Sockets Layer allows the length of the private session key generated by every encrypted transaction to be set at a variety of bit lengths (such as 40-bit, 56-bit, 128-bit, or 168-bit). A **session key** is a key used by an encryption algorithm to create cipher text from plain text during a single secure session. The longer the key, the more resistant the encryption is to attack. A Web browser that has entered into an SSL session indicates that it is in an encrypted session (most browsers use an icon in the browser status bar). Once the session is ended, the session key is discarded permanently and not reused for subsequent secure sessions.

In an SSL session, the client and server agree that their exchanges should be kept secure because they involve transmitting credit card numbers, invoice numbers, or verification codes. To implement secrecy, SSL uses public-key (asymmetric) encryption and private-key (symmetric) encryption. Although public-key encryption is convenient, it is slow compared to private-key encryption. That is why SSL uses private-key encryption for nearly all its secure communications.

Because it uses private-key encryption, SSL must have a way to get the key to both the client and server without exposing it to an eavesdropper. SSL accomplishes this by having the browser generate a private key for both to share. Then the browser encrypts the private key it has generated using the server's public key. The server's public key is stored in the digital certificate that the server sent to the browser during the authentication step. Once the key is encrypted, the browser sends it to the server. The server, in turn, decrypts the message with its private key and exposes the shared private key.

Here is how SSL works with an exchange between a browser (SSL client) and a Web server (SSL server):

1. When a client browser sends a request message to a server's secure Web site, the server sends a hello request to the browser (client). The browser responds with a client hello. The exchange of these greetings, or the handshake, allows the two computers to determine the compression and encryption standards that they both support.

2. Next, the browser asks the server for a digital certificate as a proof of identity. In response, the server sends to the browser a certificate signed by a recognized certification authority.

3. The browser checks the serial number and certificate fingerprint on the server certificate against the public key of the CA stored within the browser. Once the CA's public key is verified, the endorsement is verified. That action authenticates the Web server. The browser responds by sending its client certificate and an encrypted private session key to be used. When the server receives this information, it initiates the session, which uses the private key now shared between the browser and the Web server.

4. With the session established as secure, request messages from the browser are accepted by the Web server, which sends the necessary responses. In this secure session, the browser user can make purchases, pay bills, or trade securities without worrying about threats to the security of the information passing between the two computers.

From this point on in the session, public-key encryption is no longer used; the transmission is protected by private-key encryption. All messages sent between the client and the server are encrypted with the shared private key, also known as the session key. When the session ends, the session key is discarded.

Any new connection between a client and a secure server starts the entire process all over again, beginning with the handshake between the client browser and the server. The client and server agree to use a specific bit level of encryption (for example, 40-bit encryption or 128-bit encryption) and also agree on which specific encryption algorithm to use. Figure 10-10 illustrates the SSL handshake that occurs before a client and server exchange private-key-encoded business information for the remainder of the secure session.

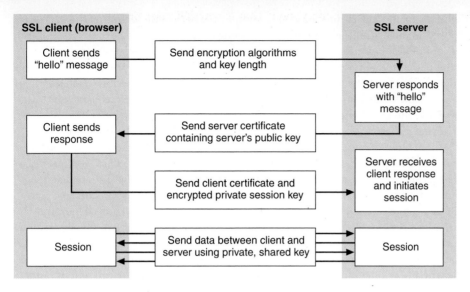

FIGURE 10-10 Establishing an SSL session

Secure HTTP (S-HTTP)

Secure HTTP (S-HTTP) is an extension to HTTP that provides a number of security features, including client and server authentication, spontaneous encryption, and request/response nonrepudiation. S-HTTP provides symmetric encryption for maintaining secret communications and public-key encryption to establish client/server authentication.

S-HTTP security is established during the initial session between a client and a server. Either the client or the server can specify that a particular security feature be required, optional, or refused. When one party stipulates that a particular security feature be required, the client or server continues the connection only if the other party (client or server) agrees to enforce the specified security. Otherwise, no secure connection is established. This process of proposing and accepting (or rejecting) various transmission conditions is called **session negotiation**. An example of this negotiation might occur in an interaction between a high-fashion clothing designer that is purchasing silk from an Asian textile mill. The points of negotiation might include:

- Designer wants the details of the transaction to remain confidential so that any eavesdropping competitors cannot learn which fabrics are to be featured in next season's designs; the designer's computer would propose that encryption be required on the transmission.
- Textile mill wants to enforce integrity on the transmission so that quantities and prices quoted to the purchaser remain intact; the mill's computer would propose that integrity be enforced on the transmission.
- Textile mill wants assurance that the purchaser is who it claims to be, rather than an imposter; the mill's computer would request nonrepudiation, a positive confirmation of the transaction.

S-HTTP differs from SSL in the way it establishes a secure session. SSL carries out a client–server handshake exchange to set up a secure communication, but S-HTTP includes security details with packet headers that are exchanged in S-HTTP. The headers define the type of security techniques, including the use of private-key encryption, server authentication, client authentication, and message integrity. Header exchanges also stipulate which specific algorithms each side supports, whether the client or the server (or both) supports the algorithm, and whether the security technique (for example, secrecy) is required, optional, or refused.

Once the client and server agree to security implementations enforced between them, all subsequent messages between them during that session are wrapped in a secure container, sometimes called an envelope. A **secure envelope** encapsulates a message and provides secrecy, integrity, and client/server authentication. In other words, it is a complete package. With it, all messages traveling on the network or Internet are encrypted so that they cannot be read. Messages cannot be altered undetectably because integrity mechanisms provide a detection code that signals a message has been altered. Clients and servers are authenticated with digital certificates issued by a recognized certification authority. The secure envelope includes all of these security features.

S-HTTP is no longer used by many Web sites. SSL has become a more generally accepted standard for establishing secure communication links between Web clients and Web servers.

You have learned how encryption provides message secrecy and confidentiality, and you have learned how digital certificates serve to authenticate a server to a client, and vice versa. In the next section, you will learn how to implement message integrity, which prevents an interloper from changing a message in transit.

Using a Hash Function to Create a Message Digest

Electronic commerce ultimately involves a client browser sending payment information, order information, and payment instructions to the Web server and that server responding with a confirmation of the order details. If an Internet interloper alters any of the order information in transit, harmful consequences can result. For instance, the perpetrator could alter the shipment address so that he or she receives the merchandise instead of the original customer. This interference is an example of an integrity violation.

Although it is difficult and expensive to prevent a perpetrator from altering a message, there are effective and efficient techniques that allow the receiver to detect when a message has been altered. To eliminate message alteration, two separate algorithms can be applied to a message. First, a hash algorithm is applied to the message. The hash value is used to create a **message digest**, which is a number that summarizes the encrypted information. The receiver of the message can calculate the message digest value independently. If the message digest values match, the receiver knows that the encrypted message was not altered in its transmission. If they do not match, the receiver can ask the sender to resend the message.

Converting a Message Digest into a Digital Signature

Hash functions are not an ideal integrity enforcement solution because the hash algorithm is public and widely known. For example, a message containing a purchase order could be intercepted, the shipping address and quantity ordered could be altered, the message digest could be regenerated, and the new message and its accompanying message digest could be sent on to the merchant. Upon receipt, the merchant would calculate the message digest value and confirm that the two message digest values match. The merchant would conclude (incorrectly) that the message had not been altered. To prevent this type of fraud, the sender can encrypt message digests using a private key.

This type of encrypted message digest is called a **digital signature**. A purchase order accompanied by a digital signature provides the merchant with positive identification of the sender and assures the merchant that the message was not altered. Because the message digest is encrypted using a public key, only the owner of the public/private key pair could have encrypted the message digest. Thus, when the merchant decrypts the message with the user's public key and calculates a matching message digest value, the result is proof that the sender is authentic. Matching the hash values proves that only the true sender could have authored the message (nonrepudiation) because only the sender's private key would yield an encrypted message that could be decrypted successfully by an associated public key. Figure 10-11 illustrates how a digital signature and a signed message are created and sent.

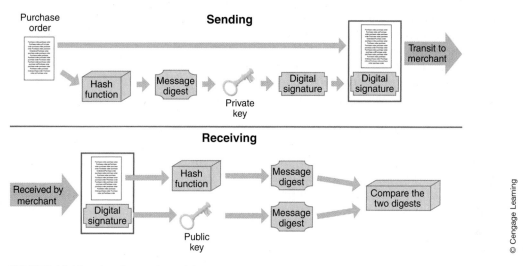

FIGURE 10-11   Sending and receiving a digitally signed message

Encrypting both the digital signature and the message itself guarantees message secrecy. Used together, public-key encryption, message digests, and digital signatures provide a high level of security for Internet transactions. Digital signatures have had the same legal status as traditional signatures in the United States, Canada, and the European Union since 2001. Today, most of the world's countries also have laws that recognize digital signatures as enforceable commitments in business transactions.

SECURITY FOR SERVER COMPUTERS

The server is the third link in the client–Internet–server electronic commerce path between the user and a Web server. Servers have vulnerabilities that can be exploited by anyone determined to cause destruction or acquire information illegally. One entry point is the Web server and its software. Other entry points include back-end programs containing data, such as a database and the server on which it runs. Although no system is completely safe, the Web server administrator's job is to make sure that security policies are documented and considered in every part of the electronic commerce operation.

Web Server Threats

Web server software, as you learned in Chapter 8, is designed to deliver Web pages by responding to HTTP requests. A Web server can compromise secrecy if it allows automatic directory listings. The secrecy violation occurs when the contents of a server's folder names are revealed to a Web browser. This can happen when a user enters a URL, such as http://www.somecompany.com/FAQ/, and expects to see the default page in the FAQ directory. The default Web page that the server normally displays is named index.htm or index.html. If that file is not in the directory, a Web server that allows

441

automatic directory listings will display all of the file and folder names in that directory. Then, visitors can click folder names at random and open folders that might not be intended for public disclosure. Careful site administrators turn off this folder name display feature. If a user attempts to browse a folder where protections prevent browsing, the Web server issues a warning message stating that the directory is not available.

One of the most sensitive files on a Web server is the file that holds Web server username–password pairs. An intruder who can access and read that file can enter privileged areas masquerading as a legitimate user. To reduce this risk, most Web servers store user authentication information in encrypted files.

The passwords that users select can be the source of a threat. Users sometimes select passwords that are guessed easily, such as their mother's maiden name, the name of a child, or their telephone number. **Dictionary attack programs** cycle through an electronic dictionary, trying every word and common name as a password.

Users' passwords, once broken, may provide an opening for entry into a server that can remain undetected for a long time. To prevent dictionary attacks, some organizations require users to create passwords that contain a combination of letters, numbers, and special characters that are unlikely to appear in an attack program's dictionary. Other organizations use their own dictionary check as a preventive measure. When a user selects a new password, the password assignment software checks the password against its dictionary and, if it finds a match, refuses to allow the use of that password. Good password assignment software checks against common words, names (including common pet names), acronyms that are commonly used in the organization, and words or characters (including numbers) that have some meaning for the user requesting the password (for example, employees might be prohibited from using their employee numbers as passwords). Figure 10-12 shows examples of passwords that range from very weak to very strong.

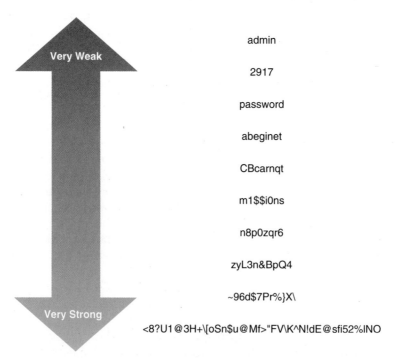

FIGURE 10-12 Examples of passwords, from very weak to very strong

There are a number of online resources that can help you create very strong passwords. One of the most respected of these is the **Gibson Research Corporation's Ultra High Security Password Generator**.

Database Threats

Electronic commerce systems store user data and retrieve product information from databases connected to the Web server. Besides storing product information, databases connected to the Web contain valuable and private information that could damage a company irreparably if disclosed or altered. Most database management systems include security features that rely on usernames and passwords. Once a user is authenticated, specific parts of the database become available to that user. However, some databases either store username/password pairs in an unencrypted table, or they fail to enforce security at all and rely on the Web server to enforce security. If unauthorized users obtain user authentication information, they can masquerade as legitimate database users and reveal or download confidential and potentially valuable information. Trojan horse programs hidden within the database system can also reveal information by changing the access rights of various user groups. A Trojan horse can even remove access controls within a database, giving all users complete access to the data—including intruders.

Other Programming Threats

Web server threats can arise from programs executed by the server. Java or C++ programs that are passed to Web servers by a client, or that reside on a server, frequently make use of a buffer. A **buffer** is an area of memory set aside to hold data read from a file or database. A buffer is necessary whenever any input or output operation takes place because a computer can process file information much faster than the information can be read from input devices or written to output devices. Programs filling buffers can malfunction and overfill the buffer, spilling the excess data outside the designated buffer memory area. This is called a **buffer overrun** or **buffer overflow** error. Usually, this occurs because the program contains an error or bug that causes the overflow. Sometimes, however, the buffer overflow is intentional. The Internet Worm of 1988 was such a program. It caused an overflow condition that eventually consumed all resources until the affected computer could no longer function.

A more insidious version of a buffer overflow attack writes instructions into critical memory locations so that when the intruder program has completed its work of overwriting buffers, the Web server resumes execution by loading internal registers with the address of the main attacking program's code. This type of attack can open the Web server to severe damage because the resumed program—which is now the attacker program—may regain control of the computer, exposing its files to disclosure and destruction by the attacking program. Good programming practices can reduce the potential damage from buffer overflows, and some computers include hardware that works with the operating system to limit the effects of buffer overflows that are intentionally programmed to create damage.

A similar attack, one in which excessive data is sent to a server, can occur on mail servers. Called a **mail bomb**, the attack occurs when hundreds or even thousands of people each send a message to a particular address. The attack might be launched by a large team of well-organized hackers, but more likely the attack is launched by one or a few hackers who have gained control over others' computers using a Trojan horse virus or some other method of turning those computers into zombies. The accumulated mail received by the target of the mail bomb exceeds the allowed e-mail size limit and can cause e-mail systems to malfunction.

Threats to the Physical Security of Web Servers

Web servers and the computers that are networked closely to them, such as the database servers and application servers used to supply content and transaction-processing capabilities to electronic commerce Web sites, must be protected from physical harm. For many companies, these computers have become repositories of important data (information about customers, products, sales, purchases, and payments). They have also become important parts of the revenue-generating function in many businesses. As key physical resources, these computers and related equipment warrant high levels of protection against threats to their physical security.

As you learned in Chapter 9, many companies use commerce service providers (CSPs) to host their Web sites. The security that CSPs maintain over their physical premises is, in many cases, stronger than the security that a company could provide for computers maintained at its own location.

Companies can take specific steps to protect their Web servers. Many companies maintain backup copies of their servers' contents at a remote location. If the Web server operation is critical to the continuation of the business, a company can maintain a duplicate of the entire Web server physical facility at a remote location. In the case of a system failure, the company's Web operations can be switched over to the backup location in less than a second. Examples of mission-critical Web servers that would warrant such a comprehensive (and expensive) level of physical security include airline reservation systems, stock brokerage firm trading systems, and bank payment clearing systems.

Some companies rely on their service providers to help with Web server security. Commerce service providers often include Web server security as an add-on service. Other companies hire smaller, specialized security service providers to handle security (see Learning from Failures—Pilot Network Services to learn more about one alternative to this approach). Having a service provider handle security can add an additional $500 to $2000 per month to the provider's standard bandwidth charges.

LEARNING FROM FAILURES

Pilot Network Services

Pilot Network Services began operations in 1993, at the dawn of commercial use of the Internet. Its goal was to build a network that would be secure for electronic commerce activities. The network it built included its own carefully monitored connections to the Internet and a database of attack signatures. Attack signatures are descriptions of the Internet traffic characteristics that indicate a cracker attack on a Web server. Pilot, as a firm specializing in security services, built an excellent collection of attack signatures and kept it updated much better than other firms that were not security specialists did at that time.

Pilot maintained the Web servers for many of its clients, and it used versions of the operating systems and Web server software that it had customized to be especially resistant to attacks. Pilot's engineers meticulously applied patches for all known points of access to the software and worked to identify new, as yet unknown, points of vulnerability—for which they immediately created and applied protective patches. For customers hosting their own servers, Pilot provided the Internet connection through its own secure network. The router between the client's network and Pilot's network and

Continued

the operating system running the Pilot network were customized to eliminate any known security loopholes.

Pilot had 24/7 monitoring of its network by computer security experts, in addition to the network technicians that any other Web hosting company would provide as part of a managed services offering. Because it offered high-quality services, its fees were considerably higher than the service charges imposed by other service providers.

Even with its high prices, Pilot had many fans among the largest companies in the United States. Pilot never had more than 300 customers, but it monitored more than 70,000 individual networks for a customer list that included General Electric, PeopleSoft, Sovereign Bancorp, The Washington Post Company, and many other major accounts. By 1999, Pilot appeared to be doing well. Its revenue had increased more than 80 percent over 1998. News releases were issued regularly announcing new customers.

In late 2000, Pilot's stock price began to fall, along with the stock prices of many companies in Internet-related businesses. Although Pilot's sales were growing, its costs were escalating at an even more rapid rate. The company had never reported a profit, and its annual losses had increased to $21.7 million in 2000. Pilot executives assured customers that the company was financially sound, but the ability of companies in Internet-related businesses to survive on the promise of future earnings had disappeared. Pilot's ability to raise the cash it needed to continue operating had vanished.

In early 2001, some Pilot customers noticed that the service was failing. Phone calls and e-mails were not being returned quickly. On the afternoon of April 25, 2001, Pilot employees received four e-mails. The first explained that telephones would be disconnected that evening. The second asked all employees to turn in their mobile phones and pagers. The third announced that the chief financial officer had resigned. The last e-mail announced that all employees were out of a job as of 4:30 p.m.

Pilot's clients, many of which found out about the collapse from the Pilot employees who had been servicing their accounts, immediately faced serious problems. Connections to the Internet vanished with no warning. The companies that had used Pilot to host their entire Web operations were in an even worse situation. A group of Pilot customers convinced AT&T (the provider of Pilot's Internet connections) to continue to carry traffic from Pilot, even though Pilot had not paid AT&T. Providian Financial, a major bank holding company and credit card processor, sent its own employees into Pilot operations centers to keep Providian's Web servers operating. Other Pilot customers that were Providian's competitors protested, concerned that their Web servers were suddenly open and vulnerable.

Several of Pilot's competitors tried to raise funding to take over the business, but all of those attempts failed, and on May 9, 2001—two weeks after the collapse—AT&T cut Internet service and Pilot was liquidated. Pilot's former customers scrambled to hire security staff, find alternative hosting firms, or find other ways to keep their Web sites operating. The lesson from this failure is that security is a critical part of an electronic commerce operation. It should be handled with the same care that a company would use to protect any physical asset. If any part of the security function is handed over to another company, that company's condition becomes an important concern and must be monitored carefully and continually.

Access Control and Authentication

Access control and authentication refers to controlling who and what has access to the Web server. Most people who work with Web servers in electronic commerce environments do not sit at a keyboard connected to the server. Instead, they access the server from a client computer. Recall that authentication is verification of the identity of the entity requesting access to the computer. Just as users can authenticate servers with which they are interacting, servers can authenticate individual users. When a server requires positive identification of a user, it requests that the client send a certificate.

The server can authenticate a user in several ways. First, if the server cannot decrypt the user's digital signature contained in the certificate using the user's public key, then the certificate did not come from the true owner. Otherwise, the server is certain that the certificate came from the owner. This procedure prevents the use of fraudulent certificates to gain entry to a secure server. Second, the server checks the timestamp on the certificate to ensure that the certificate has not expired. A server will reject an expired certificate and provide no further service. Third, a server can use a callback system in which the server software checks a user's client computer name and address against a list of authorized usernames and assigned client computer addresses before "calling back" to establish a connection. Callback systems work especially well in an intranet where usernames and client computers are controlled closely and assigned systematically. On the Internet, callback systems can be more difficult to implement, particularly if client users are mobile and work from different locations. Certificates issued by trusted CAs play a central role in authenticating client computers and their users because they provide irrefutable evidence of identity.

Usernames and passwords can also provide some element of protection. To authenticate users using passwords and usernames, the server must acquire and store a database containing users' passwords and usernames. Many Web server systems store usernames and passwords in a file. Large electronic commerce sites usually keep username/password combinations in a separate database with built-in security features.

The easiest way to store passwords is to maintain usernames in plain text and encrypt passwords using a one-way encryption algorithm. With the plain text username and encrypted password stored, the system can validate users when they log on by checking the usernames they enter against the list of usernames stored in the database. The password that a user enters when he or she logs on to a system is encrypted. Then the resulting encrypted password from the user is checked against the encrypted password stored in the database. If the two encrypted versions of the password match for the given user, the login is accepted. That is why even a system administrator cannot tell you what your forgotten password is on most systems. Instead, the administrator must assign a new temporary password that the user can change to another password. Passwords are not immune to discovery, and a person truly intent on stealing a password can often figure out a way to do so.

Note that the site visitor can save his or her username and password as a cookie on the client computer, which allows access to subscription areas of the site without entering the username and password on subsequent site visits. The trouble with that system of cookies is that the information might be stored on the client computer in plain text. If the cookie contains login and password information, then that information is visible to anyone who has access to the user's computer.

Web servers often provide access control list security to restrict file access to selected users. An **access control list (ACL)** is a list or database of files and other resources and the usernames of people who can access the files and other resources. Each file has its own access control list. When a client computer requests Web server access to a file or document that has been configured to require an access check, the Web server checks the resource's ACL file to determine if the user is allowed to access that file. This system is especially convenient to restrict access of files on an intranet server so that individuals can only access selected files on a need-to-know basis. The Web server can exercise fine control over resources by further subdividing file access into the activities of read, write, or execute. For example, some users may be permitted to read the corporate employee handbook but not allowed to update or write to the file. Only the human resources (HR) manager would have write access to the employee handbook, and that access privilege is stored along with the HR manager's ID and password in an ACL.

Firewalls

A **firewall** is software or a hardware–software combination that is installed in a network to control the packet traffic moving through it. Most organizations place a firewall at the Internet entry point of their networks. The firewall provides a defense between a network and the Internet or between a network and any other network that could pose a threat. Firewalls all operate on the following principles:

- All traffic from inside to outside and from outside to inside the network must pass through it.
- Only authorized traffic, as defined by the local security policy, is allowed to pass through it.
- The firewall itself is immune to penetration.

Those networks inside the firewall are often called **trusted**, whereas networks outside the firewall are called **untrusted**. Acting as a filter, firewalls permit selected messages to flow into and out of the protected network. For example, one security policy a firewall might enforce is to allow all HTTP (Web) traffic to pass back and forth but disallow FTP or Telnet requests either into or out of the protected network. Ideally, firewall protection should prevent access to networks inside the firewall by unauthorized users, and thus prevent access to sensitive information. Simultaneously, a firewall should not obstruct legitimate users. Authorized employees outside the firewall ought to have access to firewall-protected networks and data files. Firewalls can separate corporate networks from one another and prevent personnel in one division from accessing information from another division of the same company. Using firewalls to segment a corporate network into secure zones serves as a coarse need-to-know filter.

Large organizations that have multiple sites and many locations must install a firewall at each location that has an external connection to the Internet. Such a system ensures an unbroken security perimeter that is effective for the entire corporation. In addition, each firewall in the organization must follow the same security policy. Otherwise, one firewall might permit one type of transaction to flow into the corporate network that another excludes. Without a consistent policy, an unwanted access that occurs through a breach in one firewall can expose the information assets of the entire corporation to the threat.

Organizations should remove any unnecessary software from their firewalls. Having fewer software programs on the system should reduce the chances for malevolent software security breaches. Because the firewall computer is used only as a firewall and not as a general-purpose computing machine, only essential operating system software and firewall-specific protection software should remain on the computer. Access to a firewall should be restricted to a console physically connected directly to the firewall machine. Managers should forbid remote administration of the firewall to avoid the threat of an outside attacker gaining access to the firewall by posing as an administrator.

Firewalls are classified into the following categories: packet filter, gateway server, and proxy server. **Packet-filter firewalls** examine all data flowing back and forth between the trusted network (within the firewall) and the Internet. Packet filtering examines the source and destination addresses and ports of incoming packets and denies or permits entrance to the packets based on a preprogrammed set of rules.

Gateway servers are firewalls that filter traffic based on the application requested. Gateway servers limit access to specific applications such as Telnet, FTP, and HTTP. Application gateways arbitrate traffic between the inside network and the outside network. In contrast to a packet-filter technique, an application-level firewall filters requests and logs them at the application level, rather than at the lower IP level. A gateway firewall provides a central point where all requests can be classified, logged, and later analyzed. An example is a gateway-level policy that permits incoming FTP requests but blocks outgoing FTP requests. That policy prevents employees inside a firewall from downloading potentially dangerous programs from the outside.

Proxy server firewalls are firewalls that communicate with the Internet on the private network's behalf. When a browser is configured to use a proxy server firewall, the firewall passes the browser request to the Internet. When the Internet sends back a response, the proxy server relays it back to the browser. Proxy servers are also used to serve as a huge cache for Web pages.

One problem faced by companies that have employees working from home is that the location of computers outside the traditional boundaries of the company's physical site expands the number of computers that must be protected by the firewall. This **perimeter expansion** problem is particularly troublesome for companies that have salespeople using laptop computers to access confidential company information from all types of networks at customer locations, vendor locations, and even public locations, such as airports.

Another problem faced by organizations connected to the Internet is that their servers are under almost constant attack. Crackers spend a great deal of time and energy on attempts to enter the servers of organizations. Some of these crackers use automated programs to continually attempt to gain access to servers. Organizations often install intrusion detection systems as part of their firewalls. **Intrusion detection systems** are designed to monitor attempts to log into servers and analyze those attempts for patterns that might indicate a cracker's attack is under way.

Once the intrusion detection system identifies an attack, it can block further attempts that originate from the same IP address until the organization's security staff can examine and analyze the access attempts and determine whether they are an attack.

As more organizations rely on cloud computing for crucial production systems, the need for security in cloud environments is increasing. The development of firewalls that work with cloud computing is advancing rapidly but has lagged behind the need for these products. Instead of establishing security policies for each server, these firewalls must enforce a single set of policies across all of the servers in the cloud. One problem in cloud environments is that the servers and databases in the cloud are started up and wound down as needed. Thus, the type of identifiable servers that most firewall products are designed to protect does not exist in the same form in cloud server environments.

In addition to firewalls installed on organizations' networks, it is possible to install software-only firewalls on individual client computers. These firewalls are often called **personal firewalls**. The use of personal firewalls, such as ZoneAlarm, has become an important tool in the protection of expanded network perimeters for many companies. Many home computer users are installing personal firewalls on their home networks. You can learn more about firewall protection for your home computer at the Gibson Research Shields Up! Web site.

ORGANIZATIONS THAT PROMOTE COMPUTER SECURITY

Following the occurrence of the Internet Worm of 1988, a number of organizations were formed to share information about threats to computer systems. These organizations are devoted to the principle that sharing information about attacks and defenses for those attacks can help everyone create better computer security. Some of the organizations began at universities; others were launched by government agencies. In this section, you will learn about some of these organizations and their resources.

CERT

In 1988, a group of researchers met to study the infamous Internet Worm attack soon after it occurred. They wanted to understand how worms worked and how to prevent damage from future attacks of this type. The National Computer Security Center, part of the National Security Agency, initiated a series of meetings to figure out how to respond to future security breaks that might affect thousands of people. Soon after those meetings, the U.S. government created the Computer Emergency Response Team and housed it at Carnegie Mellon University in Pittsburgh.

The organization is now operated as part of the federally funded Software Engineering Institute at Carnegie Mellon, and it has changed its legal name from the Computer Emergency Response Team (which had been abbreviated to "CERT" by most people who wrote and talked about it) to **CERT**. CERT still maintains an effective and quick communications infrastructure among security experts so that security incidents can be avoided or handled quickly.

Today, CERT responds to thousands of security incidents each year and provides a wealth of information to help Internet users and companies become more knowledgeable about security risks. CERT posts alerts to inform the Internet community about security events, and it is regarded as a primary authoritative source for information about viruses, worms, and other types of attacks.

Other Organizations

CERT is the most prominent of these organizations and has formed relationships, such as the **Internet Security Alliance**, with other industry associations. However, CERT is not the only computer security resource. In 1989, one year after CERT was formed, a cooperative research and educational organization called the Systems Administrator, Audit, Network, and Security Institute was launched. Now known as the **SANS Institute**, this organization includes thousands of members who work in computer security consulting firms and information technology departments of companies as auditors, systems administrators, and network administrators.

Many SANS education and research efforts yield resources such as news releases, research reports, security alerts, and white papers that are available on the Web site at no cost. SANS also sells publications to generate funds that it uses for research and educational programs. The SANS Institute operates the **Infocon: SANS Internet Storm Center**, a Web site that provides current information on the location and intensity of computer attacks throughout the world.

Purdue University's Center for Education and Research in Information Assurance and Security **(CERIAS)** is a center for multidisciplinary research and education in information security. The CERIAS Web site provides resources in computer, network, and communications security and includes a section on information assurance.

The **Center for Internet Security** is a not-for-profit cooperative organization devoted to helping companies that operate electronic commerce Web sites reduce the risk of disruptions from technical failures or deliberate attacks on their computer systems. It also provides information to auditors who review such systems and to insurance companies that provide coverage for companies who operate such systems.

For current information about computer security, you can visit **CSO Online**, which carries articles that have appeared in *CSO Magazine* along with other news items related to computer security. A British publication, **Infosecurity.com**, is available online and includes articles about all types of online security issues.

Computer Forensics and Ethical Hacking

A small number of specialized consulting firms engage in the unlikely enterprise of breaking into servers and client computers at the request of the organizations that own those computers. Called **computer forensics experts** or **ethical hackers**, these computer sleuths are hired to probe PCs and locate information that can be used in legal proceedings. The field of **computer forensics** is responsible for the collection, preservation, and analysis of computer-related evidence. Ethical hackers are often hired by companies to test their computer security safeguards. They are also hired by law enforcement agencies investigating crimes and by law firms undertaking investigations on behalf of their clients.

Summary

Physical and logical computer security considerations are important issues in electronic commerce. Online businesses manage risks, often using a formal security policy document that identifies risks and countermeasures that will reduce those risks to an acceptable level. The three main elements of computer security are secrecy, integrity, and necessity. These three elements must be enforced in each of the three components of online business transactions, including: client computers, the communication channel, and server computers.

Client computer threats can be delivered as Java or JavaScript applets and as ActiveX controls. These programs and controls can be installed and run on client machines in ways that create security threats. Cookies, if not controlled and used properly, can present threats to client computers. Antivirus software is an important element in the protection of client computers.

The main communication channel used in online business is the Internet, which is especially vulnerable to attacks. Encryption provides secrecy, and several forms of encryption are available that use hash functions or more advanced algorithms. Encryption can be implemented using private-key or public-key techniques. The most effective encryption schemes use combinations of both. Integrity protections ensure that messages between clients and servers are not altered. Digital certificates provide both integrity and user authentication, which can help establish nonrepudiation in online transactions. Several Internet protocols, including Secure Sockets Layer and Secure HTTP, can be used to provide secure Internet transmissions. As wireless networks have grown to become important parts of the data communication infrastructure, security concerns have increased. Most wireless networks installed in businesses today (and many installed in homes) do have wireless encryption.

Web servers are susceptible to security threats that can come from within the server in the form of programs or that can come from outside the server. The Web server must be protected from both physical threats and Internet-based attacks on its software. Methods to protect the server include access control and authentication, which are provided by username and password login procedures and client certificates. Firewalls can be used to separate trusted inside computer networks and clients from untrusted outside networks, including other divisions of a company's enterprise network system and the Internet.

A number of organizations have been formed to share information about computer security threats and defenses. When large security outbreaks occur, the members of these organizations join together and discuss methods to locate and eliminate the threat. Computer forensics firms that undertake attacks against their clients' computers can play an important role in helping to identify security weaknesses.

Key Terms

Access control list (ACL)	Black hat hacker
Active content	Botnet
Active wiretapping	Buffer
ActiveX	Buffer overrun
Advanced Encryption Standard (AES)	Buffer overflow
Antivirus software	Certification authority (CA)
Applet	Cipher text
Asymmetric encryption	Collision
Backdoor	Computer forensics
Biometric security device	Computer forensics expert

Computer security
Countermeasure
Cracker
Cryptography
Cybervandalism
Data Encryption Standard (DES)
Decrypted
Decryption program
Delay attack
Denial attack
Denial-of-service (DoS) attack
Dictionary attack program
Digital certificate
Digital ID
Digital signature
Distributed denial-of-service (DDoS) attack
Domain name server (DNS)
Eavesdropper
Encryption
Encryption algorithm
Encryption program
Ethical hacker
Firewall
First-party cookies
Gateway servers
Hacker
Hash algorithm
Hash coding
Hash value
Integrity
Integrity violation
Intrusion detection system
Java sandbox
JavaScript
Key
Logical security
Macro virus
Mail bomb
Man-in-the-middle exploit
Masquerading
Message digest
Multivector virus
Necessity

Necessity threat
Open session
Packet-filter firewall
Perimeter expansion
Persistent cookie
Personal firewall
Phishing expeditions
Physical security
Plain text
Plug-ins
Pretty Good Privacy (PGP)
Privacy
Private key
Private-key encryption
Proxy server firewall
Public key
Public-key encryption
Remote wipe
Robotic network
Rogue app
Scripting language
Secrecy
Secure envelope
Secure Sockets Layer (SSL)
Secure Sockets Layer-Extended Validation
(SSL-EV) digital certificate
Security policy
Session cookie
Session key
Session negotiation
Signed (message or code)
Sniffer program
Spoofing
Stateless connection
Steganography
Symmetric encryption
Third-party cookies
Threat
Triple Data Encryption Standard
(Triple DES or 3DES)
Trojan horse
Trusted (network)
Untrusted (network)

Untrusted Java applet   Wireless Encryption Protocol (WEP)

Warchalking   Worm

Wardrivers   Zombie

Web bug   Zombie farm

White hat hacker

Review Questions

1. In one or two paragraphs, explain why early computer security efforts focused on controlling the physical environment in which computers operated.

2. Refer to Figure 10-1. In two paragraphs, identify and briefly describe two threats that you would place in Quadrant II and explain why you would classify them as Quadrant II threats.

3. Write a paragraph in which you identify and define the three elements of computer security.

4. In about 100 words, outline the types of information an organization would gather before writing its security policy document.

5. In about 200 words, explain why some Web sites use cookies. In your answer, discuss the reasons that cookies were first devised and why they are used today, and identify where cookies are stored.

6. Write a paragraph in which you explain the concept of a "sandbox" and describe how it is used to reduce security risks in client computers.

7. Conficker (see Figure 10-5) is a multivector worm. In about 100 words, explain why the multivector nature of this worm has made it a more severe threat than other worms have been.

8. In about 100 words, explain what a digital certificate is and why an online seller of industrial packaging products might use a digital certificate.

9. Assume you are working for an online merchant. The company keeps the computers that run its Web, database, and transaction processing servers in a room next to your office. In about 100 words, briefly describe what a biometric security device is and explain why your employer might use one or more of these devices to protect its servers.

10. Write a paragraph in which you describe the purpose and use of a sniffer program.

11. In about 200 words, describe the security threats that a company could face when it adds wireless access points (WAPs) to its network. Assume that the company occupies the six middle floors in a 12-story office building that is located in a downtown business area between two other buildings of similar height. Briefly explain how the company could reduce the risks it faces.

12. In one paragraph, explain what an encryption algorithm is and what it can do to provide increased computer security.

13. In two or three paragraphs, explain the differences between a symmetric and an asymmetric encryption system.

Exercises

1. Wilderness Trailhead, Inc. (WTI) is a retailer that offers hiking, rock-climbing, and survival gear for sale on its Web site. WTI offers about 1200 different items for sale and has about 1000 visitors per day at its Web site. The company makes about 200 sales each day on its site, with an average transaction value of $372. WTI sells products primarily through its Web site to customers in the United States and Canada. WTI ships orders from its two

warehouses: one in Vancouver, British Columbia, and another in Shoreline, Washington. WTI accepts four major credit cards and processes its own credit card transactions. It stores records of all transactions on a database server that shares a small room with the Web server computer at WTI's main offices in a small industrial park just outside Bellingham, Washington. In about 500 words, outline a security policy for the WTI database server. Be sure to consider the threats that exist because that server stores customer credit card numbers. You can use the Web Links for this exercise.

2. Many organizations rely on a firewall to prevent or deter threats to information security that arise from outside the organization. Using your favorite search engine or the resources of your library, identify the firewall issues that arise when companies use cloud computing as part of their online sales systems. In about 100 words, summarize your findings in terms of the perimeter expansion problem.

3. Third-party assurance providers such as **BBBOnline, Inc.**, and **Truste** sell their services to businesses that want to encourage Web site visitors to trust them with their personal information. Review the Web site of one or more of these third-party assurance providers and identify the security features the provider considers important in Web sites that it approves. Select two of the security features you identify and write a 200-word explanation of why the assurance provider considers these features to be important elements for preserving the privacy of site visitor information.

4. Using your library or your favorite search engine, find three Web sites that have an SSL-EV digital certificate. Note that some sites that do have SSL-EV certificates will not show the green background until you log in to the site or place an item in the site's shopping cart. For each site, write a paragraph in which you identify the CA that provides the SSL-EV certificate and explain why that site decided to incur the additional expense of buying an SSL-EV certificate. The Web Links for this exercise include links to CAs that sell SSL-EV digital certificates, which you might find useful.

Cases

C1. Bibliofind

Bibliofind was one of the first Web sites to specialize in hard-to-find and collectible books. The site featured a powerful search engine for used and rare books. The search engine's database was populated with the results of Bibliofind's daily surveys of a worldwide network of suppliers. Registered site visitors could specify the title for which they were searching, a price range, and whether they were seeking a first edition. The site also allowed visitors to build a wish list that would trigger an e-mail when a specific book on the list became available.

Bibliofind had developed a large customer list, an excellent reputation, and a solid network of rare book dealers, all of which made the company an attractive acquisition for other online bookstores. In 1999, Amazon.com bought Bibliofind, but Bibliofind continued to operate its own Web site and conduct its business as it had before the acquisition.

Several years after the Amazon.com acquisition, Bibliofind's Web site was hacked. The cracker had gained access to the company's Web server and replaced its Web pages with defaced versions. Bibliofind shut down its Web site for several days and undertook a complete review of its Web site's security. When the company's IT staff examined the server logs carefully, they found that the Web page hacking was only the tip of the iceberg. Entries in the logs showed that attackers had been accessing Bibliofind's computers for more than four months. Even worse, some of the crackers had been able to go through the Web servers to gain access to the computers that held Bibliofind customer information, including names, addresses, and credit card numbers. That information had been stored in plain text files on Bibliofind's transaction servers.

Bibliofind called in state and federal law enforcement officials to investigate the hacking incidents and sent an e-mail notification to the 98,000 customers whose private information might have been obtained by the crackers. The investigation did not result in any arrests, nor did it determine the identity of the intruders. Many of Bibliofind's customers were very upset when they learned what had happened.

A month after the hacking incident, Amazon.com moved Bibliofind into its zShops online mall (zShops was the original name of Amazon Marketplace). As an Amazon zShop, Bibliofind could process its transactions through Amazon's system and no longer needed to maintain private information about its customers on its computers; however, the company had seen its reputation seriously damaged and eventually was closed down. A successful business was ended in large part because it failed to maintain adequate security over the customer information it had gathered.

Required:

1. In about 300 words, explain how Bibliofind might have used firewalls to prevent the intruders from gaining access to its transaction servers. Be specific about where the firewalls should have been placed in the network and what kinds of rules they should have used to filter network traffic at each point.

2. In about 200 words, explain how encryption might have helped prevent or minimize the effects of Bibliofind's security breach.

3. California has a law that requires companies to inform customers whose private information might have been exposed during a security breach like the one that Bibliofind experienced. Before California enacted this law, businesses argued that the law would encourage nuisance lawsuits. In about 300 words, present arguments for and against this type of legislation.

Note: Your instructor might assign you to a group to complete this case and might ask you to prepare a formal presentation of your results to your class.

C2. Materials Equipment

You are an information technology (IT) consultant to Materials Equipment, Inc. (MEI), a major industrial equipment distributor. Its products include materials-handling machinery for assembly lines and product-packaging areas, hydraulic equipment (for moving fluids), hoses, hose fittings, and similar items. MEI has been in business for more than 70 years and sells more than $200 million worth of parts and equipment each year to its 3000 customers. MEI's customers are located all over the world, but most are in the United States, Mexico, Malaysia, China, and Singapore.

Joe Everson, MEI's director of sales, has retained you to help him with a new marketing idea. He has read about other companies that have created Web portal sites for customers, and he is interested in developing a portal site that MEI could operate with three other companies that sell products (such as bearings, seals, hoses, and hose fittings) and services (design, layout, and installation of materials-handling equipment) that are complementary to MEI products. The portal would provide MEI customers with a Web site at which they could buy MEI products, buy the products and services of the three MEI strategic partners, and obtain information about current trends in industrial equipment technologies and the application of those technologies. The portal site would also include a used equipment area in which MEI customers could list equipment for sale. Joe believes that giving customers a convenient way to liquidate old equipment will make it easier for his sales representatives to sell new equipment to those customers.

Joe has put together an internal team to examine the feasibility of the portal site, including key employees from MEI's Sales, Finance, Product Engineering, and IT Services departments. The team has identified several security issues that they want to resolve before they take the portal idea much further. Joe would like you to help the team understand two security technologies—digital certificates and encryption—and how these techniques might be used in MEI's proposed portal site.

Required:

1. Prepare two briefing reports of about 500 words each for the MEI portal team—one about digital certificates and one about encryption. Each report should explain the technology and describe one or two common applications.

2. Assume that the MEI portal project is approved and implemented. Further assume that MEI has decided to require each customer that participates in the portal to obtain a digital certificate. Write a memo of about 300 words addressed to potential participants (MEI customers) in which you explain why they must obtain a digital certificate as a condition of participation.

Note: Your instructor might assign you to a group to complete this case, and might ask you to prepare a formal presentation of your results to your class.

For Further Study and Research

Austin, R. and C. Darby. 2003. "The Myth of Secure Computing," *Harvard Business Review*, 81(6), June, 120–126.

Baldoni, R. and G. Chockler. 2012, *Collaborative Financial Infrastructure Protection: Tools, Abstractions, and Middleware*. New York: Springer.

Bank, D. and R. Richmond. 2005. "Where the Dangers Are: The Threats to Information Security That Keep the Experts Up at Night," *The Wall Street Journal*, July 18, R1.

Betts, M. 2000. "Digital Signatures Law to Speed Online B-to-B Deals," *Computerworld*, 34(26), June 26, 8.

Chickowski, E. 2009. "Is Your Information Really Safe?" *Baseline*, April, 18–23.

Chow, R., M. Jakobsson, and J. Molina. 2012. "The Future of Authentication," *IEEE Security & Privacy*, 10(1), January-February, 22–27.

Connell, S. 2004. "Security Lapses, Lost Equipment Expose Students to Possible ID Theft Loss," *The Los Angeles Times*, August 29, B4.

Costanzo, C. 2003. "Dealing with Phishing and Spoofing," *American Banker*, 168(184), September 24, 10.

Creighton, D. 2004. "Chronology of Virus Attacks," *The Wall Street Journal*, May 13. (http://online.wsj.com/article/0,,SB108362410782000798,00.html)

Curran, K., J. Doherty, A. McCann, and G. Turkington. 2011. "Good Practices for Strong Passwords," *EDPACS: The EDP Audit, Control, and Security Newsletter*, 44(5), 1–13.

DeFigueiredo, D. 2011. "The Case for Mobile Two-Factor Authentication," *IEEE Security and Privacy*, 9(5), September/October, 81–85.

DoD Directive 5215.1 CSC-STD-001-83. 1983. *Department of Defense Trusted Computer System Evaluation Criteria* (the "Orange Book"), Washington, D.C.

Dunleavey, M. 2005. "Don't Let Data Theft Happen to You," *The New York Times*, July 2, C7.

Evers, J. 2001. "Hackers Get Credit Card Data from Amazon's Bibliofind," *PC World*, March 6. (http://www.pcworld.com/news/article/0,aid,43582,00.asp)

Files, J. 2005. "For Fourth Time, Judge Seeks to Shield Indian Data," *The New York Times*, October 25, A17.

Gallagher, S. 2002. "Best Buy: May Day Mayday for Security," *Baseline*, June 7. (http://www.baselinemag.com/article2/0,3959,687,00.asp)

Glass, B. and D. Fisher. 2004. "Biometrics Security," *PC Magazine*, 23(1), January 20, 66.

Goldsborough, R. 2012. "Computer Disasters: Preparing for the Worst," *Tech Directions*, 71(6), 14.

Gorman, S. 2009. "FBI Suspects Terrorists Are Exploring Cyber Attacks," *The Wall Street Journal*, November 18, A4.

Gorman, S., E. Ramstad, J. Solomon, Y. Dreazen, R. Smith, and R. Sidel. 2009. "Cyber Blitz Hits U.S., Korea," *The Wall Street Journal*, July 9, A1, A4.

Grow, B., K. Epstein, and C. Tschang. 2008. "The New E-spionage Threat," *Business Week*, April 21, 33–41.

Hayes, F. 2002. "Thanks, Warchalkers," *Computerworld*, 36(35), August 26, 56.

Hoover, J. 2008. "What Could Slow Down the Windows Server Juggernaut?" *Information Week*, March 3, 34.

Hulme, G. 2012. "Managing the Unmanageable: Cloud Firewall Management Vendors Unleash New Wares," *CSO Online*, February 3. (http://www.csoonline.com/article/699389/managing-the-unmanageable)

Jakobsson, M., R. Chow, and J. Molina. 2012. "Authentication: Are We Doing Well Enough?" *IEEE Security and Privacy*, 10(1), January/February, 19–21.

Johnson, J. 2008. "Security Smarts: At Pacific Northwest National Laboratory, Network Defense Requires Layers of Strategic Thinking," *Information Week*, February 25, 43–46.

King, R. 2011. "Many Mobile Users Are Uneasy About Smartphone Security," *ZDNet*, October 31. (http://www.zdnet.com/blog/btl/many-mobile-users-are-uneasy-about-smartphone-security-survey/62145)

Krim, J. 2003. "WiFi Is Open, Free and Vulnerable to Hackers: Safeguarding Wireless Networks Too Much Trouble for Many Users," *The Washington Post*, July 27, A1.

Langner, R. 2011. "Stuxnet: Dissecting a Cyberwarfare Weapon," *IEEE Security and Privacy*, 9(3), May/June, 49–51.

Lee, C. 2008. "GAO Finds Data Protection Lagging," *The Washington Post*, February 26, A15.

Manes, S. 2001. "Security, Microsoft Style: No Safety Net?" *PC World*, 19(11), November, 210.

McCarthy, N. 2012. *The Computer Incident Response Planning Handbook: Executable Plans for Protecting Information at Risk*. New York: McGraw-Hill Osborne.

McCracken, H. 2004. "Microsoft's Security Problem—and Ours," *PC World*, 22(1), January, 25.

McMillan, R. 2010. "After One Year, Seven Million Conficker Infections," *PC World*, January, 44.

Menn, J. 2009. "Crippling Cyber-attacks Relied on 200,000 Computers," *Financial Times*, July 10, 6.

Nakashima, E. 2009. "Obama Set to Create A Cybersecurity Czar With Broad Mandate," *The Washington Post*, May 26, A4.

National Institute of Standards and Technology (NIST). 2001. *Federal Information Processing Standards (FIPS): Announcing the Advanced Encryption Standard (AES)*. Washington, DC: NIST. (http://csrc.nist.gov/publications/fips/fips197/fips-197.pdf)

Nerney, C. 2003. "Get It Right, Redmond," *Internet News*, May 12. (http://www.internetnews.com/commentary/article.php/2205081)

The New York Times. 2009. "Hackers Steal South Korean, U.S. Military Secrets," December 18.

Nielsen, J. 2004. "User Education Is Not the Answer to Security Problems," *Alertbox*, October 25. (http://www.useit.com/alertbox/20041025.html)

Pereira, J. 2008. "Data Theft Carried Out on Network Thought Secure," *The Wall Street Journal*, March 31, B4.

Petreley, N. 2001. "The Cost of Free IIS," *Computerworld*, 35(43), October 22, 49.

Piazza, P. 2003. "Phishing for Trouble," *Security Management*, 47(12), December, 32–33.

Rashid, F. 2011. "ZeuS Trojan Merger with SpyEye, Other Banking Malware Worry Researchers," *eWeek*, November 29. (http://www.eweek.com/c/a/Security/Zeus-Trojan-Merger-with-SpyEye-Other-Banking-Malware-Worry-Researchers-648865/)

Regan, K. 2001. "Hack Victim Bibliofind to Move to Amazon," *E-Commerce Times*, April 6. (http://www.ecommercetimes.com/story/8768.html)

Ren, K., C. Wang, and Q. Wang. 2012. "Security Challenges for the Public Cloud," *IEEE Internet Computing*, 16(1), January, 69–73.

Rivest, R. 1992. *The MD5 Message-Digest Algorithm*, IETF RFC 1321.

Rose, B. 2011. "Smartphone Security: How to Keep Your Handset Safe," *PC World*, January 10. (http://www.pcworld.com/businesscenter/article/216420/smartphone_security_how_to_keep_your_handset_safe.html)

Rosencrance, L. 2004. "Federal Audit Raises Doubts About IRS Security System," *Computerworld*, 38(36), September 6, 9.

Sang-Hun, C. and J. Markoff. 2009. "Cyberattacks Jam Government and Commercial Web Sites in U.S. and South Korea," *New York Times*, July 7, 4.

Saraswat, P. and R. Gupta. 2012. "A Review of Digital Steganography," *Journal of Pure and Applied Science & Technology*, 2(1), January, 98–106.

Sausner, R. 2009. "SSL Comes Under Fire," *Bank Technology News*, 22(9), September, 14.

Security Management, 2002. "Government Infosec Gets Failing Grade," 46(2), February, 34–35.

Shipley, G. 2001. "Growing Up with a Little Help from the Worm," *Network Computing*, 12(20), October 1, 39.

Skoudis, E. 2005. "Five Malicious Code Myths and How To Protect Yourself in 2005," *SearchSecurity.com*, January 4. (http://searchsecurity.techtarget.com/tip/1,289483,sid14_gci1041736,00.html)

Steiner, I. 2008. "eBay Changes Criteria for Sellers 'Buyer Dissatisfaction' Rate," *AuctionBytes.com*, February 8. (http://www.auctionbytes.com/cab/abn/y08/m02/i08/s02)

Strom, D. 2009. "Make E-mail Encryption Effortless," *Baseline*, December, 32–33.

Stuttard, D. and M. Pinto. 2007. *The Web Application Hacker's Handbook: Discovering and Exploiting Security Flaws*. New York: Wiley.

Thompson, J. 2012. "Smartphone Security: What You Need to Know," *TechRadar.com*, February 5. (http://www.techradar.com/news/phone-and-communications/mobile-phones/smartphone-security-what-you-need-to-know-1056995)

Tiwari, R. 2011. "Microsoft Excel File: A Steganographic Carrier File," *Digital Crime and Forensics*, 3(1), 37–52.

U.S. National Institute of Standards and Technology. 1993. *Data Encryption Standard (DES): Federal Information Processing Standards Publication 46–2*. Gaithersburg, MD: U.S. Computer Systems Laboratory.

Vaidyanathan, G. and S. Mautone. 2009. "Security in Dynamic Web Content Management Systems Applications," *Communications of the ACM*, 52(12), December, 121–125.

Vamosi, R. 2010. "New Banking Trojan Horses Gain Polish," *PC World*, January, 41–42.

Verton, D. 2002. "Mapping of Wireless Networks Could Pose Enterprise Risk," *Computerworld*, August 14. (http://computerworld.com/securitytopics/security/story/0,10801,73479,00.html)

Vijayan, J. 2001. "Corporations Left Hanging as Security Outsourcer Shuts Doors," *Computerworld*, 35(18), April 30, 13.

Vijayan, J. 2005. "Companies Scramble to Bolster Online Security," *Computerworld*, 39(10), March 7, 1, 61.

Vishwakarma, D., S. Maheshwari, and S. Joshi. 2012. "Efficient Information Hiding Using Steganography," *International Journal of Emerging Technology and Advanced Engineering*, 2(1), January, 154–159.

Wheeler, E. 2011. *Security Risk Management: Building an Information Security Risk Management Program from the Ground Up*. Waltham, MA: Syngress.

Wilshusen, G. and D. Powner. 2009. "Cybersecurity: Continued Efforts Are Needed to Protect Information Systems from Evolving Threats," *GAO Reports*, November 17, 1–20.

Wilson, T. 2008. "Before Walls Go Up, Ask What You're Really Protecting," *Information Week*, April 14, 26.

Wolfe, D. 2009. "Online Perils," *American Banker*, December 9, 5.

Zhao, J. and S. Zhao. 2010. "Opportunities and Threats: A Security Assessment of State E-government Websites," *Government Information Quarterly*, 27(1), January, 49–56.

Zissis, D. and D. Lekkas. 2012. "Addressing Cloud Computing Security Issues," *Future Generation Computer Systems*, 28(3), March, 583–592.

459

CHAPTER 11

PAYMENT SYSTEMS FOR ELECTRONIC COMMERCE

LEARNING OBJECTIVES

In this chapter, you will learn:

- The basic functions of online payment systems
- How payment cards are used in electronic commerce
- About the history and future of electronic cash
- How digital wallets work
- What stored-value cards are and how they are used in electronic commerce
- How the banking industry uses Internet technologies

INTRODUCTION

In 1991, a teenager named Max Levchin emigrated from the Ukraine to the United States. Settling in Chicago, Levchin had a burning interest in cryptography. Growing up in a Soviet police state convinced him that the ability to send coded messages that could not be read or intercepted was both important and useful. He majored in computer science at the University of Illinois and spent many hours at the school's Center for Supercomputing, pursuing his passion for making and breaking codes. When he graduated in 1998, he wanted to follow the American dream of turning his knowledge into money, so he headed for the heart of the computer industry in Palo Alto, California. Levchin's plan to build the ultimate transmission encryption scheme never did pan out, but he managed to turn his knowledge into a successful business. As cofounder and chief technical officer of **PayPal**, an online payment processing company that you will learn about in this chapter, Levchin used his expertise in cryptography and computer security to protect the firm from losses that could destroy it.

PayPal, founded in 1999, operates a service that lets people exchange money over the Internet. PayPal immediately carved itself a niche as the most popular payment system for processing auction payments on eBay. People can also use PayPal to send money to anyone who has an e-mail address, and a growing number of online stores accept PayPal in addition to (or instead of) credit cards. A number of charities accept donations through PayPal as well. These uses of PayPal—transferring money from one individual to another and as an alternative to paying by credit card at online stores—have grown rapidly in recent years. PayPal charges very small fees to business users and no fees at all to individuals, so its profit margins are small. However, it earns these small profit margins on a very large number of transactions.

One major concern for PayPal is that a single, well-organized, large-scale fraud attack could put the company out of business. Levchin's contribution to the company's success was his development of payment surveillance software that continually monitors PayPal transactions. The software searches millions of transactions as they occur every day and looks for patterns that might indicate fraud. The software notifies PayPal managers immediately when it finds something suspicious.

The software has worked well. About 1.13 percent of online credit card transactions are fraudulent, a rate that is much higher than the 0.70 percent experienced in physical stores. PayPal has kept its fraud rate below 0.50 percent. Because PayPal has kept its fraud rate so low, it can charge lower transaction fees than its competitors and still make a profit. PayPal's attention to fraud control has given it a competitive advantage over other payment processors (such as banks) and has allowed it to prosper in a very competitive business.

Since its inception, PayPal's largest customer group has been the participants (buyers and sellers) on the auction Web site eBay. As you will learn in this chapter, eBay spent three years working to establish its own payments service that could compete effectively with PayPal but finally gave up and bought PayPal for $1.4 billion. Today, PayPal offers payment services under its own name as a division of eBay.

ONLINE PAYMENT BASICS

An important function of electronic commerce sites is the handling of payments over the Internet. Most electronic commerce involves the exchange of some form of money for goods or services. As you learned in Chapter 5, many payment transactions between B2B companies are made using electronic funds transfers (EFTs). In this chapter, you will learn about a number of online payment alternatives that are available to businesses and individual consumers for B2C transactions. Online payments vary in both their size and how they are processed.

Micropayments and Small Payments

Internet payments for items costing from a few cents to approximately a dollar are called **micropayments**. Micropayment champions see many applications for such small transactions, such as paying 5 cents for an article reprint or 25 cents for a complicated literature search. However, micropayments have not been implemented very well on the Web yet. Another barrier to micropayments is a matter of human psychology. Researchers have found in a number of studies that many people prefer to buy small-value items by making regular fixed-amount payments rather than by making small payments in varying amounts, even when the small varying payments would cost less money overall. A good example of this behavior is the preference most mobile telephone users have for fixed monthly payment plans over charges based on minutes used. The comfort of knowing the exact amount of the monthly bill is more important to many people than getting the lowest price on the minutes used.

Over the past 10 years, many companies have developed systems to process micropayments. Millicent, DigiCash, Yaga, and BitPass were among the companies that entered this business and failed. Industry observers see a need for a micropayments processing system on the Web, but no company has gained broad acceptance of its system. All of the companies who entered this market used systems that either accumulated micropayments and charged them periodically to a credit card or accepted a deposit and charged the micropayments against that deposit. Some companies that offer electronic cash and bill paying services do provide micropayment capabilities as part of their services, but no company is currently devoted solely to offering micropayment services.

The payments that are between $1 and $10 do not have a generally accepted name (some industry observers use the term micropayment to describe any payment of less than $10); in this book, the term **small payments** is used to describe all payments of less than $10.

Some companies offer small payment and micropayment services through mobile telephone carriers. Buyers make their purchases using their mobile phones and the charges appear on the buyers' monthly mobile phone bill. The use of this micropayment system has been held back by the mobile carriers' substantial charges for providing the service, which can amount to 50 percent of each transaction. The company that offers the service typically takes another 10 percent, thus the buyer can end up paying more than double the actual value of the purchase. For example, a mobile phone user might pay $10 for a software application that would sell through other channels for $3 because the carrier takes a $6 fee and the payment processor takes a $1 fee.

Online Payment Methods

Cash, checks, credit cards, and debit cards are the four most common methods used in the world by consumers to pay for purchases. These four payment methods account for more than 90 percent of all consumer payments in the United States today. A small percentage of consumer payments are made by electronic transfer. The most popular consumer electronic transfers are automated payments of auto loans, insurance payments, and mortgage payments made from consumers' checking accounts.

Cash and checks are awkward or difficult to use online, so the majority of online payments (worldwide) are made using credit or debit cards (about 85 percent of the total) with alternative payment systems (predominantly PayPal) accounting for most of the remainder. Most industry analysts expect that the use of credit and debit cards will decrease somewhat as the use of alternative payment systems grows. Figure 11-1 shows forecasted forms of online payments for 2015.

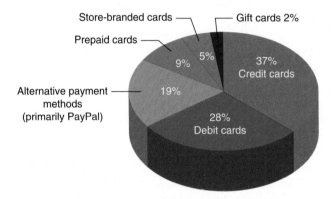

Source: Adapted from forecasts by Javelin Strategy & Research and *Internet Retailer*.

FIGURE 11-1   Forms of payment for U.S. online transactions, estimates for 2015

Online payment systems for consumer electronic commerce are still evolving. A number of proposals and implementations of payment systems currently compete for dominance. Regardless of format, electronic payments are far cheaper than mailing paper checks. Electronic payments can be convenient for customers and can save companies money. Estimates of the cost of billing one person by mail range between $1 and $1.50. Sending bills and receiving payments over the Internet can drop the transaction cost to an average of 50 cents per bill. The total savings is huge when the unit cost is multiplied by the number of customers who could use electronic payment. For example, a telephone company in a major metropolitan area might have 5 million customers, each of whom receives a bill every month. In one year, a savings of 50 cents on each of those 60 million bills adds up to about $30 million. The environmental impact is also significant. Those 60 million paper bills weigh about 1.7 million pounds. It takes 2200 trees to make that much paper—in addition to the energy consumed and the wastes generated in the paper-making process.

Online businesses must offer their customers payment options that are safe, convenient, and widely accepted. The key is to determine which choices work the best for the company and its customers. The information in this chapter will help you make those decisions. You will learn about four different payment technologies in this chapter: payment cards, electronic cash, software wallets, and smart cards (also called stored-value cards). Each technology has unique properties, costs, advantages, and disadvantages.

PAYMENT CARDS

Businesspeople often use the term **payment card** as a general term to describe all types of plastic cards that consumers (and many businesses) use to make purchases. The main categories of payment cards are credit cards, debit cards, charge cards, prepaid cards, and gift cards.

A **credit card**, such as a Visa or MasterCard, has a spending limit based on the user's credit history; a user can pay off the entire credit card balance or pay a minimum amount each billing period. Credit card issuers charge interest on any unpaid balance. Many consumers already have credit cards, or are at least familiar with how they work. Credit cards are widely accepted by merchants around the world and provide assurances for both the consumer and the merchant. A consumer is protected by an automatic 30-day period in which he or she can dispute an online credit card purchase. Online credit card purchases are similar to telephone purchases in that the card holder is not present and cannot provide proof of identity as easily as he or she can when standing at the cash register. Online and telephone purchases are often called **card not present transactions** and both include an extra degree of risk for merchants and banks.

A debit card looks like a credit card, but it works quite differently. Instead of charging purchases against a credit line, a **debit card** removes the amount of the sale from the cardholder's bank account and transfers it to the seller's bank account. Debit cards are also called **electronic funds transfer at point of sale (EFTPOS) cards**, especially outside the United States. Debit cards are issued by the cardholder's bank and usually carry the name of a major credit card issuer, such as Visa or MasterCard, by agreement between the issuing bank and the credit card issuer. By branding their debit cards (with the Visa or MasterCard name), banks ensure that their debit cards will be accepted by merchants who recognize the credit card brand names.

A **charge card**, offered by companies such as American Express, carries no spending limit, and the entire amount charged to the card is due at the end of the billing period. Charge cards do not involve lines of credit and do not accumulate interest charges. (Note: In addition to its charge card products, American Express also offers credit cards, which do have credit limits and which do accumulate interest on unpaid balances.) In the United States, many retailers, such as department stores and oil companies that own gas stations, issue their own charge cards. Cards issued by a specific retailer are sometimes called **store charge cards** or **store-branded cards**. The purchasing cards (or p-cards) that you learned about in Chapter 5 can be either credit cards or charge cards.

Some retailers offer cards that can be redeemed by anyone for future purchases. These **prepaid cards** are sometimes used by people who do not want to be tempted by a credit card to purchase more than they can afford. They can also be used to make small purchases that would be expensive for a merchant to process as credit card sales. More often, they are given to third parties as gifts. Prepaid cards sold with the intention that they be given as gifts are called **gift cards**.

Many consumers have concerns about providing their payment card numbers to vendors online, especially when the vendor is unknown to them. To address this concern, several payment card companies began offering cards with disposable numbers. These cards, sometimes called **single-use cards**, gave consumers a unique card number that was valid for one transaction only. This prevented an unscrupulous vendor from using the card number to complete unauthorized transactions on the consumer's account or selling the card number to others. Despite a flurry of interest in single-use cards when they were first introduced, issuers found that consumers did not use them. After a few years, the companies that had introduced single-use cards withdrew them from the market. The problem with single-use cards was that they required consumers to behave differently.

Not enough consumers saw a clear benefit to justify their learning how to use this new product.

Advantages and Disadvantages of Payment Cards

Payment cards have several features that make them a popular choice for both consumers and merchants in online and offline transactions. For merchants, payment cards provide fraud protection. When a merchant accepts payment cards for online payment or for orders placed over the telephone, the merchant can authenticate and authorize purchases using an interchange network. An **interchange network** is a set of connections between banks that issue credit cards, the associations that own the credit cards (such as MasterCard or Visa), and merchants' banks. You will learn more about interchange networks and how this system operates later in this chapter. For U.S. consumers, payment cards are advantageous because the Consumer Credit Protection Act limits the cardholder's liability to $50 if the card is used fraudulently. Once the cardholder notifies the card's issuer of the card theft, the cardholder's liability ends. Frequently, the payment card's issuer waives the $50 consumer liability when a stolen card is used to purchase goods. Some other countries have similar laws, but this type of protection is not common for holders of credit cards issued outside the United States. The lack of this type of protection does limit the willingness of non-U.S. consumers to use payment cards for online purchases.

Perhaps the greatest advantage of using payment cards is their worldwide acceptance. Payment cards can be used anywhere in the world, and the currency conversion, if needed, is handled by the card issuer. For online transactions, payment cards are particularly advantageous. When a consumer reaches the electronic checkout, he or she enters the payment card number and his or her shipping and billing information in the appropriate fields to complete the transaction. The consumer does not need any special hardware or software to complete the transaction.

Payment cards have one significant disadvantage for merchants when compared to cash. Payment card service companies charge merchants per-transaction fees and monthly processing fees. These fees can add up, but merchants view them as a cost of doing business. Any merchant that does not accept payment cards for purchases risks losing a significant portion of sales to other merchants that do accept payment cards. The consumer pays no direct transaction-based fees for using payment cards, but the prices of goods and services are slightly higher than they would be in an environment free of payment cards. Most consumers also pay an annual fee for credit cards and charge cards. This annual fee is much less common on debit cards.

Payment cards provide built-in security for merchants because merchants have a higher assurance that they will be paid through the companies that issue payment cards than through the sometimes slow direct invoicing process. To process payment card transactions, a merchant must first set up a merchant account. The series of steps in a payment card transaction is usually transparent to the consumer. Several groups and individuals are involved: the merchant, the merchant's bank, the customer, the customer's bank, and the company that issued the customer's payment card. All of these entities must work together for customer charges to be credited to merchant accounts (and vice versa when a customer receives a payment card credit for returned goods).

Payment Acceptance and Processing

Most people are familiar with the use of payment cards: In a physical store, the customer or a sales clerk runs the card through the online payment card terminal and the card account is charged immediately. In this type of in-person transaction, customers walk out of the store with purchases in their possession, so charging and shipment occur nearly simultaneously. Online stores and mail order stores in the United States must ship merchandise within 30 days of charging a payment card. Because the penalties for violating this law can be significant, most online and mail order merchants do not charge payment card accounts until they ship merchandise.

Processing a payment card transaction online involves two general processes, the acceptance of payment and clearing the transaction. Payment acceptance includes the steps necessary to determine that the card is valid and that the transaction will not exceed any credit limit that might exist for the card. Clearing the transaction includes all of the steps needed to move the funds from the card holder's bank account into the merchant's bank account. This section outlines the rather detailed steps involved in both of these processes.

Open and Closed Loop Systems

In some payment card systems, the card issuer pays the merchants that accept the card directly and does not use an intermediary, such as a bank or clearinghouse system. These types of arrangements are called **closed loop systems** because no other institution is involved in the transaction. American Express and Discover Card are examples of closed loop systems. Figure 11-2 shows the basic interactions among the entities involved in a closed loop payment card system.

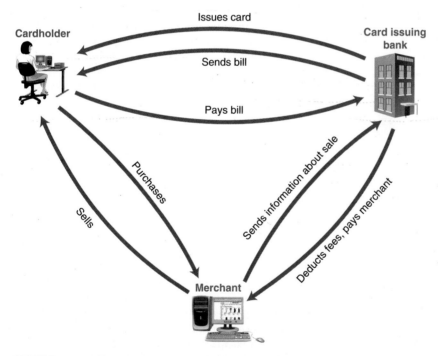

FIGURE 11-2   Closed loop payment card system

© Cengage Learning

Open loop systems add additional payment processing intermediaries to the structure of a closed loop system. Suppose an Internet shopper uses a Visa card issued by the First Bank of Woodland to purchase an item from Web Wonders, whose bank account is at the Hackensack Commerce Bank. The banking system includes one or more intermediaries (banks or other types of payment processing companies) that coordinate the transfer of funds from the First Bank of Woodland to the Hackensack Commerce Bank. Whenever additional parties, such as the intermediaries in this example, are included in payment card transaction processing, the system is called an **open loop system**. Visa and MasterCard are two of the most widely known examples of open loop systems. Many banks issue both of these cards.

Unlike American Express or Discover, neither Visa nor MasterCard issues cards directly to consumers. Visa and MasterCard are **credit card associations** that are operated by the banks who are members in the associations. These member banks, which are called **customer issuing banks** or **issuing banks**, issue credit cards to individual consumers. The issuing banks are responsible for evaluating their customers' credit standings and establishing appropriate individual credit limits. If a cardholder does not pay, the issuing bank absorbs the loss. Figure 11-3 shows the basic interactions among the entities involved in an open loop payment card system.

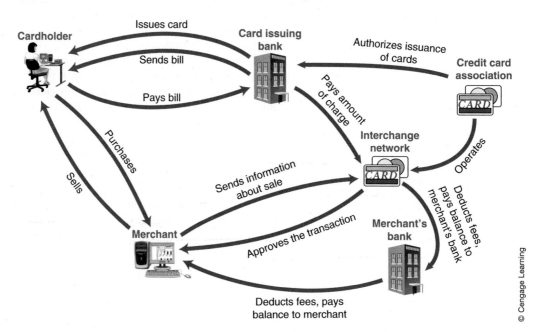

FIGURE 11-3   Open loop payment card system

Merchant Accounts

An **acquiring bank** is a bank that does business with sellers (both Internet and non-Internet) that want to accept payment cards. To process payment cards for Internet transactions, an online merchant must set up a **merchant account** with an acquiring bank. One type of merchant account is similar to a regular business checking account; the merchant's acquiring bank collects credit card receipts on behalf of the merchant from the payment card issuing bank and credits their value, net of processing fees, to the merchant's account. More commonly, a merchant account is set up to operate as a credit line rather than as a checking account. That is, the acquiring bank makes what is essentially a non-interest bearing loan to the merchant in the amount of the net credit card receipts each day. As the acquiring bank collects the proceeds of the transactions from the issuing bank, the acquiring bank reduces the balance of the non-interest bearing loan to the merchant.

A seller must provide information about its business operations to an acquiring bank before it will set up a merchant account. Typically, a new merchant must supply a business plan, details about existing bank accounts, and a business and personal credit history. The acquiring bank wants to be sure that the merchant has a good prospect of staying in business and wants to minimize its risk. If the merchant is new or is not doing well financially, the acquiring bank might ask for a deposit or personal guarantees of the owners or stockholders of the merchant. In some cases, the acquiring bank will demand that collateral (the owner's house, for example) be assigned.

The riskiness of the business also influences the acquiring bank's decision to provide a merchant account. Some types of businesses have a higher likelihood that customers will contest card charges than others. For example, a business that sells a guaranteed weight loss scheme—a business in which many customers might want their money back—might have difficulty finding an acquiring bank willing to provide a merchant account. The bank assesses the level of risk in the business based on the type of business and the credit information that is provided. Acquiring banks must estimate what percentage of sales are likely to be contested by cardholders. When a cardholder successfully contests a charge, the acquiring bank must retrieve the money it placed in the merchant account in a process called a **chargeback**. To ensure that sufficient funds are available to cover chargebacks, an acquiring bank might require a company to maintain funds on deposit in the merchant account. For example, a new or risky business that plans to make $100,000 in sales each month might be required to keep $50,000 or more on deposit in its merchant account.

In addition to chargeback deductions, the acquiring bank will deduct fees from the gross sales amount in determining the net amount to credit the merchant each day. These fees include **acquirer fees**, which are charged by the acquiring bank for providing the payment card processing service, and **interchange fees**, which are charged at rates that depend on the merchant's industry. Acquirer fees usually include a charge per month and per transaction and are set by the acquiring bank. The interchange fee rates are set by the card association itself (for example, Visa or MasterCard) and charged to the acquiring bank, which generally passes the cost along to the merchant.

One problem facing online businesses is that the level of fraud in online transactions is much higher than either in-person or telephone transactions of the same nature (that is, the same amount and the same type of good or service being purchased). Fewer than 15 percent of all credit card transactions are completed online, but those transactions are responsible for about 64 percent of the total dollar amount of credit card fraud.

According to a series of annual surveys conducted by credit card research company Cybersource, the proportion of online transactions that are fraudulent increased steadily every year from the inception of electronic commerce through 2008. Since 2008, that proportion has decreased slightly, although not enough to indicate a clear downward trend. Online fraud experts believe that this leveling off (and possible decline) in fraud losses resulted from merchants' increased use of antifraud measures. These antifraud measures include the use of fraud scoring services that provide risk ratings for individual transactions in real time, shipping only to the card billing address, and requiring **card verification numbers (CVNs)** for card not present transactions. A CVN is a three- or four-digit number that is printed on the credit card, but is not encoded in the card's magnetic strip. Having a CVN establishes that the purchaser has the card (or has seen the card) and is more likely not to be using a stolen card number. The CVN is also known by a number of different names and acronyms, including **card security code (CSC)**, **card verification data (CVD)**, **card verification value (CVV or CV2)**, **card verification value code (CVVC)**, **card verification code (CVC)**, **verification code (V-Code or V Code)**, and **card code verification (CCV)**. The next section outlines payment card authorization and payment processing options for online businesses.

Processing Payment Card Transactions

Figures 11-2 and 11-3 provide an overview of the basic processes involved in handling payment card transactions (including both credit cards and debit cards) in closed loop and open loop systems. Because most online merchants want to accept both closed loop system cards (such as American Express and Discover) and closed loop system cards (such as MasterCard and Visa), they must have internal systems that will work with both sets of processes. In addition, some online merchants accept direct deductions from customers' checking accounts. These direct deduction transactions are done through a network of banks called the **Automated Clearing House (ACH)**. Issuing banks, interchange networks, and acquiring banks use the ACH network to transfer funds to clear their card payment accounts with each other. The ACH provides a standardized funds transfer system and gives each participant a verified audit trail and nonrepudiation. These benefits are similar to those provided to EDI trading partners by a VAN, as you learned in Chapter 5. You can learn more about ACHs by following the Web Links to the **EPN**, **NACHA—The Electronic Payments Association**, and **The Clearing House**. The U.S. Federal Reserve Bank's **FedACH Services** site also has information about the operation of the ACH.

Processing payment card transactions that might be from a debit card or a credit card, that might need open loop or closed loop processing, or that might even involve the ACH directly is a complex task. Large online businesses have entire departments of highly skilled employees who build and maintain the systems needed to accomplish this work. Midsized online businesses often purchase software (separately or as part of an electronic commerce software package) that handles the processing, but they must hire skilled employees to manage the system.

Small online businesses often do not have the resources to manage this function in-house, even with purchased software. They generally rely on a service provider either to assist them in processing payment card transactions or to handle the entire function for them. These service providers are called **payment processing service providers** or **payment processors** and are usually grouped into two general types, front-end processors and back-end processors.

A **front-end processor** obtains authorization for the transaction by sending the transaction's details to the interchange network and storing a record of the approval or denial (a process which usually takes less than a second). Front-end processors (or the hardware and software that they use to obtain transaction approvals) are often called **payment gateways**. A **back-end processor** takes the transactions from the front-end processor and coordinates information flows through the interchange network to settle the transactions. The back-end processor handles chargebacks and any other reconciliation items through the interchange network and the acquiring and issuing banks, including the ACH transfers.

Some payment processors, such as IPPay, Authorize.Net, Global Payments, and FirstData, handle all elements of payment processing, including the payment gateway function, front-end processing, and back-end processing. Other companies specialize in handling just one element of the process or in a particular industry. For example, Digital River's share*it! service provides payment processing for online businesses that sell downloadable software and games.

Many payment processors work with electronic commerce software in ways that prevent a customer from realizing that a separate company is handling their credit card transactions. For online sellers with established reputations, this is beneficial because it prevents customers from worrying that another entity will be handling their credit card information. However, a number of payment processors open their Web sites in a new window to process the payment transaction. In these cases, the customer becomes aware that their payment transaction is being handled by a third party. Payment processors that operate this way include eBay's PayPal and BillMeLater services, Checkout by Amazon, Google Checkout, ClickandBuy, and Digital River's share*it! service. For smaller online sellers that do not have a well-established reputation, these payment processors can provide customers with the feeling of security that comes from having a well-recognized name such as Amazon or Google handle the card payment part of their transactions.

ELECTRONIC CASH

Although credit cards dominate online payments today, electronic cash shows promise for the future. **Electronic cash** (also called e-cash or digital cash) is a general term that describes any value storage and exchange system created by a private (nongovernmental) entity that does not use paper documents or coins and that can serve as a substitute for government-issued physical currency. Because electronic cash is issued by private entities, there is a need for common standards among all electronic cash issuers so that one issuer's electronic cash can be accepted by another issuer. This need has not yet been met. Each issuer has its own standards, and electronic cash is not universally accepted, as is government-issued physical currency.

Many stores that accept credit cards require a minimum purchase amount of $5 to $10. Similar requirements are sometimes placed on the use of debit cards. Merchants impose a minimum purchase amount because the processing fees for small purchase amounts could easily be greater than the profits on those transactions. The same is true for Internet purchases. Small purchases are not profitable for merchants that accept only credit cards for payment. The market for purchases below $10 is one potentially significant market for electronic cash. With very low fixed costs, electronic cash could allow users to spend, for example, 50 cents for an online newspaper, or 80 cents to send an electronic greeting card.

Electronic cash has another factor in its favor: Most of the world's population does not have credit cards. In the United States, adults who cannot obtain credit cards because they do not earn enough or have past debt problems and children over the age of 13 but under the age of 18 would benefit from the availability of electronic cash. Outside the United States, few people hold credit cards because they have traditionally made their purchases in cash. For them, electronic cash is a more logical next step than credit cards. Despite the many failures of electronic cash, the idea refuses to die.

Privacy and Security of Electronic Cash

Concerns about electronic payment methods include privacy and security, independence, portability, and convenience. Consumers want to know whether transactions are vulnerable and whether the electronic currency can be copied, reused, or forged. Two characteristics of physical currency are important to have in any electronic cash implementation. First, it must be impossible to spend electronic cash more than once, just as with traditional currency. Second, electronic cash ought to be anonymous, just as currency is. **Anonymous electronic cash** is electronic cash that, like bills and coins, cannot be traced back to the person who spent it. The electronic cash transaction must occur between the two parties only, and the recipient must know that the electronic currency is not counterfeit or being used in two different transactions at the same time. Perhaps the most important characteristic of cash is convenience. If electronic cash requires special hardware or software, it is not convenient for people to use. Chances are good that people will not adopt an electronic cash system that is difficult to use. A company currently in the electronic cash business is Internet Cash.

Holding Electronic Cash: Online and Offline Cash

Electronic cash can be held in online storage or offline storage. Online cash storage means that the consumer does not personally possess electronic cash. Instead, a trusted third party, such as an online bank, coordinates all transfers of electronic cash and holds the consumers' cash accounts. In an online storage system, the merchant must contact the consumer's bank to receive payment for a purchase. This helps prevent fraud by confirming that the consumer's cash is valid.

Offline cash storage is similar to money kept in a wallet. The customer holds the electronic cash and no other party is involved in the transaction. Protection against fraud is still a concern, so either hardware or software must be used to prevent fraudulent spending or double spending. **Double spending** is spending a particular piece of electronic cash twice by submitting the same electronic currency to two different vendors. When the electronic currency reaches the bank for clearance a second time, it is too late to prevent the fraudulent act. The main deterrent to double spending is the threat of detection and prosecution. The system must provide tamperproof electronic cash that can be traced back to its origin. A two-part lock provides anonymous security but signals when someone is attempting to double spend cash. When a second attempted transaction is made with the same electronic cash, the system must reveal the attempted second use and the identity of the original electronic cash holder. Electronic cash that is used correctly preserves a user's anonymity. Figure 11-4 shows a graphic representation of this double-spending detection process.

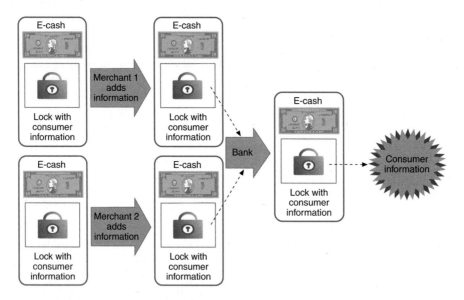

FIGURE 11-4 Detecting double spending of electronic cash

Advantages and Disadvantages of Electronic Cash

Billing for goods and services that customers purchase is part of any business. Traditional billing methods in the brick-and-mortar paradigm are costly and involve generating invoices, stuffing envelopes, buying and affixing postage to the envelopes, and sending the invoices to the customers. Meanwhile, the Accounts Payable Department must keep track of incoming payments, post accounts in the database, and ensure that customer data is current. Online sellers have many of the same payment collection inefficiencies as their brick-and-mortar cousins. Most online customers use credit cards to pay for their purchases. Electronic cash systems, though less popular than other payment methods, provide advantages and disadvantages that are unique to electronic cash.

Electronic cash transactions can be more efficient (and therefore less costly) than other methods, and that efficiency should foster more business, which eventually means lower prices for consumers. Transferring electronic cash on the Internet costs less than processing credit card transactions. Conventional money exchange systems require banks, bank branches, clerks, automated teller machines, and an electronic transaction system to manage, transfer, and dispense cash. Operating this conventional money exchange system is expensive.

Electronic cash transfers occur on the Internet, which is an existing infrastructure that uses existing computer systems. No distribution method or human oversight is required. Thus, the additional costs that users of electronic cash must incur are nearly zero. Merchants can pay other merchants in a business-to-business relationship, and consumers can pay each other. Electronic cash does not require that one party obtain an authorization, as is required with credit card transactions.

Electronic cash does have disadvantages, however, and they are significant. Using electronic cash provides no audit trail; that is, electronic cash is just like real cash in that it cannot be easily traced. Because true electronic cash is not traceable, another problem arises: money laundering. **Money laundering** is a technique used by criminals to convert money that they have obtained illegally into cash that they can spend without having it identified as the proceeds of an illegal activity. Money laundering can be accomplished by purchasing goods or services with ill-gotten electronic cash. The goods are then sold for physical cash on the open market.

Electronic cash has not been nearly as successful in the United States as it has been in Europe and Asia. In the United States, most consumers have credit cards, debit cards, charge cards, and checking accounts. These payment alternatives work well for U.S. consumers in both online and offline transactions. In most other countries of the world, consumers overwhelmingly prefer to use cash. Because cash does not work well for online transactions, electronic cash fills an important need for consumers in those countries as they conduct B2C electronic commerce. This type of need does not exist in the United States because U.S. consumers already use payment cards for traditional commerce, and these payment cards work well for electronic commerce.

One example of a successful electronic cash implementation is operated by KDD Communications (KCOM), the Internet subsidiary of Kokusai Denshin Denwa, which is Japan's largest global phone company. KCOM has its own NetCoin electronic cash system and offers electronic cash online through its NetCoin Center Web site. Shoppers can visit the site and obtain electronic cash that can be stored on their computers. Then, they can shop online for recipes or travel directories or download MP3s. Other content providers, such as Japanese newspapers, provide access to their newspaper archives and charge a small fee to retrieve articles. Japan also has a donation site where visitors can donate electronic coins to charitable organizations.

The lack of success seen by electronic cash initiatives in the United States can be blamed in part on the need that most of these systems had to be installed into consumers' Web browsers. Also, there were a number of competing technologies and no common standards were developed for all electronic cash systems. Without standards, each electronic cash alternative required its own installation and procedures, none of which were interoperable. **Interoperable software** runs transparently on a variety of hardware configurations and on different software systems.

PayPal Challengers

PayPal grew rapidly by serving the needs of buyers and sellers on auction sites such as eBay. This success and the business niche's potential for profits were noticed by a number of other companies that were eager to challenge PayPal for a share of the online payments business.

Because PayPal's early success was driven largely by its use on the eBay auction site, eBay's management team decided to compete directly against PayPal with its own payment service. In 1999, eBay purchased a small electronic payments company and, one year later, sold a 35 percent stake in that company to Wells Fargo bank. This company, Billpoint, was operated as a joint venture by eBay and Wells Fargo. Billpoint grew rapidly, but PayPal maintained the advantage it had gained as the first company to offer payment services online. PayPal continued to be the most widely used payment processing system on eBay despite Billpoint's best efforts to promote itself as a part of eBay. After unsuccessfully battling PayPal for three years, eBay finally gave up and decided to buy PayPal, as you learned in this chapter's opening case.

The profit potential of online payments also attracted the interest of several banks, whose managers thought they could use their years of experience in traditional payment processing to overcome PayPal's first-mover advantage. For example, Citibank operated its c2it payments service for several years before closing it in 2003. Other banks were similarly unsuccessful with their online payments operations.

Other financial services companies believed they could be successful in online payments, too. First Data Corporation, which owns Western Union, offered what it called electronic money orders that customers could use to settle auction transactions through its BidPay site. The company struggled to compete with PayPal for many years before closing in 2007.

PayPal performs its function very well and no other challengers have been able to find a way to do online payments any better. PayPal has prevailed not only because it was an early entrant, but because it performed as well as any of its challengers. PayPal users had no good reason to switch to any of the other providers.

DIGITAL WALLETS

As consumers are becoming more enthusiastic about online shopping, they have begun to tire of repeatedly entering detailed shipping and payment information each time they make online purchases. Filling out forms ranks high on online customers' lists of gripes about online shopping. To address these concerns, many electronic commerce sites include a feature that allows a customer to store their name, address, and credit card information on the site. However, consumers must enter their information at each site with which they want to do business. A **digital wallet** (sometimes called an **electronic wallet** or an **e-wallet**), serving a function similar to a physical wallet, is an electronic device or software that holds credit card numbers, electronic cash, owner identification, and owner contact information and provides that information at an electronic commerce site's checkout counter. Digital wallets give consumers the benefit of entering their information just once, instead of having to enter their information at every site with which they want to do business.

Increasingly, digital wallets are being promoted for purposes other than online shopping. One important purpose could be to authenticate the wallet holder's identity and credentials. For example, a digital wallet could be used to establish that a buyer of alcoholic beverages is of the appropriate age. The Prop-ID Research Project at the University of Toronto suggests that digital wallet technology broken down into three elements that include: the system (the infrastructure that accomplishes the identification), the application (the software with which the user interacts), and the device itself (if a specific device is used). Some industry observers and privacy rights activist groups are concerned about digital wallets because they give the company that issues the digital wallet access to a great deal of information about the individual using the wallet.

Software-Only Digital Wallets

Digital wallets that are software-based fall into two categories, depending on where they are stored. A **server-side digital wallet** stores a customer's information on a remote server belonging to a particular merchant or wallet publisher. For example, if you enter your information on a site such as Amazon.com and choose to store that information so you do not have to enter it when you next visit the site, Amazon.com stores your information in a server-side digital wallet.

The main weakness of server-side digital wallets is that a security breach could reveal thousands of users' personal information to unauthorized parties. Typically, server-side digital wallets employ strong security measures that minimize the possibility of unauthorized disclosure.

A **client-side digital wallet** stores a consumer's information on his or her own computer. Many of the early digital wallets were client-side wallets that required users to download the wallet software. This need to download software onto every computer used to make purchases is a chief disadvantage of client-side wallets. Server-side wallets, on the other hand, remain on a server and thus require no download time or installation on a user's computer. Before a consumer can use a server-side wallet on a particular merchant's site, the merchant must enable that specific wallet. Each wallet vendor must convince a large number of merchants to enable its wallet before it will be accepted by consumers. Thus, only a few server-side wallet vendors will be able to succeed in the market.

A disadvantage of client-side wallets is that they are not portable. For example, a client-side wallet is not available when a purchase is made from a computer other than the computer on which the wallet resides. In a client-side digital wallet, the sensitive information (such as credit card numbers) is stored on the user's computer instead of the wallet provider's central server. This removes the risk that an attack on a client-side digital wallet vendor's server could reveal the sensitive information. However, an attack on the user's computer could yield that information. Most security analysts agree that storing sensitive information on client computers is safer than storing that information on the vendor server because it requires attackers to launch many attacks on user computers, which are more difficult to identify (even though the user computers are less likely than a vendor server to have strong security features installed). It also prevents the easily identified servers of the wallet vendors from being attractive targets for such attacks.

The main weakness of server-side digital wallets is that a security breach could reveal thousands of users' personal information to unauthorized parties. Typically, server-side digital wallets employ strong security measures that minimize the possibility of unauthorized disclosure.

Microsoft Windows Live ID is a single sign-in service that includes a server-side digital wallet operated by Microsoft. Windows Live ID completes order forms automatically. All of the personal data entered into a Windows Live ID wallet is encrypted and password protected. Windows Live ID consists of four integrated services: single sign-in service (SSI), Wallet service, Kids service, and public profiles. The sign-in service allows a user to sign in at a participating Web site using his or her username and password. The Wallet service provides digital wallet functions such as secure storage and form completion of credit card and address information. When requested by a participating merchant, a consumer's secure information is released to the merchant so that the consumer does not need to enter data into a form. The Kids service helps parents protect and control their children's online privacy, and the public profiles service allows consumers to create a public page of information about themselves.

Yahoo! Wallet is a server-side digital wallet offered by Yahoo! This software-based digital wallet fills in online forms automatically with name, address, telephone, and credit card information. Yahoo! Wallet lets users store information about their major credit and charge cards, along with Visa and MasterCard debit cards. Yahoo! Wallet is accepted by thousands of Yahoo! Store merchants and can be used to pay for airplane tickets and hotel reservations booked through Yahoo! Travel. The digital wallet also works when users pay for Yahoo! services, such as premium e-mail storage or Web hosting fees. Yahoo! has the advantage of hosting a number of services and shops that it can be certain accommodate its own wallet; thus, it has a large number of merchants (including itself) that accept its wallet.

Hardware-Based Digital Wallets

The increasing prevalence of smart phones has made them candidates to become hardware-based digital wallets that can store the owner's identity credentials (such as a driver's license, medical insurance card, store loyalty cards, and other identifying documents). The smart phone can transmit portions of this identity information on command using its Bluetooth or wireless transmission capability to nearby terminals. **Near field communication (NFC)** technology, which allows for contactless data transmission over short distances, can also be used if the smart phone is equipped with a chip similar to those that have been used on payment cards (such as MasterCard's PayPass card) for a number of years.

NFC chips embedded in mobile phones are already very popular in Japan, where the devices are called *Osaifu-Keitai*, which translates approximately to "mobile wallet."

In the United States, a number of hardware-based digital wallets have been released or are in development. Google Wallet, which uses the PayPass technology that MasterCard developed for its credit cards, launched in 2011 and is available in a number of smart phones. Visa followed shortly after with its digital wallet product, V.me, and PayPal also announced that it will release a similar digital wallet that works on smart phones.

STORED-VALUE CARDS

Today, most people carry a number of plastic cards—credit cards, debit cards, charge cards, driver's license, health insurance card, employee or student identification card, and others. Most of these cards can store information electronically using either a magnetic strip or a microchip that is embedded into the card.

Magnetic Strip Cards

Most magnetic strip cards hold value that can be recharged by inserting them into the appropriate machines, inserting currency into the machine, and withdrawing the card;

the card's strip stores the increased cash value. Magnetic strip cards are passive; that is, they cannot send or receive information, nor can they increment or decrement the value of cash stored on the card. The processing must be done on a device into which the card is inserted.

Smart Cards

A **smart card** is a plastic card with an embedded microchip that can store information. Smart cards are also called **stored-value cards**. The microchip can also include a tiny computer processor that can perform calculations and storage operations right on the card. Most credit, debit, and charge cards currently store limited information on a magnetic strip. A smart card can store more than 100 times the amount of information that a magnetic strip plastic card can store. A smart card can hold private user data, such as financial facts, encryption keys, account information, credit card numbers, health insurance information, medical records, and so on.

Smart cards are safer than magnetic strip credit cards because the information stored on a smart card can be encrypted. For example, conventional credit cards show your account number on the face of the card and your signature on the back. The card number and a forged signature are all that a thief needs to purchase items and charge them against your card. With a smart card, credit theft is much more difficult because the key to unlock the encrypted information is a PIN; there is no visible number on the card that a thief can identify, nor is there a physical signature on the card that a thief can see and use as an example for a forgery.

Smart cards have been in use since the late 1990s. Popular in Europe and parts of Asia, smart cards so far have not been as successful in the United States. In Europe and Japan, smart cards are being used for telephone calls at public phones and for television programs delivered by cable to people's homes. The cards are also very popular in Hong Kong, where many retail counters and restaurant cash registers have smart card readers. The city's transportation companies—subways, buses, railways, trams, and ferries—joined together and created a smart card called the Octopus that lets commuters use one card for all of their public transportation needs. The Octopus card, which is now owned by an independent company, can be reloaded at any transportation location or at 7-Eleven stores throughout Hong Kong.

Smart cards have become more prevalent in the United States in recent years. In San Francisco, the Bay Area Metropolitan Transportation Commission created a smart card system patterned after the Octopus Card. This system, TransLink, is the first integrated ticketing system for public transportation in the United States. The transportation smart card, implemented in a 2002 pilot program, allows commuters to ride most modes of public transit available in the city, including trains, buses, cabs, and ferries, by holding the smart card near a reader device for a moment in transit vehicles or in stations. TransLink users can reload their smart cards at several retail outlets or directly from their bank accounts. The pilot program was a success and TransLink is now available to all Bay Area transit customers. More recently, a number of oil companies have issued their customers smart cards that can be swiped at the pump to pay for gasoline purchases. MasterCard has seen an increase in the use of its PayPass smart cards as well.

In the United States, the **Smart Card Alliance** promotes the benefits of smart cards. The organization promotes the widespread acceptance of multiple-application smart card technology. Its members include companies in banking, financial services, computer technology, health care, telecommunications, and a number of government agencies. The Alliance focuses on information exchange and member interaction. Every member of the Alliance recognizes that smart cards can succeed in the United States only if a critical mass of smart cards supports applications of interest to consumers. The Alliance promotes compatibility among smart cards, card reader devices, and applications.

INTERNET TECHNOLOGIES AND THE BANKING INDUSTRY

The largest dollar volume of payments in the world today are still made using paper checks. These paper checks are processed through the international banking system. The other major payment forms in use today also involve banks in one way or another. This section outlines how Internet technologies are providing new tools and creating new threats for the banking industry.

Check Processing

In the past, checks were processed physically by banks and clearinghouses. When a person wrote a check to pay for an item at a retail store, the retailer would deposit the check in its bank account. The retailer's bank would then send the paper check to a clearinghouse, which would manage the transfer of funds from the consumer's bank to the retailer's account. The paper check would then be transported to the consumer's bank, which might then send the cancelled check to the consumer. In recent years, many banks have stopped sending cancelled checks to their consumer account holders to save postage, instead providing access to PDF images of processed checks to account folders. Despite these savings, the cost of transporting tons of paper checks around the country has grown each year.

In addition to the transportation costs, another disadvantage of using paper checks is the delay that occurs between the time that a person writes a check and the time that check clears the person's bank. This delay (which is similar to the delay you learned about earlier in PayPal accounts, and which is also called float) makes it possible to write checks a few days before money is in the account to cover those checks. In effect, the bank's customer obtains the free use of funds for a few days and the bank loses the use of those funds for the same time period. Although the delay normally lasts only a few days, there are times when it can become significantly longer. Railroad and airline strikes, for example, have caused the float to be extended. The terrorist attacks of September 11, 2001 caused a significant increase in the float.

Banks have been working for years to develop technologies that will help them reduce the float. In 2004, a U.S. law went into effect that many bankers believe will eventually eliminate the float. This law, the Check Clearing for the 21st Century Act (usually referred to as **Check 21**), permits banks to eliminate the movement of physical checks entirely. In a Check 21-compliant world, the retailer can scan the customer's check. The scanned image is transmitted instantly through a clearing system and posts almost immediately to both accounts (that is, the withdrawal from the customer's account and the deposit to the retailer's account occur instantly), eliminating any float on the transaction.

You can learn more about the Check 21 law and its implementation by following this book's Web Links to the **Federal Reserve Financial Services - Check 21-Enabled Services** pages or the **American Bankers Association Check 21 Resource Center**.

NetBank

CompuBank and NetBank were two of the first Internet banks to open in the United States. They were both pure Internet banks; that is, neither was founded by an existing bank with a physical presence. After four years of operation, CompuBank had about 50,000 accounts and $64 million of deposits and was losing more than $20 million per year. NetBank had done considerably better, with 160,000 accounts and $1 billion of deposits and 10 consecutive quarters of profitability.

In early 2001, CompuBank decided to close its operations and found NetBank to be a willing purchaser of its accounts. When a bank buys accounts from another bank, it performs a series of procedures called due diligence. These **due diligence** procedures include checking the new customers' credit histories and banking records. Due diligence is usually performed before the transaction is completed and before the closing bank's customers look to the buying bank as the institution that will handle their accounts.

For a number of reasons, not all of which are clear, the due diligence process was still under way on the date that the transfer of accounts was to take place. NetBank placed holds on many accounts and sent letters to many account holders explaining that they were not acceptable customers by NetBank standards. For any bank, this would have been a difficult situation, but the nature of the two banks as Internet-only operations made things considerably worse for everyone.

Press accounts of the fiasco included stories of the problems that between 4000 and 8000 CompuBank depositors experienced. Some of the problems were small—online bill payments did not occur, debit and credit cards were rejected at stores and restaurants, and ATMs would not yield cash—while others were much larger. One couple who had kept the money to cover closing costs on a house purchase in a CompuBank account found that NetBank had placed a hold on the money. Because they could not pay the closing costs, they were forced to find another mortgage lender. In the suit they filed against NetBank, the couple asserted that the increased rate on the mortgage loan would cost them tens of thousands of dollars. Other CompuBank customers were irritated that they lost access to their money for weeks. Some customers could not determine whether the bills they had set up to be paid automatically had, in fact, been paid.

NetBank admitted failures in customer service related to the incident. Many customers who called to complain or ask for explanations experienced 45-minute waits on hold and then were transferred to the bank's Security Department, where a recording answered and asked callers to leave their Social Security numbers and wait to be called back. None of the customers reported being called back. The timing of NetBank's notification was problematic, too. Many customers reported receiving a letter from NetBank indicating that there were problems with their accounts. The letter, dated April 30, was received by the customers on or after May 14. The letter included a telephone number to call for assistance, but that number had been disconnected on May 12. Many of the unhappy customers found each other on Internet discussion boards and compared notes.

NetBank has never disclosed the number of customers it lost by its handling of this transition; indeed, it may not know. CompuBank's customers were largely experienced Internet users who chose to be part of the leading edge in handling their financial affairs. Many of them, after this experience, have sworn that they will never again do business with a bank that does not have a physical presence. The lesson from NetBank's experience is that customer service and the ability to communicate with customers become extremely important for companies that process electronic payments or are responsible for their customers' finances.

Mobile Banking

In Chapter 6, you learned about new opportunities that are emerging for businesses that want to reach customers who use smart phones and other mobile devices to connect to the Internet. In recent years, banks have begun to explore the potential of mobile commerce in their businesses, too.

In 2009, a number of banks launched sites that allow customers using smart phones to obtain their bank balance, view their account statement, or find a nearby ATM. These sites are specifically designed for the smaller screen size of smart phones and make interacting with the bank easier than using a smart phone's Web browser to view the bank's regular site. These mobile services continue to be developed. In 2012, Visa announced a partnership with Monitise that will expand the Visa processing platform, Visa DPS, to enable banks that use it to offer their customers the ability to monitor their account history and balances, transfer funds among accounts, and receive transaction alerts on their smart phones.

Many banks' future plans include offering smart phone apps that bank customers can use to transact all types of banking business, including the option of taking a picture of a check with the smart phone's camera and depositing it into their bank accounts electronically. Some vendors offer a tiny credit card reader that can be attached to a smart phone. When this device is combined with an app that runs on the smart phone, the combined hardware becomes a highly portable payment processing terminal.

CRIMINAL ACTIVITY AND PAYMENT SYSTEMS: PHISHING AND IDENTITY THEFT

Online payment systems offer criminals and criminal enterprises an attractive arena in which to operate. The average consumers who engage in online payment transactions are easy prey for expert criminals. The large amounts of money involved make online payment systems tempting targets.

In Chapter 10, you learned about the phishing expedition, which is a technique for committing fraud against the customers of online businesses. Although phishing expeditions can be launched against all types of online businesses, they are of particular concern to financial institutions because their customers expect a high degree of security to be maintained over the personal information and resources that they entrust to their online financial institutions.

Phishing Attacks

The basic structure of a phishing attack is fairly simple. The attacker sends e-mail messages (such as the one shown in Figure 11-5) to a large number of recipients who might have an account at the targeted Web site. PayPal is the targeted site in the example shown in the figure.

Date: [Date removed] 08:05:42 +0600
From: "Services PayPal" <services@paypal.com>
Subject: PayPal Account sensitive features are access limited!
To: [E-mail addresses removed]

Dear valued **PayPal** member:

PayPal is committed to maintaining a safe environment for its community of
buyers and sellers. To protect the security of your account, PayPal employs
some of the most advanced security systems in the world and our anti-fraud
teams regularly screen the PayPal system for unusual activity.

Recently, our Account Review Team identified some unusual activity in your
account. In accordance with PayPal's User Agreement and to ensure that your
account has not been compromised, access to your account was limited. Your
account access will remain limited until this issue has been resolved. This
is a fraud prevention measure meant to ensure that your account is not compromised.

In order to secure your account and quickly restore full access, we may
require some specific information from you for the following reason:

We would like to ensure that your account was not accessed by an
unauthorized third party. Because protecting the security of your account
is our primary concern, we have limited access to sensitive PayPal account
features. We understand that this may be an inconvenience but please
understand that this temporary limitation is for your protection.

Case ID Number: PP-040-187-541

We encourage you to log in and restore full access as soon as possible.
Should access to your account remain limited for an extended period of
time, it may result in further limitations on the use of your account.

However, failure to restore your records will result in account suspension.
Please update your records within 48 hours. Once you have updated your account
records, your **PayPal** session will not be interrupted and will continue as normal.

To update your **Paypal** records click on the following link:
https://www.paypal.com/cgi-bin/webscr?cmd=_login-run

Thank you for your prompt attention to this matter. Please understand that
this is a security measure meant to help protect you and your account. We
apologize for any inconvenience.

Sincerely,
PayPal Account Review Department

PayPal Email ID PP522

Accounts Management As outlined in our User Agreement, **PayPal** will
periodically send you information about site changes and enhancements.

Visit our Privacy Policy and User Agreement if you have any questions.
http://www.paypal.com/cgi-bin/webscr?cmd=p/gen/ua/policy_privacy-outside

FIGURE 11-5   Phishing e-mail message

The e-mail message tells the recipient that his or her account has been compromised and it is necessary for the recipient to log in to the account to correct the matter. The e-mail message includes a link that appears to be a link to the login page of the Web site. However, the link actually leads the recipient to the phishing attack perpetrator's Web site, which is disguised to look like the targeted Web site. The unsuspecting recipient enters his or her login name and password, which the perpetrator captures and then uses to access the recipient's account. Once inside the victim's account, the perpetrator can access personal information, make purchases, or withdraw funds at will.

When the e-mails used in a phishing expedition are carefully designed to target a particular person or organization, the exploit is called **spear phishing**. The spear phishing perpetrator must do considerable research on the intended recipient, but by obtaining detailed personal information and using it in the e-mail, the perpetrator can greatly increase the chances that the victim will open the e-mail and click the link to the phishing Web site. Spear phishers have launched attacks against employees of specific companies that include jargon and acronyms that are frequently used in the company or its industry. By using familiar language and terms, the spear phisher gains the victim's trust and is more likely to convince the victim to click the phishing link.

Phishing perpetrators are quick to capitalize on new opportunities to practice their fraud. In 2008, the U.S. government enacted an economic stimulus law that paid millions of its citizens a rebate check. Within a week of the law's passage, phishing e-mails began appearing in inboxes throughout the country. The e-mails appeared to be from the Internal Revenue Service and promised an early rebate to responders who clicked the link (to the phishing Web site) and provided details such as bank account numbers, Social Security numbers, and passwords to online accounts.

The links in phishing e-mails are usually disguised. One common way to disguise the real URL is to use the @ sign, which causes the Web server to ignore all characters that precede the @ and only use the characters that follow it. For example, a link that displays:

https://www.paypal.com@218.36.41.188/fl/login.html

looks like it is an address at PayPal. However, the @ sign causes the Web server to ignore the "paypal.com" and instead takes the victim to a Web page at the IP address 218.36.41.188. In the e-mail shown in the figure, the link appears in the victim's e-mail client software as:

https://paypal.com/cgi-bin/webscr?cmd=_login-run

but when the victim clicks the link, the browser opens a completely different URL:

http://leasurelandscapes.com/snow/webscr.dll

Instead of the URL it shows in the e-mail client, the link in the phishing e-mail actually includes the following JavaScript code:

```
<A onmouseover="window.status=`https://www.paypal.com/cgi-bin/webscr?
cmd=_login-run'; return true" onmouseout="window.status=`https://www.
paypal.com/cgi-bin/webscr?cmd=_login-run' "href="http://
leasurelandscapes.com/snow/webscr.dll">https://www.paypal.com/
cgi-bin/webscr?cmd=_login-run</A>
```

This code is invisible in many e-mail clients, so the victim might never know that the Web browser has opened a phony site. Phishers use other tricks to hide URLs, including code that opens a pop-up window that displays the financial institution's URL and positions that window so it covers the browser's address bar. Phishing perpetrators often include graphics from the Web site of the victim's financial institution in the phishing e-mail to make it even more convincing. Figure 11-6 shows a phishing e-mail that includes graphics from the Bank of America Web site. You can learn more about the details of phishing techniques by visiting the Web sites of the **Conferences on Email and Anti-Spam** and the **Anti-Phishing Working Group (APWG)**.

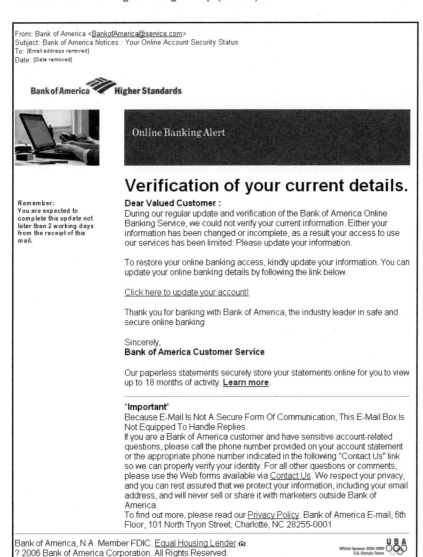

FIGURE 11-6 Phishing e-mail with graphics

Using Phishing Attacks for Identity Theft

Many perpetrators of phishing attacks are individuals working alone. However, the large amounts of illegal revenue that can be generated by combining phishing attacks with identity theft have drawn the attention of highly structured groups of criminals whose members possess a variety of specialized skills.

U.S. laws define **organized crime**, also called **racketeering**, as unlawful activities conducted by a highly organized, disciplined association for profit. The associations that engage in organized crime are often differentiated from less organized groups such as gangs and from organized groups that conduct unlawful activities for political purposes, such as terrorist organizations. Organized crime associations have traditionally engaged in criminal activities such as drug trafficking, gambling, money laundering, prostitution, pornography production and distribution, extortion, truck hijacking, fraud, theft, and insider trading. Often these activities are carried out simultaneously with legitimate business activities, which provide cover for the illegal activities.

The Internet has opened new opportunities for organized crime in its traditional types of criminal activities and in new areas such as generating spam (which you learned about in earlier chapters), phishing, and identity theft. **Identity theft** is a criminal act in which the perpetrator gathers personal information about a victim and then uses that information to obtain credit. After establishing credit accounts, the perpetrator runs up charges on the accounts and then disappears. Figure 11-7 includes a list of the types of personal information that identity thieves most want to obtain (listed in approximate order of usefulness to the criminal).

Social Security number

Driver's license number

Credit card numbers

Card verification numbers (CVNs)

Passwords (PINs)

Credit reports

Date of birth

ATM (or debit) card numbers

Telephone calling card numbers

Mortgage (or other loan) information

Telephone numbers

Home address

Employer name and address

FIGURE 11-7   Types of personal information most useful to identity thieves

Large criminal organizations can be highly efficient perpetrators of identity theft because they can exploit large amounts of personal information very quickly and efficiently. These organizations can use phishing attacks to gather personal information and then use it to perpetrate identity theft and other crimes. These criminal organizations often sell or trade information that they cannot use immediately to other organized crime entities around the world. Some of these criminal transactions are even conducted online. For example, a hacker who has planted zombie programs on a large number of computers (thus creating a **zombie farm**) might sell the right to use the zombie farm to an organized crime association that wants to launch a phishing attack (when a zombie farm is used this way, the attack is sometimes called a **pharming attack**). Individuals who commit these crimes have always posed a serious threat, but organized crime's entry into this activity increases the threat. There are two elements in phishing, the collection of the information (done by **collectors**) and the use of the information (done by **cashers**). The skills needed to perform these two activities are different. By facilitating transactions between collectors and cashers (and by participating as one or both), crime organizations have increased the efficiency and volume of phishing activity overall.

More than a million people fall victim to phishing attacks each year and as a group experience financial losses exceeding $500 million. Although the overall incidence of phishing attacks is decreasing as Internet users become aware of them, experts believe that the proportion of all phishing attacks committed by organized crime associations will continue to increase because it is so profitable.

Phishing Attack Countermeasures

In Chapter 8, you learned that several groups are working on ways to improve the Internet's mail transport protocols so that spam senders can be identified. Because spam is a key element of phishing attacks, any protocol change that improves e-mail recipients' ability to identify the source of an e-mail message will also help to reduce the threat of phishing attacks.

The most important step that companies can take today, however, is to educate their Web site users. Most online banking sites continually warn their customers that the site never sends e-mails that ask for account information or that ask the recipient to log in to their Web site and make changes to his or her account information. PayPal occasionally interrupts its own login screen sequence to insert a page that provides information about phishing attacks.

Many companies, especially those that operate financial Web sites, have contracted with consulting firms that specialize in antiphishing work. These consultants monitor the Web for new Web sites that use the company's name or logo and move quickly to shut down those sites. Most phishing perpetrators set up their entrapping Web sites (with the target company's name and logo) a few days before they launch their e-mail campaign, so this monitoring technique can be effective. Another antiphishing technique is to monitor online chat rooms that are used by criminals. By watching for offers of stolen credit card information and other phishing exploits, consultants can identify phishing schemes that are under way.

The incidence of phishing attacks has grown rapidly over the past two years and most industry analysts expect that phishing will be a problem that will plague online businesses for the near future. Phishing can be an extremely profitable criminal activity and as more companies increase their defenses, analysts expect phishing perpetrators to become even better at working around those defenses.

Summary

Online stores can accept a variety of forms of payment. Credit, debit, and charge cards (payment cards) are the most popular forms of payment on the Internet. They are ubiquitous, convenient, and easy to use. Processing payment card transactions requires that an online merchant establish a merchant account with an acquiring bank. The merchant can accomplish the card approval and transaction settlement processes using software that is included in the electronic commerce software, a separate payment processing software application, or through a payment process service provider.

Electronic cash, a form of online payment that is portable and anonymous, has been slow to catch on in the United States. Electronic cash could be useful for making micropayments because the cost of processing payment cards for small transactions is greater than the profit on such transactions.

Digital wallets provide convenience to online shoppers because they hold payment card information, electronic cash, and personal consumer identification. Software-based digital wallets eliminate the need for consumers to reenter payment card and shipping information at a site's electronic checkout counter. Instead, the digital wallet automatically fills in form information at sites that recognize the particular wallet software's technology. Hardware-based digital wallets that use a consumer's smart phone are being introduced and have great potential for making online and in-person payment card sales easier.

Stored-value cards, including smart cards and magnetic strip cards, are physical devices that hold information, including cash value, for the cardholder. Magnetic strip cards have limited capacity. Smart cards can store greater amounts of data on a microchip embedded in the card and are intended to replace the collection of plastic cards people now carry, including payment cards, driver's licenses, and insurance cards. Trials of smart cards in the United States have not been successful; however, smart cards are popular in other parts of the world.

Banks still process most monetary transactions, and a large part of the dollar volume of those transactions is still done by writing checks. Increasingly, banks are using Internet technologies to process those checks. Phishing expeditions and identity theft, especially when perpetrated by large criminal organizations, create a significant threat to online financial institutions and their customers. If not controlled, this threat could reduce the general level of confidence that consumers have in online business and hurt the growth of electronic commerce.

Key Terms

Acquirer fees	Card verification value code (CVVC)
Acquiring bank	Casher
Anonymous electronic cash	Charge card
Automated Clearing House (ACH)	Chargeback
Back-end processor	Check 21
Card code verification (CCV)	Client-side digital wallet
Card not present transaction	Closed loop system
Card security code (CSC)	Collector
Card verification code (CVC)	Credit card
Card verification data (CVD)	Credit card association
Card verification number (CVN)	Customer issuing bank (issuing bank)
Card verification value (CVV or CV2)	Debit card

Digital wallet (electronic wallet, e-wallet)
Double spending
Due diligence
Electronic cash
Electronic funds transfer at point of sale (EFTPOS) cards
Electronic wallet (e-wallet)
Front-end processor
Gift cards
Identity theft
Interchange fees
Interchange network
Interoperable software
Issuing bank
Merchant account
Micropayments
Money laundering
Near field communication (NFC)
Open loop system
Organized crime
Payment card
Payment gateway
Payment processing service provider
Payment processor
Pharming attack
Prepaid cards
Racketeering
Server-side digital wallet
Single-use card
Small payments
Smart card
Spear phishing
Store charge card (store-branded card)
Stored-value card
Verification code (V-Code or V Code)
Zombie farm

Review Questions

1. In one or two paragraphs, explain the difference between a credit card and a charge card.

2. Write a paragraph in which you describe the difficulties that can arise for merchants who want to process card not present credit card transactions.

3. In one or two paragraphs, explain the difference between a prepaid card and a gift card.

4. In one or two paragraphs, explain the difference between an open loop and a closed loop payment card processing system.

5. Write a paragraph or two in which you explain how an online merchant can use a card verification number (CVN) to reduce the risk of entering into a fraudulent credit card transaction with a customer.

6. In two or three paragraphs, distinguish between a front-end processor and a back-end processor. Be sure to include a discussion of payment gateways in your answer.

7. In one or two paragraphs, explain why electronic cash is more popular outside the United States than it has been within the United States.

8. Write a paragraph or two in which you discuss the micropayments problem and some possible solutions to it.

9. In about 100 words, outline the advantages and disadvantages of electronic cash.

10. In a paragraph or two, explain what a digital wallet is and how it can be used. Include a discussion of software-only digital wallets and those that use a physical device.

11. Write a paragraph in which you outline the key elements of the Check 21 law.

12. Many banks have decided that the best way to combat the threat of phishing attacks is to educate their customers. In about 100 words, outline the contents of a letter that a bank could use to warn its customers about possible phishing attacks. Be sure to include tips for spotting a phishing e-mail and specific advice regarding what to do (or not do) with it.

Exercises

1. Bonnie Carson has owned and managed her gift and card shop in the Central Shopping Mall for three years. She has a merchant account with her local bank to process payment card charges at her shop in the mall. Business at the mall store had been good, but Bonnie wanted to expand without paying higher rent for more floor space at the mall. A year ago, she opened an online store using a commerce service provider. Part of the monthly fee that the CSP charges covers the software needed to accept credit cards online. The CSP handles the payment processing, but the charges are much higher than those her bank charges to process the transactions for her shop in the mall. Bonnie's Web-based business is beginning to pick up and she wants to provide more payment options to her customers and is considering a payment processing service such as **PayPal** or **Checkout by Amazon** as an additional option. Identify at least three reasons that Bonnie should use such a service and at least three reasons why she should not. Prepare a briefing report for Bonnie of about 200 words summarizing your findings.

2. You are the new Manager of Online Sales for Duckworthy Rain Gear (DRG), a company that sells waterproof apparel to outdoor enthusiasts and people who must work in inclement weather. As you review the online payment processing system, you notice that DRG does not ask its customers to provide a CVN (also known as a CVV, CVV2, CSC, and so on) when they enter their credit card information on the shopping cart page of the Web site. Your prior employer always collected CVNs because they reduced the chances that a card payment transaction was fraudulent. You ask Sally Montt, the lead Web programmer, why the company is not collecting CVNs. She tells you that your predecessor established a policy of not collecting them because he said that "storing those numbers is prohibited." Using the Web Links for this exercise, your favorite search engine, or your library, evaluate the accuracy of your predecessor's statement. If storing CVNs is prohibited, determine who is responsible for the prohibition. Summarize your findings in a memo to the President of DRG in which you make a recommendation for collecting or not collecting CVNs.

3. Evan Moskowitz has formed an Internet training company called Teach-U-Comp to sell computer programming courses online. Each course costs $65, and students receive continuing education units (CEUs) based on the duration of the course and its level of difficulty. Evan expects to sell about 100 courses each month during his first few years in operation. He would like to accept credit and debit cards as payment for the courses. Evan is busy creating the online content and installing the course delivery software, so he hired you to investigate payment processing options for the site. Use your favorite search engine to identify three companies that process credit card payments for Web sites that sell downloadable digital products (such as Evan's courses). Examine the processing services offered and fees charged by these three companies and choose one that you believe would be best for Teach-U-Comp. Write a 300-word report in which you summarize your findings. In the conclusion to your report, clearly state which company you would recommend to Evan and explain why.

4. In about 200 words, outline the types of information bank customers might like to access using their smart phones. Briefly describe concerns that these customers might have because their smart phones have smaller screens than a typical computer's screen. You can use your library or your favorite search engine to conduct your research.

Cases

C1. First Internet Bank of Indiana

During the first wave of electronic commerce, many established banks opened online branches and a considerable number of new, completely online, banks were formed. Many of these online banking initiatives were closed, sold, or merged into other operations after the first wave of electronic commerce had subsided. By 2001, many notable names that had dominated the first wave were gone. For example, Bank One had closed its online subsidiary Wingspan Bank and merged its operations into its existing retail banking department. Royal Bank of Canada had done the same thing with its Security First Network Bank (generally believed to have been the first online bank). CompuBank and G&L Internet Bank were both sold to other banks, and USABancshares.com was closed in a flurry of fraud accusations and regulatory concerns.

Many early online-only banks faced similar challenges. They often bought loans instead of originating them. Purchased loans yield lower interest income because the originating bank always charges a fee or discount. They also tended to pay higher rates on customer deposits to attract new customers. These routes to rapid growth can significantly reduce profitability. Physical banks with many branches gain customers and market share because people walk or drive by a branch office and see the bank's name. New online-only banks must spend substantial sums on advertising that helps establish them as viable brands in a highly competitive market. And many well-established banks now operate online, offering customers a known brand name and the convenience of physical branches along with online banking services. Small businesses were reluctant to deal with online-only banks in the early years of their existence. Small businesses generate considerable profits for banks because they tend to borrow money at relatively high interest rates and also tend to keep large balances in their checking accounts. Thus, there were a number of challenges that made survival difficult for online-only banks.

In 2004, the U.S. Federal Deposit Insurance Corporation (FDIC) issued a report on "limited-purpose banks" (which included online-only banks) in its *Future of Banking Study* series. The FDIC report concluded that the economics of operating an online-only bank were not attractive and that very few such banks could ever expect to be successful in the long term. Despite the FDIC's gloomy outlook, a number of banks operate only online. One of those banks is the First Internet Bank of Indiana (often called First IB).

First IB was launched in early 1999. By 2001, the bank had become profitable and had more than $200 million in assets. By 2008, its assets had grown to nearly $600 million. Compared to the large international banks that dominate the financial world, $600 million is a relatively small amount (for example, the Bank of America has more than $500 billion in assets), but First IB was able to operate efficiently and with low costs because it had no physical branch offices and very few employees compared to traditional banks.

First IB invested its resources in building the best Web site it could design and then followed a process of continually adjusting the site's design and the services offered to respond to customer comments and requests. For example, First IB created a frequently asked questions (FAQ) feature that reduced customer inquiries dramatically. It was also one of the first banks to offer statements and check images online. In 2004, the bank began to make check images available online the day after the check cleared (the industry average delay at that time was four to seven days). The bank has consistently received excellent reviews of its services by online business rating agencies and in the press.

Required:

1. Create a list of 10 specific concerns that a consumer might have when considering an online-only bank. Write a paragraph for each concern that describes how First IB addresses or fails to address it.

2. Evaluate how well the design of the First IB Web site meets the needs of a potential small business customer. In about 300 words, discuss the elements of the site that work particularly well in meeting the needs of this type of site visitor. In about 300 words, outline specific changes you would make to the site to better meet the needs of a potential small business customer.

3. Assume you are a security consultant hired by First IB. The president of the bank has become concerned about the potential damage that a phishing expedition directed at First IB customers could do to the bank's reputation. In about 500 words, analyze the phishing threat that faces First IB and outline steps that First IB should take to counter the threat.

Note: Your instructor might assign you to a group to complete this case, and might ask you to prepare a formal presentation of your results to your class.

C2. The Moose Hut

Rod and Martha Nelson started The Moose Hut (TMH), a gift shop in Calgary, Alberta, more than 15 years ago. The Nelsons have capitalized on the tourist trade drawn by the Calgary Stampede, which is one of the largest rodeos in the world. The shop sells a wide range of Canadian-themed items to rodeo fans and other tourists who visit central Alberta throughout the year. TMH's offerings range from inexpensive food items, such as pure Canadian maple syrup and smoked salmon, to much more expensive handcrafted gifts, including Inuit and First Nations artwork. The company's trademark product, the Moose Mug, is one of its biggest-selling items.

Many of TMH's customers return to the store whenever they visit Calgary. TMH's line of Canada Day Party Favours is especially popular with homesick Canadians who have moved to other countries, and TMH has been selling those products by mail order for the past several years. After reviewing the sales numbers for these mail order items, Martha has decided that it might be a good idea to expand the mail order operation and begin accepting orders through a Web site. Many of the store's items have a high value-to-weight ratio and would be easy to ship to customers around the world.

TMH currently accepts only checks denominated in Canadian or U.S. currency in its mail order operation; however, taking orders on a Web site will probably require the company to be more flexible in accepting multiple payment methods. Rod and Martha asked you to help them examine payment processing alternatives for TMH's new Web business.

To be acceptable, a payment processing method needs to handle all major credit cards, perform currency conversions, and be available to a Canadian merchant. Most important is that the payment processing method must be reasonably priced. The margins on most gift items at TMH are between 10 percent and 30 percent of the selling price, but the extra costs of shipping and handling items sold through the Web site reduce those margins by another 5–10 percent of the selling price. TMH would like to keep overall payment processing costs below 4 percent of the selling price, if possible.

Required:

1. Using the Web Links for this case, identify at least three payment processing options that might be suitable for TMH. Write a report of about 400 words in which you describe each of the three payment processing options. Include specific advantages and disadvantages for each option.

2. Prepare a one-page memorandum in which you make a specific recommendation to Rod and Martha. Include an explanation of the reasons for your recommendation.

Note: Your instructor might assign you to a group to complete this case and might ask you to prepare a formal presentation of your results to your class.

For Further Study and Research

Adams, J. 2009. "New Mobile Banking Tools Get One Step Closer to Payments," *American Banker*, November 3, 14.

AFP Exchange. 2007. "Electronic Payments More Prevalent Than Three Years Ago," November, 27(9), 34.

Albornoz, L. 2007. "Accounts Payable: The Final Frontier for IT," *Computerworld*, December 17, 30–31.

American Banker. 2002. "First Internet of Indiana Turns a Profit Again," 167(95), May 17, 13.

American Banker. 2009. "Online Merchants Cut Fraud Losses," December 1, 11.

Ammons, J., G. Schneider, and A. Sheikh. 2012. "Accounting for Retailer-Issued Gift Cards: Revenue Recognition and Financial Statement Disclosures," *Journal of the International Academy for Case Studies*, 18(1), 1–8.

Berkow, J. 2011. "Smart (Phone) Money," *Financial Post*, April 23. (http://business.financialpost.com/2011/04/23/smart-phone-money/)

Berney, L. 2008. "For Online Merchants, Fraud Prevention Can Be a Balancing Act," *Cards & Payments*, February, 21(2), 22.

Bigdoli, H. and R. Phillips. 2012. *Online Banking*. Hoboken, NJ: Wiley.

Bills, S. 2009. "Consumer Demand for Mobile Banking Tools Growing Rapidly," *American Banker*, December 4. (http://www.americanbanker.com/issues/174_232/demand-mobile-tools-1004783-1.html)

Brandt, A. 2005. "Devious New Phishing Attack Outsmarts Typical Defenses," *PC World*, 23(3), March, 35.

Chang, R. 2009. "What Paying by Cellphone Will Mean for the Marketing World," *Advertising Age*, 80(33), October 5, 4, 29.

Chen, B. 2012. "A Digital Wallet Now Available on Some Smartphones," *The New York Times*, January 18. (http://bits.blogs.nytimes.com/2012/01/18/visa-digital-wallet/)

Clark, M. 2012. "Visa US Offers Mobile Services," *Near Field Communications World*, February 10. (http://www.nfcworld.com/2012/02/10/313104/visa-us-offers-mobile-services/)

Clement, A. 2011. *A Global Overview of Digital Wallet Technologies*. Toronto: University of Toronto. (http://propid.ischool.utoronto.ca/digiwallet_overview/)

Credit Card Management, 2003. "A Dubious Honor for Online Payments," 15(13), March, 14.

Credit Management. 2007. "Electronic Billing Comes of Age," December, 26.

CyberSource. 2008. *Ninth Annual Online Fraud Report: Online Payment Fraud Trends and Merchants' Response*. Mountain View, CA: CyberSource.

CyberSource. 2012. *13th Annual Online Fraud Report: Online Payment Fraud Trends, Merchant Practices and Benchmarks*. San Francisco: Visa-CyberSource.

DeCastro, M. 2009. "Mobile Takes a Breather," *American Banker*, October 29, 18–19.

Dragoon, A. 2004. "Fighting Phish, Fakes, and Frauds," *CIO Magazine*, 17(22), September 1, 33–38.

Drake, C., J. Oliver, and E. Koontz. 2004. "Anatomy of a Phishing Email," *Proceedings of the First Conference on Email and Anti-spam*. Mountain View, CA, July 30.

Fest, G. 2008. "How Will Payments Ride Rails?" *Bank Technology News*, 21(7), July, 1, 19.

Fitzgerald, K. 2009. "A Check Logjam For B2B Payments," *Cards & Payments*, 22(4), April, 20–22.

Galbraith, J. 1995. *Money: Whence it Came, Where it Went*. London: Penguin Books.

Gonsalves, A. 2009. "PayPal Unveils Plans to Open Payment Service," *InformationWeek*, November 4.

Grant, D. 2001. "Internet Banking Nightmare: Couple Sue After Access to Their Funds Was Cut Off for 10 Crucial Days," *EastSideJournal.com*, June 10.

Hernandez, W. 2009. "Noncard Payments Gaining Toehold in Bank Channel," *American Banker*, December 1, 6.

Internet Retailer. 2011. "Going Beyond Payment Cards to Drive Online Sales," February 1. (http://www.internetretailer.com/2011/02/01/going-beyond-payment-cards-drive-online-sales)

Javelin Strategy & Research. 2011. *2010 Online Retail Payments Update and Forecast*. Pleasanton, CA: Javelin Strategy & Research.

Keizer, G. 2005. "Phishing Economics 101 Reveals Collectors and Cashers," *InternetWeek*, July 29.

Kenneally, S. 2008. "Payments Cyber Roundtable: Who Moved the Payments System?" *Community Banker*, April, 17(4), 36–40.

Kingston, J. 2003. "E-Pay Overtaking Paper; Clients Want More Integration," *American Banker*, 168(81), April 29, 21.

Krim, J. 2005. "More ID May Be Required for Online Banking," *The Washington Post*, October 21, D5.

Kuykendall, L. 2003. "Citi to Pull the Plug on c2it Next Month," *American Banker*, October 1, 7.

Larkin, E. 2009. "Go Virtual for Safer Online Shopping," *PC World*, 27(11), November, 35–36.

Lewis, H. 2001. "NetBank, CompuBank Merge, Customers Get Squashed," *Bankrate.com*, May 22. (http://www.bankrate.com/bzrt/news/ob/20010521a.asp)

Livingstone, R. 2012. "Chasing the Digital Wallet," *Technology Spectator*, January 31. (http://technologyspectator.com.au/industry/financial-services/chasing-digital-wallet)

Markoff, J. 2002. "Vulnerability Is Discovered in Security for Smart Cards," *The New York Times*, May 13. (http://www.nytimes.com/2002/05/13/technology/13SMAR.html)

Marlin, S. 2003. "Who Needs Cash?" *Information Week*, December 29, 20–22.

Mearian, L. 2005. "Wells Fargo Buys into Check Image Sharing," *Computerworld*, January 14. (http://www.computerworld.com/databasetopics/data/story/0,10801,98966,00.html)

Mitchell, D. 2007. "In Online World, Pocket Change Is Not Easily Spent," *The New York Times*, August 27. (http://www.nytimes.com/2007/08/27/technology/27micro.html)

Nevius, A. 2009. "IRS Expands Electronic Payment Options," *Journal of Accountancy*, 208(6), 78.

Oehlsen, N. 2009. "Smartphone Payment Apps: Are Developers Marking the Right Call?" *Cards & Payments*, 22(8), September, 26–31.

Orr, B. 2008. "A2A Payments: Next Generation of Online Banking?" *ABA Banking Journal*, April, 100(4), 53.

Ptacek, M. 2001. "CompuBank's Demise May Signal a New Era," *American Banker*, 166(63), April 2, 16.

Ramsaran, C. 2004. "Catch of the Day: Banks Face New Phishing Scams," *Bank Systems & Technology*, December 1, 13.

Ramstad, E. 2004. "Hong Kong's Money Card Is a Hit," *The Wall Street Journal*, February 19, B3.

Ray, B. 2012. "Google Wallet PIN Security Cracked in Seconds," *The Register*, February 9. (http://www.theregister.co.uk/2012/02/09/google_wallet_pin/)

Rist, C. 2003. "Making Bank on Small Change," *Business 2.0*, 4(10), November, 56–57.

Rob, M. and E. Opara. 2003. "Online Credit Card Processing Models: Critical Issues to Consider by Small Merchants," *Human Systems Management*, 22(3), 133–142.

Roth, A. 2001. "CompuBank Merge Nettles NetBank," *American Banker*, 166(119), June 21, 1–2.

Stoneman, B. 2003. "FAQs Lighten Service Load at First Internet Bank of Indiana," *American Banker*, 168(2), January 13, 12.

Sturgeon, J. 2003. "Electronic Payments," *CFO Magazine*, 19(15), Winter, 52–53.

Tedeschi, B. 2004. "Protect Your Identity," *PC World*, 22(12), December, 107–112.

Torian, R., R. Schrader, O. Ireland, and R. Stinneford. 2008. "Current Developments in Electronic Banking and Payment Systems," *The Business Lawyer*, February, 63(2), 689–702.

Urban, M. 2005. "To Catch Phish, Banks Need Better Bait," *Bank Technology News*, 18(11), November, 57.

Wade, W. 2009. "With E-Transfers, Banks Target Gen-Y Payments," *American Banker*, December 18. (http://www.americanbanker.com/issues/174_242/e-transfers-1005381-1.html)

Wetherington, L. 2008. "The Electronic Payments Explosion," *Texas Banking*, February, 97(2), 14–17.

Wingfield, N. and J. Sapsford. 2002. "eBay to Buy PayPal for $1.4 Billion," *The Wall Street Journal*, July 9, A6.

Wolfe, D. 2008. "Mobile Micropayments to Target U.S. Teenagers," *American Banker*, December 22. (http://www.americanbanker.com/issues/173_258/-369264-1.html)

Yom, C. 2004. "Limited-purpose Banks: Their Specialties, Performance, and Prospects," *FDIC Future of Banking Study Series*, June, 1–45. Washington, D.C.: Federal Deposit Insurance Corporation (FDIC).

Yurcan, B. 2012. "Visa Rolls Out New Suite of Mobile Products for U.S. Financial Institutions," *Bank Systems & Technology*, February 9. (http://www.banktech.com/payments-cards/232600563)

PART 4

INTEGRATION

CHAPTER 12
Planning for Electronic Commerce, 497

PLANNING FOR ELECTRONIC COMMERCE

INTRODUCTION

AlliedSignal (now part of **Honeywell**) is a diversified manufacturing and technology business selling products in the aerospace, automotive, chemicals, fibers, and plastics industries. In 1999, the company had more than 70,000 employees and annual sales exceeding $15 billion. Although some of AlliedSignal's products used new technologies or helped other firms create new technologies, many of the products were commodity items that were manufactured and sold just as they had been for decades. In early 1999, AlliedSignal's CEO, Larry Bossidy, called together the heads of the company's business units for a one-day conference to develop strategic plans for electronic commerce at the company. He invited Michael Dell, chairman and CEO of Dell Computers, and John Chambers, CEO of Cisco Systems, to speak about their companies' electronic commerce implementation successes.

At the end of the day, Bossidy gave the business unit heads their marching orders. They were to take what they had learned and create a strategy for implementing electronic commerce in their business units—in two months. Bossidy told the room full of surprised managers that, although most of their business units were at or near the top of their industries, the Internet would change everything. He believed that the kinds of electronic commerce strategies that had worked so well for Dell and Cisco in the computer industry could also work in many of AlliedSignal's businesses. He wanted to make sure that AlliedSignal was the first to exploit those strategies and any other Internet-enabled business ideas that the managers could devise. In two months, managers reported back with strategies that included multiple online projects, including Web sites for selling products, providing customer service, improving corporate infrastructure, managing supply chains, coordinating logistics, holding auctions, and creating virtual communities. These plans were evaluated in the company's regular annual budget process, and the best ones were chosen for funding and immediate implementation.

In a matter of months, one of the largest industrial enterprises in the world had drastically altered its course, setting sail for the uncharted waters of the first wave of electronic commerce. In the years since, AlliedSignal has gone through many changes, including a merger with Honeywell. The initiatives it undertook as a result of this first electronic commerce strategic planning session were important in making the company an attractive merger candidate. Today, as part of Honeywell, the businesses that were formerly AlliedSignal are using a wide range of Internet technologies in a variety of their supply chain management and purchasing functions.

IDENTIFYING BENEFITS AND ESTIMATING COSTS OF ELECTRONIC COMMERCE INITIATIVES

The ability of companies to plan, design, and implement cohesive electronic commerce strategies makes the difference between success and failure for the majority of them. The tremendous leverage that firms can gain by being the first to do business a new way on the Web has caught the attention of top executives in many industries. The keys to successful implementation of any information technology project are planning and execution. This chapter provides some useful guidelines for those readers who will manage the planning, implementation, and continuing operations of electronic commerce initiatives. A successful business plan for an electronic commerce initiative should include activities that identify the initiative's specific objectives and link those objectives to business strategies (strategies that you learned about in Chapters 3, 4, 5, and 6).

In setting the objectives for an electronic commerce initiative, managers should consider the strategic role of the project, its intended scope, and the resources available for executing it. In this section, you will learn how to identify objectives and link those business objectives to business strategies. In later sections of this chapter, you will learn about Web site development strategies and how to manage the implementation of an electronic commerce initiative.

Identifying Objectives

Businesses undertake electronic commerce initiatives for a wide variety of reasons. Objectives that businesses typically strive to accomplish through electronic commerce include: increasing sales in existing markets, opening new markets, serving existing customers better, identifying new vendors, coordinating more efficiently with existing vendors, or recruiting employees more effectively.

Organizations of different sizes will have different objectives for their electronic commerce initiatives. Decisions regarding resource allocations for electronic commerce initiatives should consider the expected benefits and costs of meeting the objectives. These decisions should also consider the risks inherent in the electronic commerce initiative and compare them to the risks of inaction—a failure to act could concede a strategic advantage to competitors.

Linking Objectives to Business Strategies

Businesses use tactics called **downstream strategies** to improve the value that the business provides to its customers. Alternatively, businesses can pursue **upstream strategies** that focus on reducing costs or generating value by working with suppliers or inbound shipping and freight service providers.

In earlier chapters of this book, you learned about many of the things that companies are doing on the Web. The Web is an attractive sales channel for many firms; however, companies use electronic commerce to do much more than sell. They can use the Web to complement their business strategies and improve their competitive positions. Electronic commerce opportunities can inspire businesses to undertake activities such as:

- Building brands
- Enhancing existing marketing programs
- Selling products and services
- Selling advertising
- Developing a better understanding of customer needs
- Improving after-sale service and support
- Purchasing products and services
- Managing supply chains
- Operating auctions
- Building or using virtual communities to maintain relationships with customers and suppliers

The success of these activities can be difficult to measure. In the first wave of electronic commerce, many companies engaged in these activities on the Web without setting specific, measurable goals. In the late 1990s, companies that had good ideas could find plenty of investors and start a business activity on the Web. These early activities were often highly speculative. Successes and failures were measured in broad strokes. A company would either become a leader in its industry (perhaps after being acquired by a larger company) or would disappear into bankruptcy—all within a few short years.

In the second wave of electronic commerce, companies started taking a closer look at the benefits and costs of their electronic commerce initiatives before committing resources to them. It became necessary for online business ideas to have specific objectives for benefits to be achieved and costs to be incurred. Companies began creating pilot Web sites to test their online business ideas and then released production Web sites to handle full implementations. Companies started specifying clear goals that their pilot tests had to meet before they would launch new Web sites in their full production versions.

In the third wave, companies are moving beyond a conceptualization of online business as a Web site that communicates to individual users running Web browser software on their computers. The pervasiveness of smart phones and tablet devices puts the power of a Web browser into many more hands in many more locations. It also changes the nature of online communication. Messaging between a Web client running on a fixed location computer and a Web browser is a communication from one point to another, much like a land-line telephone. Web clients running on multiple devices (some of which might be used simultaneously by a single user) make the types of communication and interaction richer and able to accomplish a wider array of tasks. The ease of acquiring the benefits of a technology is also increasing. For example, a company might never own its own microblogging-based social media tool (such as Twitter), but it can certainly use the tool Twitter provides to participate in a virtual community in ways that cement its relationships with customers, suppliers, and even its own employees. The most profound change in the third wave, however, is likely to be the increase in electronic commerce activities by smaller businesses. These firms can use the existing communication infrastructure of the world's Facebooks, Twitters, and similar social media tools to get information out to potential customers very effectively without investing large amounts of money in their own Web infrastructures. Some experts even suggest that small businesses might be better off investing their promotional resources in social media than in traditional Web sites.

Identifying and Measuring Benefits

Some benefits of electronic commerce initiatives are obvious, tangible, and easy to measure. These include such things as increased sales or reduced costs. Other benefits are intangible and can be much more difficult to identify and measure, such as increased customer satisfaction. When identifying benefits, managers should try to set objectives that are measurable, even when those objectives are for intangible benefits. For example, success in achieving a goal of increased customer satisfaction might be measured by counting the number of first-time customers who return to the site and buy.

Companies that create Web sites to build brands or enhance their existing marketing programs can set goals in terms of increased brand awareness, which they can measure with market research surveys and opinion polls. Companies that sell goods or services online can measure increases in sales volume. One complication that can occur when measuring either brand awareness or sales is that the increases can be caused by other things that the company is doing at the same time or by a general improvement in the economy. A good marketing research staff or outside consulting firm can help a company sort out the specific effects of their online marketing or sales initiatives. Marketing research staff or outside consultants can also help a firm set and evaluate its specific goals for online business initiatives.

Companies that want to use Web sites to improve customer service or after-sale support might set goals of increased customer satisfaction or reduced costs of providing customer service or support. For example, Philips Lighting wanted to use the Web to provide an ordering system for its smaller customers that did not use EDI. The primary goal for this initiative was to reduce the cost of processing smaller orders. Philips had identified that responding to inventory availability and order status inquiries accounted for over half the cost of processing smaller orders. Customers that placed small orders often called or sent faxes asking for this information. Philips built a pilot Web site and invited a number of its smaller customers to try it. The company found that customer service phone calls from the test group of customers dropped by 80 percent. Based on that measurable increase in efficiency, Philips decided to invest in additional hardware and personnel to staff a version of the Web site that could handle virtually all of its smaller customers. The reduction in the cost of handling small orders justified the additional investment.

Companies can use a variety of similar measurements to assess the benefits of other electronic commerce initiatives. Supply chain managers can measure supply cost reductions, quality improvements, or faster deliveries of ordered goods. Auction sites can set goals for the number of auctions, the number of bidders and sellers, the dollar volume of items sold, the number of items sold, or the number of registered participants. The ability to track such numbers is usually built into auction site software. Virtual communities and Web portals measure the number of visitors and try to measure the quality of their visitors' experiences.

Some sites use online surveys to gather this data; however, most settle for estimates based on the length of time each visitor remains on the site and how often visitors return. A summary of benefits and measurements that companies can make to assess the value of those benefits (these measurements are often called **metrics**) appears in Figure 12-1.

Electronic commerce initiatives	Common measurements of benefits provided
Build brands	Surveys or opinion polls that measure brand awareness, changes in market share
Enhance existing marketing programs and create new marketing programs	Change in per-unit sales volume, frequency of customer contact, conversion (to buyers) rate
Improve customer service	Customer satisfaction surveys, quantity of customer complaints, customer loyalty
Reduce cost of after-sale support	Quantity and type (telephone, fax, e-mail) of support activities, change in net support cost per customer
Improve supply chain operation	Cost, quality, and on-time delivery of materials or services purchased, overall reduction in cost of goods sold
Hold auctions	Quantity of auctions, bidders, sellers, items sold, registered participants; dollar volume of items sold; participation rate
Provide portals, social networks, and virtual communities	Number of visitors, number of return visits per visitor, duration of average visit, participation in online discussions

FIGURE 12-1   Measuring the benefits of electronic commerce initiatives

No matter how a company measures the benefits provided by its Web site, it usually tries to convert the raw activity measurements to dollars. Having the benefits measured in dollars lets the company compare benefits to costs and compare the net benefit (benefits minus costs) of a particular initiative to the net benefits provided by other projects. Although each activity provides some value to the company, it is often difficult to measure that value in dollars. Usually, even the best attempts to convert benefits to dollars yield only rough approximations.

Identifying and Estimating Costs

At first glance, the task of identifying and estimating costs may seem much easier than the task of setting benefits objectives. However, many managers have found that information technology project costs can be just as difficult to estimate and control as the benefits of those projects. Because Web development uses hardware and software technologies that change even more rapidly than those used in other information technology projects, managers often find that their experience does not help much when they are making estimates. Most changes in the cost of hardware are downward, but the increasing sophistication of software often requires more of the newer, less-expensive hardware. This often yields a net increase in overall hardware costs. The more sophisticated software often costs more than the amount originally budgeted, too. Even though electronic commerce initiatives are often completed within a shorter time frame than many other information technology projects, the rapid changes in Web technology can quickly destroy a manager's best-laid plans.

Total Cost of Ownership

In addition to hardware and software costs, the project budget must include the costs of hiring, training, and paying the personnel who will design the Web site, write or customize the software, create the content, and operate and maintain the site. Many organizations now track costs by activity and calculate a total cost for each activity. These cost numbers, called **total cost of ownership (TCO)**, include all costs related to the activity. Increasing some costs can reduce other costs, so most managers find the TCO of a project to be a more appropriate focus for their cost control efforts than the individual elements of the project's cost.

The TCO of an electronic commerce implementation includes the costs of hardware (server computers, routers, firewalls, and load-balancing devices), software (licenses for operating systems, Web server software, database software, and application software), design work outsourced, salaries and benefits for employees involved in the project, and the costs of maintaining the site once it is operational. A good TCO calculation would, for example, include assumptions about how often the site would need to be redesigned in the future.

Opportunity Costs

For many companies, one of the largest and most significant costs associated with electronic commerce initiatives is the opportunity lost by not undertaking such an initiative. The foregone benefits that a company could have obtained from an electronic commerce initiative that they chose not to pursue are costs. Managers and accountants use the term **opportunity cost** to describe such lost benefits from an action not taken.

Opportunity costs of not undertaking an online business initiative could include the value of customers never obtained, sales not made, suppliers not identified, or cost reductions not achieved in the company's supply chain. Although opportunity costs never show up in the accounting records, they are real and avoidable losses. Good managers try to think of opportunity costs whenever they make business decisions of any kind.

Web Site Costs

Since companies began setting up Web sites, information technology research firms (such as International Data Corporation and Gartner) and management consulting firms (such as Booz & Company and McKinsey & Company) have regularly estimated the costs of implementing various types of online business operations. Although the total dollar amounts required to create and operate a Web site have varied over the years (and across specific types of businesses), the relative proportion of startup costs has remained surprisingly stable. About 10 percent of the cost is for computer hardware, another 10 percent is for software, and about 80 percent of the cost is for labor (including both internal labor and the cost of outside consultants). The annual cost of operating an online business Web site generally ranges between 50 and 200 percent of the initial cost of the site.

As you learned in Chapter 9, a small online store can be placed in operation for under $5000, and a typical small to midsize online business operation with full transaction and payment processing capabilities usually requires an initial investment between $50,000 and $1 million. In fact, surveys of smaller companies showed that their expenditures on construction of new electronic commerce Web sites average $80,000.

Current estimates of the cost to launch electronic commerce sites for larger companies, especially those that must be integrated with existing business operations, are substantially higher. Figure 12-2 summarizes recent industry estimates for the cost of creating and operating online Web sites for various sizes of businesses.

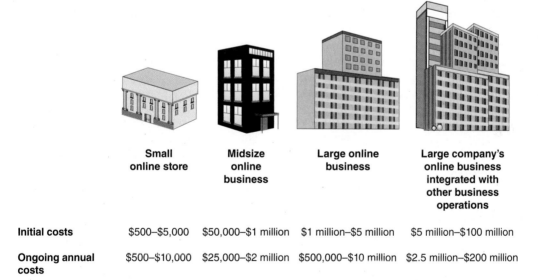

	Small online store	Midsize online business	Large online business	Large company's online business integrated with other business operations
Initial costs	$500–$5,000	$50,000–$1 million	$1 million–$5 million	$5 million–$100 million
Ongoing annual costs	$500–$10,000	$25,000–$2 million	$500,000–$10 million	$2.5 million–$200 million

FIGURE 12-2   Estimated costs for business Web sites

Many industry observers have noted that costs are generally heading downward. Startup firms increasingly find they can get their operations launched for dollar amounts that are in the low end of the range in each category. Lower costs for broadband access and computer hardware play a major role, but the most significant trend is that the cost of developing and maintaining software to run an online business (a cost that includes a substantial labor component) is decreasing.

Sarah Lacy, a journalist who writes about high-tech companies, compared one of the Internet's first successful startups, Netscape, with more recent startup companies. She noted that Netscape needed more than $40 million to buy equipment, bandwidth, and to pay people to build the software it needed just to get started in the early 1990s. Kevin Rose started his online business, Digg, with an investment of under $500,000 in 2004.

When considering hosting options for a Web site, businesses need to ensure that their needs will be met. The hosting option must support the functions that the business wants to carry out on the site and have the ability to handle any increases in volume that the business anticipates. It must also be reliable and secure, with provisions for backup and recovery of important business information. Of course, the cost of the hosting option is always a factor as well. The most important factors to evaluate when selecting a hosting service are shown in Figure 12-3.

Feature	Typical measures
Functionality	Bandwidth, number of different operating systems and databases supported, disk space, number of e-mail accounts allowed, number and type of software provided (for Web site contruction, traffic analysis, and so on)
Reliability	Guaranteed uptime percentage, guaranteed speed of service reinstatement when it does fail
Scalability	Ease of expansion of bandwidth, disk space, additional software (database, traffic analysis, and so on) that can be added to an account as it grows
Security	Employee background checks, features that provide physical protection of the facilities (fences, alarms, guards, security cameras, and so on) and protection against online intrusions (firewalls, network security software and devices)
Backup and recovery	Frequency of backups, automation of backups, off-site storage of backup media
Cost	Initial and ongoing charges for setup and operation, additional charges for specific software and other features

FIGURE 12-3   Important Web hosting service features

Funding Online Business Startups

In the early days of the Web, many businesses were started by individuals who knew something about computers and technology and who had an idea for a business. Although most of those early businesses failed, some of them were successes. For example, both eBay and Yahoo! were started by computer enthusiasts who decided they might be able to make a little money with their hobbies.

Because many online business initiatives are startup companies (rather than ideas launched by existing businesses), the traditional ways businesses finance expansions (borrowing from a bank or offering bonds or stock to investors) are not available to them. Banks are reluctant to lend money on the strength of a good idea alone, and the stock and bond markets are limited to companies with long track records of profitability. Most startup businesses of any kind are funded out of the founders' savings, along with investments or loans from friends and relatives.

As business interest in the Web grew in the late 1990s, many online startups were attractive to investors who wanted a chance to make some fast money in what had become the Internet boom. A person would come up with an idea for an online business and pitch it to a group of businesspersons who had money. These investors, often called **angel investors**, would fund the initial startup. In return for their capital, angel investors would become stockholders in the business and would often own more of the business than the founder. Typical funding by angel investors would be between a few hundred thousand dollars and a few million dollars, which is substantially more than most entrepreneurs can raise from relatives and friends. Angel investors hoped that the business would grow rapidly so that in a short time they could sell their interest in the company at a profit to the next round of investors, called venture capitalists.

Venture capitalists are very wealthy individuals, groups of wealthy individuals, or investment firms that look for small companies that are about to grow rapidly. They invest large amounts of money (between a million and a few hundred million dollars) hoping that in a few years the company will be large enough to sell stock to the public in an event called an **initial public offering (IPO)**. In the IPO, the venture capitalists take their profits and once again search for a new small company in which to invest.

The supply of angel investors and venture capitalists (and their willingness to invest in new startups) has waxed and waned with the booms and busts of online business activity. It has always been easier to find money for electronic commerce initiatives when business is good than when it is declining.

This system of financing startup and initial growth of online businesses has both benefits (it provided access to large amounts of capital early in the life of the business) and costs (angel investors and venture capitalists got most of the profits and put great pressure on the business to grow rapidly) for the founders of those businesses. With the high costs of launching online business Web sites in the first wave of electronic commerce, business founders had few alternatives. Now that the costs of creating an online business have gone down, the number of founders who can avoid venture capitalists and even angel investors is increasing. By relieving the pressure to grow rapidly, online entrepreneurs can be more creative and have a chance to learn from their mistakes. Industry observers expect this trend toward more and smaller online ventures to continue as the cost of creating an online business continues to fall.

Comparing Benefits to Costs

Most companies have procedures that call for an evaluation of any major expenditure of funds. These major investments in equipment, personnel, and other assets are called **capital projects** or **capital investments**. The techniques that companies use to evaluate proposed capital projects range from very simple calculations to complex computer simulation models. However, no matter how complex the technique, it always reduces to a comparison of benefits and costs. If the benefits exceed the costs of a project by a comfortable margin, the company invests in the project.

A key part of creating a business plan for electronic commerce initiatives is the process of identifying potential benefits (including intangibles such as employee satisfaction and company reputation), identifying the total costs required to generate those benefits, and evaluating whether the value of the benefits exceeds the total of the costs. Companies should evaluate each element of their electronic commerce strategies using this cost/benefit approach. A representation of the cost/benefit approach appears in Figure 12-4.

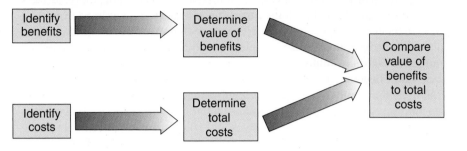

FIGURE 12-4 Cost/benefit evaluation of electronic commerce strategy elements

Return on Investment (ROI)

You might have learned techniques for capital project evaluation, such as the payback method, the net present value method, or the internal rate of return method, in your accounting or finance courses. These evaluation approaches are called **return on investment (ROI)** techniques because they measure the amount of income (return) that will be provided by a specific current expenditure (investment). ROI techniques provide a quantitative expression of whether the benefits of a particular investment exceed their costs (including opportunity costs). They can also mathematically adjust for the reduced value of benefits that the investment will return in future years (benefits received in future years are worth less than those received in the current year).

Although most companies evaluate the anticipated value of electronic commerce initiatives in some way before approving them, many companies see these projects as absolutely necessary investments. Thus, businesses might not subject these initiatives to the same close examination and rigid requirements as other capital projects. These companies fear being left behind as competitors stake their claims in the online marketspace. The value of early positioning in a new market is so great that many companies are willing to invest large amounts of money with few near-term profit prospects.

Newspaper Web sites are one example of an industry's willingness to incur losses to establish an online presence. In the first wave of electronic commerce, there were few profitable newspaper sites (such as Gannet's **USA Today** and *The Wall Street Journal's* **WSJ.com** sites). Most newspaper sites took many years to become profitable, and a significant number remain unprofitable today. As you learned earlier in this book, newspaper sites have experimented with various ways to generate revenue, such as charging for subscriptions, charging for access to certain content, or charging for access to archived articles. Despite their continuing losses, most newspaper companies believe that they cannot afford to ignore the long-term potential of the Web. These companies estimate that the opportunity costs of not being present on the Web (for example, the loss of future profits to be earned from the Web site or the risk of losing market share to competitors) are greater than the losses they are experiencing in their online operations.

In the second wave of electronic commerce, more companies began taking a harder look at Web-related expenditures. Many companies have turned to ROI as the measurement tool for evaluating new electronic commerce projects because that is what they used for other IT projects in the past. ROI is a simple-to-understand tool that is easily applied; however, managers should be careful when using it to evaluate online business initiatives. ROI has some built-in biases that can lead managers to make poor decisions.

First, ROI requires that all costs and benefits be stated in dollars. Because it is usually easier to quantify costs than benefits, ROI measurements can be biased in a way that gives undue weight to costs. Second, ROI focuses on benefits that can be predicted. Many electronic commerce initiatives have returned benefits that were not foreseen by their planners. The benefits developed after the initiatives were in place. For example, Cisco Systems created online customer forums to allow customers to discuss product issues with each other. The main benefits from this initiative were to reduce customer service costs and increase customer satisfaction regarding the availability of product information; however, the forums turned out to be a great way for Cisco engineers to get feedback from customers on new products that they were developing. This second use was not foreseen by the project's planners and has become the most important and beneficial outcome of the customer forums. An ROI analysis would have missed this benefit completely.

Yet another weakness of ROI is that it tends to emphasize short-run benefits over long-run benefits. The mathematics of ROI calculations do account for both correctly, but short-term benefits are easier to foresee, so they tend to get included in the ROI calculations. Long-term benefits are harder to imagine and harder to quantify, so they tend to be included less often and less accurately in the ROI calculation. This biases ROI calculations to weigh short-term costs and benefits more heavily than long-term costs and benefits, which can lead managers who rely on ROI measures to make incorrect decisions. You can learn more about this topic at the **CIO Budget** and the *Computerworld* **ROI Knowledge Center** Web pages.

In the third wave, companies undertake highly sophisticated analyses of any planned online business activity. For example, a bank that is planning to launch mobile banking services would develop ROI estimates for each element of the implementation, including the Web site for mobile users, any apps that would be offered for various smart phone operating systems, and social media promotions that would entice users to switch their accounts to the bank or use more bank services (and thus generate more fee revenue for the bank).

STRATEGIES FOR DEVELOPING ELECTRONIC COMMERCE WEB SITES

When companies first established presences on the Web, the typical Web site was a static brochure that was not updated frequently with new information and seldom included any business transaction processing capabilities. The next generation of Web sites included transaction processing and a variety of other automated business processing capabilities. Web sites became important parts of companies' information systems infrastructures. These transaction processing capabilities were eventually enhanced with personalization (in which sites customized the Web site's presentation to each specific user) and the customer relationship management features you learned about in earlier chapters. More recently, Web sites became integrated with social media networks such as Facebook and Twitter. Companies also added separate Web sites with formatting that provides specific functionality for mobile devices with smaller screens, such as smart phones and mobile tablet devices. This evolution of Web site functions is shown in Figure 12-5.

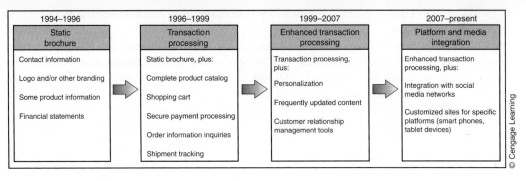

1994–1996	1996–1999	1999–2007	2007–present
Static brochure	**Transaction processing**	**Enhanced transaction processing**	**Platform and media integration**
Contact information	Static brochure, plus:	Transaction processing, plus:	Enhanced transaction processing, plus:
Logo and/or other branding	Complete product catalog	Personalization	Integration with social media networks
Some product information	Shopping cart	Frequently updated content	Customized sites for specific platforms (smart phones, tablet devices)
Financial statements	Secure payment processing	Customer relationship management tools	
	Order information inquiries		
	Shipment tracking		

© Cengage Learning

FIGURE 12-5   Evolution of Web site functions

This transformation occurred rapidly in most companies. Because the change in the focus of Web sites happened so fast, many businesses had difficulty changing the way they developed and managed their Web sites rapidly enough. Although the scope of business Web sites has expanded, many businesses have struggled to change the way they manage their Web sites to match the dynamic business applications they have become. Some large companies have been successful in developing tools that they use to manage their software development projects. As more companies begin to see their Web sites as collections of software applications, they are starting to use these tools to manage the development and maintenance of their Web sites. This evolution in the management of companies' Web presences as dynamic business applications will likely continue to grow in the near future.

Many companies have found it challenging to develop new information systems and Web sites that can help them create new markets or reconfigure their supply chains. In the past, companies that have been successful in devising new ways of working with their customers and suppliers have had years to complete the reconfiguration of their supply chains. The Internet has changed markets and marketing channels throughout many supply chains very quickly. A number of companies have been able to respond to these changes rapidly by using alternatives to traditional systems development methods, including the incubator and fast venturing approaches that you will learn about later in this chapter.

Internal Development vs. Outsourcing

Although many companies would like to think that they can avoid electronic commerce site development problems by finding a contractor who can handle the entire project for them, savvy leaders realize that they cannot. No matter what kind of electronic commerce initiative a company is contemplating, the initiative's success depends on how well it is integrated into and supports the activities in which the business is already engaged. Using internal people to lead all projects helps to ensure that the company's specific needs are addressed and that the initiative is congruent with the goals and the culture of the organization. Outside consultants are seldom able to learn enough about an organization's culture to accomplish these objectives. However, few companies are large enough or have sufficient in-house expertise to launch an electronic commerce project without some external help. The key to success is finding the right balance between outside and inside support for the project. Hiring another company to provide the outside support for all or part of the project is called **outsourcing**.

The Internal Team

The first step in determining which parts of an electronic commerce project to outsource is to create an internal team that is responsible for the project. This team should include people with enough knowledge about the Internet and its technologies to know what kinds of things are possible. Team members should be creative thinkers who are interested in taking the company beyond its current boundaries, and they should be people who have distinguished themselves in some way by doing something very well for the company. If they are not already recognized by their peers as successful individuals, the project may suffer from lack of credibility.

Some companies make the mistake of appointing as electronic commerce project leader a technical wizard who does not know much about the business and is not well-known throughout the company. Such a choice can greatly increase the likelihood of failure. Business knowledge, creativity, and the respect of the firm's operating function managers are all much more important than technical expertise in establishing successful electronic commerce. Project leaders need a good sense of the company's goals and culture to manage an implementation effectively.

Measuring the achievements of this internal team is very important. The measurements do not have to be monetary. Achievement can be expressed in whatever terms are appropriate to the objectives of the initiative. Customer satisfaction, number of sales leads generated, and reductions in order-processing time are examples of metrics that can provide a sense of the team's level of accomplishment. The measurements should show how the project is affecting the company's ability to provide value to the consumer. Many consultants advise companies to set aside between 5 percent and 10 percent of a project's budget for quantifying the project's value and measuring the achievement of that value.

Increasingly, companies are recognizing the value of the accumulated mass of employees' knowledge about the business and its processes. The value of an organization's pool of this type of knowledge is called **intellectual capital**. In the past, many companies ignored the value of intellectual capital because these human assets did not appear in the accounting records or financial statements.

Leif Edvinsson pioneered the use of human capital measures at Skandia Group, a large financial services company in Sweden. In addition to acknowledging employees' competencies, Edvinsson's measures include the value of customer loyalty and business partnerships as part of a company's intellectual capital. This networking approach to evaluating intellectual capital shows promise as a tool for assessing and tracking the value of internal teams and their connections to external consultants. These measurements are now being adapted for use in measuring systems development efforts. You can learn more about the use of human capital measurements by reading the books by Edvinsson and Max Boisot, another proponent of human capital measurement, which are included in the For Further Study and Research section at the end of this chapter.

The internal team should hold ultimate and complete responsibility for the electronic commerce initiative, from the setting of objectives to the final implementation and operation of the site. The internal team decides which parts of the project to outsource, to whom those parts are outsourced, and what consultants or partners the company needs to hire for the project. Consultants, outsourcing providers, and partners can be extremely important early in the project because they often develop skills and expertise in new technologies before most information systems professionals.

Early Outsourcing

In many electronic commerce projects, the company outsources the initial site design and development to launch the project quickly. The outsourcing team then trains the company's information systems professionals in the new technology before handing the operation of the site over to them. This approach is called **early outsourcing**. Because operating an electronic commerce site can rapidly become a source of competitive advantage for a company, it is best to have the company's own information systems people working closely with the outsourcing team and developing ideas for improvements as early as possible in the life of the project.

Late Outsourcing

In the more traditional approach to information systems outsourcing, the company's information systems professionals do the initial design and development work, implement the system, and operate the system until it becomes a stable part of the business operation. Once the company has gained all the competitive advantage provided by the system, the maintenance of the electronic commerce system can be outsourced so that the company's information systems professionals can turn their attention and talents to developing new technologies that will provide further competitive advantage. This approach is called **late outsourcing**. Although for years late outsourcing has been the standard for allocating scarce information systems talent to projects, electronic commerce initiatives lend themselves more to the early outsourcing approach.

Partial Outsourcing

In both the early outsourcing and late outsourcing approaches, a single group is responsible for the entire design, development, and operation of a project—either inside or outside the company. This typical outsourcing pattern works well for many information systems projects. However, electronic commerce initiatives can benefit from a partial outsourcing approach, too. In **partial outsourcing**, which is also called **component outsourcing**, the company identifies specific portions of the project that can be completely designed, developed, implemented, and operated by another firm that specializes in a particular function.

Many smaller Web sites outsource their e-mail handling and response functions. Customers expect rapid and accurate responses to any e-mail inquiry they make of a Web site with which they are doing business. Many companies send the customer an automatic order confirmation by e-mail as soon as the order or credit card payment is accepted. A number of companies provide e-mail auto response functions on an outsourcing basis.

Another common example of partial outsourcing is an electronic payment system. Many vendors are willing to provide complete customer payment processing. These vendors provide a site that takes over when customers are ready to pay and returns the customers to the original site after processing the payment transaction.

One of the most common elements of electronic commerce initiatives that companies outsource using this approach is Web hosting activity. Web hosting service providers are usually willing to accommodate requests for a variety of service levels. Small businesses can rent space on an existing server at the ISP's location. Larger companies can purchase the server hardware and have the service provider install and maintain it at the service provider's location. The service provider has the continuous staffing and expertise needed to keep an electronic commerce site up and running 24 hours a day, seven days a week (this kind of service is often called **24/7 operation**).

A number of service providers offer services beyond basic Internet connectivity to companies that want to do business on the Web. Many of these services were described earlier as candidates for partial outsourcing strategies and include automated e-mail response, transaction processing, payment processing, security, customer service and support, order fulfillment, and product distribution.

Nordisk Aviation

Nordisk Aviation is a subsidiary of the Norwegian Norsk Hydro Group. It designs, manufactures, and repairs air cargo containers for both freight and passenger baggage for major airlines throughout the world and for freight carriers such as FedEx and UPS. It also designs and sells handling systems and pallets that work with the containers. The company has annual sales of more than $100 million and employs more than 150 people at its locations around the world.

Nordisk was a strong believer in using the outsourcing approach for its IT projects—its IT Department included only two people. These two IT staff members worked as the overseers of every IT design and implementation project for the company. They also managed the ongoing IT services provided to Nordisk by other companies.

In late 2000, Manfred Gollent, the president of Nordisk, decided it was time to upgrade the company's Web site—which had been operating as an information site for several years—to include portal features that would allow Nordisk customers to check order status and learn about current developments in container and container-handling systems design. The logical approach for Nordisk was to find a company to which it could outsource the project.

The two members of Nordisk's IT staff went to work finding suitable Web developers. The previous Web developer had disappeared; they were unable to find any trace of the person who had created the existing Web site. The developer had created the Web site so that it used a number of programs to deliver dynamic pages. Unfortunately, the developer had given Nordisk only the executable code and not the actual programs. He also did not provide Nordisk with any documentation of the programs.

When the Web site was initially created, it was not an important strategic project for Nordisk. The IT staff members, who were busy with other important projects, did not ensure that the application code and documentation were received. Nordisk had to hire a company to rebuild the site completely to obtain the additional portal functions it wanted to add to the site. The lesson from the Nordisk case is that even when a company is outsourcing virtually all of its Web development, it must have procedures in place to ensure that the project is internally managed and documented.

New Methods for Implementing Partial Outsourcing

Although partial outsourcing has been used in IT management for many years, new ways of implementing it were developed specifically for Web businesses. The next two sections describe two of these new implementation approaches: incubators and fast venturing.

Incubators

An **incubator** is a company that offers startup companies a physical location with offices, accounting and legal assistance, computers, and Internet connections at a very low monthly cost. Sometimes, the incubator offers seed money, management advice, and marketing assistance as well. In exchange, the incubator receives an ownership interest in the company, typically between 10 percent and 50 percent.

When the company grows to the point that it can obtain venture capital financing or launch a public offering of its stock, the incubator sells all or part of its interest and reinvests the money in a new incubator candidate. One of the first Internet incubators was Idealab, which helped companies such as CarsDirect.com, Overture, and Tickets.com get their starts. Today, Idealab focuses on its own internally generated ideas rather than soliciting ideas from outside entrepreneurs, but it still operates as an incubator.

Some companies have created internal incubators. A number of companies used internal incubators in the past to develop technologies that the companies planned to use in their main business operations. Most of these were unsuccessful and, ultimately, were shut down. Employees in internal incubators found it difficult to maintain an entrepreneurial spirit when they knew that the technology they were developing would ultimately be taken away and controlled by the parent company.

More recently, companies such as Matsushita Electric's U.S. Panasonic division started internal incubators to help launch new companies that will grow to become important strategic partners. The business ideas developed in the incubator are eventually launched as separate companies that assume ownership of the assets used to create the ideas or products. The incubator development team also stays on as the managers of the new company. These strategic partner incubators have yielded much better results than old-style technology development incubators.

Fast Venturing

Often, large companies struggle to emulate the entrepreneurial spirit of smaller companies as they launch their Internet business initiatives. Many of these companies are trying to expand the internal incubator model and create an effective support system for new business and technology ideas, such as electronic commerce initiatives. One approach that is becoming popular is called fast venturing.

In **fast venturing**, an existing company that wants to launch an electronic commerce initiative joins external equity partners and operational partners that can offer the experience and skills needed to develop and scale up the project very rapidly. Equity partners are usually banks or venture capitalists that sometimes offer money, but are more likely to offer experience gained from guiding other startups that they have funded. Operational partners are firms, such as systems integrators and consultants, that have experience in moving projects along and scaling up prototypes. The roles of each participant in fast venturing are described in Figure 12-6.

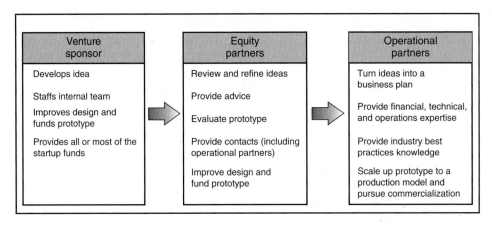

FIGURE 12-6   Elements of fast venturing

Fast venturing starts with the venture sponsor, an existing company that wants to launch the electronic commerce initiative but that does not have experience in starting new businesses. Then, equity partners which are entities that have provided startup money to new ventures in the past and have developed knowledge about operating new ventures, provide advice to the venture sponsor. The operational partners are people and companies that previously have built Web business sites. Thus, they can provide expertise in the technologies and business practices needed to create a successful operating electronic commerce site. The figure shows the flow of information and activities over time, starting with the venture sponsor, whose activities lead to the work done by the equity partners, who then pass the project on to the operational partners for implementation.

MANAGING ELECTRONIC COMMERCE IMPLEMENTATIONS

The best way to manage any complex electronic commerce implementation is to use formal management techniques. Project management, project portfolio management, specific staffing, and postimplementation audits are methods businesses use to efficiently administer their electronic commerce projects.

Project Management

Project management is a collection of formal techniques for planning and controlling the activities undertaken to achieve a specific goal. Project management was developed by the U.S. military and the defense contractors that worked with the military in the 1950s and the 1960s to develop weapons and other large systems. Not only was defense spending increasing in those years, but individual projects also were becoming so large that it became impossible for managers to maintain control over them without some kind of assistance.

The project plan includes criteria for cost, schedule, and performance—it helps project managers make intelligent trade-off decisions regarding these three criteria. For example, if it becomes necessary for a project to be completed early, the project manager can compress the schedule by either increasing the project's cost or decreasing its performance.

Today, project managers use specific application software called **project management software** to help them oversee projects. Commercial project management software products, such as **Oracle Primavera** and **Microsoft Project**, give managers an array of built-in tools for managing resources and schedules. The software can generate charts and tables that show, for example, which parts of the project are critical to its timely completion, which parts can be rescheduled or delayed without changing the project finish date, and where additional resources might be most effective in speeding up the project. **Open Workbench**, **OpenProj**, and **Redmine** are open-source project management software packages that offer many of the same features as the leading commercial products.

In addition to managing the people and tasks of the internal team, project management software can help the team manage the tasks assigned to consultants, technology partners, and outsourced service providers. By examining the costs and completion times of tasks as they are completed, project managers can learn how the project is progressing and continually revise the estimated costs and completion times of future tasks.

Information systems development projects have a well-deserved reputation for running out of control and ultimately failing. They are much more likely to fail than other types of projects, such as building construction projects. The main causes for information systems project failures are rapidly changing technologies, long development times, and changing customer expectations. Because of this vulnerability, many teams rely on project management software to help them achieve project goals.

Although electronic commerce certainly uses rapidly changing technologies, the development times for most electronic commerce projects are relatively short—often they are accomplished in under six months. This gives both the technologies and the expectations of users less time to change. Thus, electronic commerce initiatives are, in general, more successful than other types of information systems implementations.

You can learn more about project management by reading the references listed in the For Further Study and Research section at the end of this chapter, or by clicking the Web Link for the Project Management Institute, a not-for-profit organization devoted to the promotion of professional project management practices.

Project Portfolio Management

Larger organizations often have many IT implementation projects going on simultaneously—a number of which could be electronic commerce implementations or updates. A company's top technology manager is its **chief information officer (CIO)**. CIOs of some larger companies now use a portfolio approach to managing these multiple projects. **Project portfolio management** is a technique in which each project is monitored as if it were an investment in a financial portfolio. The CIO records the projects in a list (usually using spreadsheet or database management software) and updates the list regularly with current information about each project's status. By managing each project as a portfolio element, project portfolio managers can make trade-offs between cost, schedule, and quality across projects as well as within individual projects. This gives the organization more flexibility in allocating resources to achieve the best set of benefits from all of the projects in the timeliest manner.

Project management software performs a function similar to this for the tasks within a project, but most project management software packages are designed to handle individual projects and do not do a very good job of consolidating activities across multiple projects. Also, the information used in project portfolio management differs somewhat from the information used to manage specific projects. Project management software tracks the details of how each project is accomplishing its specific goals. In project portfolio management, the CIO assigns a ranking for each project based on its importance to the strategic goals of the business and its level of risk (probability of failure).

To develop these rankings, managers can use any of the methods that are commonly used to evaluate the risk of making investments in business assets. Indeed, using these tools helps the IT function explain electronic commerce projects as investments in assets, which makes it easier for other top managers in the company to understand the business characteristics of these projects. You can learn more about project portfolio management by reading the Berinato article cited in the For Further Study and Research section at the end of this chapter.

Staffing for Electronic Commerce

Regardless of whether the internal team decides to outsource parts of the design and implementation activity, it must determine the staffing needs of the electronic commerce initiative. The general areas of staffing that are most important to the success of an electronic commerce initiative include:

- Chief information officer (CIO)
- Business managers
- Project managers
- Project portfolio managers

- Account managers
- Applications specialists
- Web programmers
- Web graphics designers
- Content creators
- Content managers or editors
- Social networking administrators
- Online marketing managers
- Customer service reps
- Systems administrators
- Network operators
- Database administrators

In addition to implementing IT projects, the CIO is responsible for overseeing all of the information systems and related technological elements required to undertake and operate online business activities. The CIO's perspective is strategic and the person holding this position often serves as an important advocate for online business initiatives. CIOs frequently have an undergraduate degree in computer information systems (or a similar field) and a graduate degree in business or information technology management. They must have many years of experience in increasingly responsible management positions.

The business management function should include internal staff. The **business manager** should be a member of the internal team that sets the objectives for the project. The business manager is responsible for implementing the elements of the business plan and reaching the objectives set by the internal team. If revisions to the plan are necessary as the project proceeds, the business manager develops specific proposals for plan modifications and additional funding and presents them to the internal team and top management for approval.

The business manager should have experience and knowledge related to the business activity that is being implemented on the electronic commerce site. For example, if business managers are assigned to a retail consumer site, they should have experience managing a retail sales operation.

In addition to including the business manager, the business management function in large electronic commerce initiatives may include other individuals who carry out specialized functions, such as project management or account management, that the business manager does not have time to handle personally. A **project manager** is a person with specific training or skills in tracking costs and the accomplishment of specific objectives in a project. Many project managers are certified by organizations such as the Project Management Institute (which you learned about earlier in this chapter) and have skills in the use of project management software.

The **project portfolio manager** is usually promoted from the ranks of the project managers and has the responsibility for tracking all ongoing projects and managing them as a portfolio. This is the person who makes the trade-offs in cost, schedule, and quality across projects and balances the needs of the organization with the resources devoted to all projects.

An **account manager** keeps track of multiple Web sites in use by a project or keeps track of the projects that will combine to create a larger Web site. Most larger projects will have a test version, a demonstration version, and a production version of the Web site located on different servers. The test version is the "under construction" version of a Web site. Because most sites are frequently updated with new features and content, the test version gives the company a place to make sure that each new feature works before exposing it to customers. The demonstration version has features that have passed testing and must be demonstrated to an internal audience (for example, the Marketing Department) for approval. The production version is the full operating version of the site that is available to customers and other visitors. The account manager supervises the location of specific Web pages and related software installations as they are moved from test to demonstration to production. In smaller projects, the business manager handles the project and account management functions.

As more vendors provide packaged software solutions for electronic commerce, such as those you learned about in Chapter 9, companies need information systems staff that can install and maintain the software. Most large businesses have **applications specialists** who maintain accounting, human resources, and logistics software. Similarly, electronic commerce sites that buy software to handle catalogs, payment processing, and other features need applications specialists to maintain the software. Although the installation of these software packages can be outsourced, most companies prefer to train internal staff to install the software so they can be better prepared to manage the software when the site becomes operational.

Web sites have evolved from static HTML to more complex designs built with dynamic Web page generation technologies and XML data integration. As Web sites have become more complicated, the need for **Web programmers**, who design and write the underlying code for dynamic database-driven Web pages, has increased. Good Web programmers are familiar with several different dynamic Web page generation technologies and are highly skilled in at least one of them. Many Web programmers also have database manipulation and query skills, such as the ability to write SQL or PHP code.

Because the Web is a visual medium, the role of graphic elements on individual Web pages is important. A company must either retain the services of a graphics design firm, a Web design firm that includes graphics designers, or must hire employees with graphics design skills. A **Web graphics designer** is a person trained in art, layout, and composition and who also understands how Web pages are constructed. The Web graphics designer, or design team for larger sites, must ensure that the Web pages on the site are visually appealing, easy to use, and make consistent use of graphics elements from page to page.

Most larger sites and many smaller sites include content created specifically for the Web site. Other sites adapt content from existing sources within the company for use on the Web site, or purchase content to use on the site. These activities require that the company hire **content creators** to write original content and **content managers** or **content editors** to purchase existing material and adapt it for use on the site.

A relatively new addition to the online business team is the **social networking administrator**, who is responsible for managing the virtual community elements of the Web operation. These administrators might have backgrounds in technology, sales, customer service, or in widely diverse fields such as sociology or anthropology. They must coordinate all of the technologies that make the site work as a social network in ways that create value for the organization.

Although many organizations operate their online marketing function out of their traditional marketing departments, employees with the position of **online marketing manager** specialize in the specific techniques used to build brands and increase market share using the Web site and other online tools, such as e-mail marketing. These managers often have an extensive background in marketing and combine that with knowledge of technologies that allow them to manage the organization's online marketing function.

The Web offers businesses a unique opportunity to reach out to their customers. Thus, business-to-consumer and business-to-business sites that want to capitalize on that opportunity must include a customer relationship management function. **Customer service** personnel help design and implement customer relationship management activities in the electronic commerce operation. They can, for example, issue and administer passwords, design customer interface features, handle customer e-mail and telephone requests for service or follow-up action, and conduct telemarketing for the site. Companies strive to provide the best possible service to satisfy the demands of their customers. The increasing power of customers to organize and express their expectations on the Web is a natural extension of the increase in consumerism that has occurred over the past two decades.

Some companies outsource parts of their customer relationship management operation to independent call centers. A **call center** is a company that handles incoming customer telephone calls and e-mails for other companies. Using a call center often makes sense for smaller companies that do not have the volume of customer inquiries to justify creating an internal call center operation. Some call centers work with a variety of businesses; others focus on one specialty area. For example, a specialized call center might contract with software manufacturers to provide installation help for their software products. Call center employees who are skilled in helping customers install one software package are often able to learn how to support other software packages very quickly.

A systems administrator who understands the server hardware and operating system is an essential part of a successful electronic commerce implementation. The **systems administrator** is responsible for the system's reliable and secure operation. If the site operation is outsourced to a service provider, the service provider supplies this function. If the site is hosted by the company, it needs to devote at least one person to this job. In addition, the internal system administrator needs sufficient staff to maintain full 24/7 operation and site security. These **network operations** staff functions include load estimation and load monitoring, resolving network problems as they arise, designing and implementing fault-resistant technologies, and managing any network operations that are outsourced to service providers or telephone companies.

Most electronic commerce sites require a **database administration** function to support activities such as transaction processing, order entry, inquiry management, or shipment logistics. These activities require either an existing database into which the site is being integrated, or a separate database established for the electronic commerce initiative. It is important to have a database administrator who can effectively manage the design and implementation of this function.

Postimplementation Audits

After an electronic commerce site is successfully launched, most of the project's resources are devoted to maintaining and improving the site's operations. However, an increasing number of businesses are realizing the value of a postimplementation audit. A **postimplementation audit** (also called a **postaudit review**) is a formal review of a project after it is up and running.

The postimplementation audit gives managers a chance to examine the objectives, performance specifications, cost estimates, and scheduled delivery dates that were established for the project in its planning stage and compare them to what actually happened. In the past, most project reviews focused on identifying individuals to blame for cost overruns or missed delivery dates. Because many external forces in technology projects can overwhelm the best efforts of managers, this blame identification approach was generally unproductive, as well as uncomfortable, for the managers on the project. Today, the postimplementation audit is used by most organizations to gather lessons learned from both successful and unsuccessful projects. These lessons can be accumulated and, over time, used to create and update a set of standard best practices for the organization.

A postimplementation audit allows the internal team, the business manager, and the project manager to raise questions about the project's objectives and provide their "in-the-trenches" feedback on strategies that were set in the project's initial design. By agreeing beforehand not to lay blame, the company obtains valuable information that it can use in planning future projects and gives the participants a meaningful learning experience.

The audit should result in a comprehensive report that analyzes the project's overall performance, how well the project was administered, whether the organizational structure was appropriate for the project, and the specific performance of the project team(s). Each section of the report should compare actual results to the project's objectives. Many companies modify their project management organization structure after completing each project based on the contents of postaudit review reports. Many companies also include a confidential section in the report that evaluates each team member's performance on the project. Summaries of member performance can help managers decide which employees should be included in future team projects.

Change Management

Any information system project involves change, and change can be upsetting to people. As employees of an organization become accustomed to their specific duties, many of them draw comfort from their knowledge and develop a sense of security because they know their jobs well and are good at doing them. When changes are introduced into a workplace, employees become concerned about their abilities to cope with the changes and with their ability to continue to do good work. They often become worried that they might lose their jobs. These concerns can lead to increased stress that can be damaging to morale and work performance.

Management researchers have developed strategies for **change management**, which is the process of helping employees cope with these changes. Change management techniques include communicating the need for change to employees, including employees in the decision processes leading up to the change, allowing employees to participate in the planning for the change, and other tactics designed to help employees feel that they are a part of the change. This helps employees overcome the feelings of powerlessness that can lead to stress and reduced work performance.

Summary

This chapter provides an overview of key elements that are typically included in business plans for electronic commerce implementations. The first step is setting overall goals for the implementation. More specific objectives derive from these overall goals and include planned benefits and planned costs. The benefit and cost objectives should be stated in measurable terms, such as dollars or quantities, and they should be linked to the organization's business strategies. Before undertaking any online business initiative, companies should evaluate the initiative's estimated costs and benefits. Some costs, such as opportunity costs, can be difficult to identify and estimate.

Businesses use a number of evaluation techniques; however, most businesses calculate projects' ROI to gauge their value. The benefits of electronic commerce projects can be harder to define and quantify than the benefits expected from most other IT projects, so managers should be careful when using quantitative measures, such as ROI, to evaluate electronic commerce projects.

Companies must decide how much, if any, of an electronic commerce project to outsource. Forming an internal team that includes knowledgeable individuals from within the company is a good first step in developing an outsourcing strategy. The internal team develops the specific project objectives and is responsible for meeting those objectives. The internal team can select from specific strategies, such as using incubators or fast venturing, and should supervise the staffing of any part of the project that is to be developed internally.

Project management is a formal way to plan and control specific tasks and resources used in a project. It provides project managers with a tool they can use to make informed trade-offs among the project elements of schedule, cost, and performance. Large organizations are beginning to use project portfolio management techniques to track and make trade-offs among multiple ongoing projects. Electronic commerce initiatives are usually completed within a short time frame and thus are less likely to run out of control than other information systems development projects.

The company must staff the electronic commerce initiative regardless of whether portions of the project are outsourced. Critical staffing areas include business management, application specialists, customer service staff, systems administration, network operations staff, and database administration. A good way for all participants to learn from project experiences is to conduct a postimplementation audit that compares project objectives to the actual results.

Key Terms

24/7 operation	Content editor
Account manager	Content manager
Angel investor	Customer service
Applications specialist	Database administration
Business manager	Downstream strategies
Call center	Early outsourcing
Capital investment	Fast venturing
Capital project	Incubator
Change management	Initial public offering (IPO)
Chief information officer (CIO)	Intellectual capital
Component outsourcing	Late outsourcing
Content creator	Metrics

Network operations
Online marketing manager
Opportunity cost
Outsourcing
Partial outsourcing
Postimplementation audit (postaudit review)
Project management
Project management software
Project manager
Project portfolio management

Project portfolio manager
Return on investment (ROI)
Social networking administrator
Systems administrator
Total cost of ownership (TCO)
Upstream strategies
Venture capitalist
Web graphics designer
Web programmer

Review Questions

1. Name three benefit objectives that a business might decide to measure in an electronic commerce initiative, and then write one paragraph about each in which you explain how the business might measure its accomplishment of that objective.

2. Write a paragraph in which you explain the differences between upstream and downstream business strategies.

3. In one paragraph, outline the types of costs that would be included in the total cost of ownership of an online business initiative.

4. Write a paragraph in which you explain the concept of opportunity cost, including at least one example of such a cost.

5. In one or two paragraphs, summarize how the cost of launching an online business has changed over the years. Be sure to include an explanation of why such costs have changed.

6. Write a paragraph in which you explain the advantages and disadvantages of using ROI to evaluate online business proposals.

7. In two or three paragraphs, explain why the head of the business management function of an electronic commerce initiative should be an employee of the company implementing the project, even if most of the work is outsourced.

8. In one paragraph, explain why late outsourcing is seldom used in electronic commerce projects.

9. Write a paragraph in which you explain the concept of partial outsourcing.

10. Write two or three paragraphs in which you describe the most important functions to look for in a project management software package.

11. In about 100 words, explain why IT projects (such as Web site development or redesign) are less likely to be delivered on time and within budget than large building construction projects. Include a discussion of how project management software can help IT project managers achieve their goals.

12. In one or two paragraphs, explain why blame identification is not an important element of a postimplementation review.

Exercises

1. Your friend, Samantha Arturo, has what you think is a great idea for a Web site that will post ideas for inventions submitted by creative thinkers around the world. Samantha wants to solicit feedback on the ideas using social media and then fund the development of the ideas that get the most positive feedback. She realizes that she will need to raise substantial capital to launch the site and fund the development of each invention. In fact, she is thinking that she will need several hundred thousand dollars during the first three years before any of the inventions start to generate significant profits. Her Uncle Dave has offered to invest $100,000 in the business to help get things going. She has taken her idea to several banks, but they all told her that her business was too speculative for them to lend her large amounts of money. Samantha has heard about angel investors and venture capitalists and asks you for more information about them. Use your favorite search engine or the resources of your library to learn more about angel investors and venture capitalists and then write about 100 words in which you explain to Samantha what angel investors and venture capitalists are, how much money she is likely to be able to raise from each, and what the advantages and disadvantages of using them are.

2. The Grover Cams Company manufactures cams and other components for diesel engines. As Web site manager for Grover, you created a Web site that allows your smaller customers to order directly from Grover instead of through their local distributors. The site also includes information about the company's history, its financial statements, and detailed downloadable technical specifications for most of its main products. You would like to add features to the Web site that allow mobile users to obtain information and order parts using their smart phones or tablet devices. You also have some ideas for using social media along with the Web site to generate interest among potential customers. You created a capital budget proposal for adding these features and submitted it to Grover's board of directors last month. The board always calculates and evaluates a capital project's return on investment before approving it. Grover's CIO, Tom Buckles, told you that the project did not provide a high enough financial return for the board to approve it; however, some of the board members thought the project could improve Grover's future strategic position. Therefore, the board is willing to consider nonmonetary factors as a basis for approving the project. Tom would like to take your project back to the board next month with a solid proposal that includes nonmonetary factors. He asks you to write a memo that outlines nonmonetary factors that could be important to Grover's future strategic position. In addition to considering the discussion in this chapter, you can use your library and the Web Links for this exercise as you prepare your memo.

3. You are working for International Delicacies, which has become successful selling unusual food and gift items through its mail order catalog. Most customers call the toll-free telephone number on the catalog, but some still send in orders by mail. Your manager, Jagdish Singh, wants to add an online store that will complement the company's existing mail order and telephone sales channels. He wants you to lead the internal team for the project. Write a memo to Jagdish of about 300 words in which you outline the steps you will take to staff the internal team, make decisions about internal development versus outsourcing, and choose a hosting service. Be sure to include an evaluation of whether an incubator or a fast venturing strategy might make sense for this project.

4. As manager of networks and computing operations for Fashion Land, a retailer of women's clothing and accessories, you have seen the business grow from seven stores in Kansas City to over 100 stores located throughout the Midwest. Fashion Land's marketing research team realizes that the majority of its target customer group—females between the ages of 13 and 35—are regular users of the Web. The researchers have asked you for help in

developing an electronic commerce initiative for Fashion Land. Alone, or in a team assigned by your instructor, do the following:

a.   Outline a business strategy for Fashion Land's online business initiative. The outline should include a list of specific objectives and the costs and benefits of accomplishing each objective. The outline should also include recommendations regarding what to outsource and what staff should be hired.

b.   Prepare a memo that lists and briefly describes the major hardware, software, security, payment processing, advertising, international, legal, and ethics issues that might arise in the development of this electronic commerce site.

Cases

C1. Idealab

Bill Gross started his first company (a solar-powered device manufacturer) when he was 15 years old. After graduating from Caltech, he started a software company, GNP, that he later sold to Lotus (the spreadsheet software pioneer that is now part of IBM). Gross had made a considerable amount of money and was interested in exploring better ways of getting ideas converted into profitable businesses. He became fascinated by the idea of business incubators about the same time he became fascinated with the business potential of the Internet. In 1996, he pooled some of his wealth with contributions from several partners to create Idealab.

Idealab was one of the first companies to provide an incubator that was open to individual entrepreneurs. Idealab provided venture capital and gave entrepreneurs a place to work and develop their business ideas alongside other entrepreneurs. In the first wave of electronic commerce, Idealab was very successful. Although many of its incubated companies eventually failed, enough of them succeeded that Idealab was able to fund several generations of new businesses through its operations. In its first year, it supported 10 new businesses, including the very successful CitySearch Web site. In its second year, Idealab helped create another 10 businesses, including the successful sites Shopping.com, Tickets.com, and WeddingChannel.com. In subsequent years, Idealab incubated companies such as NetZero, Cooking.com, CarsDirect.com, Picasa, and GoTo.com (which later became Overture and was eventually acquired by Yahoo!). Not all of Idealab's companies were successful, however. One of the most dramatic failures of the first wave of electronic commerce, eToys, had been an Idealab company. Idealab had more winners than losers, though; by early 2000, the company had more than $4 billion in assets.

In 2000, Gross devised a new strategy that would go beyond Idealab's original purpose as an incubator. He developed a plan to compete with Amazon.com using existing Idealab companies. His plan was to combine about 10 of the companies in the incubator (including specialty retailer Eve.com and online jewelry store Ice.com) and promote them (using large amounts of money that would be raised from outside investors) as a single marketplace under the name Big.com. However, just as Gross began raising money to support the launch of this new marketplace, the pool of dot-com investment funds dried up. The new combined company quickly failed. Eve.com and Big.com no longer exist. The founders of Ice.com bought their company back from Idealab and moved it to their home in Montreal (where the company is now operating profitably). Within a few months, the failure of Big.com and the lower stock market valuations of Idealab's holdings reduced the value of the company's assets from $4 billion to $200 million.

Idealab's investors were upset by Gross' change in strategy and by the drop in their company's value. In January 2002, 44 of them sued Gross and other Idealab managers for $750 million. The suit alleged mismanagement of the funds invested and further alleged that

Gross had used Idealab funds to pay personal expenses. Eighteen months later, a court held that the allegations were without merit and the suit was dismissed. Gross was once again able to devote his time to operating Idealab as an incubator.

Gross laid off more than two-thirds of Idealab's employees and stopped accepting outside venture capital. Idealab no longer provides incubator space for entrepreneurs who have developed ideas on their own. The company only funds ideas generated by the Idealab management team. Idealab's asset value has rebounded somewhat and is now between $400 million and $800 million. Idealab still incubates a number of online businesses along with startups in the areas of social media management, solar power generation, and data management.

Required:

1. In its first three years of operation, Idealab recruited entrepreneurs to its incubator who had business experience, but who did not know much about the Internet. In about 300 words, explain what benefits Idealab was able to provide to these entrepreneurs and why the incubator environment was beneficial to them.

2. In about 200 words, analyze Idealab's 2000 decision to change its focus from being an incubator to merging its companies in an attempt to compete with Amazon.com. In your analysis, discuss whether the decision was a strategic error or just a case of bad timing.

523

3. In about 200 words, explain why you think Gross decided to devote Idealab's resources to the development of internally generated ideas in 2003. Be sure to consider whether this change will help Idealab succeed in the second wave of electronic commerce.

Note: Your instructor might assign you to a group to complete this case, and might ask you to prepare a formal presentation of your results to your class.

C2. Davis Humanics

Davis Humanics (DH) is a company founded in 1982 that provides human resources services to about 7000 companies with a total of nearly 100,000 employees. These services include payroll processing, tax filing, health insurance and claims management, and retirement plan management. DH has annual sales of $2 billion and about 1000 employees. In 1997, DH launched a Web site that has since grown to include a variety of tools for connecting with customers. DH has expanded rapidly and has clients of all sizes, ranging from smaller companies with fewer than 50 employees to Fortune 500 companies.

As DH grows, it is having trouble maintaining a consistent quality of service. Account managers each must handle more clients, and it is becoming difficult for those account managers to maintain a high degree of personal contact with the human resources executives who control DH's contracts. In the past, account managers worked with a small set of client contact people, but now account managers must work with more people, many of whom they have never met. In addition to account managers, client personnel have regular contacts with DH operations staff (who handle input tasks), DH systems staff (who help customize the interfaces between DH systems and client systems), and DH professional staff (lawyers, actuaries, and human resources professionals who consult with DH clients and their legal counsel regarding the operation of their retirement and benefits plans).

Because DH's clients are so different in size and how they operate, DH has to be flexible in handling input data. For example, DH's payroll-processing service allows clients many different ways to send in time card data. The largest clients arrange for customized computer-to-computer transfer of information. Some large clients use EDI transfers. Most medium and smaller-sized clients e-mail or fax the time card information, but a significant number of them mail paper lists that DH must scan into its systems. The health insurance claims-handling operation is even more troublesome. In addition to having clients send information in various formats, the insurance companies demand that information be submitted in specific formats, each of which is different.

The complexity of DH's operations is growing as rapidly as the company adds new clients. Sandi Higbee, DH's director of Operations, asks for your help in outlining a Web-based customer relationship management (CRM) system that will help manage the account managers' ever-increasing levels of customer contact. Sandi reviewed the products offered by several leading CRM vendors and believes that one might work as a base product, but no matter which product is chosen, she believes that substantial customization will be necessary because DH's operations are so complex and different from most companies that sell products or simple services to customers. A good CRM system for DH would need to monitor all types of customer interactions with DH account managers, operations staff, systems staff, and professional staff. In addition, the system's Web interface should allow DH clients to access parts of the CRM system so they can track DH's follow-up on their work requests and pending inquiries.

DH evaluates all capital projects, including IT projects, using ROI. Sandi is worried about this because she believes that many of the benefits of this CRM project will be hard to quantify. On the other hand, the costs of the CRM project (software and hardware purchase and cost of consultants who will customize the CRM software to meet DH's specific needs) will be very easy to quantify and will be large. Sandi expects the vendor-consultant teams to submit bids of between $1 million and $2 million for this project.

Required:

1. Prepare an outline of the benefits that DH might expect to obtain from this CRM project. Use categories to organize your list of benefits; for example, you might identify benefits that will accrue to DH's account managers, operations staff, IT staff, and professional staff. Because DH's clients will also benefit, you might be able to identify benefits that will accrue to DH's Marketing and Sales departments or to DH's New Product Development department. Be sure to include any long-term benefits that you think might occur after the CRM system has been in place for several years.

2. Estimate the dollar value of each benefit you identified in the first part of your answer.

3. Prepare a one-page memorandum to the DH board of directors in which you argue against using ROI as the primary method for evaluating this project. Keep in mind that these directors have little time to review your arguments and are very much inclined to use ROI for all project evaluations.

Note: Your instructor might assign you to a group to complete this case, and might ask you to prepare a formal presentation of your results to your class.

For Further Study and Research

Abdel-Hamid, T. and S. Madnick. 1991. *Software Project Dynamics: An Integrated Approach*. Englewood Cliffs, NJ: Prentice Hall.

Abdel-Hamid, T., K. Sengopta, and C. Sweet. 1999. "The Impact of Goals on Software Project Management: An Experimental Investigation," *MIS Quarterly*, 23(4), December, 531–555.

Anthes, G. 2008. "What's Your Project Worth?" *Computerworld*, 42(11), March 10, 29–31.

Aragon, L. 2004. "Idealab: Bubble Fund Finds Itself Back at Square One," *Venture Capital Journal*, 44(6), June, 20.

Armour, P. 2010. "Return at Risk: Calculating the True Likely Cost of Projects," *Communications of the ACM*, 53(9), September, 23–25.

Bannan, K. 2004. "Entrepreneur Learns Why It's Best to Optimize Site Before It Launches," *B to B*, 89(15), December 13, 19.

Beal, V. 2012. "Can Facebook Replace Traditional Business Websites?" *InfoWorld*, February 2. (http://www.infoworld.com/d/applications/can-facebook-replace-traditional-business-websites-185579)

Betts, M. 2011. "Banks Can Reap 300% ROI From Advanced Smartphone Services, Study Says," *Computerworld*, February 21. (http://www.computerworld.com/s/article/354870/Banks_Can_Reap_Big_Profits_From_Mobile_Services)

Blazier, A. 2003. "Far from Dead, Idealab Continues to Build for Future," *San Gabriel Valley Tribune*, July 12, C1.

Boyer, A. 2012. "Social Media for Small Businesses with 'No Time,'" *BlogWorld*, February 25. (http://www.blogworld.com/2012/02/25/social-media-for-small-businesses-with-no-time/)

Brandel, M. 2008. "Xtreme ROI," *Computerworld*, 42(7), February 11, 30–33.

Brodie, T. 2012. "What Small Business Can Learn From Super Bowl Ads," *The Globe and Mail*. February 3. (http://www.theglobeandmail.com/report-on-business/small-business/sb-tools/small-business-briefing/what-small-business-can-learn-from-super-bowl-ads/article2325325/)

Buderi, B. 2005. "Conquering the Digital Haystack: New Startups Are Changing the Way People Search the Web," *Inc.*, January, 34–35.

Cendrowski, S. 2012. "Nike's New Marketing Mojo," *Fortune*, February 27, 81–88.

Cerpa, N. and J. Verner. 2009. "Why Did Your Project Fail?" *Communications of the ACM*, 52(12), December, 130–134.

Edvinsson, L. and M. Malone. 1997. *Intellectual Capital: Realising Your Company's True Value by Finding its Hidden Brainpower*. New York: HarperCollins.

Fisher, E. and R. Reuber. 2011. "Social Interaction Via New Social Media: (How) Can Interactions on Twitter Affect Effectual Thinking and Behavior?" *Journal of Business Venturing*, 26(2), January, 1–18.

Fisher, T. 2009. "ROI in Social Media: A Look at the Arguments," *Journal of Database Marketing & Customer Strategy Management*, 16(3), September, 189–195.

Fleming, Q. and J. Koppelman. 2003. "What's Your Project's Real Price Tag?" *Harvard Business Review*, 81(9), September, 20–21.

Grimes, A. 2004. "Court Deals Blow to Investors' Suit Against Idealab," *The Wall Street Journal*, June 30, B6.

Haeussler, C., H. Patxelt, and S. Zahra. 2012. "Strategic Alliances and Product Development in High Technology New Firms: The Moderating Effect of Technological Capabilities," *Journal of Business Venturing*, 27(2), March, 217–233.

Hannay, C. 2012. "Toughest to Track: How to Measure Social Media Success," *The Globe and Mail*, February 16. (http://www.theglobeandmail.com/report-on-business/small-business/digital/web-strategy/toughest-to-track-how-to-measure-social-media-success/article2339488/)

Havenstein, H. 2007. "IT Execs Seek New Ways to Justify Web 2.0," *Computerworld*, 41(33), August 13, 14–15.

Hellweg, E. and S. Donahue. 2000. "The Smart Way to Start an Internet Company," *Business 2.0*, March 1, 64–66.

Jepson, K. 2009. "How Two Credit Unions Are Achieving Banner ROI On Their Web Sites," *Credit Union Journal*, September 21, 1–15.

Kambil, A., E. Eselius, and K. Monteiro. 2000. "Fast Venturing: The Quick Way to Start a Web Business," *Sloan Management Review*, 41(4), Summer, 55–67.

Keefe, P. 2003. "Backing Up ROI," *Computerworld*, 37(12), March 24, 22.

Keen, J. and R. Joshi. 2011. *Making Technology Investments Profitable: ROI Roadmap From Business Case to Value Realization*. Second edition. Hoboken, NJ: Wiley.

Keen, P. 2000. "Six Months—or Else," *Computerworld*, 34(15), April 10, 48.

Keil, M. and D. Robey, 1999. "Turning Around Troubled Software Projects: An Exploratory Study of the De-Escalation of Commitment to Failing Courses of Action," *Journal of Management Information Systems*, 15(4), 63–87.

Keil, M., P. Cule, K. Lyytinen, and R. Schmidt. 1998. "A Framework for Identifying Software Project Risks," *Communications of the ACM*, 41(11), November, 76–83.

Kerzner, H. 2009. *Project Management: A Systems Approach to Planning, Scheduling, and Controlling*. Tenth Edition. New York: John Wiley & Sons.

Lacy, S. 2008. *Once You're Lucky, Twice You're Good: The Rebirth of Silicon Valley and the Rise of Web 2.0*. New York: Gotham Press.

Leung, L. 2003. "Managing Offshore Outsourcing," *Network World*, 20(49), December 8, 59.

Madachy, R. 2008. *Software Process Dynamics*. Hoboken, NJ: Wiley.

Mathiassen, L. and T. Tuunanen. 2011. "Managing Requirements Risks in IT Projects," *IT Professional*, 13(6), November–December, 40–47.

Mathieu, R. and R. Pal. 2011. "The Selection of Supply Chain Management Projects: A Case Study Approach," *Operations Management Research*, 4(3–4), December, 164–181.

McConnell, S. 1996. *Rapid Development: Taming Wild Software Schedules*. Redmond, WA: Microsoft Press.

Metz, R. 2012. "The Startup Whisperer," *Business Impact*, February, 16.

Murthi, S. 2002. "Managing the Strategic IT Project," *Intelligent Enterprise*, 5(18), November 15, 49–52.

Nocera, J. and E. Florian. 2001. "Bill Gross Blew Through $800 Million in Eight Months (and He's Got Nothing to Show for It): Why Is He Still Smiling?" *Fortune*, 143(5), March 5, 70–77.

O'Leary, S., K. Sheehan, and S. Lentz. 2011. *Small Business Smarts: Building Buzz With Social Media*. Santa Barbara, CA: Praeger/Greenwood.

Petrecca, L. 2012. "Small Businesses Use Social Media to Grow," *USA Today*, February 16. (http://www.usatoday.com/money/smallbusiness/story/2012-02-16/small-business-social-media-outreach-smachburger/53122300/1)

Pentina, I. and R. Hasty. 2009. "Effects of Multichannel Coordination and E-Commerce Outsourcing on Online Retail Performance," *Journal of Marketing Channels*, 16(4), 359–374.

Phillips, J., W. Brantley, and P. Phillips. 2012. *Project Management ROI*. Hoboken, NJ: Wiley.

Ramsey, C. 2000. "Managing Web Sites as Dynamic Business Applications," *Intranet Design Magazine*, June. (http://idm.internet.com/articles/200006/wm_index.html)

Rivard, S. and R. Dupré. 2009. "Information systems project management in PMJ: A brief history," *Project Management Journal*, 40(4), December, 20–30.

Sacks, D. 2005, "The Accidental Guru," *Fast Company*, January, 64–71.

Sawhney, M. 2002. "Damn the ROI, Full Speed Ahead: 'Show Me the Money' May Not Be the Right Demand for E-Business Projects," *CIO*, 15(19), July 15, 36–38.

Schonfeld, E. 2007. "The Startup King's New Gig," *Business 2.0*, 8(9), October, 68.

Schwalbe, K. 2007. *Information Technology Project Management*. Fifth Edition. Boston, MA: Course Technology.

Schwalbe, K. 2009. *Introduction to Project Management*. Second Edition. Boston, MA: Course Technology.

Stewart, T. 1999. "Larry Bossidy's New Role Model: Michael Dell," *Fortune*, 139(7), April 12, 166–167.

Tan, B., N. Tang, and P. Forrester. 2004. "Application of Quality Function Deployment for e-Business Planning," *Production Planning & Control*, 15(8), December, 802–815.

Taylor, H., E. Artman, and J. Woelfer. 2012. "Information Technology Project Risk Management: Bridging the Gap Between Research and Practice." *Journal of Information Technology*, 27, 17–34.

Teo, T. and T. Koh. 2010. "Lessons From Multi-Agency Information Management Projects: Case of the Online Business Licensing Service Project, Singapore," *International Journal of Information Management*, 30(1), February, 85–93.

United States Department of Justice Inspector General. 2002. *Audit Report No. 03–09: Federal Bureau of Investigation's Management Of Information Technology Investments*. Washington, D.C.: U.S. Department of Justice.

United States General Accounting Office. 2002. *Desktop Outsourcing: Positive Results Reported, But Analyses Could Be Strengthened*. Washington, D.C.: U.S. General Accounting Office.

Warren, L., D. Patton, and D. Bream. 2009. "Knowledge Acquisition Processes During the Incubation of New High Technology Firms," *International Entrepreneurship and Management Journal*, 5(4), 481–495.

Weinberg, B. 2011. "Social Spending: Managing the Social Media Mix," *Business Horizons*, 54(3), May–June, 275–282.

Wysocki, B. 2000. "U.S. Incubators Help Japan Hatch Ideas," *The Wall Street Journal*, June 12, A1.

Wysocki, B. 2009. *Effective Project Management: Traditional Agile, Extreme*. Fifth Edition. Indianapolis: Wiley.

Yourdon, E. and P. Becker. 1997. *Death March: The Complete Software Developer's Guide to Surviving "Mission Impossible" Projects*. Upper Saddle River, NJ: Prentice Hall.

527

GLOSSARY

24/7 operation The operation of a site or service 24 hours a day, seven days a week.

802.11a, 802.11b, 802.11g, 802.11n Various updates to an improved version of Wi-Fi introduced in 2002; capable of transmitting data at speeds up to 54 Mbps. 802.11n rates are 300–450 Mbps.

Acceptance An expression of willingness to take an offer, including all of its stated terms.

Access control list (ACL) A list of resources and the usernames of people who are permitted access to those resources within a computer system.

Account aggregation A feature of online banks that allows a customer to obtain bank, investment, loan, and other financial account information from multiple Web sites and to display it all in one location at the bank's Web site.

Account manager A person who keeps track of multiple Web sites in use by a project or keeps track of the projects that combine to create a larger Web site.

Accredited Standards Committee X12 (ASC X12) A committee that develops and maintains uniform EDI standards in the United States.

Acquirer fees Fees charged by an acquiring bank for providing payment card processing services.

Acquiring bank Synonymous with merchant bank, which is a bank that does business with merchants who want to accept credit cards.

Acquisition cost The total amount of money that a site spends, on average, to draw one visitor to the site.

Active ad A Web ad that generates graphical activity that "floats" over the Web page itself instead of opening in a separate window.

Active content Programs that are embedded transparently in Web pages that cause action to occur.

Active Server Pages (ASP) Applications that generate dynamic content within Web pages using either Jscript code or Visual Basic.

Active wiretapping An integrity threat that exists when an unauthorized party can alter a message.

ActiveX An object, or control, that contains programs and properties that are put in Web pages to perform particular tasks.

Activity A task performed by a worker in the course of doing his or her job.

Ad view A Web site visitor page request that contains an advertisement.

Ad-blocking software A program that prevents banner ads and pop-up ads from loading.

Addressable media Advertising efforts sent to a known addressee; these include direct mail, telephone calls, and e-mail.

Advance fee fraud A scam in which the perpetrator offers to share the proceeds of some large payoff with the victim if the victim will make a "good faith" deposit or provide some partial funding first. The perpetrator then disappears with the deposit.

Advanced Encryption Standard (AES) The encryption standard designed to keep government information secure using the Rijndael algorithm. It was introduced in February 2001 by the National Institute of Standards and Technology (NIST).

Advertising-subscription mixed revenue model A revenue model in which subscribers pay a fee and accept some level of advertising.

Advertising-supported revenue model A revenue model in which Web sites provide free content along with advertising or messages provided by other companies that pay the Web site operator for delivering the advertising or messages.

Affiliate marketing An advertising technique in which one Web site (called an "affiliate") includes descriptions, reviews, ratings, or other information about products that are sold on another Web site. The affiliate site includes links to the selling site, which pays the affiliate site a commission on sales made to visitors who arrived from a link on the affiliate site.

Affiliate program broker A company that serves as a clearinghouse or marketplace for sites that run affiliate programs and sites that want to become affiliates.

AJAX (asynchronous JavaScript and XML) A development framework that can be used to create interactive Web sites that look like applications running in a Web browser.

Amazon law State laws that require online retailers to collect and remit sales taxes on sales they make in their states, even though the online retailers do not have nexus with the state.

American National Standards Institute (ANSI) The coordinating body for electrical, mechanical, and other technical standards in the United States.

Analytical processing A technique that examines stored information and looks for patterns in the data that are not yet known or suspected; also called data mining.

Anchor tag The HTML tag used to specify hyperlinks.

Angel investors Investors who fund the initial startup of an online business. In return for their capital, angel investors become stockholders in the business and often own more of the business than the founder. Typical funding by angel investors is between a few hundred thousand dollars and a few million dollars.

Animated GIF Animated Web ad graphics that grab a visitor's attention.

Anonymous electronic cash Electronic cash that cannot be traced back to the person who spent it.

Anonymous FTP A protocol that allows users to access limited parts of a remote computer using FTP without having an account on the remote computer.

Antivirus software Software that detects viruses and worms and either deletes them or isolates them on the client computer so they cannot run.

Applet A program that executes within another program; it cannot execute directly on a computer.

Application integration The coordination of all of a company's existing systems to each other and to the company's Web site.

Application program (application, application software) A program that performs a specific function, such as creating invoices, calculating payroll, or processing payments received from customers.

Application program interface (API) A general name for the ways programs interconnect with each other.

Application server A middle-tier software and hardware combination that lies between the Internet and a corporate back-end server.

Application service provider (ASP) A Web-based site that provides management of applications such as spreadsheets, human resources management, or e-mail to companies for a fee.

Application software Synonymous with application, which is a program that performs a specific function.

Applications specialist The member of an electronic commerce team who is responsible for maintenance of software that performs a specific function, such as catalog, payment processing, accounting, human resources, and logistics software.

Apps Application software that is sold for use on mobile phones.

AS2 (Applicability Statement 2) A specification based on the HTTP rules for Web page transfers.

AS3 (Applicability Statement 3) A more secure version of AS2.

Ascending-price auction A type of auction in which bidders publicly announce their successively higher bids until no higher bid is forthcoming; also called an English auction.

ASP.NET Microsoft-developed server-side dynamic Web page-generation technology.

Asymmetric connection An Internet connection that provides different bandwidths for each direction.

Asymmetric digital subscriber line (ADSL) Internet connections using the DSL protocol with bandwidths from 16 to 640 Kbps upstream and 1.5 to 9 Mbps downstream.

Asymmetric encryption Synonymous with public-key encryption, which is the encoding of messages using two mathematically related but distinct numeric keys.

Asynchronous transfer mode (ATM) Internet connections with bandwidths of up to 622 Gbps.

Atom Publishing Protocol A blogging application that simplifies the blog publishing process and makes its functions available as a Web service so other computers can interact with blog content.

Attachment A data file (document, spreadsheet, or other) that is appended to an e-mail message.

Auction consignment services Companies that take an item and create an online auction for that item, handle the transaction, and remit the balance of the proceeds after deducting a fee. These services are performed on behalf of people and small businesses who want to use an online auction but do not have the skills or the time to become a seller.

Auctioneer The person who manages an auction.

Authority to bind The ability of an individual to commit his or her company to a contract.

Automated clearing house (ACH) One of several systems set up by banks or government agencies, such as the U.S. Federal Reserve Board, that process high volumes of low dollar amount electronic fund transfers.

Backbone routers Computers that handle packet traffic along the Internet's main connecting points; they can each handle more than 50 million packets per second.

Backdoor An electronic hole in electronic commerce software left open by accident or intentionally that allows users to run the program without going through the normal authentication procedure for access to the program.

Back-end processor A banking service provider that takes transactions from the front-end processor and coordinates information flows through the interchange network to settle transactions. The back-end processor handles chargebacks and any other reconciliation items through the interchange network and the acquiring and issuing banks, including the ACH transfers.

Bandwidth The amount of data that can be transmitted in a fixed amount of time. Also, the number of simultaneous site visitors that a Web site can accommodate without degrading service.

Banner ad A small rectangular object on a Web page that displays a stationary or moving graphic and includes a hyperlink to the advertiser's Web site.

Banner advertising network An organization that acts as a broker between advertisers and Web sites that carry ads.

Banner exchange network An organization that coordinates ad sharing so that other sites run your ad and your site runs other exchange members' ads.

Base 2 (binary) A number system in which each digit is either a 0 or a 1, corresponding to a condition of either "off" or "on." Also known as a binary system.

Bayesian revision A statistical technique in which additional knowledge is used to revise earlier estimates of probabilities.

Behavioral segmentation The creation of a separate experience for customers based on their behavior.

Benchmarking Testing that compares hardware and software performances.

Bid An offer of a certain price made on an item that is up for auction.

Bidder A potential buyer at an auction; one who places bids.

Bill presentment A Web site feature that allows customers to view and pay bills online.

Biometric security device A security device that uses an element of a person's biological makeup to confirm identification. These devices include writing pads that detect the form and pressure of a person writing a signature, eye scanners that read the pattern of blood vessels in a person's retina, and palm scanners that read the palm of a person's hand (rather than just one fingerprint).

Black hat hackers Hackers who use their skills for harmful purposes.

Black list spam filter Software that looks for From addresses in incoming messages that are known to be spammers. The software can delete the message or put it into a separate mailbox for review.

Blade server A server configuration in which small server computers are each installed on a single computer board and then many of those boards are installed into a rack-mounted frame.

Blog Synonymous with Web log, which is a Web site on which people post their thoughts and invite others to add commentary.

Bluetooth A wireless standard that is used for short distances and lower bandwidth connections.

Bonded warehouse A secure location where incoming international shipments can be held until customs requirements are satisfied or until payment arrangements are completed.

Border router The computers located at the border between the organization and the Internet that decide how best to forward each packet of information as it travels on the Internet to its destination. Synonymous with gateway computer and gateway router.

Bot (robot) A program that automatically searches the Web to find Web pages that might be interesting to people.

Botnet A robotic network that can act as an attacking unit, sending spam or launching denial-of-service attacks against specific Web sites. Synonymous with zombie farm.

Brand Customers' perceptions of the attributes of a product or service, including name, history, and reputation.

Brand leveraging A strategy in which a well-established Web site extends its dominant positions to other products and services.

Breach of contract The failure of one party to comply with the terms of a contract.

Broadband Connections that operate at speeds of greater than about 200 Kbps.

Browser-wrap acceptance Synonymous with Web-wrap acceptance, which is the compliance with EULA conditions with which a user agrees through the act of using a Web site.

Buffer An area of a computer's memory that is set aside to hold data read from a file or database.

Buffer overrun, buffer overflow An error that occurs when programs filling buffers malfunction and overfill the buffer, spilling the excess data outside the designated buffer memory area. Also called buffer overflow.

Bulk mail Electronic junk mail that can include solicitations, advertisements, or e-mail chain letters. Also called spam or unsolicited commercial e-mail.

Bulletin board system (BBS) Computers that allow users to connect through modems (using dial-up connections through telephone lines) to read and post messages in a common area.

Business logic Rules of a particular business.

Business manager The member of an electronic commerce team who is responsible for implementing the elements of the business plan and reaching the objectives set by the internal team. The business manager should have experience in and knowledge of the business activity being implemented in the site.

Business model A set of processes that combine to yield a profit.

Business process offshoring The distribution of nonmanufacturing business activities to international suppliers.

Business process patent A patent that protects a specific set of procedures for conducting a particular business activity.

Business processes The activities in which businesses engage as they conduct commerce.

Business rules The way a company runs its business.

Business-to-business (B2B) Transactions conducted between businesses on the Web.

Business-to-consumer (B2C) Transactions conducted between shoppers and businesses on the Web.

Business-to-government (B2G) Business transactions conducted with government agencies, such as paying taxes and filing required reports.

Business unit A unit within a company that is organized around a specific combination of product, distribution channel, and customer type. Synonymous with strategic business unit.

Byte An 8-bit number (in most computer applications).

Call center A company that customer handles telephone calls and e-mails for other companies.

Cannibalization The loss of traditional sales of a product to its electronic counterpart.

Capital investment A major outlay of funds made by a company to purchase fixed assets such as property, a factory, or equipment.

Capital project Synonymous with capital investment.

Card not present transaction A credit card transaction in which the card holder is not at the merchant's location and the merchant does not see the card. Includes mail order, online, and telephone sales.

Card verification number (CVN, card code verification (CCV), card verification data (CVD), card verification value (CVV or CV2), card verification code (CVC), card verification value code (CVVC), card security code (CSC), verification code (V-Code or V Code)) A three- or four-digit number that is printed on the credit card, but is not encoded in the card's magnetic strip, which establishes that the purchaser has the card (or has seen the card) and is likely not using a stolen card number.

Cascading Style Sheets (CSS) An HTML feature that allows designers to apply multiple predefined page display styles to Web pages.

Casher The participant in a phishing scam who uses the acquired information.

Catalog On electronic commerce sites, a listing of goods or services that may include photographs and descriptions, often stored in a database.

Catalog model A revenue model in which the seller establishes a brand image and then uses the strength of that image to sell through printed catalogs mailed to prospective buyers. Buyers place orders by mail or by calling the seller's toll-free telephone number.

Cause marketing An affiliate marketing program that benefits a charitable organization.

Centralized architecture A server structure that uses a few very large and fast computers.

Certification authority (CA) A company that issues digital certificates to organizations or individuals.

Challenge-response A content-filtering security technique that requires an unknown sender to reply to a challenge presented in an e-mail. These challenges are designed so that a human can respond easily, but a computer would have difficulty formulating the response.

Change management The process of helping employees cope with changes in the workplace.

Channel conflict The problem that arises when a company's sales in one sales outlet interfere with its sales in another sales outlet; for example, when sales through the company's Web site interfere with sales in that company's retail store.

Channel cooperation A strategy that coordinates sales and credit among various sales outlets, including online, catalog, and brick-and-mortar sales.

Charge card A payment card with no preset spending limit. The entire amount charged to the card must be paid in full each month.

Chargeback The process in which a merchant bank retrieves the money it placed in a merchant account as a result of a cardholder successfully contesting a charge.

Check 21 A U.S. law that permits banks to replace the physical movement of checks with transmission of scanned images.

Chief information officer (CIO) An organization's top technology manager; responsible for overseeing all of the business's information systems and related technological elements.

Cipher text Text that is composed of a seemingly random assemblage of bits. Cipher text is what messages become after they are encrypted.

Circuit A specific route between source and destination along which data travels.

Circuit switching A way of connecting computers or other devices that uses a centrally controlled single connection. In this method, which is used by telephone companies to provide voice telephone service, the connection is made, data is transferred, and the connection is terminated.

Click Synonymous with click-through.

Clickstream Data about site visitors.

Click-through The loading of an advertiser's Web page that results from a visitor clicking an advertisement on another Web page.

Click-wrap acceptance A user's compliance with a site's EULA or its terms and conditions through clicking a button on the Web site.

Client-level filtering An e-mail content filtering technique in which the filtering software is placed on the individual user's computer.

Client/server architecture A combination of client computers running Web client software and server computers running Web server software.

Client-side digital wallet An electronic or digital wallet that stores a consumer's information on the consumer's own computer.

Client-side scripting The generation of active content through software on the browser.

Closed architecture The use of proprietary communication protocols by computer manufacturers in the early days of computing, preventing computers made by different manufacturers from being connected to each other. Also called proprietary architecture.

Closed loop system A payment card arrangement involving a consumer, a merchant, and a payment card company (such as American Express or Discover) that processes transactions between the consumer and merchant without involving banks.

Closing tag The second half of a two-sided HTML tag; it is identified by a slash (/) that precedes the tag's name.

Cloud computing The practice of replacing a company's investment in computing equipment by selling Internet-based access to its own computing hardware and software.

ColdFusion Adobe's server-side dynamic page-generation technology.

Collector In a phishing attack, the computer that collects data from the potential victim.

Collision The occurrence of two messages resulting in the same hash value; the probability of this happening is extremely small.

Co-location (collocation, colocation) An Internet service arrangement in which the service provider rents a physical space to the client to install its own server hardware.

Colon hexadecimal (colon hex) The shorthand notation system used for expressing IPv6 addresses that uses eight groups of 16 bits ($8 \times 16 = 128$). Each group is expressed as four hexadecimal digits and the groups are separated by colons.

Commerce service provider (CSP) A Web host service that also provides commerce hosting services on its computer.

Commodity item A product or service that has become so standardized and well-known that buyers cannot detect a difference in the offerings of various sellers; buyers usually base their purchase decisions for such products and services solely on price.

Common Gateway Interface (CGI) A standard way of interfacing external applications with Web servers.

Common law The part of English and U.S. law that is established by the history of law.

Communication modes Ways of identifying and reaching customers.

Company A business engaged in commerce; synonymous with firm.

Component outsourcing Synonymous with partial outsourcing; the outsourcing of the design, development, implementation, or operation of specific portions of an electronic commerce system.

Component-based application system A business logic approach that separates presentation logic from business logic.

Computer forensics The field responsible for the collection, preservation, and analysis of computer-related evidence to be used in legal proceedings.

Computer forensics expert An individual hired to access client computers to locate information that can be used in legal proceedings.

Computer network Any technology that allows people to connect computers to each other.

Computer security The protection of computer resources from various types of threats.

Computer virus Synonymous with virus, which is software that attaches itself to another program and can cause damage when the host program is activated.

Configuration table Information about connections that lead to particular groups of routers, specifications on which connections to use first, and rules for handling instances of heavy packet traffic and network congestion.

Conflict of laws A situation in which federal, state, and local laws address the same issues in different ways.

Consideration The bargained-for exchange of something valuable, such as money, property, or future services.

Constructive notice The idea that citizens should know that when they leave one area and enter another, they become subject to the laws of the new area.

Consumer-to-business An industry term for electronic commerce that occurs in general consumer auctions; bidders at a general consumer auction might be businesses.

Consumer-to-consumer (C2C) A category of electronic commerce that includes individuals who buy and sell items among themselves.

Content creator A person who writes original content for a Web site.

Content editor A person who purchases and adapts existing material for use on a Web site.

Content management software Software used by companies to control the large amounts of text, graphics, and media files used in business.

Content manager Synonymous with content editor.

Contextual advertising An advertising technique in which ads are placed in proximity to related content.

Contract An agreement between two or more legal entities that provides for an exchange of value between or among them.

Contract purchasing Direct materials purchasing in which the company negotiates long-term contracts for most of the materials that it will need. Also called replenishment purchasing.

Conversion The transition of a first-time visitor to a customer.

Conversion cost The total amount of money that a site spends, on average, to induce one visitor to make a purchase, sign up for a subscription, or (on an advertising-supported site) register.

Conversion rate Used in advertising to calculate the percentage of recipients that respond to an ad or promotion.

Cookies Bits of information about Web site visitors created by Web sites and stored on client computers.

Copy control An electronic mechanism for providing a fixed upper limit to the number of copies that one can make of a digital work.

Copyright A legal protection of intellectual property.

Cost per thousand (CPM) An advertising pricing metric that equals the dollar amount paid to reach 1000 people in an estimated audience.

Countermeasure A physical or logical procedure that recognizes, reduces, or eliminates a threat.

Cracker A technologically skilled person who uses his or her skills to obtain unauthorized entry into computers or network systems, usually with the intent of stealing information or damaging the information, the system's software, or the system's hardware.

Crawler Synonymous with spider, which is the first part of a search engine, which automatically and frequently searches the Web to find pages and updates its database of information about old Web sites.

Credit card A payment card that has a spending limit based on the cardholder's credit limit. A minimum monthly payment must be made against the balance on the card, and interest is charged on the unpaid balance.

Credit card associations Member-run organizations that issue credit cards to individual consumers. Also called customer issuing banks.

Cryptography The science that studies encryption, which is the hiding of messages so that only the sender and receiver can read them.

Culture The combination of language and customs that are unique to a particular population.

Customer-centric The Web site development approach of putting the customer at the center of all site designs.

Customer issuing banks (issuing banks) Member-run organizations that issue credit cards to individual consumers. Also called credit card associations.

Customer life cycle The five stages of customer loyalty.

Customer portal A corporate Web site designed to meet the needs of customers by offering additional services such as private stores, part number cross-referencing, product-use guidelines, and safety information.

Customer relationship management (CRM) Synonymous with technology-enabled relationship management, it is the obtaining and use of detailed customer information.

Customer relationship management (CRM) software Software that collects data on customer activities; this data is then used by managers to conduct analytical activities.

Customer service The people within an electronic commerce team who are responsible for managing customer relationships in the electronic commerce operation.

Customer value The cost that a customer pays for a product, minus the benefits the customer gains from the product.

Customs broker A company that arranges the payment of tariffs and compliance with customs laws for international shipments.

Customs duty (duty) A tax levied on a product as it enters a country.

Cyberbullying Threats, sexual remarks, or pejorative comments transmitted on the Internet or posted on Web sites.

Cybersquatting The practice of registering a domain name that is the trademark of another person or company with the hope that the trademark owner will pay huge amounts of money for the domain rights.

Cybervandalism The electronic defacing of an existing Web site page.

Data Encryption Standard (DES) An encryption standard adopted by the U.S. government for encrypting sensitive information.

Data-grade lines The quality of telephone wiring in most urban and suburban areas; made more carefully of higher grade copper than voice-grade lines so they can better carry data.

Data mining A technique that examines stored information and looks for patterns in the data that are not yet known or suspected. Also called analytical processing.

Data warehouse In a CRM system, the database containing information about customers, their preferences, and their behavior.

Database The storage element of a search engine.

Database administration The person or team that is responsible for defining the data elements in an organization's database design and the operation of its database management software.

Database manager (database management software) Software that stores information in a highly structured way.

Database server The server computer on which database management software runs.

Dead link A Web link that when clicked displays an error message instead of a Web page.

Debit card A payment card that removes the amount of the charge from the cardholder's bank account and transfers it to the seller's bank account.

Decentralized architecture A server structure that uses a large number of less-powerful computers and divides the workload among them.

Decrypted Information that has been decoded. The opposite of encrypted.

Decryption program A procedure to reverse the encryption process, resulting in the decoding of an encrypted message.

Dedicated hosting A Web hosting option in which the hosting company provides exclusive use of a specific server computer that is owned and administered by the hosting company.

Deep Web Information that is stored in databases and is accessible to users through Web interfaces.

Defamatory A statement that is false and injures the reputation of a person or company.

Delay attack A computer attack that disrupts normal computer processing.

Demographic information Characteristics that marketers use to group visitors, including address, age, gender, income level, type of job held, hobbies, and religion.

Demographic segmentation The grouping of customers by characteristics such as age, gender, family size, income, education, religion, or ethnicity.

Denial-of-service (DoS) attack (denial attack) A computer attack that disrupts normal computer processing or denies processing entirely.

Descending-price auction Synonymous with Dutch auction, which is an open auction in which bidding starts at a high price and drops until a bidder accepts the price.

Dictionary attack program A program that cycles through an electronic dictionary, trying every word in the book as a password.

Digital certificate (digital ID) An attachment to an e-mail message or data embedded in a Web page that verifies the identity of a sender or Web site.

Digital content revenue model A revenue model in which a business sells subscriptions for access to the information it owns.

Digital ID See digital certificate.

Digital rights management Software that limits the number of copies that can be made of an audio file.

Digital signature An encryption message digest.

Digital Subscriber Line (DSL) Telephone-line ISP connectivity that is a higher grade than standard 56K connectivity.

Digital wallet (electronic wallet, e-wallet) A software utility that holds credit card information, owner identification and address information, and provides this data automatically at electronic commerce sites; electronic wallets can also store electronic cash.

Digital watermark A digital code or stream embedded undetectably in a digital image or audio file.

Direct connection EDI The form of EDI in which EDI translator computers at each company are linked directly to each other through modems and dial-up telephone lines or leased lines.

Direct materials Materials that become part of the finished product in a manufacturing process.

Disintermediation The removal of an intermediary from a value chain.

Distributed architecture Synonymous with decentralized architecture, which is a server structure that uses a large number of less-powerful computers and divides the workload among them.

Distributed database system A database within a large information system that stores the same data in many different physical locations.

Distributed denial-of-service (DDoS) attack A simultaneous attack on a Web site (or a number of Web sites) from all of the computers in a botnet.

Distributed information system A large information system that stores the same data in many different physical locations.

Distribution (place) The need to have products or services available in many different locations.

Domain name The address of a Web page, it can contain two or more word groups separated by periods. Components of domain names become more specific from right to left.

Domain name hosting A service that permits the purchaser of a domain name to maintain a simple Web site so that the domain name remains in use.

Domain name ownership change The changing of owner information maintained by a public domain registrar in the registrar's database to reflect the new owner's name and business address.

Domain name parking Synonymous with domain name hosting, which is a service that permits the purchaser of a domain name to maintain a simple Web site so that the domain name remains in use.

Domain name server (DNS) A computer on the Internet that maintains directories that link domain names to IP addresses.

Dot-com A company that operates only online.

Dotted decimal The IP address notation in which addresses appear as four separate numbers separated by periods.

Double auction A type of auction in which buyers and sellers each submit combined price-quantity bids to an auctioneer. The auctioneer matches the sellers' offers (starting with the lowest price, then going up) to the buyers' offers (starting with the highest price, then going down) until all of the quantities are sold.

Double-spending The spending of the same unit of electronic cash twice by submitting the same electronic currency to two different vendors.

Download To receive a file from another computer.

Downstream bandwidth (downlink bandwidth) The connection that occurs when information travels to your computer from your ISP.

Downstream strategies Tactics that improve the value that a business provides to its customers.

Due diligence Background research procedures.

Dutch auction A form of open auction in which bidding starts at a high price and drops until a bidder accepts the price.

Dynamic catalog An area of a Web site that stores information about products in a database.

Dynamic content Nonstatic information constructed in response to a Web client's request.

Dynamic page A Web page whose content is shaped by a program in response to a user request.

Early outsourcing The hiring of an external company to do initial electronic commerce site design and development. The external team then trains the original company's information systems professionals in the new technology, eventually handing over complete responsibility of the site to the internal team.

Eavesdropper A person or device who is able to listen in on and copy Internet transmissions.

EDI compatible Firms that are able to exchange data in specific standard electronic formats with other firms.

EDI for Administration, Commerce, and Transport (EDIFACT) The 1987 publication that summarizes the United Nations' standard transaction sets for international EDI.

EDIINT (Electronic Data Interchange-Internet Integration or EDI-INT) A set of protocols for the exchange of data (EDI, XML, and other formats) over the Internet.

Effect The impact of an action.

E-government The use of electronic commerce by governments and government agencies to perform business-like activities.

Electronic business (e-business) Another term for electronic commerce; sometimes used as a broader term for electronic commerce that includes all business processes, as distinguished from a narrow definition of electronic commerce that includes sales and purchase transactions only.

Electronic cash A form of electronic payment that is anonymous and can be spent only once.

Electronic commerce (e-commerce) Business activities conducted using electronic data transmission over the Internet and the World Wide Web.

Electronic customer relationship management (eCRM) Synonymous with technology-enabled relationship management, it is the obtaining and use of detailed customer information.

Electronic data interchange (EDI) Exchange between businesses of computer-readable data in a standard format.

Electronic funds transfer (EFT) Electronic transfer of account exchange information over secure private communications networks.

Electronic funds transfer at point of sale (EFTPOS) cards Another term for debit cards, usually used outside the United States.

Electronic mail (e-mail) Messages that are exchanged among users using particular mail programs and protocols.

Electronic wallet (e-wallet, digital wallet) A software utility that holds credit card information, owner identification, and address information, and provides this data automatically at electronic commerce sites; electronic wallets can also store electronic cash.

E-mail client software Programs used to read and send e-mail.

E-mail server A computer that is devoted to handling e-mail.

EMV standard A single standard for the handling of payment card transactions developed cooperatively by Visa, MasterCard, and MasterCard Europe.

Encapsulation The process that occurs when VPN software encrypts packet contents and then places the encrypted packets inside an IP wrapper in another packet.

Encryption The coding of information using a mathematical-based program and secret key; it makes a message illegible to casual observers or those without the decoding key.

Encryption algorithm The logic that implements an encryption program.

Encryption program A program that transforms plain text into cipher text.

End-user license agreement A contract that the user must accept before installing software.

English auction A type of auction in which bidders publicly announce their successively higher bids until no higher bid is forthcoming.

Enterprise application integration The coordination of all of a company's existing systems to each other and to the company's Web site.

Enterprise-class software Commerce software used by large-scale electronic commerce businesses.

Enterprise resource planning (ERP) Business software that integrates all facets of a business, including planning, manufacturing, sales, and marketing.

Entity body The part of a message from a client that contains the HTML page requested by the client and passes bulk information to the server.

E-procurement The use of Internet technologies in a company's purchasing and supply management functions.

E-procurement software Software that allows a company to manage its purchasing function through a Web interface.

Escrow service An independent third party who holds an auction buyer's payment until the buyer receives the purchased item and is satisfied that it is what the seller represented it to be.

E-sourcing The use of Internet technologies in the activities a company undertakes to identify vendors that offer materials, supplies, and services that the company needs.

Ethical hacker A computer security specialist hired to probe computers and computer networks to assess their security; can also be hired to locate information that can be used in legal proceedings.

Extensible Hypertext Markup Language (XHTML) A new markup language proposed by the WC3 that is a reformulation of HTML version 4.0 as an XML application.

Extensible Markup Language (XML) A language that describes the semantics of a page's contents and defines data records on a page.

Extensible Stylesheet Language (XSL) A language that formats XML code for viewing in a Web browser.

Extranet A network system that extends a company's intranet and allows it to connect with the networks of business partners or other designated associates.

Fair use The approved limited use of copyright material when certain conditions are met.

False positive An e-mail message that is incorrectly rejected by an e-mail filter as being spam when it is actually valid e-mail.

Fan Someone who follows a company's discussion activity on a social media site.

Fan base A collection of fans.

Fast venturing The joining of an existing company that wants to launch an electronic commerce initiative with external equity partners and operational partners who provide the experience and skills needed to develop and scale up the project very rapidly.

Fee-for-service revenue model A revenue model in which payment is based on the value of the service provided.

Fee-for-transaction revenue model A revenue model in which businesses charge a fee for services based on the number or size of the transactions they process.

File Transfer Protocol (FTP) A protocol that enables users to transfer files over the Internet.

Finger An Internet utility program that runs on UNIX computers and allows a user to obtain limited information about other network users.

Firewall A computer that provides a defense between one network (inside the firewall) and another network (outside the firewall, such as the Internet) that could pose a threat to the inside network. All traffic to and from the network must pass through the firewall. Only authorized traffic, as defined by the local security policy, is allowed to pass through the firewall. Also used to describe the software that performs these functions on the firewall computer.

Firm A business engaged in commerce.

First-mover advantage The benefit a company can gain by introducing a product or service before its competitors.

First-party cookie A cookie that is placed on the client computer by the Web server site.

First-price sealed-bid auction A type of auction in which bidders submit their bids independently and privately, with the highest bidder winning the auction.

Fixed-point wireless A data transmittal service that uses a system of repeaters to forward a radio signal from an ISP to customers.

Forum selection clause A statement within a contract that dictates that the contract will be enforced according to the laws of a particular state; signing a contract with a forum selection clause constitutes voluntary submission to the jurisdiction named in the forum selection clause.

Four Ps of marketing The essential issues of marketing: product, price, promotion, and place.

Fourth-generation (4G) wireless technology Wireless technology that offers download speeds up to 12 Mbps and upload speeds up to 5 Mbps.

Fractional T1 High-bandwidth telephone company connections that operate at speeds between 128 Kbps and 1.5 Mbps in 128-Kbps increments.

Frame relay A routing technology.

Freight forwarder A company that arranges shipping and insurance for international transactions.

Front-end processor A banking service provider that obtains authorization for a transaction by sending the transaction's details to the interchange network and storing a record of the approval or denial.

Full-privilege FTP A protocol that allows users to upload files to and download files from a remote computer using FTP.

Funnel model of customer acquisition, conversion, and retention A method of evaluating specific marketing strategy elements.

Gateway computers Synonymous with routers, which are computers that determine the best way for data packets to move forward.

Gateway server A firewall that filters traffic based on applications requested by clients on the trusted network.

Generalized Markup Language (GML) An early markup language resulting from efforts to create standard formatting styles for electronic documents.

Generic top-level domain (gTLD) The main top-level domain names, including .com, .net, .edu, .gov, .mil, .us, and .org.

Geographic segmentation The grouping of customers by location of home or workplace.

Gift card A prepaid card sold to be given as a gift.

Graphical user interface (GUI) Computer program control functions that are displayed using pictures, icons, and other easy-to-use graphical elements.

Green computing The reduction of the environmental impact of large computing installations.

Group purchasing site (group shopping site) A type of auction Web site that negotiates with a seller to obtain lower prices on an item as individual buyers enter bids on that item.

Hacker A dedicated programmer who writes complex code that tests the limits of technology; usually meant in a positive way.

Hash algorithm A security utility that mathematically combines every character in a message to create a fixed-length number (usually 128 bits in length) that is a condensation, or fingerprint, of the original message.

Hash coding The process used to calculate a number from a message.

Hash value The number that results when a message is hash coded.

Hexadecimal (base 16) A number system that uses 16 digits.

Hierarchical business organization Firms that include a number of levels with cumulative responsibility. These organizations are typically headed by a top-level president or officer. A number of vice presidents report to the president. A larger number of middle managers report to the vice presidents.

Hierarchical hyperlink structure A hyperlink structure in which the user starts from a home page and follows links to other pages in whatever order they wish.

High-speed DSL (HDSL) An Internet connection service that provides 768 Kbps of symmetric bandwidth.

Home page In a hierarchical Web page structure, the introductory page of a Web site. Synonymous with start page.

Hot spot A wireless access point (WAP) that is open to the public.

HTML extensions Developer-created Web page features that only work in certain browsers.

Hyperlink A type of tag that points to another location within the same or another HTML document. Also called a hypertext link.

Hypertext A system of navigating between HTML pages using links.

Hypertext elements HTML text elements that are related to each other within one document or among several documents.

Hypertext link (hyperlink) A type of tag that points to another location within the same or another HTML document.

Hypertext Markup Language (HTML) The language of the Internet; it contains codes attached to text that describe text elements and their relation to one another.

Hypertext Preprocessor (PHP) A Web programming language that can be used to write server-side scripts that generate dynamic Web pages.

Hypertext server Synonymous with Web server, which is a computer that is connected to the Internet and that stores files written in HTML that are publicly available through an Internet connection.

Hypertext Transfer Protocol (HTTP) The Internet protocol responsible for transferring and displaying Web pages.

Idea-based networking The act of participating in Web communities that are based on the connections between ideas.

Idea-based virtual community A Web community based on the connections between ideas.

Identity theft A criminal act in which the perpetrator gathers personal information about a victim and then uses that information to obtain credit in the victim's name. After establishing credit accounts, the perpetrator runs up charges on the accounts and then disappears.

IEEE An organization that creates wireless networking specifications; originally named the Institute of Electrical and Electronic Engineers.

Impact sourcing Offshoring that is done to benefit training or charitable activities in less-developed parts of the world. Also called smart sourcing.

Implied contract An agreement between two or more parties to act as if a contract exists, even if no contract has been written and signed.

Implied warranty A promise to which the seller can be held even though the seller did not make an explicit statement of that promise.

Impression The loading of a banner ad on a Web page.

Income tax Taxes that are levied by national, state, and local governments on the net income generated by business activities.

Incubator A company that offers start-up businesses a physical location with offices, accounting and legal assistance, computers, and Internet connections at a very low monthly cost.

Independent exchange A vertical portal that is not controlled by a company that was an established buyer or seller in the industry.

Independent industry marketplace A vertical portal that is focused on a specific industry.

Index A list containing every Web page found by a spider, crawler, or bot.

Indirect connection EDI The form of EDI in which each company transmits and receives EDI messages through a value-added network.

Indirect materials Materials and supplies that are purchased by a company in support of the manufacturing of an item, but not directly used in the production of the item.

Industry Multiple firms selling similar products to similar customers.

Industry consortia-sponsored marketplace A marketplace formed by several large buyers in a particular industry.

Industry marketplace A vertical portal that is focused on a single industry.

Industry value chain The larger stream of activities in which a particular business unit's value chain is embedded.

Initial public offering (IPO) The original sale of a company's stock to the public.

Inline text ad A text ad consisting of text in an article or story that is displayed as a hyperlink and that leads to an advertiser's Web site.

Integrated Services Digital Network (ISDN) High-grade telephone service that uses the DSL protocol and offers bandwidths of up to 128 Kbps.

Integrity The category of computer security that addresses the validity of data; confirmation that data has not been modified.

Integrity violation A security violation that occurs whenever a message is altered while in transit between sender and receiver.

Intellectual capital The value of the accumulated mass of employees' knowledge about a business and its processes.

Intellectual property A general term that includes all products of the human mind, including tangible and intangible products.

Intentional tort A tortious act in which the seller knowingly or recklessly causes injury to the buyer.

Interactive Mail Access Protocol (IMAP) A newer e-mail protocol with improvements over POP.

Interactive marketing unit (IMU) ad format The standard banner sizes that most Web sites have voluntarily agreed to use.

Interchange fees Fees charged by a card association to an acquiring bank that are usually passed to the merchant.

Interchange network A set of connections between banks that issue credit cards, the associations that own the credit cards (such as MasterCard or Visa), and merchants' banks.

Internet, internet A global system of interconnected computer networks. An internet (small "i") is a group of computer networks that have been interconnected.

Internet access provider (IAP) Synonymous with Internet service provider.

Internet backbone Routers that handle packet traffic along the Internet's main connecting points.

Internet EDI EDI on the Internet.

Internet host A computer that is directly connected to the Internet.

Internet Protocol (IP) Within TCP/IP, the protocol that determines the routing of data packets. See TCP/IP.

Internet Protocol version 4 (IPv4) The version of IP that has been in use for the past 20 years on the Internet; it uses a 32-bit number to identify the computers connected to the Internet.

Internet Protocol version 6 (IPv6) The protocol that will replace IPv4.

Internet service provider (ISP) A company that sells Internet access rights directly to Internet users.

Internet2 A successor to the Internet used for conducting research; it offers bandwidths in excess of 1 Gbps.

Interoperability The coordination of a company's information systems so that they all work together.

Interoperable software Software that runs transparently on a variety of hardware and software configurations.

Interstitial ad An intrusive Web ad that opens in its own browser window, instead of the page that the user intended to load.

Intranet An interconnected network of computers operated within a single company or organization.

Intrusion detection system A part of a firewall that monitors attempts to log in to servers and analyzes those attempts for patterns that might indicate a cracker's attack is under way.

IP address The 32-bit number that represents the address of a particular location (computer) on the Internet.

IP tunneling The creation of a private passageway through the public Internet that provides secure transmission from one extranet partner to another.

IP wrapper The outer packet in the encapsulation process.

Jailbreaking Modifying an Apple iPhone's operating system.

Java sandbox A Web browser security feature that limits the actions that can be performed by a Java applet that has been downloaded from the Web.

Java servlet An application that runs on a Web server and generates dynamic content.

JavaScript A scripting language developed by Netscape to enable Web page designers to build active content.

JavaServer pages (JSP) A server-side scripting program developed by Sun Microsystems.

Judicial comity An accommodation by a court in one country in which it voluntarily enforces another country's laws or court judgments when no strict requirement to do so exists.

Jurisdiction A government's ability to exert control over a person or corporation.

Key A number used to encode or decode messages.

Knowledge management The intentional collection, classification, and dissemination of information about a company, its products, and its processes.

Knowledge management (KM) software Software that helps companies collect and organize information, share the information among users, enhance the ability of users to collaborate, and preserve the knowledge gained for future use.

Late outsourcing The hiring of an external company to maintain an electronic commerce site that has been designed and developed by an internal information systems team.

Law of diminishing returns The characteristic of most activities to yield less value as the amount of consumption increases.

Leaderboard ad Web site banner ad that is designed to span the top or bottom of a Web page.

Leased line A permanent telephone connection between two points; it is always active.

Legitimacy The idea that those subject to laws should have some role in formulating them.

Life-cycle segmentation The use of customer life cycle stages to identify groups of customers that are in each stage.

Linear hyperlink structure A hyperlink structure that resembles conventional paper documents in which the user reads pages in serial order.

Link checker A site management tool that examines each page on the site and reports any URLs that are broken, that seem to be broken, or that are in some way incorrect.

Link rot The undesirable situation of a site that contains a number of links that no longer work.

Liquidation broker An agent that finds buyers for unusable and excess inventory.

Load-balancing switch A piece of network hardware that monitors the workloads of servers attached to it and assigns incoming Web traffic to the server that has the most available capacity at that instant in time.

Local area network (LAN) A network that connects workstations and PCs within a single physical location.

Localization A type of language translation that considers multiple elements of the local environment, such as business and cultural practices, in addition to local dialect variations in the language.

Localized advertising Online advertising in which ads are generated in response to a search for products or services in a specific geographic area.

Lock-in effect The inherent greater value to customers of existing companies than new sites.

Log file A collection of data that shows information about Web site visitors' access habits.

Logical security The protection of assets using nonphysical means.

Long-arm statute A state law that creates personal jurisdiction for courts.

Long Term Evolution (LTE) A 4G wireless technology that offers download speeds up to 12 Mbps and upload speeds up to 5 Mbps.

Machine translation Language translation that is done by software; such translation can reach speeds of 400,000 words per hour.

Macro virus A virus that is transmitted or contained inside a downloaded file attachment; it can cause damage to a computer and reveal otherwise confidential information.

Mail bomb A security attack in which many computers (hundreds or thousands) each send a message to a particular address, exceeding the recipient's allowable mail limit and causing mail systems to malfunction; the computers are often under the surreptitious control of a third party.

Mail-order model Synonymous with catalog model.

Mailing list An e-mail address that forwards messages to certain users who are subscribers.

Maintenance, repair, and operating (MRO) Commodity supplies, including general industrial merchandise and standard machine tools, that are used in a variety of industries.

Mall-style commerce service provider A CSP that provides small businesses with an Internet connection, Web site creation tools, and little or no banner advertising clutter.

Managed service provider (MSP) A Web site hosting service firm; synonymous with ASP and CSP.

Man-in-the-middle exploit A message integrity violation in which the contents of the e-mail are changed in a way that negates the message's original meaning.

Many-to-many communications A model of communications in which a number of entities communicate with a number of other entities.

Many-to-one communications model A model of communications in which a number of entities communicate with a single other entity.

Market A real or virtual space in which potential buyers and sellers come into contact with each other and agree on a medium of exchange (such as currency or barter).

Market segmentation The identification by advertisers of specific subsets of their markets that have common characteristics.

Marketing channel Each different pathway that a business uses to reach its customers.

Marketing mix The combination of elements that companies use to achieve their goals for selling and promoting their products and services.

Marketing strategy A particular marketing mix that is used to promote a company or product.

Marketspace A market that occurs in the virtual world instead of in the physical world.

Markup tags (tags) Web page code that provides formatting instructions that Web client software can understand.

Masquerading Pretending to be someone you are not (for example, by sending an e-mail that shows someone else as the sender) or representing a Web site as an original when it is an imposter. Synonymous with spoofing.

Mass media The method of contacting potential customers through the distribution of broadcast, printed, billboard, or mailed advertising materials.

Meetup An in-person meeting between people who are acquainted through a blog.

Merchandising The combination of store design, layout, and product display intended to create an environment that encourages customers to buy.

Merchant account An account that a merchant must hold with a bank that allows the merchant to process payment card transactions.

Mesh routing A version of fixed-point wireless that directly transmits Wi-Fi packets through hundreds of short-range transceivers that are located close to each other.

Message digest The number that results from the application of an encryption algorithm to plain text information.

Metalanguage A language that comprises a set of language elements and can be used to define other languages.

Metrics Measurements that companies use to assess the value of site visitor activity.

Microblog A Web site such as Twitter that functions as a very informal blog site with entries (messages, or tweets) that are limited to 140 characters in length.

Microlending The practice of lending very small amounts of money to people who are starting or operating small businesses, especially in developing countries.

Micromarketing The practice of targeting very small and well-defined market segments.

Micropayments Internet payments for items costing very little—usually $1 or less.

Middleware Software that handles connections between electronic commerce software and accounting systems.

Minimum bid In an English auction, the price for an item at which the auctioning begins.

Minimum bid increment The amount by which one bid must exceed the previous bid.

Mobile ads Advertising messages that appear as part of mobile apps.

Mobile apps Programs that run on wireless devices such as smartphones and tablets.

Mobile commerce (m-commerce) Resources accessed using devices that have wireless connections, such as stock quotes, directions, weather forecasts, and airline flight schedules.

Mobile wallet A mobile phone that operates as a credit card.

Monetizing The conversion of existing regular site visitors seeking free information or services into fee-paying subscribers or purchasers of services.

Money laundering A technique used by criminals to convert money that they have obtained illegally into cash that they can spend without having it identified as the proceeds of an illegal activity.

Multipurpose Internet Mail Extension (MIME) An e-mail protocol that allows users to attach binary files to e-mail messages.

Multivector virus A virus that can enter a computer system in several different ways.

Näive Bayesian filter E-mail filtering software that classifies messages based on learned patterns indicated by the e-mail user's categorization of incoming mail. The filter eventually learns to recognize spam and filter it out.

Name changing (typosquatting) A problem that occurs when someone registers purposely misspelled variations of well-known domain names. These variants sometimes lure consumers who make typographical errors when entering a URL.

Name stealing Theft of a Web site's name that occurs when someone, posing as a site's administrator, changes the ownership of the domain name assigned to the site to another site and owner.

Near field communication (NFC) Contactless wireless transmission of data over short distances.

Necessity The category of computer security that addresses data delay or data denial threats.

Necessity threat The disruption of normal computer processing or denial of processing. Also called delay, denial, or denial-of-service threat (DoS).

Negligent tort A tortious act in which the seller unintentionally provides a harmful product.

Net bandwidth The actual speed information travels, taking into account traffic on the communication channel at any given time.

Netbook A small notebook computer with wireless connectivity but with less computing functionality than a full-featured notebook.

Network access points (NAPs) The four primary connection points for access to the Internet backbone in the United States.

Network access providers The few large companies that are the primary providers of Internet access; they, in turn, sell Internet access to smaller Internet service providers.

Network Address Translation (NAT) device A computer that converts private IP addresses into normal IP addresses when they forward packets to the Internet.

Network Control Protocol (NCP) Used by ARPANET in the early 1970s to route messages in its experimental wide area network.

Network economic structure A business structure wherein firms coordinate their strategies, resources, and skill sets by forming a long-term, stable relationship based on a shared purpose.

Network effect An increase in the value of a network to its participants, which occurs as more people or organizations participate in the network.

Network operations Web site staff whose responsibilities include load estimation and monitoring, resolving network problems as they arise, designing and implementing fault-resistance technologies, and managing any network operations that are outsourced to ISPs, CSPs, or telephone companies.

Network specification The set of rules that equipment connected to a network must follow.

Newsgroup A topic area in Usenet where people read and post articles.

Nexus The association between a tax-paying entity and a governmental taxing authority.

Nigerian scam (419 scam) A scam in which the victim receives an e-mail from a Nigerian government official requesting assistance in moving money to a foreign bank account.

Nonrepudiation Verification that a particular transaction actually occurred; this prevents parties from denying a transaction's validity or its existence.

Notice The expression of a change in rules (usually, legal or cultural rules) typically represented by a physical boundary.

N-tier architecture Higher-order client-server architectures that have more than three tiers.

Occasion segmentation Behavioral segmentation that is based on things that happen at a specific time or occasion.

Octet An 8-bit number.

Offer A declaration of willingness to buy or sell a product or service; it includes sufficient details to be firm, precise, and unambiguous.

Offshoring Outsourcing that is done by organizations outside the country.

One-to-many communication model A model of communications in which one entity communicates with a number of other entities.

One-to-one communication model A model of communications in which one entity communicates with one other entity.

One-to-one marketing A highly customized approach to offering products and services that match the needs of a particular customer.

Online community Synonymous with virtual community, which is an electronic gathering place for people with common interests.

Online marketing manager An employee who specializes in the specific techniques used to build brands and increase market share using the Web site and other online tools, such as e-mail marketing.

Ontology A set of standards that defines, in detail, the structure of a particular knowledge domain; in the Semantic Web, it defines the relationships among RDF standards and specific XML tags.

Open architecture The philosophy behind the Internet that dictates that independent networks should not require any internal changes to be connected to the network, packets that do not arrive at their destinations must be retransmitted from their source network, routers do not retain information about the packets they handle, and no global control exists over the network.

Open auction (open-outcry auction) An auction in which bids are publicly announced (such as an English auction).

Open EDI EDI conducted on the Internet instead of over private leased lines.

Open loop system A payment card arrangement involving a consumer and his or her bank, a merchant and its bank, and a third party (such as Visa or MasterCard) that processes transactions between the consumer and merchant.

Open-outcry double auction A double auction in which buy and sell offers are announced publicly. Typically conducted in exchange floor or trading pit environments for items of known quality, such as securities or graded agricultural products, that are regularly traded in large quantities.

Open session A continuous connection that is maintained between a client and server on the Internet.

Open-source software Software that is developed by a community of programmers who make the software available for download and use at no cost.

Opening tag An HTML tag that precedes the text that a tag affects.

Opportunity cost Lost benefits from an action not taken.

Optical fiber A data transmission cable that uses glass fibers to achieve bandwidths up to 10 Gbps.

Opt-in A personal information collection policy in which the company collecting the information does not use the information for any other purpose (or sell or rent the information) unless the customer specifically chooses to allow that use.

Opt-in e-mail The practice of sending e-mail messages to people who have requested information on a particular topic or about a specific product.

Opt-out A personal information collection policy in which the company collecting the information assumes that the customer does not object to the company's use of the information unless the customer specifically chooses to deny permission.

Organized crime Unlawful activities conducted by a highly organized, disciplined association for profit. Also called racketeering.

Orphan file A file on a Web site that is not linked to any page.

Outsourcing The hiring of another company to perform design, implementation, or operational tasks for an information systems project.

Packet-filter firewall A firewall that examines all data flowing back and forth between a trusted network and the Internet.

Packets The small pieces of files and e-mail messages that travel over the Internet.

Packet-switched A network in which packets are labeled electronically with their origin, sequence, and destination addresses. Packets travel from computer to computer along the interconnected networks until they reach

their destination. Each packet can take a different path through the interconnected networks, and the packets may arrive out of order. The destination computer collects the packets and reassembles the original file or e-mail message from the pieces in each packet.

Page view A page request made by a Web site visitor.

Page-based application system Application server software that returns pages generated by scripts that include the rules for presenting data on the Web page with the business logic.

Paid placement (sponsorship) The purchasing of a top listing in results listings for a particular set of search terms.

Partial outsourcing The outsourcing of the design, development, implementation, or operation of specific portions of an electronic commerce system.

Participatory journalism The practice of inviting readers to help write an online newspaper.

Patent An exclusive right to make, use, and sell an invention granted by a government to the inventor.

Payment card A general term for plastic cards used instead of cash to make purchases, including credit cards, debit cards, and charge cards.

Payment processing service provider, payment processor A third-party company that handles payment card processing for online businesses.

Pay-per-click model A revenue model in which an affiliate earns payment each time a site visitor clicks a link to load the seller's page.

Pay-per-conversion model A revenue model in which an affiliate earns payment each time a site visitor is converted from a visitor into either a qualified prospect or a customer.

Pay wall A digital control mechanism that limits the number of times a visitor may visit a site to a specific number of visits before the user must pay for continued access.

Per se defamation A legal cause of action in which a court deems some types of statements to be so negative that injury is assumed.

Perimeter expansion The increase in firewall limits beyond traditional borders caused by telecommuting.

Permission marketing A marketing strategy that only sends specific information to people who have indicated an interest in receiving information about the product or service being promoted.

Persistent cookie A cookie that exists indefinitely.

Personal area network (PAN) A small, low-bandwidth Bluetooth network of up to 10 networks of eight devices each. It is used for tasks such as wireless synchronization of laptop computers with desktop computers and wireless printing from laptops, PDAs, or mobile phones. Synonymous with piconet.

Personal contact A method of identifying and reaching customers that involves searching for, qualifying, and contacting potential customers.

Personal firewall A software-only firewall that is installed on an individual client computer.

Personal jurisdiction A court's authority to hear a case based on the residency of the defendant; a court has personal jurisdiction over a case if the defendant is a resident of the state in which the court is located.

Personal shopper An intelligent agent program that learns a customer's preferences and makes suggestions.

Pharming attack The use of a zombie farm, often by an organized crime association, to launch a massive phishing attack.

Phishing expedition A masquerading attack that combines spam with spoofing. The perpetrator sends millions of spam e-mails that appear to be from a respectable company. The e-mails contain a link to a Web page that is designed to look exactly like the company's site. The victim is encouraged to enter his or her username, password, and sometimes credit card information.

Physical security Tangible protection devices such as alarms, guards, fireproof doors, fences, and vaults.

Piconet A small, low-bandwidth Bluetooth network of up to 10 networks of eight devices each. It is used for tasks such as wireless synchronization of laptop computers with desktop computers and wireless printing from laptops, PDAs, or mobile phones. Synonymous with personal area network.

Ping (Packet Internet Groper) A program that tests the connectivity between two computers connected to the Internet.

Place (distribution) The need to have products or services available in many different locations.

Plain old telephone service (POTS) The network that connects telephones; it provides a reliable data transmission bandwidth of about 56 Kbps.

Plain text Normal, unencrypted text.

Platform neutrality The ability of a network to connect devices that use different operating systems.

Plug-in An application that helps a browser to display information (such as video or animation) but is not part of the browser.

Pop-behind ad A pop-up ad that is followed very quickly by a command that returns the focus to the original browser window, resulting in an ad that is parked behind the user's browser waiting to appear when the browser is closed.

Pop-up ad An ad that appears in its own window when the user opens or closes a Web page.

Portal A Web site that serves as a customizable home base from which users do their searching, navigating, and other Web-based activity. Synonymous with Web portal.

Post Office Protocol (POP) The protocol responsible for retrieving e-mail from a mail server.

Postimplementation audit (postaudit review) A formal review of a project after it is up and running.

Power A form of control over physical space (such as a state) and the people and objects that reside in that space.

Prepaid card A purchased card that contains a limited value and that can be used for making purchases from retailers.

Presence The public image conveyed by an organization to its stakeholders.

Pretty Good Privacy (PGP) A popular technology used to implement public-key encryption to protect the privacy of e-mail messages.

Price The amount a customer pays for a product.

Primary activities Activities that are required to do business: design, production, promotion, marketing, delivery, and support of products or services.

Privacy The protection of individual rights to nondisclosure of information.

Private company marketplace A marketplace that provides auctions, requests for quotes postings, and other features to companies that want to operate their own marketplace.

Private IP addresses A series of IP numbers that have been set aside for subnet use and are not permitted on packets that travel on the Internet.

Private key A single key that is used to encrypt and decrypt messages. Synonymous with symmetric key.

Private-key encryption The encoding of a message using a single numeric key to encode and decode data; it requires both the sender and receiver of the message to know the key, which must be guarded from public disclosure.

Private network A private, leased-line connection between two companies that physically links their individual computers or intranets.

Private store A password-protected area of a Web site that offers individual customers negotiated price reductions on a limited selection of products and other customized features.

Private valuation The amount a bidder is willing to pay for an item that is up for auction.

Procurement The business activity that includes all purchasing activities plus the monitoring of all elements of purchase transactions.

Product The physical item or service that a company is selling.

Product-based structure A business organization based on product categories.

Product disparagement A statement that is false and injures the reputation of a product or service.

Project management Formal techniques for planning and controlling activities undertaken to achieve a specific goal.

Project management software Application software that provides built-in tools for managing people, resources, and schedules.

Project manager A person with specific training or skills in tracking costs and the accomplishment of specific objectives in a project.

Project portfolio management A technique in which each project is monitored as if it were an investment in a financial portfolio.

Project portfolio manager An employee who is responsible for tracking all ongoing projects and managing them as a portfolio.

Promotion Any means of spreading the word about a product.

Property tax Taxes levied by states and local governments on the personal property and real estate used in a business.

Proprietary architecture The use of vendor-specific communication protocols by computer manufacturers in the early days of computing, preventing computers made by different manufacturers from being connected to each other. Also called closed architecture.

Prospecting The part of personal contact selling in which the salesperson identifies potential customers.

Protocol A collection of rules for formatting, ordering, and error-checking data sent across a network.

Proxy bid In an electronic auction, a predetermined maximum bid submitted by a bidder.

Proxy server firewall A firewall that communicates with the Internet on behalf of the trusted network.

Psychographic segmentation The grouping of customers by variables such as social class, personality, or their approach to life.

Public key One of a pair of mathematically related numeric keys, it is used to encrypt messages and is freely distributed to the public.

Public-key encryption The encoding of messages using two mathematically related but distinct numeric keys.

Public marketplace A vertical portal that is open to new buyers and sellers just entering an industry.

Public network An extranet that allows the public to access its intranet or when two or more companies link their intranets.

Purchasing card (p-card) Payment cards that give individual managers the ability to make multiple small purchases at their discretion while providing cost-tracking information to the procurement office.

Pure dot-com A company that operates only online; also called dot-com.

Python A scripting language that can be used in dynamic Web page generation.

Racketeering Unlawful activities conducted by a highly organized, disciplined association for profit. Also called organized crime.

Radio frequency identification device (RFID) Small chips that include radio transponders; they can be used to track inventory as it moves through an industry value chain.

Rational branding An advertising strategy that substitutes an offer to help Web users in some way in exchange for their viewing an ad.

Real-time location systems Tracking systems that use bar codes to monitor inventory movements and ensure that goods are shipped as quickly as possible.

Reintermediation The introduction of a new intermediary into a value chain.

Remote server administration Control of a Web site by an administrator from any Internet-connected computer.

Remote wipe Removing personal information from a lost or stolen mobile device by clearing all of the data stored on the device, including e-mails, text messages, contact lists, photos, videos, and any type of document file.

Repeat visits Subsequent visits a Web site visitor makes to a particular page.

Repeater A transmitter-receiver device used in a fixed-point wireless network to forward a radio signal from the ISP to customers. Synonymous with transceiver.

Replenishment purchasing Direct materials purchasing in which the company negotiates long-term contracts for most of the materials that it will need. Also called contract purchasing.

Representational State Transfer (REST) A principle that describes the way the Web uses networking architecture to identify and locate Web pages and the elements (graphics) that make up those Web pages.

Request header The part of an HTTP message from a client to a server that contains additional information about the client and more information about the request.

Request line The part of an HTTP message from a client to a server that contains a command, the name of the target resource (without the protocol or domain name), and the protocol name and version.

Request message The HTTP message that a Web client sends to request a file or files from a Web server.

Reserve price (reserve) The minimum price a seller will accept for an item sold at auction.

Resource description framework (RDF) A set of standards for XML syntax.

Response header field In a client/server transmission, the field that follows the response header line and returns information describing the server's attributes.

Response header line The part of a message from a server to a client that indicates the HTTP version used by the server, status of the response, and an explanation of the status information.

Response message The reply that a Web server sends in response to a client request.

Response time The amount of time a server requires to process one request.

RESTful applications (REST) Web services that are built on the REST model.

RESTful design The use of the REST model in building Web services.

Retained customer A customer who returns to a site one or more times after making his or her first purchase.

Retention costs The costs of inducing customers to return to a Web site and buy again.

Return on investment (ROI) A method for evaluating the potential costs and benefits of a proposed capital investment.

Revenue model The combination of strategies and techniques that a company uses to generate cash flow into the business from customers.

Reverse auction (seller-bid auction) A type of auction in which sellers bid prices for which they are willing to sell items or services.

Reverse bid The process in which an auction customer seeks products by describing an item or service in which he or she is interested, and then entertains responses from merchants who offer to supply the item at a particular price.

Reverse link checker A Web site management program that checks on sites with which a company has entered a link exchange program and ensures that link exchange partners are fulfilling their obligation to include a link back to the company's Web site.

Rich media ad A Web ad that generates graphical activity that "floats" over the Web page itself instead of opening in a separate window. Also called an active ad.

Rich media objects Programming components of attention-grabbing Web banner ads.

Right of publicity A limited right to control others' commercial use of an individual's name, image, likeness, or identifying aspect of identity.

Roaming The shifting of Wi-Fi devices from one WAP to another without requiring intervention by the user.

Robot (bot) A program that automatically searches the Web to find Web pages that might be interesting to people.

Robotic network A network that can act as an attacking unit, sending spam or launching denial-of-service attacks against specific Web sites. Synonymous with botnet or zombie farm.

Rooting Modifying an Android smartphone's operating system.

Router A computer that determines the best way for data packets to move forward to their destination.

Router computers (routing computers) The computers that decide how best to forward each packet of information as it travels on the Internet to its destination. Synonymous with gateway computers and routers.

Routing algorithm The program used by a router to determine the best path for data packets to travel.

Routing table Synonymous with configuration table, which is information about connections that lead to particular groups of routers, specifications on which connections to use first, and rules for handling instances of heavy packet traffic and network congestion.

Ruby on Rails A Web programming development framework for creating dynamic Web pages that present users with an interface similar in appearance to application software running in a Web browser.

Scalable A system's ability to be adapted to meet changing requirements.

Scripting language A programming language that provides scripts, or commands, that are executed.

Sealed-bid auction An auction in which bidders submit their bids independently and are usually prohibited from sharing information with each other.

Search engine Web software that finds other pages based on key word matching.

Search engine optimization (search engine positioning, search engine placement) The combined art and science of having a particular URL listed near the top of search engine results.

Search engine placement broker A company that aggregates inclusion and placement rights on multiple search engines and then sells those combination packages to advertisers.

Search term sponsorship The option of purchasing a top listing on results pages for a particular set of search terms. Also called paid placement or sponsorship.

Search utility The part of a search engine that finds matching Web pages for search terms.

Second-price sealed-bid auction A type of auction in which bidders submit their bids independently and privately; the highest bidder wins the auction but pays only the amount bid by the second-highest bidder.

Secrecy The category of computer security that addresses the protection of data from unauthorized disclosure and confirmation of data source authenticity.

Secure envelope A security utility that encapsulates a message and provides secrecy, integrity, and client/server authentication.

Secure Sockets Layer (SSL) A protocol for transmitting private information securely over the Internet.

Secure Sockets Layer-Extended Validation (SSL-EV) digital certificate A more secure certificate for which a certification authority must confirm the legal existence of the organization by verifying the organization's registered legal name and other facts.

Security policy A written statement describing assets to be protected, the reasons for protecting the assets, the parties responsible for protection, and acceptable and unacceptable behaviors.

Segment Also called a market segment; a subset of a company's potential customer pool that has common demographic characteristics.

Self-hosting A system of Web hosting in which the online business owns and maintains the server and all its software.

Semantic Web A project initiated by Tim Berners-Lee intended to blend technologies and information to create a next-generation Web in which words on Web pages are tagged (using XML) with their meanings.

Server A powerful computer dedicated to managing disk drives, printers, or network traffic.

Server architecture The different ways that servers can be connected to each other and to related hardware such as routers and switches.

Server farm A large collection of electronic commerce Web site servers.

Server-level filtering An e-mail content filtering technique in which the filtering software resides on the mail server.

Server software The software that a server computer uses to make files and programs available to other computers on the same network.

Server-side digital wallet An electronic or digital wallet that stores a customer's information on a remote server that belongs to a particular merchant or to the wallet's publisher.

Server-side scripting A Web page response approach in which programs running on the Web server create Web pages before sending them back to the requesting Web clients as parts of response messages.

Service mark A distinctive mark, device, motto, or implement used to identify services provided by a company.

Session cookie A cookie that exists only until you shut down your browser.

Session key A key used by an encryption algorithm to create cipher text from plain text during a single secure session.

Session negotiation When establishing S-HTTP security, the process of proposing and accepting (or rejecting) various transmission conditions.

Sexting The illegal practice of sending sexually explicit messages or photos using a mobile phone.

Shared hosting A Web hosting arrangement in which the hosting company provides Web space on a server computer that also hosts other Web sites.

Shill bidder An individual employed by a seller or auctioneer who makes bids on behalf of the seller, sometimes artificially inflating an item's price. Shill bidders may be prohibited by the rules of a particular auction.

Shipping profile The collection of attributes, including weight and size, that affect how easily a product can be packaged and delivered.

Shopping cart An electronic commerce utility that keeps track of items selected for purchase and automates the purchasing process.

Short message service (SMS) A protocol used to transmit short text messages to cell phones and other wireless devices.

Shrink-wrap acceptance A buyer's acceptance of the conditions of the EULA, demonstrated by removing the shrink wrap from the product box.

Signature Any symbol executed or adopted for the purpose of authenticating a writing.

Signed (message or code) The status of a message or Web page when it contains an attached digital certificate.

Simple Mail Transfer Protocol (SMTP) A standardized protocol used by a mail server to format and administer e-mail.

Simple Object Access Protocol (SOAP) A message-passing protocol that defines how to send marked up data from one software application to another across a network.

Single-use card A payment card with disposable numbers, which gives consumers a unique card number that is valid for one transaction only.

Site map On a hierarchically structured Web site, a page that contains a map or listing of the Web pages in their hierarchical order.

Site sponsorship The opportunity for an advertiser to sponsor part or all of a Web site to promote its products, services, or brands. Site sponsorships are more subtle than banner or pop-up ads.

Skyscraper ad A large banner ad on the side of a Web page that remains visible as the user scrolls down through the page.

Small payment Any payment of less than $10.

Smart card A plastic card with an embedded microchip that contains information about the card owner.

Smartphone A mobile phone that includes a functional Web browser and a full keyboard.

Smart sourcing Offshoring that is done to benefit training or charitable activities in less developed parts of the world. Also called impact sourcing.

Sniffer program A program that taps into the Internet and records information that passes through a router from the data's source to its destination.

Snipe The act of placing a winning bid in an online auction at the last possible moment.

Sniping software Auction software that observes auction progress until the last second or two of the auction clock, and then places a bid high enough to win the auction.

Social commerce The use of interpersonal connections online to promote or sell goods and services.

Social media Web sites that allow participants to exchange ideas and report news and information updates to each other.

Social networking administrator An employee who is responsible for managing the virtual community elements of the Web operation.

Social networking site A Web site that individuals and businesses can use to conduct social interactions online.

Social shopping The practice of bringing buyers and sellers together in a social network to facilitate retail sales.

Software agent A program that performs information gathering, information filtering, and/or mediation on behalf of a person or entity. Synonymous with intelligent software agent.

Sourcing The part of procurement devoted to identifying suppliers and determining the qualifications of those suppliers.

Spam (unsolicited commercial e-mail or bulk mail) Electronic junk mail.

Spear phishing A phishing expedition in which the e-mails are carefully designed to target a particular person or organization.

Spend The total dollar amount of the goods and services that a company buys during a year.

Spider The first part of a search engine, it automatically and frequently searches the Web to find pages and updates its database of information about old Web sites.

Sponsored top-level domain (sTLD) A top-level domain for which an organization other than ICANN is responsible.

Spoofing Synonymous with masquerading, which is pretending to be someone you are not (for example, by sending an e-mail that shows someone else as the sender) or representing a Web site as an original when it is an imposter.

Spot market A loosely organized market within a specific industry.

Spot purchasing Direct materials purchasing that occurs within a spot market.

Stakeholders The various entities involved in a business; these include customers, suppliers, employees, stockholders, neighbors, and the general public.

Standard Generalized Markup Language (SGML) An old, complex text markup language used to create frequently revised documents that need to be printed in various formats.

Start page In a hierarchical Web page structure, the introductory page of a Web site. Synonymous with home page.

Stateless connection A connection between a client and server over the Internet in which each transmission of information is independent; no continuous connection is maintained.

Static catalog A simple list of products written in HTML and displayed on a Web page or a series of Web pages.

Static page A Web page that displays unchanging information retrieved from a disk.

Statistical modeling A technique that tests theories that CRM analysts have about relationships among elements of customer and sales data.

Statute of Frauds State law that specifies that contracts for the sale of goods worth more than $500 and contracts that require actions that cannot be completed within one year must be created by a signed writing.

Statutory law That part of British and U.S. law that comprises laws passed by elected legislative bodies.

Steganography The hiding of information (such as commands) within another piece of information.

Stickiness The ability of a Web site to keep visitors at its site and to attract repeat visitors.

Sticky The condition of having stickiness.

Stockout A loss of sales suffered by a retailer when it does not have specific goods on its shelves that customers want to buy.

Store charge card (store-branded card) A charge card issued by a specific retailer.

Stored-value card Either an elaborate smart card or a simple plastic card with a magnetic strip that records currency balance, such as a prepaid phone, copy, subway, or bus card.

Strategic alliance The coordination of strategies, resources, and skill sets by companies into long-term, stable relationships with other companies and individuals based on shared purposes.

Strategic business unit (SBU) A unit within a company that is organized around a specific combination of product, distribution channel, and customer type.

Strategic partners The entities taking part in a strategic alliance.

Strategic partnership Synonymous with strategic alliance.

Streamlined Sales and Use Tax Agreement (SSUTA) An agreement between U.S. states that would simplify state sales taxes by making the various state tax codes more congruent with each other while allowing each state to set its own rates.

Style sheet A set of instructions used for Web page formatting. It is stored in a separate file and lets designers apply specific formatting styles to a page.

Subject-matter jurisdiction A court's authority to decide a dispute between entities based on the issue of dispute.

Subnetting The use of reserved private IP addresses within LANs and WANs to provide additional address space.

Sufficient jurisdiction A court's ability to hear a matter if it has both subject-matter jurisdiction and personal jurisdiction.

Supply alliances Long-term relationships among participants in the supply chain.

Supply chain The part of an industry value chain that precedes a particular strategic business unit. It includes the network of suppliers, transportation firms, and brokers that combine to provide a material or service to the strategic business unit.

Supply chain management The process of taking an active role in working with suppliers and other participants in the supply chain to improve products and processes.

Supply chain management (SCM) software Software used by companies to coordinate planning and operations with their partners in the industry supply chains of which they are members.

Supply management Synonymous with procurement, which is the business activity that includes all purchasing activities plus the monitoring of all elements of purchase transactions.

Supply web An industry value chain that includes many participants that are interconnected in a web or network configuration.

Supporting activities Secondary activities that back up primary business activities. These include human resource management, purchasing, and technology development.

SWOT analysis Evaluation of the strengths and weaknesses of a business unit, and identification of the opportunities presented by the markets of the business unit and threats posed by competitors of the business unit.

Symmetric connection An Internet connection that provides the same bandwidth in both directions.

Symmetric encryption The encryption of a message using a single numeric key to encode and decode data. Synonymous with private-key encryption.

Systems administrator A member of an electronic commerce team who understands the server hardware and software and is responsible for the system's reliable and secure operation.

T1 High-bandwidth Internet connections that operate at 1.544 Mbps.

T3 High-bandwidth Internet connections that operate at 44.736 Mbps.

Tablet device A small computing device with wireless connectivity that is larger than a mobile phone but smaller than most laptop and notebook computers.

Tags (markup tags) Web page code that provides formatting instructions that Web client software can understand.

Tariff A tax levied on products as they enter the country; also called duty or customs duty.

TCP/IP The set of protocols that provide the basis for the operation of the Internet. The TCP protocol includes rules that computers on a network use to establish and break connections. The IP protocol determines routing of data packets.

Technology-enabled customer relationship management Synonymous with technology-enabled relationship management.

Technology-enabled relationship management The business practice of obtaining detailed information about a customer's behavior, preferences, needs, and buying patterns and using that information to set prices, negotiate terms, tailor promotions, add product features, and provide other customized interactions.

Teergrubing A antispamming approach in which the receiving computer launches a return attack against the spammer, sending e-mail messages back to the computer that originated the suspected spam.

Telecommuting An employment arrangement in which the employee logs in to the company computer from an off-site location through the Internet instead of traveling to an office.

Telework Synonymous with telecommuting.

Telnet A program that allows users to log on to a computer and access its contents from a remote location.

Telnet protocol The set of rules used by Telnet programs.

Terms of service (ToS) Rules and regulations intended to limit the Web site owner's liability for what a visitor might do with information obtained from the site.

Text ad A short promotional message that does not use any graphic elements and is usually placed along the top or right side of a Web page.

Text markup language A language that specifies a set of tags that are inserted into the text.

Third-generation (3G) wireless technology Wireless mobile phone technology that offers download speeds up to 2 Mbps and upload speeds up to 800 Kbps and also uses the SMS protocol to send and receive text messages.

Third-party cookie A cookie that originates on a Web site other than the site being visited.

Third-party logistics (3PL) provider A transportation or freight company that operates all or most of a customer's material movement activities.

Threat An act or object that poses a danger to assets.

Three-tier architecture A client/server architecture that builds on the two-tier architecture by adding applications and their associated databases that supply non-HTML information to the Web server on request.

Throughput The number of HTTP requests that a particular hardware and software combination can process in a unit of time.

Tier-one suppliers The capable suppliers that work directly with and have long-term relationships with businesses.

Tier-three suppliers Suppliers that provide components and raw materials to tier-two suppliers.

Tier-two suppliers Suppliers that provide components and raw materials to tier-one suppliers.

Top-level domain (TLD) The last part of a domain name; the most general identifier in the name.

Tort An action taken by a legal entity that causes harm to another legal entity.

Total cost of ownership (TCO) Business activity costs including the costs of hiring, training, and paying the personnel who will design the Web site, write or customize the software, create the content, and operate and maintain the site. TCO also includes hardware and software costs.

Touchpoint Online and offline customer contact points.

Touchpoint consistency The provision of similar levels and quality of service in all of a company's interactions with its customers, whether those interactions occur in person, on the telephone, or online.

Tracert A route-tracing program that sends data packets to every computer on the path (Internet) between one computer and another computer and clocks the packets' round-trip times, providing an indication of the time it takes a message to travel from one computer to another and back, pinpointing any data traffic congestion, and ensuring that the remote computer is online.

Trade name The name (or a part of that name) that a business uses to identify itself.

Trademark A distinctive mark, device, motto, or implement that a company affixes to the goods it produces for identification purposes.

Trademark dilution The reduction of the distinctive quality of a trademark by alternative uses.

Trading partners Businesses that engage in EDI with one another.

Transaction An exchange of value.

Transaction costs The total of all costs incurred by a buyer and seller as they gather information and negotiate a transaction.

Transaction processing Processes that occur as part of completing a sale; these include calculation of any discounts, taxes, or shipping costs and transmission of payment data (such as a credit card number).

Transaction server The computer on which a company runs its accounting and inventory management software.

Transaction sets Formats for specific business data interchanges using EDI.

Transaction taxes Sales taxes, use taxes, excise taxes, and customs duties that are levied on the products or services that a company sells or uses.

Transceiver A transmitter-receiver device used in a fixed-point wireless network to forward a radio signal from the ISP to customers. Synonymous with repeater.

Transmission Control Protocol The protocol that includes rules that computers on a network use to establish and break connections. See TCP/IP.

Trial visit The first visit a Web site visitor makes to a particular page.

Trigger word A key word used to jog the memory of visitors and remind them of something they want to buy on the site.

Triple Data Encryption Standard (3DES) A robust version of the Data Encryption Standard used by the U.S. government that cannot be cracked even with today's supercomputers.

Trojan horse A program hidden inside another program or Web page that masks its true purpose (usually destructive).

Trusted (network) A network that is within a firewall.

Tweet A short message sent from one Twitter user to another.

Two-tier client/server architecture A client/server architecture in which only a client and server are involved in the requests and responses that flow between them over the Internet.

Typosquatting (name changing) A problem that occurs when someone registers purposely misspelled variations of well-known domain names. These variants sometimes lure consumers who make typographical errors when entering a URL.

Ultimate consumer orientation A focus on the needs of the consumer who is at the end of an industry value chain.

Ultra Wideband (UWB) A wireless communication technology that provides wide bandwidth (up to about 480 Mbps in current versions) connections over short distances (30 to 100 feet).

Uniform Resource Locator (URL) Names and abbreviations representing the IP address of a particular Web page. Contains the protocol used to access the page and the page's location. Used in place of dotted quad notations.

Universal ad package The four most common standard Web ad formats.

Universal Description, Discovery and Integration (UDDI) specification The set of protocols that identify locations of Web services and their associated WSDL descriptions.

Unsolicited commercial e-mail (UCE) Electronic junk mail that can include solicitations, advertisements, or e-mail chain letters. Also called spam or bulk mail.

Untrusted (network) A network that is outside a firewall.

Untrusted Java applet A Java applet that is not known to be secure.

Upload bandwidth Synonymous with upstream bandwidth.

Upstream bandwidth The connection that occurs when you send information from your connection to your ISP.

Upstream strategies Tactics that focus on reducing costs or generating value by working with suppliers or inbound logistics.

URL broker A business that sells or auctions domain names that it believes others will find valuable.

Usability testing The testing and evaluation of a company's Web site for ease of use by visitors.

Usage-based market segmentation Customizing visitor experiences to match the site usage behavior patterns of each visitor or type of visitor.

Use tax A tax levied by a state on property used in that state that was not purchased in that state.

Usenet (User's News Network) One of the first mailing lists; it allows subscribers to read and post articles within topic areas.

Usenet newsgroup Message posting areas on Usenet computers in which interested persons (primarily from the education and research communities) can discuss those topic areas.

Value chain A way of organizing the activities that each strategic business unit undertakes to design, produce, promote, market, deliver, and support the products or services it sells.

Value system Synonymous with industry value chain.

Value-added network (VAN) An independent company that provides connection and EDI transaction forwarding services to businesses engaged in EDI.

Venture capitalist A very wealthy individual or investment firm that invests in small companies that are about to grow rapidly. By investing large amounts of money (between a million and a few hundred million dollars), venture capitalists attempt to help these growing companies become large enough to sell stock to the public.

Vertical integration The practice of an existing firm replacing one of its suppliers with its own strategic business unit that creates the supplied product.

Vertical portal (vortal) A vertically integrated Web information hub focusing on an individual industry.

Vicarious copyright infringement The violation of an organization's rights that occurs when a company capable of supervising the infringing activity fails to do so and obtains a financial benefit from the infringing activity.

Vickrey auction Synonymous with second-price sealed-bid auction. Named for William Vickrey, who won the 1996 Nobel Prize in Economics for his studies of the properties of this auction type.

Viral marketing Tactics that rely on existing customers to tell other persons—the company's prospective customers—about the products or services they have enjoyed using.

Virtual community An electronic gathering place for people with common interests.

Virtual company A strategic alliance occurring among companies that operate on the Internet.

Virtual host Multiple servers that exist on a single computer.

Virtual learning network A virtual community used for distance learning.

Virtual model A graphic image built from customer measurements and physical traits on which customers can try clothes. Typically found on sites selling clothing and accessories.

Virtual private network (VPN) A network that uses public networks and their protocols to transmit sensitive data using a system called "tunneling" or "encapsulation."

Virtual server Synonymous with virtual host.

Virus Software that attaches itself to another program and can cause damage when the host program is activated.

Visit The request of a Web site visitor for a page from a Web site.

Voice-grade line Telephone wiring that costs less than lines designed to carry data, is made of lower-grade copper, and was never intended to carry data. These lines can only carry limited bandwidth—usually less than 14 Kbps.

Warchalking The practice of placing a chalk mark on a building that has an easily entered wireless network.

Wardrivers Network attackers who drive around in cars using their wireless-equipped laptop computers to search for unprotected wireless network access points.

Warranty disclaimer A statement indicating that the seller will not honor some or all implied warranties.

Web See World Wide Web.

Web 2.0 Technologies that include software that allow users of Web sites to participate in the creation, editing, and distribution of content on a Web site owned and operated by a third party.

Web APIs Techniques for interconnection of programs with each other over the Web.

Web browser (Web browser software) Software that lets users read HTML documents and move from one HTML document to another using hyperlinks.

Web bug A tiny, invisible Web page graphic that provides a way for a Web site to place cookies.

Web catalog revenue model A revenue model of selling goods and services on the Web wherein the seller establishes a brand image that conveys quality and uses the strength of that image to sell through catalogs mailed to prospective buyers. Buyers place orders by mail or by calling the seller's toll-free telephone number.

Web client computer A computer that is connected to the Internet and is used to download Web pages.

Web client software Software that sends requests for Web page files to other computers.

Web community Synonymous with virtual community.

Web directory A listing of hyperlinks to Web pages that is organized into hierarchical categories.

Web EDI EDI on the Internet.

Web graphics designer A person trained in art, layout, and composition who also understands how Web pages are constructed and who ensures that the Web pages are visually appealing, are easy to use, and make consistent use of graphics elements from page to page.

Web log A Web site on which people post their thoughts and invite others to add commentary. Synonymous with blog.

Web portal Synonymous with portal, which is a Web site that serves as a customizable home base from which users do their searching, navigating, and other Web-based activity.

Web programmer A programmer who designs and writes the underlying code for dynamic database-driven Web pages.

Web server A computer that receives requests from many different Web clients and responds by sending HTML files back to those Web client computers.

Web server software Software that makes files available to other computers on the Internet.

Web services A combination of software tools that let application software in one organization communicate with other applications over a network using the SOAP, UDDI, and WSDL protocols.

Web Services Description Language (WSDL) A language that describes the characteristics of the logic units that make up specific Web services.

Web-wrap acceptance The compliance with EULA conditions with which a user agrees through the act of using a Web site.

White hat hackers Hackers who use their skills for positive purposes.

White list spam filter Software that looks for From addresses in incoming messages that are known to be good addresses.

Wide area network (WAN) A network of computers that are connected over large distances.

Wi-Fi (wireless Ethernet, 802.11b, 802.11a, 802.11g, 802.11n) The most common wireless connection technology for use on LANs; it can communicate through a wireless access point connected to a LAN to become a part of that LAN.

Winner's curse A psychological phenomenon that causes bidders to become caught up in the excitement of competitive bidding and bid more than their private valuation.

Wire transfer Synonymous with electronic funds transfer, which is the electronic transfer of account exchange information over secure private communications networks.

Wireless access point (WAP) A device that transmits network packets between Wi-Fi-equipped computers and other devices that are within its range.

Wireless Application Protocol (WAP) A protocol that allows Web pages formatted in HTML to be displayed on devices with small screens, such as PDAs and mobile phones.

Wireless Encryption Protocol (WEP) A set of rules for encrypting transmissions from wireless devices.

Wireless Ethernet The most common wireless connection technology for use on LANs.

Worldwide Interoperability for Microwave Access (WiMAX) A 4G wireless technology that offers download speeds up to 12 Mbps and upload speeds up to 5 Mbps.

World Wide Web (Web) The subset of Internet computers that connects computers and their contents in a specific way, and that allows for easy sharing of data using a standard interface.

World Wide Web Consortium (W3C) A not-for-profit group that maintains standards for the Web.

Worm A virus that replicates itself on other machines.

Writing A tangible representation of the terms of a contract.

XML parser A program that can format an XML file so it can appear on the screen of a computer, a wireless PDA, a mobile phone, or other device.

XML vocabulary A set of XML tag definitions.

Yankee auction A type of English auction that offers multiple units of an item for sale and allows bidders to specify the quantity of items they want to buy.

Zombie A program that secretly takes over another computer for the purpose of launching attacks on other computers. Zombie attacks can be difficult to trace to their perpetrators.

Zombie farm A group of computers on which a hacker has planted zombie programs.

Note: **Bold** page numbers indicate where a key term is defined in the text.

1&1 Internet, 388
"1-by-1 GIFs," **415**
1-Click feature, 164, 299–300
3DES, **435**
3G wireless technology, **90**
4G wireless technology, **90**
7-Eleven, 38
20th Century Fox, 114
24/7 operation, **510**
43 Things, 251
419 scam, **307**
802.11 connections, **88–89**
911 attacks, 427, 478
911Gifts.com, 153–155
1996 Nobel Prize in Economics, 261

A

Abacus Direct Corporation, 315
ABC, 114
About.com, 299
Abrams, Jonathan, 248
academic content, 112
acceptances, **291–293**
access control, 445–447
accessibility, 137–139
account aggregation, **122**, 131
account managers, **515–516**
accounting software, 381, 382–383
Accredited Standards Committee X12
    (ASC X12), 219
ACH (Automated Clearing House), **225–226**, 470
ACL (access control list), **446–447**
ACLU (American Civil Liberties Union), 134–135
acquirer fees, **469**
acquiring banks, **468–469**
acquisition, customer, 167–**168**, 169–170
Across Asia on the Cheap, 147
active ads, 174–175
active content, **415–416**
Active Server Pages (ASP), 337, 383
active wiretapping, **431**
ActiveX controls, 415, **417–418**
activities, **6**
ad views, **176**
AdAge.com, 178
ad-blocking software, **174**
AdDesigner.com, 171–172

addressable media, **142**
Adleman, Leonard, 434
administration activities, 31, 213–214
Adobe ColdFusion, 337, 383
Adobe Dreamweaver, 84, 355, 378–379
Adobe Flash, 138, 336, 418
Adobe PDF, 138–139
Adobe SiteCatalyst, 354
Adobe Systems, Beta Computers v., 293
AdReady, 172
AdSense, 191
ADSL (asymmetric digital subscriber line), **85**
advance fee fraud, **306**
advanced content filtering, 347–348
Advanced Encryption Standard (AES), **435**
Advanced Research Projects Agency (ARPA), 56
Advanced Rotocraft Technology, 192
advertising. *See also* marketing
    ad formats, 171–175
    banner ads, **171**–173
    case examples, 197–200
    changing strategies in, 127–130
    contextual, **191**
    cost and effectiveness, 176–178
    customer loyalty model, 170–171
    in first and second waves, 12, 14
    in late 1990s, 4
    localized, **191**
    in mobile apps, 175, 257
    regulation, 304–306
    revenue models, 114–119
    site sponsorships, **175**–176
    on social networking sites, 252–253
    statistics, 189–190
    text ads, **173**–174
.aero (sponsored TLD), 69
AES (Advanced Encryption Standard), **435**
affiliate marketing, **185**–186
affiliate program brokers, **186**
Affluence Models database, 167
Africa, 38
Agilent, 269
AJAX (asynchronous JavaScript and XML), **337**
Akamai, 114
Al Qaeda, 427
alcoholic beverages, 290, 305–306
Alertbox, 140
Algeria, 39
algorithms, encryption, **433**–434
Alibaba.com, 274–275, 289
AlliedSignal, 497–498
Allwall.com, 192
AltaVista, 3

Amazon laws, 319–320
Amazon.com
    1-Click feature, 164, 299–300
    affiliate program, 185
    Askville, 253
    auction site, 265
    Bibliofind, 454–455
    as case example, 45–48
    Checkout, 471
    cookies, use of, 315
    as CSP, 389
    data storage service, 128
    electronic books, 112–113
    hacker attacks on, 431, 432
    Kindle Fire, 258
    in late 1990s, 3
    Marketplace, 47
    music, 113
    payments from publishers, 310
    sales taxes, 319–320
    strategic alliances, 131
    third wave opportunities and, 15
    tracking system, 230
    traffic statistics, 252
    video, 114
    wine store, 290
    zShop, 455
American Bankers Association Check 21 Resource
    Center, 478
American Civil Liberties Union (ACLU), 134–135
American Express, 259, 465, 467
American National Standards Institute (ANSI), 219
American Packaging Machinery, 241–242
American Psychological Association, 112
American Registry for Internet Numbers (ARIN), 63
American Society of Mechanical Engineers, 193
Ameritrade, 312
Amphire Solutions, 233
analytical processing, **182–183**
anchor tags, **77**
Anderson, Chris, 126
Andreessen, Marc, 67–68
Android, 256–257, 258, 259, 428–429
angel investors, **505**
animated GIFs, 171, 172
Annals of International Business, 149
Anonymizer, 431
anonymous electronic cash, **471**
anonymous FTP, **354**
ANSI (American National Standards Institute), 219
Antigua, 307
Anti-Phishing Working Group (APWG), 483
antispam tactics by individual users, 344–345
antivirus software, **420**, 423, 428
AOL, 128, 252, 352, 431
Apache HTTP/Web Server, 334, 341–342
Apache Software Foundation, 337
Apartments.com, 266
API (application program interface), **385**
APNIC (Asia-Pacific Network Information
    Center), 63
App Inventor, 258
Apple Computer
    app store, 257, 311, 428
    iPad, 138, 258
    iPhone, 138, 255–257
    iPod, 299

iTunes, 13, 113, 114, 299
    offshoring, 208
    QuickTime, 418
    traffic statistics, 252
applets, **415**, 416–417
Applicability Statement 2 (AS2), 225
Applicability Statement 3 (AS3), 225
application integration, **383–384**
application program interface (API), **385**
application programs/software, **383–384**
application servers, **383–384**
application service providers (ASPs), **373**
applications specialists, **516**
apps, mobile, **175**, 257–259
APWG (Anti-Phishing Working Group), 483
Arab Spring of 2011, 39
architectures
    client/server, **65**, **335**, 338–340
    Web server hardware, 358–361
Argentina, 42
ARIN (American Registry for Internet Numbers), 63
ARIN Whois, 63
ARPA (Advanced Research Projects Agency), 56
ARPANET, 56, 62, 342–343
Arrington, Michael, 250–251
Art.com, 192
Artuframe, 192
As Time Goes By (Freeman and Louçã), 11
AS2 (Applicability Statement 2), 225
AS3 (Applicability Statement 3), 225
ASC X12 (Accredited Standards
    Committee X12), 219
ascending-price auctions, **260**
Asia, 10
Asia-Pacific Network Information Center
    (APNIC), 63
ASP and ASP.NET, 337
ASPs (application service providers), **373**
Association for Computing Machinery, 112
Association for the Study of International Business
    (ASIB), 148–150
asymmetric connections, **84**
asymmetric digital subscriber line (ADSL), **85**
asymmetric encryption, **434–437**
asynchronous JavaScript and XML (AJAX), **337**
asynchronous transfer mode (ATM)
    connections, 87
At Home Corporation, 186
AT&T, 86, 445
The Atlantic Monthly, 66
ATM (asynchronous transfer mode)
    connections, 87
Atom Publishing Protocol, **387**
attachments, **343**
Auction Universe, 266
AuctionBytes, 270
auctioneer, **259**
AuctionHawk, 271
auctions
    auction-related services, 270–271
    case example, 276
    history and overview, **259**
    types of, 259–262
    Web site categories, 263–269
Audible, 112
audits, postimplementation, **517–518**
authentication, 445–447

authority to bind, **295**
Authorize.Net, 470
Autobytel, 124
Automated Clearing House (ACH), **225–226**, **470**
automobile sales sites
  brand image, 134
  combined with traditional commerce, 18
  revenue model, 124
Automotive Industry Action Group, 240
AutoTrader.com, 118
awareness, **166**

B

B2B. *See* business-to-business (B2B)
B2C (business-to-consumer)
  growth in online sales, 10–11
  overview, **6–8**
B2G, **7–8**
backbone routers, **60**
backdoors, **430**
back-end processors, **470**
bandwidth, **84–85**
Bank, Grameen, 254
Bank of America, 312
Bank One, 489
banking
  account aggregation, 122, 131
  fee-for-transaction model, 121–122
  Internet technologies and, 478–480, 489
  mobile, 258, 480
banner ads, **171–173**
banner advertising networks, **172**
banner exchange networks, **172**
bar codes, 229, 230
Barbados, 307
bargainers, **165**
Barnes & Noble, 112–113, 300
Barron's, 112
barrydiller.com, 302
base 2 number system, **63**
base 16 (hexadecimal), **64**
basic content filtering, **345–346**
BATF (Bureau of Alcohol, Tobacco, and Firearms),
  305–306
Bay Area TransLink, 477
Bayesian revision, **347–348**
BBC Worldwide, 147
BBSs (bulletin board systems), 247
Bechtel, 269
behavioral segmentation, **163–165**
benchmarking, **357**
Berkeley Technology Law Journal, 289
Berkman Center for Internet & Society, 289
Berkman Center UDRP Opinion Guide, 302
Berners-Lee, Tim, 66–67, 72, 91
Best Buy, 109, 164, 299, 433
Beta Computers v. Adobe Systems, 293
Bethlehem Steel, 233
Better Business Bureau, 270, 305
Betty's Crystal, 276
Beverly Hills Internet, 247
Bezos, Jeff, 45–46, 142, 300
BIB NET, 231
Bibliofind, 454–455
BidPay, 474

bids and bidders, **259**
BigCommerce, 378
bill presentment service, **122**
Billings, Hilary, 154
BillMeLater, 471
BillPoint, 474
binary number system, **63**
Bing, 117
biometric technologies, 15, **428**
Bitnet, 56
BitPass, 463
black hat hackers, **411**
black list spam filter, **345**
BlackBerry, 54, 255–257
Blackboard, 251
blade servers, **356**
Blockbuster, 114
Blogbus, 40
blogs, **160**, 250–251
Blue Nile, 132
Blue Spike, 303
Bluefly, 397
BlueMountainArts, 186
Bluetooth, **88**
Boeing, 228, 269
Boisot, Max, 509
bonded warehouses, **41**
bookmarks manager, social, 251
books, electronic, 13, 112–113
Books-on-Tape, 112
border routers, **59**
borders and jurisdiction, 283–290
Bossidy, Larry, 497–498
Boston College, 312
botnets, **416**, 432
brand, product's, 18, **155**
branding
  elements of, 183–184
  emotional vs. rational, **184–185**
  image and Web presence, 133–134
Brazil, 38, 365
breach of contract, **287**
Brin, Sergey, 4
Britannica.com, 129
broadband connections, 12, 14, **85**
broadcast media, **142**
Broadvision, 383, 393
browser mode, visitors in, 163–164
browsers. *See also* client security
  HTML documents and, **67**
  plug-ins, **418–419**
  usability and, 137
  as Web clients, **65**, 335
browser-wrap acceptance, **293**
buffer overrun/overflow, **443**
Bugbear, 419–420, 422
bulk mail, **178–179**. *See also* spam
bulletin board systems (BBSs), 247
bullying, 308
Bureau of Alcohol, Tobacco, and Firearms (BATF),
  305–306
Bureau of Consumer Protection Business Center,
  304–305
Bureau of Public Debt, 215
Bush, Vannevar, 66
business content, 112
business logic, **383**

business managers, **515**
business models, **15**–16
business process offshoring, **208**
business process patents, 299–300
business processes, 7, 16
business rules, **382**
business units, **26**
Business.com, 192
business-to-business (B2B). *See also* electronic data interchange (EDI)
    case examples, 239–242, 274–275
    e-Government, **215**–216
    electronic marketplaces and portals, **232**–236
    growth in online sales, 10–11
    introduction, 205–207
    logistics activities, 208, 212–213
    overview, **6–8**, **10**
    purchasing activities, 208–212
    supply chain management, **226**–232
    supply webs, **217**
    support activities, 208, 213–214
business-to-business auctions, 267–269
business-to-consumer (B2C)
    growth in online sales, 10–11
    overview, **6–8**
business-to-government (B2G), **7–8**
"Buy It Now," 300
BuyDomains.com, 193
Buyer Behavior Indicator database, 167
buyer mode, visitors in, 164
BuyUSA, 42
bytes, **63**

C

C2C (consumer-to-consumer), **7–8**
c2it, 474
cable channels, 114
cable modems, 85
Caesar, Julius, 409
CA.gov, 8, 215–216
call centers, **517**
callback systems, 446
Calvin Klein, 132
cannibalization, **130**–131
CAN-SPAM law, 348–351
capital projects/investments, **505**
CAPTCHA Project, 346
Carbonite, 128
card code verification (CCV), **469**
card not present transactions, **465**
card security code (CSC), **469**
card verification numbers (CVNs) and variants, **469**
cards, payment. *See* payment cards
CareerBuilder.com, 118
cars, used, 18
Cars.com, 266
Carson, Ellen, 326
Cart32, 430
CAs (certification authorities), **426**–427
Cascading Style Sheets (CSS), **78**
case examples
    Alibaba.com, 274–275
    Amazon.com, 45–48
    American Packaging Machinery, 241–242

Association for the Study of International Business, 148–150
Bibliofind, 454–455
Davis Humanics, 523–524
Ellasaurus Products Enterprises, 326
First Internet Bank of Indiana, 489–490
Hal's Woodworking, 48–49
Harley-Davidson, 239–241
Hyderabad, Internet access in, 98
Idealab, 522–523
Ingersoll-Rand Club Car Division, 402–403
Lonely Planet, 147–148
Materials Equipment, 455–456
Microsoft and open source software, 364–366
Montana Mountain Biking, 199–200
The Moose Hut, 490–491
Nissan.com, 324–326
Old Metamora, 276
Oxfam, 197–199
Portable Fun Instruments, 98–99
Random Walk Shoes, 366–367
Web Services for State Government, 403–404
cash, as payment method, 464
cash, electronic, **471**–474
cashers, **485**
Castells, Manuel, 28
catalog display software, **375**–377
catalog revenue model, **107**–111
Catchings, Bill, 314
cause marketing, **185**
CBS, 114
CCV (card code verification), **469**
CDNow, 47
Cengage Learning, 425, 426
censorship, 38–40
Center for Education and Research in Information Assurance and Security (CERIAS), 417, 449
Center for Internet Security, 450
Center for Responsive Politics, 136
centralized architecture, **358**–359
Cerf, Vinton, 62
CERIAS (Center for Education and Research in Information Assurance and Security), 417, 449
Cerias - Java Security Page, 417
CERN, 66, 91
CERT (Computer Emergency Response Team), 449
certification authorities (CAs), **426**–427
CGI (Common Gateway Interface), 337
challenge-response filtering, **346**–347
Chambers, John, 497
Chanel, 132
change management, **518**
channel conflicts, **130**–131
channel cooperation, **131**
charge cards, **465**
chargebacks, **469**
Charles Schwab, 120
chat feature, 111
Check 21 (Check Clearing for the 21st Century Act), **478**
check processing, 478
CheckPointHR, 214
checks, as payment method, 464
Chemdex, 232–233
Cheviot, Ohio Web site, 216
Chevrolet Nova, 37–38
Chicago Board Options Exchange, 261

chief information officers (CIOs), **514**, 515
child elements, **82**
children
    communications with, 315–317
    Ellasaurus, 326
    online crimes involving, 308
Children's Internet Protection Act (CIPA), 316
Children's Online Privacy Protection Act (COPPA), 316–317
Children's Online Protection Act (COPA), 316
China
    during 2008–2009 recession, 10
    censorship in, 39–40
    culture and government, 285–286
    Microsoft and, 365–366
    mobile phones in, 90
    Pepsi's advertising campaign in, 38
    U.S. corporations in, 289
Chinapage.com, 274
ChoicePoint, 312
Christie's, 259
Chrome, 65
Chrysler, 159
Chung, Jen, 250
Cigarbid.com, 266
CIO Budget, 507
CIOs (chief information officers), **514**, 515
CIPA (Children's Internet Protection Act), 316
cipher text, **433**
circuit switching, **59**
circuits, **58**–59
Cisco, 19, 234, 269, 507
CISG (Contracts for the International Sale of Goods), 293
Citibank, 474
Clampi, 420, 423
Claritas, 167
Clark, James, 68
classified ads, 117–118, 266
Classified Ventures, 266
"clear GIFs," **415**
The Clearing House, 470
ClickandBuy, 471
clickstream, **180**
click-throughs, **176**–177
click-wrap acceptance, **293**
client security
    active content, **415**–416
    cookies and Web bugs, 413–415
    digital certificates, **424**–427
    for mobile devices, 428–429
    physical security, 428
    steganography, **427**–428
    viruses, worms, antivirus software, 419–424
client-level filtering, **345**
client/server architectures, **65**, **335**, 338–340
client-side digital wallets, **475**
client-side scripting, **336**
closed architecture, **62**
closed loop payment card systems, **467**
closing tags, **73**
clothing, challenge with color, 111
cloud computing, **398**, 448
C-NET, 117
CNN, 250
Coase, Ronald, 23, 24
Coca Cola, 133–134

Code Red, 419, 421, 424
ColdFusion, 337
Coldwater Creek, 260
collectors, **485**
collisions, **434**
co-location, **373**
colon hexadecimal, **64**
colors and culture, 38
Commerce Clause, 290
Commerce Server, Microsoft, 392
commerce service providers (CSPs), **373**, 387–389, 444
Commission Junction, 186
commitment, 166, **167**
commodity items, **17**–18
Common Gateway Interface (CGI), 337
common law, **300**
Common Object Request Broker Architecture (CORBA), 384
communication channel security
    digital signatures, 440–441
    encryption solutions, 433–440
    integrity threats, 431
    message digests, 440–441
    necessity threats, 431–432
    overview, 429
    physical threats, 432
    secrecy threats, 429–431
    wireless threats, 432–433
communication modes, **142**–144
communities, **247**–248. See also social media and networks
Comodo, 426
companies, **22**–23
Compaq, 34
Competitive Advantage, 29, 31
complexity, marketing and, 158–159
Component Object Model (COM), 384
component outsourcing, **510**–513
component-based application systems, **383**–384
CompuBank, 479, 489
CompuPay, 214
Compuserv, 247
CompuServe, 57
Computer Crime & Intellectual Property Section, 306
Computer Emergency Response Team (CERT), 449
computer forensics, **450**
computer networks, **55**
computer security, **410**. See also security
Computer Security Institute, 309
computer viruses. See viruses
"Conditions of Use," 295–296
Conferences on Email and Anti-Spam, 483
Conficker, 420, 423
configuration tables, **59**
conflict of laws, **290**
connection options, 84–91
connectors, **165**
considerations, **291**
consignment services, auction, **271**
consortia-sponsored marketplaces, **235**
ConstantContact, 179
Constitution, U.S., 290, 296
constructive notice, 286
consumer auctions, 263–267
Consumer Credit Protection Act, 466

Consumer Leasing Act of 1976, 282
ConsumerReports.org, 119
consumer-to-business auctions, **263**
consumer-to-consumer (C2C), **7–8**
content creators, **516**
content editors, **516**
content filtering, 345–348
content management software, **395**
content managers, **516**
contextual advertising, **191**
contract purchasing, **211**
contracts, **287**, 291–296
Contracts for the International Sale of Goods
    (CISG), 293
"Controlling the Assault of Non-Solicited
    Pornography and Marketing," 348–351
Convergence Center, 70
conversion, customer, **168**, 169–170
conversion rate, **179**
Cookie Central, 415
cookies
    DoubleClick, **315**, 414
    security and, 413–415
    shopping carts and, 379
COPA (Children's Online Protection Act), 316
Cope Today, 126
COPPA (Children's Online Privacy Protection Act),
    316–317
copy control, **303**
copyright issues, **297–299**
CORBA (Common Object Request Broker
    Architecture), 384
Cornell Law School, 292
Cornell University, 247–248
cost per thousand (CPM), **176–177**
cost/benefit evaluation, 506
Costco, 107, 132
countermeasures, **410**
country TLDs, 69–70
Cover Pages: XML Applications and Initiatives, 83
CPA Directory, 125
crackers, **410**. *See also* hackers
craigslist, 118, 251
crawlers, **188**
credit card associations, **468**
credit card number theft, 429, 430
credit cards, 464, **465**. *See also* payment cards
Cricket Jr. (sniping software), 271
crimes, 306–309, 480–485
CRM. *See* customer relationship management
cruise vacation sites, 123
Crutchfield, 164
cryptography, **433**
CSC (card security code), **469**
CSO Online, 450
CSPs (commerce service providers), **373**, 387–389,
    444
CSS (Cascading Style Sheets), **78**
Cubic, 269
culture
    concerns regarding, 21
    international commerce and, **37–40**
    laws, ethics, and, 283–286, 313
CUNA Mutual Group, 385
currencies, 12
customer behavior, segmentation using, 163–165

customer issuing banks, **468**
customer portal sites, **234**, 235–236
customer relationship management
    acquisition, conversion, retention, **167–168**
    case example, 402–403
    elements of typical, 183
    funnel model, **169–170**
    relationship intensity in each stage, 165–167
    software, **396–398**
    technology-enabled, 180–183
    Web 2.0 and, 13
customer service
    job description for, **517**
    personal touch, 111
    trust and loyalty and, 139–140
customer value, **156**
customer-based marketing, 157
customer-centric Web site design, 140–**141**
CustomerVision, 395
customs brokers, 41
customs duty, 320
CVNs (card verification numbers) and variants, **469**
cyberbullying, **308**
Cybergold, 299
CyberSource, 469
cybersquatting, **300–301**
cybervandalism, **431**

D

DaimlerChrysler, 324
data analysis software, 354
Data Encryption Standard (DES), **435**
Data Interchange Standards Association (DISA), 219
The Data Mine, 183
data mining, **182–183**, 389
data storage service, 128
data warehouses, **182–183**
Data Warehousing Information Center, 183
database, search engine, **188**
database administrators, **517**
database management software, **382**
database servers, **338**
databases, threats to, 443
data-grade lines, **85**
data-type definitions (DTDs), **82**
Datsun, 324
Davis, Robin, 200
Davis Humanics, 523–524
DDoS (distributed denial-of-service) attacks, **432**
dead links, **354**
debit cards, 464, **465**. *See also* payment cards
decentralized architecture, **358–359**
DecisionStep, 111
decryption, **433**
dedicated hosting, **373**
Deep Crack, 435
deep Web, 68–69
defamation, **303**
delay attacks, **431–432**
del.icio.us, 251
Dell, Michael, 497
Dell Computer
    as business example, 8
    FTC and, on leasing ad, 281–282

hardware, 355
market segmentation and, 162, 165
private stores, 234, 236
supply chain management, 227, 228–229
SWOT analysis by, 33–34
Delta Air Lines, 192
demographic information, **115**
demographic segmentation, 160–161
demonstration version, **516**
denial-of-service (DoS) attacks, **431**–432
Department for Work and Pensions, 215
Department of Homeland Security (DHS), 215
Department of Transportation (DOT), 305–306
DES (Data Encryption Standard), **435**
descending-price auctions, **260**–261
destinationCRM.com, 183
DHS (Department of Homeland Security), 215
dial-up modems, 12
dictionary attack programs, **442**
differentiation, **183**–184
Digg, 504
DigiCash, 463
DigiCert, 426
Digi-Key, 212
Digimarc, 303
digital cash, **471**–474
digital certificates/IDs, **424**–427
Digital Divide Data, 207
Digital Equipment Corporation (DEC), 184
digital products
in first and second waves, 13, 14
logistics and, 156
opportunity to sell online, 19
physical products vs., 126
piracy, **13**, 14, 364–366
revenue models for, **112**–119, 126
VAT on sales of, 320
Digital Rights Management (DRM), **113**
Digital River, 470, 471
digital signatures
converting message digests into, **440**–441
forgery and, 295
legality for contract purposes, **293**
Digital Subscriber Line (DSL), **85**
digital wallets, **474**–476
Digital Watermarking, Stenography and, 428
digital watermarks, **302**
Diller, Barry, 302
diminishing returns, **28**
DiPonetti, Mario, 148–150
direct connection EDI, **223**–224
Direct Marketing Association (DMA), 311–312
direct materials, **211**
Directive on the Protection of Personal Data, 313
DISA (Data Interchange Standards Association), 219
disclaimers, warranty, **294**
Disco Virtual, 20–21
discount retailers, 107–108
Discover Card, 467
disintermediation, **120**–125
Disney Online, 316
disposable card numbers, **465**–466
distance learning, 20
distributed architecture, **358**–359
distributed database systems, **382**
distributed denial-of-service (DDoS) attacks, **432**
distributed information systems, **382**

distribution, as marketing element, **156**
DMA (Direct Marketing Association), 311–312
DNSs (domain name servers), **431**
Dobkins, Jake, 250
dodge.com, 324
domain name ownership change, **302**
domain name parking/hosting, **193**
domain name servers (DNSs), **431**
domain names
case examples, 324–326
intellectual property issues, 300–302
main discussion, **69**–70
naming issues, 191–193
Donnelley Marketing, 167
Donovan, Hal, 48–49
DoS (denial-of-service) attacks, **431**–432
DOT (Department of Transportation), 305–306
dot-com boom, bust, rebirth, 10–11
dot-com businesses, **5**
dotted decimal notation, **63**
double auctions, **261**–262
double spending, **472**
DoubleClick, 172, 315
Dow Jones, 112
Download.com, ActiveX page at, 417
downstream bandwidth, **85**
downstream strategies, **499**
DRM (Digital Rights Management), **113**
Dropbox, 128
Drudge Report, 115
drug products, advertising, 305–306
DS0 lines, **87**
DSL (Digital Subscriber Line), **85**
DSW Shoe Warehouse, 312
DTDs (data-type definitions), **82**
due diligence, **479**
Duke University, 56
Dutch auctions, **260**–261, **262**
duty, customs, **320**
DVDs, suitability for e-commerce of, 18
dynamic catalogs, **375**
dynamic content/pages, **336**–337

E

E*Trade Financial, 120
early outsourcing, 509–**510**
Eastman Kodak, 140
eavesdroppers, **410**
eBay. *See also* PayPal
bargainers on, 165
as C2C example, 8
in China, 289
competition, 264–265, 266, 267
as CSP, 389
illegal items on, 310–311
in late 1990s, 3
masquerading attacks on, 431
mechanics of, 263–264
patent case, 300
stolen goods and, 309
Stores, 389
third wave opportunities and, 15
traffic statistics, 252
Web servers at, 359–360

e-books, 13, 112–113
EBSCO Information Services, 112, 149–150
e-business, **5**
e-cash, **471**–474
Eccles, David, 271
e-commerce. *See* electronic commerce
E-Commerce Usability, 140
eCompanies, 192
economic forces and e-commerce, 22–29
Eddie Bauer, 131
eDeposit, 270
EDI. *See* electronic data interchange (EDI)
EDI compatible, **217**
EDI for Administration, Commerce, and Transport (EDIFACT), 219
EDIINT (Electronic Data Interchange-Internet Integration), 225
Edmunds.com, 124
Edvinsson, Leif, 509
effects, **285**
EFTPOS (electronic funds transfer at point of sale) cards, **465**
EFTs (electronic funds transfers), **8**, 225–226
e-Government, **215**–216
Egypt, 39
EJBs (Enterprise JavaBeans), 384
electronic books, 13, 112–113
electronic business, **5**
electronic cash, **471**–474
electronic commerce. *See also* payment systems; security; software; Web servers; Web sites and presences; *specific topics*
    analyzing business unit opportunities, 29–34
    business models and processes, 15–18
    case examples, 45–49. *See also under specific topics*
    categories, 6–8
    cautions and concerns, 20–22
    defined, **5**
    development and growth, 8–9
    dot-com boom, bust, rebirth, 10–11
    economic forces and, 22–29
    first wave, 11–14, 15–16, 499–500, 506
    history, 3–5
    international. *See* international electronic commerce
    merchandising, **16**–17
    opportunities, 18–20
    planning for. *See* planning
    product/process suitability, 17–18
    second wave, 11–14, 15, 500, 506–507
    third wave, 14–15, 500, 507
Electronic Communications Privacy Act of 1986, 311
electronic CRM, **180**–183
electronic data interchange (EDI)
    B2B sales and, 10
    case examples, 239–242
    direct and indirect connections, 223–225
    history, 218–219
    overview, 9, 217–218
    payments, 225–226
    process, paper-based vs., 220–223
Electronic Data Interchange-Internet Integration (EDIINT), 225
Electronic Frontier Foundation, 435
electronic funds transfer at point of sale (EFTPOS) cards, **465**

electronic funds transfers (EFTs), **8**, 225–226
electronic mail. *See* e-mail
electronic wallets, **474**–476
Ellasaurus Products Enterprises, 326
E-LOAN, 124
Elsop LinkScan, 355
e-mail. *See also* spam
    benefits and drawbacks, 343
    emergence of, 56, 342–343
    encryption, 435
    in first and second waves, 12, 14
    mail bomb, 443
    marketing via, 178–180
    network effect, 29
    outsourcing, 180, 510
    phishing, **431**, 480–485
    protocols, **64**–65
    responsiveness via, 140
    secrecy and, 429–430
    security and, 408–409, 419, 423, 428
e-mail addresses, collection of, 311
e-mail client software, **64**
e-mail servers, **64**, **338**
eMarketer, 178
The eMarketplace, 235
emotional branding, **184**–185
employees
    change management for, **518**
    staffing, 21, 514–517
employment sites, 118
eMusic, 113
encapsulation, **61**
encryption solutions, **433**–440
Encyclopaedia Britannica, 128
end-user license agreements (EULAs), **292**–293
Engelbart, Douglas, 66
engineering.org, 193
English auctions, **259**–260, **262**
enterprise application integration, 383–384
Enterprise JavaBeans (EJBs), 384
enterprise resource planning (ERP) software, **384**
enterprise-class software, **393**–398
entity bodies, **339**
Entrust, 426
Ephemeral Web-Based Applications, 138
EPN, 470
Epocrates, 258
ePowerSellers, 271
e-procurement, **6**, **234**–235
equity partners, 512
Ericsson, 214
ERP (enterprise resource planning) software, **384**
eS-Books, 38
escrow services, **265**, 270
Escrow.com, 270
e-sourcing, **209**
ESPN, 119
e-Steel, 235
Ethernet, **88**–89
ethical hackers, **450**
ethics
    communications with children, 315–317
    culture and, 283–284, 313
    online business practices and, 310–311
    privacy rights and obligations, 311–315
Etsy, 251
EU (European Union), 284

EULAs (end-user license agreements), **292**–293
Europe, smart cards in, 477
European Laboratory for Particle Physics, 66
European Union (EU), 284, 313, **320**
event tickets, 121
Everson, Joe, 455–456
e-wallets, **474**–476
Excel spreadsheet format, 138–139
Expedia, 122
exploration, **166**
Extensible Business Reporting Language (XBRL), 82
Extensible Hypertext Markup Language (XHTML), **71**
Extensible Stylesheet Language (XSL), **83**
extranets, **61**–62
Extreme Pizza, 188

F

Facebook. *See also* social media and networks
    in Egypt, 39
    green computing, 357
    history, 245–246
    marketing through social media, 160, 186–188
    traffic statistics, 252
    as Web 2.0 example, 13, 15
Factiva, 112
Failure of Corporate Websites, 136
fair use, **297**–298, 299
familiarity, **166**
fans and fan bases, **188**
fast venturing, **512**–513
Faust, Bill, 184
FBI, 309
FDA (Food and Drug Administration), 305–306
FDIC (Federal Deposit Insurance Corporation), 489
FedACH Services, 470
Federal Bureau of Investigation, 309
Federal Reserve Bank, 307
Federal Reserve Board, 282
Federal Reserve Financial Services - Check
    21-Enabled Services, 478
Federal Trade Commission. *See* U.S. Federal Trade
    Commission
FedEx, 213, 357
fee-for-content revenue models, 112–119
fee-for-service revenue models, **125**–126, 253
fee-for-transaction revenue models, **119**–125
Fielding, Roy, 387
File Transfer Protocol (FTP), **353**–354, 437
filtering software, **316**
finance activities, 31, 213–214
Financial Management Service (FMS), 215
financial services, fee-for-transaction model, 121–122
Finger command, **352**
fingerprint readers, 428
firearms, 305–306, 311
Firefox, 65, 68, 75, 414, 415–416, 427
firewalls, **447**–448
firms, **22**–23
First Amendment, right of publicity and, 296
First Data Corporation, 474
First Internet Bank of Indiana, 122, 489–490
FirstData, 470
first-mover advantage, **13**, 14
first-party cookies, **414**
first-price sealed-bid auctions, **261**, **262**
Fisher, Marshall, 227

FitFinder, 138
fixed-point wireless, **89**
Flash: 99% Bad, 138
Flash Usability Challenge, 138
float, **478**
food, advertising, 305–306
Food and Drug Administration (FDA),
    305–306
Ford Motor Company, 159, 324–325
forged identities, 295
forum selection clause, **287**–288
four Ps of marketing, **155**–156
fourth-generation (4G) wireless technology, **90**
Fox, 114
frame relay connections, 87
France, 40, 285
fraud
    ChoicePoint incident, 312
    credit card, 469
Free the Grapes, 290
Freedom Forum, 434
Freedom House, 39
Freeman, Chris, 11
freight forwarders, **41**
FreshDirect, 20
Friendster, 248
front-end processors, **470**
fruit, 20
FTC. *See* U.S. Federal Trade Commission
FTP (File Transfer Protocol), **353**–354, 437
full-privilege FTP, **354**
Fundación Invertir, 42
funding, 12, 14
funnel model, **169**–170
Future of Banking Study, 489

G

G&L Internet Bank, 489
gambling, 307–308
games, 125
Gartner, Inc., 424
Gate.com, 388
Gates, Bill, 365
Gateway, 355
gateway computers, **59**
gateway servers, **448**
The General, 121
General Agreement on Trade and Services, 307
general consumer auctions, 263–266
General Electric, 9, 26, 63
General Motors, 37–38, 192, 324
General Services Administration, 269
Generalized Markup Language (GML), **72**
generic top-level domains (gTLDs), **69**–70
GEnie, 247
GeoCities, 247
geographic segmentation, **160**–161
GeoTrust, 426
Gibson Research Corporation, 443, 448
GIFs, clear, **415**
gift cards, **465**
gift shops, 153–155
Gilt, 267, 371–372
global electronic commerce. *See* international
    electronic commerce
Global Payments, 470

global positioning satellite (GPS), 213, 258
Global Trust and Culture, 37
GMAC Mortgage, 124
Gmail, 64
gm.com, 324
GML (Generalized Markup
    Language), **72**
GNP, 522
Godin, Seth, 179
GoIndustry Dove Bid, 268
Golden Shield Project, 40
Gollent, Manfred, 511
Google
    advertising model on, 117
    AdWords and AdSense, 191
    Android, 256
    Answers, 253–254
    Checkout, 471
    in China, 289
    Chrome, 65
    data storage service, 128
    early years of, 4
    eBookstore, 112
    first-mover advantage and, 13
    Gmail, 64
    green computing, 357
    IPO Dutch auction, 261
    local search service, 191
    music, 113
    Orkut, 248, 249
    Scholar, 150
    targeted ads by, 12
    text ads, 173
    traffic statistics, 252
    Urchin, 354
    video, 114
    Wallet, 259, 476
    Web Server, 341–342
Google+, 160, 186, 248
Gordon Brothers Group, 268
Gothamist, 250
Goto.com, 4
government. *See also* legal environment
    culture and, 38–40
    legitimacy and, 285–286
GPS (global positioning satellite),
    213, 258
Graham-Cumming, John, 348
Grainger.com, 8, 212, 234
Grand Haven, Michigan, 89
graphical user interface (GUI), **67**
graphics, security and, 418
graphics designers, **516**
GreatDomains, 193
GREE, 248
green computing, **356–357**
Green Dam Youth Escort, 40
Grocery Gateway, 20–21
Gross, Bill, 522–523
group purchasing/shopping sites, **267**
Groupon, 267
GSN.com, 125
gTLDs (generic top-level domains),
    69–70
GUI (graphical user interface), **67**
Gupta, Yash, 98–99

H

hackers
    ethical, **450**
    incidents involving, 312, 380,
        454–455
    overview, **410–411**
Hal's Woodworking, 48–49
Handspring, 54
hardware, Web server, 355–361
hardware-based digital wallets, 476
Harley-Davidson, 239–241
Harris Corporation, 357
Harvard Business Review, 227
Harvard Law School, 289, 302
hash coding, **434**, 436
hashing and message digests, 440
HBO, 114
HDSL (high-speed DSL), **85**
Hello Kitty, 326
Hewlett-Packard, 34, 325, 355, 357
hexadecimal (base 16), **64**
Hickory Farms, 107
hierarchical business organizations, **22–26**
hierarchical hyperlink structure, **76–77**
Higbee, Sandi, 524
high-speed DSL (HDSL), **85**
HitExchange, 172
Hockenstein, Jeremy, 207
Hollywood studios, 114
Home Depot, 48, 109, 157
home pages, **76**
HomeGrocer, 20
Honeywell, 497, 498
Hong Kong, smart cards, 477
Horizon Blue Cross Shield of New Jersey, 312
hosting. *See* Web hosting
hot spots, **89**
HotBot, 3
HowStuffWorks, 116
HREF (hypertext reference) property, **77**
HTML
    editors, 83–84
    links, 75–78
    overviews, **67**, **71**, **72–73**
    style sheets, **78**
    tags, 73–75
"HTML 5," 73, 114, 138
HTML extensions, **72**
HTTP (Hypertext Transfer Protocol), **66**
HTTPS, 437
The Huffington Post, 115
Hulu, 114
human resource activities, 31, 213–214
Human Rights Watch, 38
hybrid hyperlink structure, **76–77**
Hyderabad case example, 98–99
hyperlinks, **67**, 75–78
hypertext, **66**
hypertext elements, **72**
Hypertext Markup Language. *See* HTML
Hypertext Preprocessor (PHP), 337, 383
hypertext reference (HREF) property, **77**
hypertext servers, **67**
Hypertext Transfer Protocol (HTTP), **66**
Hyves, 249

I Can Has Cheezburger, 252
i2 Technologies, 396
IAB (Interactive Advertising Bureau), 171, 178
IAPs (Internet access providers), **84**
IBM
 B2B marketplace product by, 233
 in Brazil, 365
 content management, 395
 DB2, 382
 as Dell competitor, 34
 on e-business, 5
 knowledge management, 395
 reverse auctions and, 269
 spam-sending computers and, 351
 Tivoli, 383
 use of computers by Lands' End, 334
 WebSphere Commerce, 391–392, 393
ICANN (Internet Corporation for Assigned Names
 and Numbers), 69, 193
ICANN UDRP Proceedings, 302
ICANNWatch, 70
Ice.com, 132
idea-based networking, **251**
Idealab, 511, 522–523
identity theft, 483–485
Identity Theft Resource Center, 312
IEEE, **88**
IETF (Internet Engineering Task Force), **64**, 352
IIS (Internet Information Server), 341–342, 424
Illinois, 307
ILOVEYOU virus, 419, 421
IMAP (Interactive Mail Access Protocol), **65**
IMAP Connection, 65
impact sourcing, **208**
implied contracts, **291**
implied warranties, **293**–294
import tariffs, **320**
impressions, **176**
IMU (interactive marketing unit) ad formats, **171**
income taxes, **317**, 318–319
incubators, **511**–512, 522–523
independent exchanges, **232**–233
independent industry marketplaces, **232**–233, 235
index, search engine, **188**
indexing programs, **354**
India, 38, 54, 90, 98
indirect connection EDI, **224**–225
indirect materials, **211**
Industrial Revolution, 11
industries, **29**
industry consortia-sponsored marketplaces, **235**
industry marketplaces, **232**–233
industry value chains, **31**–33, **209**
Infocon: SANS Internet Storm Center, 449
Informatica, 383
Information Security Policy World, 412
InformationWeek/Accenture, 309
Infosecurity.com, 450
infrastructure issues, 40–42
Ingersoll-Rand Club Car Division, 402–403
initial public offering (IPO), **505**
inline text ads, **173**–174
Inside Supply Management, 210
Institute for Supply Management (ISM), 210–211
Institute of Practitioners in Advertising (IPA), 178

insurance brokers, 120–121
Insurance.com, 121
Insure.com, 192
InsWeb, 121
Integrated Services Digital Network (ISDN), **85**
integrity and integrity violations, **411**
integrity threats, 431
Intel, 11
intellectual capital, **509**
intellectual property
 advertising regulation, 304–306
 copyright issues, **297**–299
 deceptive trade practices, 304
 defamation, **303**
 domain names, 300–302
 overview, **296**–297
 patent issues, **299**–300
 protecting online, 302–303
 trademark issues, **300**
intentional torts, **288**
Interactive Advertising Bureau (IAB), 171, 178
Interactive Corp, 252
Interactive Mail Access Protocol (IMAP), **65**
interactive marketing unit (IMU) ad formats, **171**
interchange fees, **469**
interchange networks, **466**
Internal Revenue Service (IRS), 318–319
internal social networking, 254
internal teams, 509
International Business Today, 149
international commerce, jurisdiction in, 288–290
international electronic commerce
 cultural issues, **37**–40
 infrastructure issues, 40–42
 language issues, 12, 14, 36–37
 overview, 34–35
 parties in typical transaction, 41–42
 trust issues, 35–36
International Organization for Standardization
 (ISO), 72
Internet. *See also* Web
 case examples, 98–99
 commercial use of, 57
 connection options, 84–91
 emergence of new uses for, 56
 growth of, 57–58
 introduction, 53–**55**
 origins of, 55–56
 packet-switched networks, 58–62
 protocols, 62–66
Internet access providers (IAPs), **84**
Internet backbone, **60**
Internet Cash, 471
Internet Corporation for Assigned Names and
 Numbers (ICANN), 69, 193
Internet EDI, **225**
Internet Engineering Task Force (IETF), **64**, 352
Internet Explorer, 65, 68, 72, 75, 81, 427
Internet Governance Project, 70
Internet hosts, **57**–58
Internet Information Server (IIS), 341–342, 424
Internet Protocol (IP), **62**–63
Internet Protocol version 4 (IPv4), **63**–64
Internet Protocol version 6 (IPv6), **64**
Internet Security Alliance, 449
Internet service providers (ISPs), **57**
Internet Society, 64

Internet2, **91**–92
internets, **55**
INTERNTCO Corp., Barry Diller v., 302
interoperability, **383**
interoperable software, **473**
Interpol, 309
Intershop Enfinity, 391
interstitial ads, **174**
intranets, **61**–62
intrusion detection systems, **448**
Intuit, 19
inventory management software, 382
iOS operating system, 256, 258
IP (Internet Protocol), **62**–63
IP addresses, hiding, 430–431
IP Addressing, **63–64**
IP tunneling, **61**
IP wrapper, **61**
IPA (Institute of Practitioners in Advertising), 178
iPad, 258
iPhone, 255–257
IPO (initial public offering), **505**
iPod, 299
IPPay, 470
IPv4 (Internet Protocol version 4), **63**–64
IPv6 (Internet Protocol version 6), **64**
IRS (Internal Revenue Service), 318–319
ISDN (Integrated Services Digital Network), **85**
ISM (Institute for Supply Management),
    210–211
ISO (International Organization for
    Standardization), 72
iSold It, 271
ISPs (Internet service providers), **57**
issuing banks, **468**
ITSRx, 111
iTunes, 13, 113, 114, 299
Ivory Soap, 184

J

jailbreaking, **257**
Janah, Leila, 206
Janet, 56
Japan, 37, 38, 473, 476, 477
Jasig, 251
Java applets, **415**, 416–417
Java sandbox, **417**
Java Server Pages (JSP), 337, 383
Java servlets, **337**
JavaScript, 336, 415, 417
J.B. Hunt, 213
J. C. Penney, 107
JDA Software, 396
jewelry, 18, 20, 132
job descriptions, 514–517
Johnson, Sarah, 48–49
J.P. Morgan Chase & Co., 385
Jscript, 337
JSP (Java Server Pages), 337, 383
judicial comity, **289**
Juicy Couture, 162
jurisdiction
    crimes and, 306–309
    main discussion, 283–290
JustBeads.com, 266

K

Kahn, Robert, 62
Kanoodle, 191
Kayak, 117
KCOM (KDD Communications), 473
Keynote Systems, 334
keys
    digital certificates and, 426
    encryption and, 433–440
Kindle Fire, 258
Kindle readers, 112–113, 258
Kiva, 254
KM (knowledge management) software, **395**
Kmart, 107
KMWorld, 214
knowledge management, **214**
knowledge management (KM) software, **395**
Kohl's, 230
Kokusai Denshin Denwa, 473
Kozmo, 169

L

LabTech Software, 355
Lacy, Sarah, 504
The Ladders, 118
Lands' End, 111, 132, 333–334
language issues, 12, 14, 36–37
Language Translation Services, 36
LANs (local area networks), **58**
late outsourcing, **510**
law of diminishing returns, **28**
Law on the Web, 125
Lawrence, Amy, 366–367
laws. *See* legal environment
leaderboard ads, **171**
leased lines, **60**, 87
Lee Jeans, 138
Lefkofsky, Eric, 267
legal content, 112
legal environment. *See also* intellectual property
    borders and jurisdiction, 283–290
    case examples, 324–326
    conflict of laws, **290**
    contracts, 291–296
    crime, terrorism, warfare, 306–310
    ethical issues. *See* ethics
    introduction, 21, 281–283
    spam and, 348–351
    taxation, 317–320
legal services, 125
LegalXML, 82
legitimacy, **285**–286
Leonhardt, Ted, 184
LetsBuyIt.com, 267
Levi Strauss & Company, 130–131
LexisNexis, 112
Libya, 39
life-cycle segmentation, **167**
lighttpd, 341
Lilly Pulitzer, 132
lincoln.com, 324
linear hyperlink structure, **76**–77
link rot, **354**
link-checking utilities, **354**–355

LinkedIn, 248
LinkShare, 186
LinxCop, 355
liquidation brokers, **268**
Literary Machines, 66
live chats, 111
LivingSocial, 267
LL Bean, 107
load-balancing, **360–361**
local area networks (LANs), **58**
localization, **37**
localized advertising, **191**
lock-in effect, **265**–266
log files, **354**
logical security, **410**
Logility, 396
logistics, 208, 212–213
Lonely Planet, 147–148
Long Term Evolution (LTE), **90**
long-arm statutes, **288**
LookSmart, 190, 261
lottery tickets, 307–308
Lotus, 522
Louçã, Francisco, 11
"love bug," 419
Lowe's, 48
loyalty
    stages of, 165–167, 170–171
    usability and, 139–140
LTE (Long Term Evolution), **90**
luxury goods, 132
Lycos, 3

M

Ma, Jack, 274–275
machine translation, **37**
macro viruses, **419**
magazines
    advertising, 117–119, 127
    inline text ads, **173**–174
magnetic strip cards, 476–477
mail bomb, **443**
mailing lists, **56**
mail-order model, **107**
maintenance, repair, and operating (MRO)
    supplies, **211**–212
Malaga, Michael, 86
mall-style CSPs, **388**–389
managed service providers (MSPs), **373**
management, 513–518
man-in-the-middle exploits, **411**
Manugistics, 396
many-to-many communication model, 143, **144**
many-to-one communication model, 143, **144**
marketing. *See also* advertising
    affiliate, **185**–186
    case examples, 197–200
    cause, **185**
    creating and maintaining brands, 183–188
    customer behavior and relationship intensity,
        163–170
    domain naming issues, 191–193
    e-mail, 178–180
    introduction, 153–155
    media selection, 158–160

search engine positioning, 188–191
    segmentation, 160–162
    strategies, **155**–157
    technology-enabled CRM, 180–183
    using multiple channels, **109**–110
    viral, **186**–188
marketing mix, **155**
MarketMile, 233
markets, **22**–26
marketspace, **181**–183
markup languages
    HTML. *See* HTML
    overview, 70–72
    XML, 78–83
markup tags, **70**, 73–75, 82
marshmallows and taxes, 319
Martindale.com, 125
Mason, Andrew, 267
masquerading, **431**
mass media approach
    main discussion, **142**–143
    measuring effectiveness, 176
    micromarketing vs., **160**
    trust and, 158–159
MasterCard
    credit cards, 465
    interchange networks and, 466, 468, 469
    PayPass, 476, 477
    phone reader, 259
Materials Equipment, 455–456
materials-tracking technologies, 229–231
MathML, 82
Matsushita Electric, 512
Maytag, 130–131, 140
McAfee Virus Information, 423
McCool, Rob, 342
MCI Mail, 57
McKinsey & Company, 164–165
McMaster-Carr, 212
m-commerce. *See* mobile commerce
    and devices
media choice, 158–160
medical services, 125–126
meetups, **250**
Memex, 66
Mercata, 267
MercExchange, 300
merchandising, **16**–17
merchant accounts, **468**–469
Mercury Technologies, 325
mercury.com, 325
Merriam-Webster's Collegiate Dictionary, 129
mesh routing, **89**
message digests, **440**
meta tags, **188**
metalanguages, **71**
MetalSite, 233–234
metrics, **501**
MI Analyst, 128
Michelin North America, 231
microblogs, **250**
microlending, **254**
micromarketing, **160**
Micron Technology, 281–282
MicroPatent, 309
micropayments, **463**
MicroPlace, 254

Microsoft
    Access, 382
    ASP, 337, 383
    B2B marketplace product by, 233
    COM, 384
    Commerce Server, 392
    Excel spreadsheet format, 138–139
    Expedia, 122
    Expression Web, 378–379
    FrontPage, 378–379
    Hotmail, 64
    IIS, 341–342, 424
    Internet Explorer, 65, 68, 72, 75, 81, 427
    MSN Games, 125
    MSN Money, 385
    MSN Music, 113
    open source software and, 364–366
    Outlook, 64
    Project, 513
    server-side digital wallet, 475–476
    SharePoint, 395
    Silverlight, 418
    Slate, 127
    spam and, 352
    SQL Server, 382
    traffic statistics, 252
    viruses and, 419–420, 424
    Windows Live ID, 475–476
    Windows Phone, 256
Microsoft Safety & Security Center, 424
Middle East, 38–39, 40
middleware, **382**–383
Millicent, 463
MIME (Multipurpose Internet Mail Extensions), **65**
Mindcraft, 357
minimum bid, **260**
minimum bid increment, **264**
Minneapolis Web site, 216
minors. See children
mixed revenue models, 118–119, 253
mixi, 248
mobile commerce and devices
    ads, **175**
    apps, **175**, 257–259
    banking, 480
    business strategies and, 500
    connectivity, 89–90
    designing Web sites for, 141
    as digital wallets, 476
    emergence of, **14**
    Internet-capable, early years of, 53–54
    micropayments, **463**
    operating systems, 255–257
    payment apps, 258–259
    security, 428–429
    social networking and, 255–259
    tablet devices, 258
mobile telephone networks, 89–90
mobile wallets, **258**
modems, 85
monetizing, **253**
money laundering, **473**
Monitise, 480
Monster.com, 118
Montana Mountain Biking, 199–200
Montgomery Ward, 107
Moodle, 251

The Moose Hut, 490–491
Morocco, 39
mortgage loan sites, 124–125
Mosaic, 67–68
The Motley Fool, 253
Motorola, offshoring, 208
movies. See videos
Mozilla Firefox, 65, 68, 75, 414, 415–416, 427
Mozilla Thunderbird, 64
MP3s. See music
MSN Games, 125
MSN Money, 385
MSN Music, 113
MSPs (managed service providers), **373**
Multipurpose Internet Mail Extensions (MIME), **65**
multivector viruses, **419**
music
    copyright and, 298–299
    fee-for-content model, 113
    piracy, 13
Muslim countries, 38, 39
Mutual Mobile, 257
My Conversation with Jeff Bezos, 300
My Virtual Model, 111, 334
MySpace, 248, 316
MySQL, 382

N

NACHA - The Electronic Payments Association,
    470
naïve Bayesian filter, 347–348
Nakagawa, Eric, 252
name changing, **301**
name stealing, **302**
NAPs (network access points), **57**
Napster, 298–299
NAT (Network Address Translation) devices, **64**
National Association of Purchasing Management,
    210
National Center for Supercomputing Applications
    (NCSA), 342
National Conference of State Legislatures, 320
National Governors Association, 320
National Retail Federation, 309
National Science Foundation (NSF), 57
Nationwide Building Society, 385
Nazi memorabilia, 285
NBC, 114
NCP (Network Control Protocol), **62**
NCSA (National Center for Supercomputing
    Applications), 342
near field communication (NFC), **476**
necessity threats, **431**–432
negligent torts, **288**
Nelson, Anne, 403–404
Nelson, Rod and Martha, 490
Nelson, Ted, 66
net bandwidth, **84**
NetBank, 479
netbooks, **90**
NetCoin, 473
Netcraft, 341–342
Netflix, 114
NetMechanic, 355
Netscape, 68, 72, 417, 504

network access points (NAPs), **57**
network access providers, 57
Network Address Translation (NAT) devices, **64**
Network Associates, 435
Network Control Protocol (NCP), **62**
network economic structures, **27**–28, 217
Network Economics, 217
network effects, **28**–29
network operations staff, **517**
New Orleans Web site, 216
The New Pioneers, 19
New York, 307, 319
New York Post, 191
New York Stock Exchange, 261–262
The New York Times, 118–119, 129, 257, 310, 418
Newark.com, 212
NewHomeNetwork.com, 266
Newmark, Craig, 251
newsgroups, **56**
newspapers
    inline text ads, **173**–174
    participatory journalism, **250**
    revenue models, 117–119, 129–130
    ROI and, 506
Nexus, **317**–318
NFC (near field communication), **476**
nginx, 341
Nick Jr., 326
Nielsen, 253
Nielsen, Jakob, 136, 138, 140
Nielsen//NetRatings, 189
Nigerian scam, **307**
Nimda, 419, 422
Nissan Computer Corp., 325
Nissan.com, 324–326
Nokia, 54, 256
nonrepudiation, **225**
Nook reader, 112–113
Nordisk Aviation, 511
North Africa, 38–39
North Korea, 40, 408
Northern Light, 128
NorthPoint Communications, 86
"Norton Secured, powered by VeriSign," 426
not-for-profit organizations, 134–136, 206–207, **208**
notice, 286
NSF (National Science Foundation), 57
n-tier architectures, **340**
NTT DoCoMo, 255, 258

O

Oasis, 83
Obama, Barack, 408
Object Management Group, 384
object tag, **78**
obligations, privacy rights and, 311–315
occasion segmentation, **163**
octets, **63**
Octopus, 477
Odnoklassniki, 249
offers, **291**–292
Office Depot, 212
offshoring, **208**. *See also* business-to-business (B2B)

Old Metamora, 276
one-to-many communication model, **143**
one-to-one communication model, 143, **144**
one-to-one marketing, **162**
OnGuardOnline.gov Spam site, 349–350
online advertising. *See* advertising
online auctions. *See* auctions
online banking. *See* banking
online communities, **247**
online marketing managers, **516**
Online Publishers Association, 178
online warfare, 309–310
ontologies, **92**
open architecture, **62**
open EDI, **225**
open loop payment card systems, 467–**468**
open or open-outcry auctions, **260**
open sessions, **413**
Open Source Initiative, 251
open source software
    Android, 256
    Microsoft and, 364–366
    MySQL, 382
    project management, 513
    virtual learning networks, **251**
Open Workbench, 513
opening tags, **73**
OpenMarket, 394
open-outcry double auctions, **261**, **262**
OpenProj, 513
operating expenses, 389–390
operating systems, mobile, 255–257
operational partners, 512
opportunity costs, **502**–503
optical fiber connections, 87
opt-in e-mail, **179**
opt-in vs. opt-out approach, **313**–314
opt-out mechanism, 348–349
Oracle
    B2B marketplace product by, 233
    content management, 395
    CRM, 397–398
    E-Business Suite, 393
    ERP software, 384
    hardware, 355
    iPlanet, 341
    Primavera, 513
    software, 382
"Orange Book," 409
Orbitz, 122–123
O'Reilly, Tim, 300
organized crime, **484**
Orkut, 248, 249
orphan files, **355**
Osaifu-Keitai, 476
Ottawa Wireless, 89
outsourcing. *See also* business-to-business (B2B)
    defined, **208**
    e-mail processing, 180, 510
    staffing, 514–517
    Web site development, **508**–512
Overstock.com, 107
overstocks, 132
Overture, 191
Owens Corning, 268–269
Oxfam, 197–199

P

Pace University Law School CISG Database, 293
Packet Internet Groper, **352**
packet-filter firewalls, **447**
packets, **59**
packet-switched networks, 58–62
Page, Lawrence, 4
page views, **176**
page-based application systems, **383**
paid placements, **189**, 191
Palestinian Authority, 39
Palm, 255–257
PANs (personal area networks), **88**
Paramount, 114
partial outsourcing, **510**–513
participatory journalism, **250**
passive RFID tags, **230**–231
passwords, 442–443, 446
patent issues, 299–300
pay wall, **130**
Paychex, 208
Pay.gov, 215
payment cards
    acceptance and processing, 466–471
    advantages and disadvantages, 466
    merchant accounts, 468–469
    open and closed loop systems, 467–468
    overview, **465**–466
payment gateways, **470**
payment processors, **470**
payment systems
    banking and Internet, 478–480
    basics, 463–464
    cards. *See* payment cards
    case examples, 489–491
    criminal activity and, 480–485
    digital wallets, **474**–476
    electronic cash, **471**–474
    outsourcing, 510
    stored-value cards, **476**–478
PayPal
    2015 forecasted forms of payment, 464
    browser window, 471
    challengers to, 474
    eBay and, 265, 289, 462–463
    hacker attacks on, 431
    main discussion, 461–463
    mobile wallet announcement, 476
PayPass, 476, 477
pay-per-click model, **185**
pay-per-conversion model, **185**–186
payroll processing services, 214
PC Week, 314
p-cards, **211**, 465
PDF, 138–139
PDG Software, 380
Peapod, 20
Pentagon, 407
People's Republic of China. *See* China
Peppers, Don, 179
Pepsi, 38, 133–134
per se defamation, **303**
perceived value, **184**
performance evaluation for Web servers, 357–358
perimeter expansion, **448**

perishable goods, 20
Perl, 337
permission marketing, **179**
persistent cookies, 413–414
personal area networks (PANs), **88**
personal contact model, 142–144, 159
personal firewalls, **448**
personal jurisdiction, **287**–288
personal shopper feature, **111**
personal touch, 111
Peru, 365
PETCO, 22
PetFoodDirect.com, 22
Pets.com, 22
Petzinger, Thomas, 19
PGP (Pretty Good Privacy), **435**
PGP International, 435
pharming attacks, **484**
Philippines, 419
Philips Lighting, 501
phishing expeditions, **431**, 480–485
PHP (Hypertext Preprocessor), 337, 383
physical security
    for clients, 428
    for communication channels, 432
    defined, **410**
    for Web servers, 444
piconets, **88**
Pilot Network Services, 444–445
Ping, **352**
piracy, **13**, 14, 364–366
place, as marketing element, **156**
plain old telephone service (POTS), **85**
plain text, **433**
planning
    case examples, 522–524
    comparing benefits to costs, 505–507
    developing Web sites, 507–513
    funding startups, 504–505
    identifying and estimating costs,
        502–504
    identifying and measuring benefits,
        500–502
    identifying objectives, 499
    linking objectives to business strategies,
        499–500
    managing implementations, 513–518
platform neutrality, **335**
plug-ins, 418–419
poker, 307
political parties and organizations, 135–136, 250
POP (Post Office Protocol), **65**
pop-behind ads, **174**
POPFile, **348**
pop-up ads, **174**
pornographic material, 307
Portable Fun Instruments, 98–99
portals
    advertising models on, **116**–117
    vertical, **232**–236
Porter, Michael, 16, 29, 31
positions, staff, 514–517
Post Office Protocol (POP), **65**
postaudit reviews, 517–518
postimplementation audits, **517**–518
POTS (plain old telephone service), **85**

Pottery Barn, 154
power, **284–285**
prepaid cards, **465**
presence, **132**
Pretty Good Privacy (PGP), **435**
price, as marketing element, **155–156**
Price Watch, 271
Priceline.com, 266–267, 299
primary activities, **29–31**
privacy, secrecy vs., **429**
Privacy Council, 429
Privacy Rights Advocacy Groups, 314
privacy rights and obligations, 311–315
private company marketplaces, **234–235**
private ip addresses, **64**
private keys, **434**
private networks, **60**
private stores, **234**, 235–236
private valuations, **259**
private-key encryption, **435–437**
PRIZM, 167
Procter & Gamble, 184, 192
procurement, 6, **209**
Procurement Center of the State Council, 366
Prodigy, 247
product disparagement, **303**
product-based marketing, **156–157**
production version, **516**
product/process suitability, **17–18**
products, as marketing element, **155**
professional services, 125–126
ProFlowers, 155
ProgrammableWeb, 387
programmers, **516**
programming threats, 443
Progressive, 105–106, 121
ProHosting.com, 388
project management, **513–514**
Project Management Institute, 514
project managers, **515**
project portfolio management, **514**
project portfolio managers, **515**
Promedix, 233
promotion, as marketing element, **156**
property taxes, **317**
Prop-ID Research Project, 475
proprietary architecture, **62**
ProQuest, 112
prospecting, **142**
protocols, **62–66**
Provide Commerce, 155
Providian Financial, 445
proxy bids, **264**
proxy server firewalls, **448**
Prudential, 142
psychographic segmentation, **161**
public marketplaces, **232–233**
public networks, **60**
Public Security Bureau, 39–40
public-key encryption, **434–437**
purchasing
    activities, 208–212
    paper-based vs. EDI, 220–223
purchasing cards, **211**, 465
pure dot-com businesses, **5**
Python, **337**

Q

Qaddafi, Muammar, 39
QQ.com, 248
Quicken, 432
QuickTime, 418
Quotesmith, 121

R

racketeering, **484**
radio frequency identification devices (RFIDs), 15, **230–231**
Raisch, Warren, 235
Random Walk Shoes, 366–367
RapidSSL.com, 426
rational branding, **185**
Rayport, Jeffrey, 181
Raytheon, 269
RDFs (resource description frameworks), **92**
real estate, 18, 124–125, 312
RealPlayer, 418
real-time location systems (RTLS), 230
Realtor.com, 124
RedEnvelope, 154–155, 168
Redmine, 513
refdesk.com, 117
registrars, domain name, **193**
Regulation M, 282
reintermediation, **120–124**
relevance, **184**
remote server administration software, **355**
remote wipes, **428**
Renren, 248
repeat visits, **176**
repeaters, **89**
replenishment purchasing, **211**
Representational State Transfer (REST), **387**
request messages and headers, **339**
Research in Motion (RIM), 54
Réseaux IP Européens (RIPE), 63
reserve price, **260**
resource description frameworks (RDFs), **92**
response messages and headers, **339**
response time, **357**
REST (Representational State Transfer), **387**
RESTful design and applications, **387**
retailers, discount, 107–108
retention, customer, **168**, 169–170
return on invesment (ROI), 21, **506–507**
Return Path, 179
revenue models
    advertising, 114–119
    case examples, 147–150
    changing strategies in, 127–130
    fee-for-content, 112–119
    fee-for-service, **125–126**
    fee-for-transaction, **119–125**
    free for many, fee for a few, 126
    issues, 130–132
    overview, **16**, 106–107
    for social networking sites, 251–254
    web catalog, 107–111

reverse auctions
  business-to-business, 268–269
  consumer, 266–267
  in general, **262**
reverse bids, **266**
reverse link checkers, **355**
RFID Journal, 230
RFIDs (radio frequency identification devices), 15, 230–231
Rhapsody, 113
Rheingold, Howard, 247
rich media ads, **174**–175
rich media objects, 171, 172
right of publicity, **296**, 303
RIM (Research in Motion), 54
RIPE (Réseaux IP Européens), 63
risk management model, 410
Rivest, Ronald, 434
roaming, **89**
robotic networks, **416**, 432
robots, **188**
Roebuck, Alvah, 107
rogue apps, **428**
ROI (return on investment), 21, **506**–507
ROI Knowledge Center, 507
root element, **82**
rooting, **257**
Rose, Kevin, 504
RosettaNet, 82
Ross-Dove Company, 268
routiners, **165**
routing algorithms, **59**
routing computers, **59**
routing tables, **59**
Royal Ahold, 20
Royal Bank of Canada, 489
RSA Public Key Cryptosystem, 434
RTLS (real-time location systems), 230
Ruby on Rails, **337**
Ryder Supply Chain, 213

S

Sabre, 122
Safeway, 20
sales taxes, 319–320, 381
SalesCart, 378
Salesforce.com, 13, 398
Salon Core, 127
Salon.com, 127, 247
Samasource, 206–207
San Francisco TransLink, 477
Sanriotown, 316–317
SANS Institute, 449
SAP, 233, 384, 398
satellite connections, 87
Saudia Arabia, 39
SBUs (strategic business units), **26**
scalable, defined, **373**
Scandinavian countries, 286
schemas, XML, **82**
Schneider National, 213
SCM. *See* supply chain management
Scotland, 293
Scottrade Mobile, 257
scripting languages, **415**

sealed-bid auctions, **261**, 262
Search Engine Land, 191
search engine placement brokers, **190**–191
search engines
  indexing software and, **354**
  in late 1990s, 3–5
  parts of, **188**–189
  positioning, 4–5, 189–191
search term sponsorships, **189**
search utility, **188**
Sears, Roebuck & Company, 9, 107, 157, 333
second-price sealed-bid auctions, **261**, **262**
secrecy, **411**
secrecy threats, 429–431
secure envelopes, **440**
Secure HTTP, **439**–440
Secure Sockets Layer (SSL), **437**–439
Secure Sockets Layer-Extended Validation (SSL-EV) digital certificates, **426**–427
security. *See also* client security; communication channel security
  breaches, 312, 407–408, 454–455
  case examples, 454–456
  establishing policies, **411**–413
  organizations promoting, 449–450
  overview, 408–413
  server, 441–448
Security First Network Bank, 489
security policies, **411**–413
segmentation
  behavioral and occasion, **163**–165
  life-cycle, **167**
  market, **160**–162
self-hosting, **373**
seller-bid auctions, **262**
Semantic Web, **91**–92
separation, 166, **167**
server architectures, **357**
server farms, **358**
server security, 441–448
server software, **337**
server-level filtering, **345**
servers. *See also* Web servers
  hardware, 355–356
  multiple meanings of, **337**–338
server-side digital wallets, **475**–476
server-side scripting, **336**–337
service marks, **300**
session cookies, **413**–414
session keys, **437**
session negotiation, **439**
sexting, **308**
SGML (Standard Generalized Markup Language), **71**, 72
ShadowSurf.com, 431
Shamir, Adi, 434
share*it! 470, 471
shared hosting, **373**
Shari's Berries, 155
Shields Up! 448
shill bidders, **259**
shipping costs, 381
shipping profiles, **18**, 22
Shneiderman, Ben, 140
shopper mode, visitors in, 164
shopping, social, **251**

shopping carts
    Cart32 backdoor, 430
    cultures and, 38
    software, 377–380
    visitors in buyer mode and, **164**
ShopSite, 379
ShopTogether, 111
short message service (SMS), **89**, 255
Showtime, 114
shrink-wrap acceptance, **293**
Shriver, Betty, 276
S-HTTP, **439**–440
Siebel Systems, 397–398
Siemens, 420
signatures, **293**. *See also* digital signatures
signed messages, **425**
Silicon Graphics, 68
Simple Mail Transfer Protocol (SMTP), **65**
Simple Object Access Protocol (SOAP), **386**–387
simplifiers, **165**
Singapore, 39, 40, 285–286
Singapore Government Online, 215
SinglePoint, 128
Singleton, Jerry, 199–200
single-use cards, **465**–466
SITA, 69
site maps, **76**
site sponsorships, **175**–176
Six Degrees, 248
Skandia Group, 509
skyscraper ads, **171**
Slate, 127
small payments, **463**
Smart Card Alliance, 477–478
smart cards, **477**–478
smart phones. *See* mobile commerce and devices
smart sourcing, **208**
SMS (short message service), **89**, 255
SMTP (Simple Mail Transfer Protocol), **65**
sniffer programs, **429**–430
snipes and sniping software, **271**
SOAP (Simple Object Access Protocol), **386**–387
social commerce, **15**
social media and networks. *See also* auctions
    blogs, 250–251
    business strategies and, 500
    children and, 316
    criminals and, 309
    cyberbullying, **308**
    early communities, 247–248
    emergence of, **248**–249
    idea-based, 251
    marketing through, **160**, **186**–188
    mobile commerce, 255–259
    revenue models, 251–254
    for shoppers, 251
    virtual learning, 251
social networking administrators, **516**
social shopping, **251**
Society of American Travel Writers Silver Award, 147
Softbank, 38
software. *See also* browsers
    auction, 271
    case examples, 364–366, 371–372
    e-commerce. *See* software, e-commerce
    e-mail, **64**–65
    filtering, **316**

project management, 513, 514
Web hosting alternatives, 373
for Web servers, 340–342. *See also* Web servers
software, e-commerce
    basic functions of, 374–381
    case examples, 402–404
    for large businesses, 392–398
    for midsize to large businesses, 390–392
    other software and, 381–387
    for small and midsize companies, 387–390
software agents, **91**–92
software-only digital wallets, **475**–476
Sohu.com, 39
Solar Turbines, 269
Solaris, 334
Sony, 114, 269
Sony Online Entertainment, 125
Sotheby's, 259, 265
South Korea, 12, 408, 432
Southwest Airlines, 191–192
spam
    basic content filtering, 345–346
    individual user antispam tactics, 344–345
    legal solutions, 348–351
    overview, **178**–179
    phishing and, 485
    statistics, 343–344
    technical solutions, 351–352
Spam and Open Relay Blocking System, 345
Spamhaus Project, 345–346
spear phishing, **482**
specialists (auctioneers), **261**–262
specialty consumer auctions, 266
spend, **210**
spiders, **188**
sponsored top-level domains (sTLDs), **69**
sponsorships, search term, **189**
sponsorships, site, **175**–176
spoofing, **431**
sports programming, market segments for, 161
sportsters, **165**
spot market, **211**
spot purchasing, **211**
SpyEye, 420–421, 423
SRI International, 56
SSL (Secure Sockets Layer), **437**–439
SSL-EV (Secure Sockets Layer-Extended
    Validation) digital certificates, **426**–427
SSUTA (Streamlined Sales and Use Tax
    Agreement), 320
staffing, 21, 514–517
stakeholders, **132**
stalking, 308
Standard Generalized Markup Language (SGML),
    71, 72
Standard Performance Evaluation Corporation, 357
Stanford Copyright & Fair Use, 298
Staples, 157, 212
start pages, **76**
startups, funding, 504–505
Starz, 114
state sales taxes, 319–320
stateless connections, **413**
static catalogs, **375**
static pages, **336**
statistical modeling, **182**–183
Statute of Frauds, **293**

statutory law, **300**
steganography, **427**–428
stickiness, **115**, 252–253
Sting, 302
sTLDs (sponsored top-level domains), **69**
stock brokerage firms, 120
stockouts, **230**
stolen goods, 309
store charge cards, **465**
store-branded cards, **465**
stored-value cards, 476–478
Storm virus, 420, 422
strategic alliances/partnerships, **27**, **131**
strategic business units (SBUs), **26**
Streamlined Sales and Use Tax Agreement
    (SSUTA), 320
StubHub, 121
style sheets, **78**
subject-matter jurisdiction, **287**
subnetting, **64**
subscription model
    fee-for-content, 112–119
    transitions from/to, 127–128
sufficient jurisdiction, **287**
suitability of products for e-commerce, 17–18
Sullivan, Danny, 191
Summer, Gordon, 302
Sun Microsystems, 334, 337
supply alliances, **226**, 232
Supply Chain Council, 227
supply chain management
    auctions and, 269
    case example, 239–241
    main discussion, **226**–232
    software, **396**
supply management, **6**
supply webs, **217**
support. *See* customer service
supporting activities, **29**–31, 208, 213–214
surf vacation sites, 123–124
surfers, **165**
Sviokla, John, 181
Swebapps, 258
SWOT analysis, **33**–34
Symantec, 426, 435
Symantec Intelligence Reports, 352
Symantec Security Response, 423
Symbian, 256–257
symmetric connections, **84**
symmetric encryption, **435**–437
Syracuse University, 70
systems administrators, **517**

T

T. Rowe Price, 140
T1 lines, **87**
T3 lines, **87**
tablet devices, **14**, 54, 90, 258
tags, **70**, 73–75, 82
Talbots, 109, 162
Target, 47, 107, 131, 230
tariffs, **320**
TaskCity, 258
taxation, 317–320, 381
TCO (total cost of ownership), **502**

TCP (Transmission Control Protocol), **62**–63
TCP/IP, **62**–63
TD Ameritrade, 120
Teak Wood Patio Furniture, 375–376
Tealeaf, 397
TechCrunch, 251
technical content, 112
technologies. *See also* Internet; *specific*
    technologies
    in first and second waves, 12–14
    integration issues, 21
    materials-tracking, 229–231
    in third wave, 15
    usability and, 137
    video delivery, 113–114
technology development activities, 31,
    213–214
technology-enabled CRM, **180**–183
teergrubing, **351**–352
telecommuting/telework, **7**, 20, 27, 61
telephone connections, voice-grade, 85
television
    advertising, 115
    fee-for-content model, 114
    market segmentation, 161
Telnet, **353**, 437
Tereshchuk, Myron, 309
terms of service (ToS) agreements, **295**–296
terrorism, 309–310
Tesco, 20–21
test version, **515**–516
testing, usability, **140**
text ads, **173**–174
text chats, 111
text markup languages, **70**
Thaler, William, 260
Thawte, 426
The Wall Street Journal, 112
Theglobe.com, 247–248
Theil, Peter, 246
therapy services, 126
TheStreet.com, 253
third-generation (3G) wireless technology, **90**
third-party cookies, **414**
third-party logistics (3PL) providers, **213**
Thorn Tree, 148
threats
    database, 443
    integrity, 431
    necessity, 431–432
    overview, **410**
    physical, communication channels, 432
    physical, Web servers, 444
    programming, 443
    secrecy, 429–431
    Web server, 441–443
three-tier architecture, **339**–340
throughput, **357**
Ticketmaster, 121
TicketsNow, 121
tier-*N* suppliers, **226**
Tiffany & Co., 184
Time Warner, 68, 312
Times Mirror, 266
TLDs (top-level domains), **69**–70
tobacco products, advertising, 305–306
Tohan, 38

Tomlinson, Ray, 56, 343
Top 10 Information Architecture Mistakes, 136
top-level domains (TLDs), 69–70
torts, **287**–288
ToS (terms of service) agreements, **295**–296
total cost of ownership (TCO), **502**
touchpoints, **166**, 182
Toys"R"Us, 46–47
Trace Center, 137
Tracert, **352**–353
tracking technologies, 229–231
trade names, **300**
trademark dilution, **304**
trademark issues, **300**
Trader Publishing, 118
trading partners, 9
trading pits, **261**
training activities, 214
transaction costs, 23–27
transaction processing, 380–381
transaction servers, **338**
transaction taxes, **317**, 320
transactions, defined, 6
transceivers, **89**
transfer taxes, **317**
TransLink, 477
Transmission Control Protocol (TCP), **62**–63
transportation, advertising, 305–306
travel sites, revenue models, 117, 122–123
Travelocity, 122
TreasuryDirect, 215
Treo, 54
trial visits, **176**
Tribe.net, 248
trigger words, **164**
TriNet, 208
Triple DES, **435**
Tripod, 247
Trojan horses, **416**, 420–423, 428, 443
trust
    international commerce and, 35–36
    marketing and, 158–160
    in supply chain, 226, 229, 232
    usability and, 139–140
Trusted Computer System Evaluation Criteria, 409
trusted networks, **447**
TrustPass, 275
Tuenti, 249
Tunisia, 39
TurboTax, 19
TV.com, 114
Twitter, 15, 160, 186–188, 248, 250
two-tier client/server architecture, **338**
typosquatting, **301**

U

UCC (Uniform Commercial Code), 292
UCLA Online Institute for Cyberspace Law and
    Policy, 289–290
Uclue, 254
UDDI (Universal Description, Discovery, and
    Integration Specification), **386**
UDRP (Uniform Domain Name Dispute Resolution
    Policy), 301–302

UIGEA (Unlawful Internet Gambling Enforcement
    Act), 307
ultimate consumer orientation, **231**–232
Ultra High Security Password Generator, 443
Ultra Wideband (UWB), **88**
UN/EDIFACT, 219
Unfair Contract Terms European Union Directive,
    293
Uniform Commercial Code (UCC), 292
Uniform Domain Name Dispute Resolution Policy
    (UDRP), 301–302
Uniform Resource Locators (URLs), **66**
United Arab Emirates, 39
United Kingdom, 56, 92
United Nations, UN/EDIFACT, 219
United Nations Convention on CISG, 293
United States. *See U.S. entries*
Universal, 114
universal ad package (UAP), **171**
Universal Description, Discovery, and Integration
    Specification (UDDI), **386**
University of California, Berkeley, 217
University of California, Los Angeles, 56
University of California, Santa Barbara, 56
University of Illinois, 67–68, 342
University of Louisville Medical School, 258
University of Maryland Human-Computer
    Interaction Lab, 140
University of North Carolina, 56
University of Texas Copyright Crash Course, 298
University of Toronto, 475
University of Utah, 56
UNIX, 334, **352**
Unlawful Internet Gambling Enforcement Act
    (UIGEA), 307
unsolicited commercial e-mail (UCE), **178**–179.
    See also spam
untrusted Java applets, **417**
untrusted networks, **447**
uPortal, 251
UPS, 213
upstream bandwidth, **84**
upstream strategies, **499**
Urchin from Google, 354
URL brokers, **193**
URLs (Uniform Resource Locators), **66**
URLzone, 420, 423
U.S. Anticybersquatting Consumer Protection Act,
    301, 325
U.S. CAN-SPAM law, 348–351
U.S. Commercial Service, 42
U.S. Constitution, 290, 296
U.S. Department of Defense, 55–57, 72, 409
U.S. Department of Homeland Security, 309
U.S. Department of Justice, 306
U.S. Department of the Treasury, 307
U.S. Federal Deposit Insurance Corporation (FDIC),
    489
U.S. Federal Trade Commission
    advertising regulation by, 304–306
    Dell Computer and Micron Technology and,
        281–282
    privacy rights and, 311, 315
    on spam, 348–351
U.S. government, use of EDI by, 9
U.S. income taxes, 318–319

U.S. Navy, 269
U.S. Panasonic, 512
U.S. Pentagon, 407
U.S. state sales taxes, 319–320
USA Today, 506
USABancshares.com, 489
usability, 136–141
usage-based market segmentation, **163–165**
use taxes, **318**–319
used cars, 18
used vehicle sites, 118
Usenet, **56**, **247**
"User Agreement," 295–296
User's News Network, **56**
utility programs, 352–355
UWB (Ultra Wideband), **88**

V

VacationsToGo.com, 123
Value Added Tax (VAT), **320**
value chains, **29**–33
value systems, **31**
value-added networks (VANs), 9, 223–225
ValueClick, 172
value-to-weight ratio, 18, 22
Van Name, Mark, 314
Vanorder, Gil, 324
VANs (value-added networks), 9, 223–225
VAT (Value Added Tax), **320**
VBScript, 337, 415
V-Code (verification code), **469**
vegetables, 20
Vendio, 271
Ventro, 232–233, 235
venture capitalists, **505**
venture sponsor, 512
Verance, 302–303
verification code (V-Code), **469**
VeriSign, 426
Verizon, 86
vertical integration, **25**–26
vertical portals, **232–236**
Veterans Affairs Medical Center, 258
vicarious copyright infringement, **298**
Vickrey, William, 261
Vickrey auctions, **261**
video chats, 111
videos
    fee-for-content model, 113–114
    rich media ads, 175
    YouTube, 13, 248
viral marketing, **186–188**
virtual communities, **19**, **247**
The Virtual Community, 247
virtual companies, **27**
virtual learning networks, **251**
virtual model feature, **111**
virtual private networks (VPNs), **61**
virtual servers/hosts, **356**
Virtual Vineyards, 37
Virtual Works, 324
viruses
    main discussion, 419–424

on mobile devices, 428
overview, **343**
Visa
    credit cards, 465
    digital wallet, 476
    DPS, 480
    interchange networks and, 466, 468, 469
    mobile banking, 480
    phone reader, 259
visits, **176**
VKontakte, 249
V.me, 476
voice-grade lines, **85**
voice-grade telephone connections, 85
Volkswagen, 324
Voltrank, 172
Volusion, 378
vortals, **232–236**
VPNs (virtual private networks), **61**
vw.com and vw.net, 324

W

W3C (World Wide Web Consortium), **71**, 72–73
W3C Getting Started with HTML, 75
W3C HTML 5, 73
W3C HTML Working Group, 72
W3C Semantic Web, 92
W3C Web Accessibility Initiative, 137
W3C XHTML Version 1.0 Specification, 71
W3C XML Pages, 83
The Wall Street Journal, 118–119
Wal-Mart, 8, 9, 107–108, 225, 230
Walt Disney, 114
WANs (wide area networks), **58**
WAP (Wireless Application Protocol), 256
WAPs (wireless access points), **88**, 432
warchalking, **433**
Ward, Aaron Montgomery, 107
wardrivers, **433**
warfare, 309–310
Warner Brothers, 114
warranties and disclaimers, **293–294**
WaveHunters.com, 124
We Love Etsy, 251
Web. *See also* Internet; markup languages
    communication on, 142–143
    defined, **55**
    emergence of, 66–70
    page request and delivery protocols, 65–66
Web 2.0, **13**, 15
Web APIs, **385**
Web browser software, **65**
Web browsers. *See* browsers
Web bugs, **415**
web catalog revenue model, **107–111**
Web client computers, **65**
Web client software, **65**
Web communities, **247**
Web directories, **116–117**, **189**
Web EDI, **225**
Web graphics designers, **516**

Web hosting
  alternatives, 373
  evaluation criteria, 504
  outsourcing, 510
Web logs, **160**, 250–251
Web portals. *See* portals
Web programmers, **516**
Web server software, **65**
Web servers. *See also* e-mail
  basics, **335–340**
  case examples, 333–334, 366–367
  defined, **67**
  hardware, 355–361
  performance evaluation, 357–358
  software, 340–342
  threats, 441–443
  utility programs, 352–355
  XML and, 83
Web services, **384–387**, 403–404
Web Services Description Language
  (WSDL), **386**
Web site design, customer-centric, 140–**141**
Web sites and presences. *See also* software
  brand image, 133–134
  case examples, 147–150
  communicating with customers, 142–144
  costs, 503–504
  creating effective presence, **132–136**
  evolution of, 507–508
  identifying goals, 133–136
  for mobile devices, 257
  strategies for developing, 507–513
  usability, 136–141
WebSphere Commerce, 391–392, 393
WebTrends, 354
WebVan, 20
WebWord.com, 138
Web-wrap acceptance, **293**
WELL, 247
Wells Fargo, 474
WEP (Wireless Encryption Protocol), **432**
Wescorp, 258
Wheeler, Tony and Maureen, 147
white hat hackers, **411**
white list spam filter, 346
Whole Earth Review, 247
wide area networks (WANs), **58**
Wi-Fi, **88**–89
Wikimedia Foundation, 252
Wikipedia, 13
Williamson, Oliver, 25
Williams-Sonoma, 154
WiMAX (Worldwide Interoperability for Microwave
  Access), **90**
Windows Live ID, 475–476
Windows Phone, 256
WindowsSecurity.com, 412
wine industry, sales regulation, 290
Winebid, 266
Wine.com, 37
Wingspan Bank, 489
winner's curse, **260**
WIPO (World Intellectual Property Organization),
  301–302
wire transfers, **8**
Wired Magazine, 126

wireless access points (WAPs), **88**, 432
Wireless Application Protocol
  (WAP), 256
wireless connections, 87–90
Wireless Encryption Protocol (WEP), **432**
wireless networks, threats to, 432–433
World Intellectual Property Organization (WIPO),
  301–302
World Trade Organization, 307
World Wide Web. *See* Web
World Wide Web Consortium (W3C), **71**, 72–73
Worldwide Interoperability for Microwave Access
  (WiMAX), **90**
worms, **419–424**, 432, 443
writings, **293**
WSDL (Web Services Description
  Language), **386**
WSJ.com, 506
Wu, Juliet, 365
W.W. Grainger, 8, 212, 234

X

Xanadu, 66
XBRL (Extensible Business Reporting
  Language), 82
Xdrive Technologies, 128
XHTML (Extensible Hypertext Markup
  Language), **71**
Xiaobo, Liu, 40
Xing, 249
XML
  editors, 84
  main discussion, 78–83
  Semantic Web and, 91–92
  SGML and, **71**
  in Web services, 386–387
XML parsers, **83**
XML Spy, 84
XML vocabularies, **83**
XSL (Extensible Stylesheet Language), **83**

Y

Yaga, 463
Yahoo!
  advertising model on, 116–117
  Answers, 253
  antivirus scan on e-mails, 423
  Art.com, agreement with, 192
  auction site, 264–265, 285
  botnet attack on, 432
  domain names, 192
  Games, 69
  GeoCities, 247
  Japan, 38
  in late 1990s, 4
  Mail, 64, 423
  Mobile, 257
  Nazi memorabilia, 285
  Overture, 190–191
  premium e-mail service, 126
  Small Business Merchant
    Solutions, 388

social networking elements, 253
spam and, 352
third wave opportunities and, 15
traffic statistics, 252
Wallet, 476
Web Directory, 297
Yankee auctions, **260**
Yemen, 39
Yesmail, 179
Yodlee, 122, 131
YouTube, 13, 248
YoYoDyne, 179
Yunus, Muhammad, 254

Z

Zeus, 420–421, 423
Zhivago, Kristin, 141
Zhivago Marketing Partners, 141
Zimmerman, Phil, 435
zombies and zombie farms, **416**, 484
ZoneAlarm, 448
Zotob, 420, 422
Zuckerberg, Mark, 245–246
Zynga, 257